Lectures on Mathematics for Economic and Financial Analysis

Giorgio Giorgi • Bienvenido Jiménez •
Vicente Novo

Lectures on Mathematics for Economic and Financial Analysis

Springer

Giorgio Giorgi
Department of Economics and Management
University of Pavia
Pavia, Italy

Bienvenido Jiménez
Department of Applied Mathematics
National University of Distance Education
Madrid, Spain

Vicente Novo
Department of Applied Mathematics
National University of Distance Education
Madrid, Spain

ISBN 978-3-031-83341-0 ISBN 978-3-031-83339-7 (eBook)
https://doi.org/10.1007/978-3-031-83339-7

This Springer imprint is published by the registered company Springer Nature Switzerland AG
The registered company address is: Gewerbestrasse 11, 6330 Cham, Switzerland

For Professor Angelo Guerraggio

Giorgio Giorgi

For my wife Elena and my children Roberto, Cristina, and Elena

Bienvenido Jiménez

For my wife María Antonia and my daughters Noa and Marta.

Vicente Novo

Preface

This book stems from the long didactic experience of its authors in teaching Mathematics and its applications to Economics, Finance, Operations Research, Engineering, Statistics, etc. Indeed, it is expected that students attending the above courses (or related ones) have a good knowledge of several important mathematical tools, especially some specific topics of Linear Algebra, Calculus of functions of several variables, Dynamical Systems, Dynamic Optimization, etc. Therefore, the main purpose of the present book is to help students (undergraduates and also postgraduates) to acquire some mathematical tools which will give them access to the literature relevant in the abovementioned courses and, in case, to conduct successfully a research project or a final thesis. Hence, the present book is conceived for students, not for professors, even if we hope that also some professors will read some sections, perhaps making some constructive criticism. We emphasize again that this book is designed to be a "tools box....."

We have given the proofs of several results, but only those proofs that are not too intricate. However, in our opinion, it is fruitful the fact that students become acquainted with the various proof techniques. We have inserted many examples, some of them with an economic meaning. The book is endowed with a quite essential list of bibliographical references, so that the willing reader has the possibility to deepen the treatment of the various subjects. Altogether the book is the result of the author's long didactic experience: the famous English mathematician G. H. Hardy declared that young people must find new results and old people must write books! We hope that our efforts will be of some utility for students and in general for those readers interested in mathematics for economic and financial analysis. We are grateful to all who will point out errors, inaccuracies, or other questions. As almost nobody reads the preface of the books, we prefer to stop here, with the dedication of our book to all its readers.

Pavia, Italy — Giorgio Giorgi
Madrid, Spain — Bienvenido Jiménez
Madrid, Spain — Vicente Novo
November 2024

Acknowledgments

This work, for the second and third author, was partially supported by Ministerio de Ciencia e Innovación and Agencia Estatal de Investigación (Spain) under project with reference PID2020-112491GB-I00/AEI/10.13039/501100011033.

Competing Interests The authors have no competing interests to declare that are relevant to the content of this manuscript.

Contents

List of Figures

Chapter 1
Topics of Linear Algebra

1.1 The Structure of a Linear Space

In this first section of the present chapter we wish to introduce a rather abstract and general notion, which however is of great importance not only in Linear Algebra and in its various applications, but also in other branches of pure and applied mathematics. It is the notion of *linear space* or *vector space* (more precisely: *linear space over the real field;* it is possible to consider also the complex field, but this will not be done in the present section).

Let us consider a set S, made of elements not a priori specified, but endowed with two "binary" (i.e. involving two "ingredients") operations: an *internal* operation, said *sum* or *addition* and denoted by the usual symbol +, and an *external* operation, said *product* or *multiplication* of an element of S by a real number (i.e. a real scalar quantity). The elements of S are also called "vectors", even if in the sequel we shall mainly use this name for the elements of $\mathbb{R}^n$ (Euclidean n-space) of also for the elements of $\mathbb{C}^n$ (complex n-space). The set S is called a *linear space* or a *vector space* (or, more precisely, *a linear (vector) space over the real field)* if the following two sets of axioms are satisfied.

(a) With respect to the addition.

A1. If $x, y \in S$ also $x + y \in S$. In other words, S is "closed" (in an algebraic sense) with respect to the operation of addition between its elements. That is, by summing any two elements of S we still remain in S.

A2. If $x, y \in S$, then $x + y = y + x$, that is the addition is *commutative.*

A3. If $x, y, z \in S$, then $(x + y) + z = x + (y + z)$, that is the *associative* property holds.

A4. There exists in S a unique element, said *zero element* or *zero vector* and denoted by $[0]$ or by $\mathbf{0}$ or by $\underline{0}$, etc., such that $x + [0] = [0] + x = x, \forall x \in S$.

G. Giorgi et al., *Lectures on Mathematics for Economic and Financial Analysis*,
https://doi.org/10.1007/978-3-031-83339-7_1

A5. For every $x \in S$ there exists a unique element, said the *opposite* of x, and denoted by $-x$, such that $x+(-x) = (-x)+x = [0]$. The sum $x+(-x)$ is usually written as $x - x$.

(b) With respect to the multiplication by a (real) scalar.

M1. If $x \in S$ and $\lambda \in \mathbb{R}$, then $\lambda x \in S$. In other words, S is "closed" (in an algebraic sense) with respect to the product by real scalars. That is, if we multiply any element of S by any real number, we still remain in S.

M2. The product is *distributive with respect to the addition of scalars*: $(\alpha + \beta)x = \alpha x + \beta x, \forall \alpha, \beta \in \mathbb{R}, \forall x, y \in S$.

M3. The product is *distributive with respect to the addition of elements of* S : $\lambda(x + y) = \lambda x + \lambda y, \forall \lambda \in \mathbb{R}, \forall x, y \in S$.

M4. The following *associative property* holds: $\alpha(\beta x) = (\alpha\beta)x, \forall \alpha, \beta \in \mathbb{R}$, $\forall x \in S$.

M5. There exists a unique *neutral* element with respect to the multiplication by scalars: it is (obviously) the number $1 : 1x = x, \forall x \in S$.

It is also said that a linear space has the structure of a *commutative group* with respect to the sum. Hence, any set of elements which satisfy the above axioms **A1-A5** and **M1-M5** is a linear (or vector) space over the real field. It is immediate to verify that the said axioms generate other properties, for example: $-x = (-1)x$, $0x = [0]$, and $\lambda\,[0] = [0]$.

Given any linear space, the zero element $[0]$ will always belong to the said space. An important example of linear space is the set of all real vectors with n components, i.e. the space $\mathbb{R}^n$, encountered in some previous courses of Mathematics. However, there are many other sets whose elements satisfy the axioms **A1-A5** and **M1-M5**: for example the set $\mathbb{C}^n$ of all complex vectors with n components (see further, Sect. 1.5), the set of all (real) matrices of order (m, n), the set of all continuous functions $f : \mathbb{R} \longrightarrow \mathbb{R}$ defined on the same domain $T \subset \mathbb{R}$, the set of all polynomials, the set of all polynomials with degree less or equal than n, etc. It is *not* a linear space, for example, the set of all polynomials with a *given* degree n. The reason is simple: if we consider, e.g., the following two polynomials of degree 5:

$$(a)\; x^5 + 2x^2 + 1; \quad (b)\; -x^5 + 3x^3 - x$$

and we perform the sum, we obtain

$$3x^3 + 2x^2 - x + 1,$$

i.e. a polynomial of degree 3. Hence we violate the axiom **A1.**

Definition 1.1.1 A (nonempty) subset L of a linear space S is said to be a *linear subspace of* S if the elements of L satisfy the axioms **A1-A5** and **M1-M5**, with respect to L.

It follows that L, subset of the linear space S, is itself a linear subspace if and only if

(1) $x, y \in L \Longrightarrow x + y \in L$,
(2) $x \in L,\ \lambda \in \mathbb{R} \Longrightarrow \lambda x \in L$.

We can also summarize the said two properties with a unique implication, i.e.

$$x, y \in L,\ \alpha_1, \alpha_2 \in \mathbb{R} \Longrightarrow \alpha_1 x + \alpha_2 y \in L.$$

It follows that, for example, in $\mathbb{R}^3$ we have the following linear subspaces: the origin of $\mathbb{R}^3$, any straight line passing through the origin, every plane passing through the origin, the whole space $\mathbb{R}^3$. For example, the set

$$L_1 = \left\{x \in \mathbb{R}^2 : x_1 > 0,\ x_2 > 0\right\}$$

is *not* a linear subspace of $\mathbb{R}^2$, as the origin of $\mathbb{R}^2$ does not belong to L_1. The set

$$L_2 = \left\{x \in \mathbb{R}^2 : x_1 + x_2 = k,\ k \in \mathbb{R}\right\}$$

is a linear subspace of $\mathbb{R}^2$ only if $k = 0$. The linear space formed only by the zero element, is also said "trivial linear space". If $L \neq \{[0]\}$ (L linear subspace), then L contains infinite elements: indeed, if $x \in L$, then also $\lambda x \in L,\ \forall \lambda \in \mathbb{R}$.

1.2 Basic Notions on Vectors of $\mathbb{R}^n$ and on Topology in $\mathbb{R}^n$

As already observed, the Euclidean space $\mathbb{R}^n$ is an important example of a linear space. This particular linear space will be the subject of the present section and also of several next sections and chapters. Many issues on *real vectors* should be known by the reader from previous courses of Mathematics. We recall that the *zero vector* of $\mathbb{R}^n$ will be denoted by

$$[0] = \begin{bmatrix} 0 \\ 0 \\ 0 \\ \vdots \\ 0 \end{bmatrix}.$$

Usually the vectors will be considered as *column vectors*. The *transpose* of x will be denoted by $x^\top$, i.e. $x^\top$ is the *row vector* whose components are the same of x, but written as a row. The *sum vector* of $\mathbb{R}^n$ will be denoted by $e = [1, 1, 1, \ldots, 1]^\top$.

It is possible to introduce in $\mathbb{R}^n$ a partial ordering by comparing two vectors $x \in \mathbb{R}^n$ and $y \in \mathbb{R}^n$ in the following way:

- $x = y$, if $x_i = y_i, \ \forall i = 1, \ldots, n$ ($x \neq y$ in the opposite case).
- $x > y$, if $x_i > y_i, \ \forall i = 1, \ldots, n$.
- $x \geqq y$ (x greater than or equal to y), if $x_i \geqq y_i, \ \forall i = 1, \ldots, n$. For example, $[5, 2]^\top \geqq [1, -6]^\top$, but also $[5, 2]^\top \geqq [5, 2]^\top$.
- $x \geq y$ (x *quasi-greater than* y), if $x \geqq y$, but $x \neq y$. In other words: $x_i \geqq y_i, \ \forall i = 1, \ldots, n$, and there exists and index i such that $x_i > y_i$.

The relations $x \leqq y, \ x \leq y, \ x < y$ are introduced in a similar way.

Obviously the following implications hold true:

$$x > y \Longrightarrow x \geq y \Longrightarrow x \geqq y.$$

If, with $[0] \in \mathbb{R}^n$, we have

- $x \geqq [0]$, we say that x is a *nonnegative vector;*
- $x \geq [0]$, we say that x is a *semipositive vector* (some authors use also the term"quasipositive vector");
- $x > [0]$, we say that x is a *positive vector.*

Similarly, we define *nonpositive vectors* ($x \leqq [0]$), *seminegative vectors* ($x \leq [0]$), *negative vectors* ($x < [0]$).

The operations of sum of two (or more) vectors x and y of $\mathbb{R}^n$ is introduced as

$$x + y = w \in \mathbb{R}^n, \text{ being } w_k = x_k + y_k, \ \forall k = 1, \ldots, n.$$

The operation of multiplication of $x \in \mathbb{R}^n$ by a scalar $\alpha \in \mathbb{R}$, is defined as

$$\alpha x = z \in \mathbb{R}^n, \text{ being } z_k = \alpha x_k, \ k = 1, \ldots, n.$$

We recall that the *standard vectors* of $\mathbb{R}^n$ (called also *basic vectors, unit vectors*) are:

$e^1 = [1, 0, 0, \ldots, 0]^\top, e^2 = [0, 1, 0, \ldots, 0]^\top, \ldots e^n = [0, 0, \ldots, 0, 1]^\top$.

Consider now the set $L \subset \mathbb{R}^n$; for what previously said, L is a *linear or vector subspace* of $\mathbb{R}^n$ if:

(*a*) $x, y \in L \Longrightarrow x + y \in L$;
(*b*) $\lambda \in \mathbb{R}, x \in L \Longrightarrow \lambda x \in L$.

Now, let L be a linear subspace of $\mathbb{R}^n$, let $x^1, x^2, \ldots, x^m$ be vectors of L and let $\lambda_1, \lambda_2, \ldots, \lambda_m$ be real scalars. The expression

$$\lambda_1 x^1 + \lambda_2 x^2 + \cdots + \lambda_m x^m = \sum_{i=1}^{m} \lambda_i x^i = \bar{x}$$

is called a *linear combination* (l. c.) of $x^1, x^2, \ldots, x^m$, with *coefficients* $\lambda_1, \lambda_2, \ldots, \lambda_m$. It is clear that $\bar{x} \in L$.

A way to build a linear subspace of $\mathbb{R}^n$ (maybe coinciding with the whole $\mathbb{R}^n$) is to consider the set X of *all linear combinations* of some given vectors $x^1, x^2, \ldots, x^k$:

$$X = \left\{ x : x = \sum_{i=1}^{k} \lambda_i x^i \right\}.$$

It is easy to prove that X is a linear subspace of $\mathbb{R}^n$:

(1) It holds the "closure property" with respect to the addition. Let us consider two vectors of X,

$$x = \sum_{i=1}^{k} \lambda_i x^i; \quad y = \sum_{i=1}^{k} \alpha_i x^i.$$

Their sum is

$$x + y = \sum_{i=1}^{k} \lambda_i x^i + \sum_{i=1}^{k} \alpha_i x^i = \sum_{i=1}^{k} (\lambda_i + \alpha_i) x^i$$

i.e. their sum is a linear combination of the two given vectors.

(2) It holds the "closure property" with respect to the product by scalars. From

$$\alpha x = \alpha \left(\sum_{i=1}^{k} \lambda_i x^i \right) = \sum_{i=1}^{k} (\alpha \lambda_i) x^i$$

we deduce a similar conclusion.

In the said case we say that the set of vectors $\{x^1, x^2, \ldots, x^k\}$ is a *spanning set or set of generators* for X and that X is the subspace of $\mathbb{R}^n$ *spanned (or generated)* by the said vectors; this fact is also denoted as

$$X = \text{span} \left\{ x^1, x^2, \ldots, x^k \right\}.$$

It is also said that $x^1, x^2, \ldots, x^k$ are a "system of generators" for X and that $x^1, x^2, \ldots, x^k$ *span* (or *generate*) the linear subspace X.

For example, we can get any vector $x \in \mathbb{R}^n$ (i.e. the whole $\mathbb{R}^n$) as a linear combination of the standard vectors $e^1, e^2, \ldots, e^n$. Indeed we have

$$\lambda_1 e^1 + \lambda_2 e^2 + \cdots + \lambda_n e^n = [\lambda_1, \lambda_2, \ldots, \lambda_n]^\top .$$

If $x \in \mathbb{R}^n$ is a column vector, its *transpose,* $x^\top$, is a row vector.

We recall now another basic concept of the Linear Algebra: the *linear independence and linear dependence* of a set of vectors of $\mathbb{R}^n$. Let be given the following vectors of a linear subspace L of $\mathbb{R}^n$: $x^1, x^2, \dots, x^m$. We build their linear combination: $z = \lambda_1 x^1 + \lambda_2 x^2 + \cdots + \lambda_m x^m = \sum_{i=1}^m \lambda_i x^i$. It is evident that if we choose the coefficients $\lambda_1 = \lambda_2 = \cdots = \lambda_m = 0$, we get $z = [0]$. If the said choice of coefficients (i.e. all zero coefficients) is the *unique* which produces the zero vector $[0]$, then we say that the m given vectors are *linearly independent* (l. i.) or that their set is linearly independent. If there exists a choice of coefficients $\lambda_1, \lambda_2, \dots, \lambda_m$, *not all zero,* such that

$$\sum_{i=1}^{m} \lambda_i x^i = [0]$$

we say that the m given vectors are *linearly dependent* (l. d.) or that their set is linearly dependent.

For example, two proportional vectors of $\mathbb{R}^n$ are linearly dependent: from x and αx, $\alpha \neq 0$, we have obviously $(-\alpha)x + 1(\alpha x) = [0]$. Furthermore, it is immediate to verify that if among a set of vectors there is the zero vector $[0]$, then the said vectors are linearly dependent. The standard vectors of $\mathbb{R}^n$, i.e. the vectors $e^1, e^2, \dots, e^n$ are linearly independent. Indeed: $\lambda_1 e^1 + \lambda_2 e^2 + \cdots + \lambda_n e^n = [\lambda_1, \lambda_2, \dots, \lambda_n]^\top$, and this vector is the zero vector if and only if $\lambda_1 = \lambda_2 = \cdots = \lambda_n = 0$. We have the following important theorems on linear dependent and independent vectors.

Theorem 1.2.1 *If and only if m vectors of a linear subspace L of $\mathbb{R}^n$ are linearly dependent, one at least of them is given by the linear combination of the remaining $(m-1)$ vectors. Hence, the said vectors are linearly independent if and only if none of them is given by the linear combination of the remaining vectors.*

Proof Let be given the linearly dependent vectors $x^1, x^2, \dots, x^m$ of the linear subspace L of $\mathbb{R}^n$. Then, there exists at least a coefficient $\lambda_h \neq 0$ $(1 \leqq h \leqq m)$ such that

$$\lambda_1 x^1 + \lambda_2 x^2 + \cdots + \lambda_m x^m = [0].$$

From this equality it is possible to deduce

$$\sum_{i=1}^{m} \frac{\lambda_i}{\lambda_h} x^i = \frac{\lambda_1}{\lambda_h} x^1 + \cdots + x^h + \cdots + \frac{\lambda_m}{\lambda_h} x^m = [0]$$

from which

$$x^h = \sum_{\substack{i=1 \\ i \neq h}}^{m} \left(-\frac{\lambda_i}{\lambda_h} \right) x^i.$$

Vice-versa, if $x^h = \sum_{\substack{i=1 \\ i \neq h}}^m \lambda_i x^i$, then

$$\sum_{\substack{i=1 \\ i \neq h}}^{m} \left(-\frac{\lambda_i}{\lambda_h}\right) x^i - x^h = [0],$$

hence, with $\lambda_h = -1$, the vectors $x^1, x^2, \ldots, x^m$ are linearly dependent. □

Taking Theorem 1.2.1 into account, we can therefore state that a set of vectors (of $\mathbb{R}^n$) is linearly dependent if:

(1) At least one of them is the zero vector $[0]$;
(2) If a vector is multiple of another vector (for example, if two vectors are equal);
(3) If a vector is given by the sum (or by the difference) of other vectors.

Theorem 1.2.2 *If, among the vectors $x^1, x^2, \ldots, x^m$ of L there are $k < m$ linearly dependent vectors, then the given m vectors are linearly dependent.*

Proof Let us suppose, without loss of generality, that the k dependent vectors are the *first* k vectors. Then, there exists at least a coefficient $\lambda_h \neq 0$, with $1 \leqq h \leqq m$, such that $\lambda_1 x^1 + \lambda_2 x^2 + \cdots + \lambda_h x^h + \cdots + \lambda_k x^k = [0]$, with $\lambda_h \neq 0$. If we put $\lambda_{k+1}, \lambda_{k+2}, \ldots, \lambda_m = 0$, we get

$$\lambda_1 x^1 + \lambda_2 x^2 + \cdots + \lambda_m x^m = [0]$$

with coefficients λ_i, $i = 1, \ldots, m$, not all zero. □

Theorem 1.2.3 *If $x^1, x^2, \ldots x^m$ are linearly independent vectors of L, then any subset of the same set of vectors is a linearly independent set (in other words $k < m$ vectors of the given vectors, are linearly independent).*

Proof Absurdly suppose that k vectors of the given m vectors are linearly dependent. Then, by Theorem 1.2.2, any set containing the said set of k vectors is formed by linearly dependent vectors, against the assumption that the given m vectors are linearly independent. □

We have recalled the notion of a *spanning set* for a linear subspace L of $\mathbb{R}^n$. It must be observed that a nontrivial linear subspace can have many (also infinite) spanning sets of vectors. It is not difficult to be persuaded of this fact. We have seen that, for example, the two basic vectors of $\mathbb{R}^2$, e^1 and e^2 generate the whole $\mathbb{R}^2$. It is then obvious that the triplet

$$e^1 = \begin{bmatrix} 1 \\ 0 \end{bmatrix}, \; e^2 = \begin{bmatrix} 0 \\ 1 \end{bmatrix}, \; x = \begin{bmatrix} \alpha \\ \beta \end{bmatrix}, \quad \alpha, \beta \in \mathbb{R}$$

again generates $\mathbb{R}^2$: it is sufficient to consider the linear combination $\lambda_1 e^1 + \lambda_2 e^2 + \lambda_3 x$, with $\lambda_3 = 0$.

Let us consider the following example, perhaps a bit less trivial. Let us show that the vectors

$$\begin{bmatrix}1\\1\end{bmatrix}, \begin{bmatrix}0\\1\end{bmatrix}, \begin{bmatrix}\alpha\\\beta\end{bmatrix}, \alpha, \beta \in \mathbb{R}, \ \alpha, \beta \neq 0,$$

generate, by linear combination, the whole $\mathbb{R}^2$ (being α and β arbitrary, we have therefore infinite spanning sets for $\mathbb{R}^2$). It must be, with $x_1, x_2 \in \mathbb{R}$,

$$\begin{bmatrix}x_1\\x_2\end{bmatrix} = \lambda_1 \begin{bmatrix}1\\1\end{bmatrix} + \lambda_2 \begin{bmatrix}0\\1\end{bmatrix} + \lambda_3 \begin{bmatrix}\alpha\\\beta\end{bmatrix} = \begin{bmatrix}\lambda_1 + \alpha\lambda_3\\\lambda_1 + \lambda_2 + \beta\lambda_3\end{bmatrix}.$$

The system

$$\begin{cases} x_1 = \lambda_1 + \alpha\lambda_3 \\ x_2 = \lambda_1 + \lambda_2 + \beta\lambda_3 \end{cases}$$

has obviously infinite solutions. Let us consider, e.g., $\lambda_1 = t \in \mathbb{R}$. It results ($\alpha \neq 0$ by assumptions)

$$\lambda_3 = \frac{x_1 - t}{\alpha}; \ \lambda_2 = x_2 - t - \beta\frac{x_1 - t}{\alpha}.$$

Let us check the results:

$$t\begin{bmatrix}1\\1\end{bmatrix} + (x_2 - t - \frac{\beta}{\alpha}(x_1 - t))\begin{bmatrix}0\\1\end{bmatrix} + \frac{x_1 - t}{\alpha}\begin{bmatrix}\alpha\\\beta\end{bmatrix} =$$

$$= \begin{bmatrix}t + x_1 - t\\ t + x_2 - t - \frac{\beta}{\alpha}(x_1 - t) + \frac{\beta}{\alpha}(x_1 - t)\end{bmatrix} = \begin{bmatrix}x_1\\x_2\end{bmatrix}:$$

We have therefore found infinite spanning sets of $\mathbb{R}^2$.

Now we wish to recall another basic notion, in order to find those special spanning sets, of a given linear subspace of $\mathbb{R}^n$, which have the *minimal number of elements.* With reference to the above example, let us try to "cut down" the system of generators we have found, by removing the last vector, thus by considering only the first two vectors:

$$\begin{bmatrix}1\\1\end{bmatrix}; \begin{bmatrix}0\\1\end{bmatrix}.$$

One realizes at once that these vectors are able to generate, by linear combination, the whole $\mathbb{R}^2$. Indeed, we have to find λ_1 and λ_2 such that

$$\begin{bmatrix} x_1 \\ x_2 \end{bmatrix} = \lambda_1 \begin{bmatrix} 1 \\ 1 \end{bmatrix} + \lambda_2 \begin{bmatrix} 0 \\ 1 \end{bmatrix}.$$

We have

$$\begin{bmatrix} x_1 \\ x_2 \end{bmatrix} = \begin{bmatrix} \lambda_1 \\ \lambda_1 + \lambda_2 \end{bmatrix},$$

from which we obtain the unique solution $\lambda_1 = x_1, \ \lambda_2 = x_2 - x_1$.

We point out that, while there were infinite ways to get all vectors of $\mathbb{R}^2$ by considering the three vectors of the first example, here, if we fix the pair $x_1, \ x_2$, we have an unique way to obtain this pair by using the two vectors

$$\begin{bmatrix} 1 \\ 1 \end{bmatrix}, \ \begin{bmatrix} 0 \\ 1 \end{bmatrix}.$$

If we start from the pair

$$\begin{bmatrix} 1 \\ 1 \end{bmatrix}, \ \begin{bmatrix} 2 \\ 2 \end{bmatrix},$$

again obtained from the three vectors previously considered, by removing the second vector and choosing $\alpha = \beta = 2$ in the third vector, we have

$$\lambda_1 \begin{bmatrix} 1 \\ 1 \end{bmatrix} + \lambda_2 \begin{bmatrix} 2 \\ 2 \end{bmatrix} = \begin{bmatrix} \lambda_1 + 2\lambda_2 \\ \lambda_1 + 2\lambda_2 \end{bmatrix} = \begin{bmatrix} \gamma \\ \gamma \end{bmatrix}, \ \gamma \in \mathbb{R}.$$

Hence, in this case (note that the two vectors we have chosen are linearly dependent) we are able only to generate those vectors of $\mathbb{R}^2$ with equal components (i.e. we generate the bisecting line of the first and third quadrant of the Cartesian plane). In this case we do not generate the whole $\mathbb{R}^2$ but only a proper linear subspace of $\mathbb{R}^2$.

Let be given a linear subspace $L \subset \mathbb{R}^n$ and a system of its generators $x^1, x^2, \ldots, x^m$, that is

$$\operatorname{span}\left\{x^1, x^2, \ldots, x^m\right\} = L.$$

We are interested in finding a spanning set of L containing the minimum number of vectors. It is immediate to note that this "minimal" spanning set must be formed by linearly independent vectors. Indeed, if this is not the case, one at least of them could be expressed as a linear combination of the remaining $(m-1)$ vectors and the number of the generators of L would be less of a unit at least, as any other vector of L could be obtained as a linear combination of the remaining $(m-1)$ vectors. For example, if the removed vector is x^k, $1 \leqq k \leqq m$, we get

$$x^k = \sum_{i \neq k} \alpha_i x^i$$

and any vector $\bar{x}$ of L can then be expressed as

$$\begin{aligned}\bar{x} &= \sum_{i \neq k} \lambda_i x^i + \lambda_k x^k = \sum_{i \neq k} \lambda_i x^i + \lambda_k \sum_{i \neq k} \alpha_i x^i \\ &= \sum_{i \neq k} (\lambda_i + \lambda_k \alpha_i) x^i .\end{aligned}$$

On the other hand, thanks again to Theorem 1.2.1, *no linearly independent* vectors of the given m vectors can be removed, as in this case the removed vector could not be obtained as a linear combination of the remaining vectors (Theorem 1.2.1). *Hence the minimal number of the generators of* $L \subset R^n$ *coincides with the maximal number of linearly independent vectors of* L. If k is this number, this is said the *dimension* of L and it is said that L is a linear subspace of $\mathbb{R}^n$ of dimension equal to k (we are treating finite-dimensional linear spaces):

$$\dim(L) = k.$$

Definition 1.2.4 Let be given a linear subspace L of $\mathbb{R}^n$. A *basis* of L (more precisely: a basis of Hamel) is a set of vectors $x^1, x^2, \ldots, x^k$ of L such that:

(1) The vectors $x^1, x^2, \ldots, x^k$ are linearly independent;
(2) The same vectors span L : span $\{x^1, x^2, \ldots, x^k\} = L$.

Note that a spanning set of a linear subspace of $\mathbb{R}^n$ is not necessarily a basis of the same subspace, whereas the vice-versa is always true. Indeed, from a spanning set it could be possible to remove some elements without destroying the property that the remaining vectors are able to generate the whole linear subspace; from a basis this is never possible (a basis is formed by linearly independent vectors!). A "pictorial" analogue of the concept of "basis": with the basic colours red, blue, yellow and black it is possible to generate any nuance of any colour. Furthermore, we have the following result.

Theorem 1.2.5

(1) If the linear subspace L of $\mathbb{R}^n$ admits a basis of k vectors, any other basis of L is formed by k vectors.
(2) Any set of k linearly independent vectors of L is a basis for L.
(3) The representation of any given vector x of L as a linear combination of the elements of a given basis L is unique, i.e. we have a unique choice of multipliers.

From property 1 it follows that the vectors of a basis (of a linear subspace L of $\mathbb{R}^n$) are not a specific result of a particular choice, but they are a characteristics of that linear subspace : In short... a basis is equivalent to another one! We can therefore also say that $\dim(L)$ is the number of vectors which form any basis of L. Hence the *dimension* of a linear subspace L of $\mathbb{R}^n$ is:

- The number of vectors which form any basis of L or
- The minimal number of the generators of L or
- The maximal number of linearly independent vectors which is possible to find in L.

In particular, the property 3 of Theorem 1.2.5 is quite easy to prove: we want to prove that there exists a unique set of coefficients $\lambda_1, \lambda_2, \ldots, \lambda_k$ such that, with $\bar{x}$ any vector of L, it holds

$$\bar{x} = \sum_{i=1}^{k} \lambda_i x^i.$$

Let us show that two different choices of coefficients of the linear combination do not exist. Absurdly suppose that

$$\bar{x} = \sum_{i=1}^{k} \lambda_i x^i = \sum_{i=1}^{k} \lambda_i' x^i,$$

with at least one $\lambda_i \neq \lambda_i'$. Without loss of generality, let us suppose that $\lambda_k \neq \lambda_k'$. Then, from $\sum_{i=1}^{k} \lambda_i x^i = \sum_{i=1}^{k} \lambda_i' x^i$, we get

$$x^k = \frac{1}{\lambda_k - \lambda_k'} \sum_{i=1}^{k-1} (\lambda_i - \lambda_i') x^i = \sum_{i=1}^{k-1} \gamma_i x^i$$

where

$$\gamma_i = \frac{(\lambda_i - \lambda_i')}{(\lambda_k - \lambda_k')}, \text{ for } k = 1, 2, \ldots, k-1.$$

Hence we have that x^k is given by the linear combination of the vectors $x^1, x^2, \ldots, x^{k-1}$, but this is against the assumptions that the vectors of a basis are linearly independent.

Definition 1.2.6 The *standard basis or natural basis or canonical basis* of $\mathbb{R}^n$ is the basis formed by the n standard vectors $e^i,\ i = 1, \ldots, n$.

Indeed, the said vectors are linearly independent, as previously remarked, and they span, by linear combination, every vector of $\mathbb{R}^n$, as previously remarked. Hence we can say that the dimension of $\mathbb{R}^n$ is just n and write

$$\dim(\mathbb{R}^n) = n.$$

We have to remark that not always it is possible, given a general linear space S, to find a basis of the same, formed by a finite number of elements. An example of a linear space S which does not admit a basis formed by a finite number of elements is the set of all polynomials, with real coefficients or also with complex coefficients, in one variable x (real or complex). This is a linear space, however it is indeed absurd to think that it is possible to find a finite number of polynomials $P_1(x), P_2(x), \ldots, P_s(x)$ which are able to generate *all polynomials* of any degree. These linear spaces, quite important in Mathematics, are called *infinite-dimensional linear spaces*. We shall not develop the questions related to this kind of linear spaces.

We recall that with $x \in \mathbb{R}^n$ and $y \in \mathbb{R}^n$, the *scalar product* or *inner product* of x and y is the number

$$x^\top y = x_1 y_1 + x_2 y_2 + \cdots + x_n y_n = \sum_{i=1}^{n} x_i y_i .$$

It is customary, also for practical reasons, to consider $x^\top$ as a row vector and y as a column vector, even if this is not necessary in the above definition. The reason of this convention will appear clear when we shall recall the definition of product of two matrices. We recall that the scalar product does not satisfy the "product cancellation law": in other words, we can have $x^\top y = 0$, with $x \neq [0]$, $y \neq [0]$. For example, with $e^1, e^2 \in \mathbb{R}^2$, we have $(e^1)^\top e^2 = 0$. When we have, as in the last example, $x^\top y = 0$, with x and y *nonzero vectors* of $\mathbb{R}^n$, the vectors x and y are said to be *orthogonal*. Several authors define "orthogonal" two vectors $x, y \in \mathbb{R}^n$, such that $x^\top y = 0$, without requiring $x \neq [0]$, $y \neq [0]$. Obviously, with this definition, $[0] \in \mathbb{R}^n$ is orthogonal to any vector of $\mathbb{R}^n$.

As already remarked, the standard vectors of $\mathbb{R}^n$ are orthogonal, in the sense that $(e^i)^\top e^j = 0, \ \forall i \neq j$.

Theorem 1.2.7 *If $x^1, x^2, \ldots, x^m$ are nonzero vectors of $\mathbb{R}^n$, two by two orthogonal, then they are linearly independent.*

Proof Let us consider the relation

$$\lambda_1 x^1 + \lambda_2 x^2 + \cdots + \lambda_m x^m = [0],$$

where $x^1, x^2, \ldots, x^m$ are two by two orthogonal. Let us compute the scalar product of the said linear combination by the vector x^i, $1 \leqq i \leqq m$. We obtain

$$(x^i)^\top (\lambda_1 x^1 + \lambda_2 x^2 + \cdots + \lambda_m x^m) = 0.$$

Hence, as the nonzero vectors are two by two orthogonal, we have $\lambda_i (x^i)^\top x^j = 0$, with $j \neq i$, but being $(x^i)^\top x^i > 0$, it follows $\lambda_i = 0$. As the index i is arbitrary, it follows $\lambda_1 = \lambda_2 = \cdots = \lambda_m = 0$. □

Given any vector $x \in \mathbb{R}^n$, its *Euclidean norm,* denoted by $\|x\|$, is the nonnegative quantity

$$\|x\| = \sqrt{x^\top x} = \sqrt{\sum_{i=1}^{n} (x_i)^2}.$$

When n is, respectively, $1, 2, 3,$ the said quantity becomes the modulus of the real number x, the lenght of the vector $x \in \mathbb{R}^2$, the length of the vector $x \in \mathbb{R}^3$. The norm is a type of *distance* and hence it verifies some basic properties of the distance:

1. $\|x\| \geqq 0, \ \forall x \in \mathbb{R}^n.$
2. $\|x\| = 0$ if and only if $x = [0]$.
3. $\|\alpha x\| = |\alpha| \ \|x\|, \ \forall \alpha \in \mathbb{R}, \forall x \in \mathbb{R}^n.$
4. $\|x + y\| \leqq \|x\| + \|y\|, \ \forall x, y \in \mathbb{R}^n.$

The last property is said "triangular inequality". In dimensions 2 or 3 it becomes the well known theorem of Geometry which states that the lenght of a side of a triangle is not greater of the sum of the lengths of the remaining two sides. The triangular inequality is a consequence of another inequality:

$$\left| x^\top y \right| \leqq \|x\| \ \|y\|,$$

known as "Cauchy-Schwarz inequality". If x and y are two nonzero vectors of $\mathbb{R}^n$ it makes sense to write the quotient

$$\frac{x^\top y}{\|x\| \ \|y\|},$$

which is a number between -1 and 1, owing to the Cauchy-Schwarz inequality. This quotient is the *cosine* of the angle between x and y and it is denoted by $\cos(\widehat{xy})$. It results that two nonzero vectors are orthogonal if and only if the cosine of the angle formed by the said vectors is zero. When $n = 2$, the above quotient becomes the usual notion of cosine defined in Trigonometry. For instance, if

$$x^\top = [1, 2, 0, 1]; \quad y^\top = [0, -1, -2, -4],$$

we have $\|x\| = \sqrt{6}$; $\|y\| = \sqrt{21}$; $x^\top y = -6$, and hence

$$\cos(\widehat{xy}) = \frac{-6}{\sqrt{6}\sqrt{21}} = -\frac{\sqrt{14}}{7}.$$

The formula of the cosine between two nonzero vectors of $\mathbb{R}^n$ is capable of an interesting interpretation, in mathematical statistics terms. Suppose that the elements of the vectors $x, y \in \mathbb{R}^n$, $(x \neq [0], y \neq [0])$ describe the deviations between n statistical observations and their respective expected values. If we denote by X and Y the two statistical variables, with elements X_s, Y_s, and by m_X and m_Y their expected values, we have

$$x_s = X_s - m_X; \quad y_s = Y_s - m_Y, \ s = 1, \ldots, n.$$

We can rewrite the formula of the cosine in the following way:

$$\frac{\left[\sum_{s=1}^{n} (X_s - m_X)(Y_s - m_Y)\right]/n}{\sqrt{\sum_{s=1}^{n} (X_s - m_X)^2/n} \cdot \sqrt{\sum_{s=1}^{n} (Y_s - m_Y)^2/n}},$$

i.e., in statistical terms,

$$\frac{Cov(X, Y)}{\sigma(X)\sigma(Y)}.$$

Hence, we have obtained the so-called *linear correlation coefficient* between the statistical variables X and Y ($Cov(X, Y)$, $\sigma(X)$, $\sigma(Y)$ denote, respectively, the *covariance* between X and Y and their *standard deviations*). If the correlation coefficient is equal to zero, the statistical variables X, Y are said to be *uncorrelated* or *orthogonal.*

A *nonzero* vector x of $\mathbb{R}^n$ is said to be *normalized,* if its norm is equal to one: $\|x\| = 1$. It is always possible to normalize a vector $x \neq [0]$: it is sufficient to consider $y = \frac{1}{\|x\|}x$. The vector y turns out to be normalized; indeed,

$$\|y\| = \left\| \frac{1}{\|x\|} x \right\| = \frac{1}{\|x\|} \|x\| = 1.$$

We note that the standard vectors of $\mathbb{R}^n$, i.e. $e^1, e^2, \ldots, e^n$, are normalized.

Remark 1.2.8

(1) Other notations for the scalar product between two vectors x and y are (x, y); $\langle x, y \rangle$.
(2) Normalized vectors are also called *versors;* in particular, the standard vectors e^i, $i = 1, \ldots, n$, of $\mathbb{R}^n$, are also called "versors of the axes".
(3) Normalized and orthogonal vectors are also said to be *orthonormal.* If a basis of a certain linear subspace $L \subset \mathbb{R}^n$ is formed by orthonormal vectors, it is said to be an *orthonormal basis.* Note that the standard vectors e^i, $i = 1, \ldots, n$, of $\mathbb{R}^n$ are an orthonormal basis for $\mathbb{R}^n$. The so-called "Gram-Schmidt procedure" ensures that, starting from any basis for $\mathbb{R}^n$, it is always possible to build an orthonormal basis for the same space. See, e.g., [17, 20].

If we have two vectors $x, y \in \mathbb{R}^n$, their *Euclidean distance* is defined as the nonnegative quantity

$$d(x, y) = \|x - y\| = \sqrt{\sum_{i=1}^{n}(x_i - y_i)^2}.$$

There are other possible distances in $\mathbb{R}^n$, for example:
$d(x, y) = \max\limits_{i=1,\ldots,n} \{|x_i - y_i|\}$, $d(x, y) = \sum_{i=1}^{n} |x_i - y_i|$, etc.

However, we shall usually consider only the Euclidean distance. It results therefore that the norm $\|x\|$ is nothing but the Euclidean distance of x from the zero vector $[0]$, i.e. from the origin of $\mathbb{R}^n$. No surprise if the distance verifies properties similar to the ones of the norm:

(*a*) $d(x, y) \geqq 0$, $\forall x, y \in \mathbb{R}^n$, being $d(x, y) = 0$ if and only if $x = y$.
(*b*) $d(x, y) = d(y, x)$ (symmetric property).
(*c*) $d(x, y) \leqq d(x, z) + d(z, y)$, $\forall x, y, z \in \mathbb{R}^n$ (triangular inequality).

By means of the Euclidean distance we can introduce the notion of *open neighborhood* or *open ball* of radius $\varepsilon > 0$ of a point $x^* \in \mathbb{R}^n$: it is the set of all vectors $x \in \mathbb{R}^n$ for which the Euclidean distance from x^* is less than ε, that is, denoting this set by $N(x^*, \varepsilon)$ or $B(x^*, \varepsilon)$ or other notations,

$$N(x^*, \varepsilon) = \left\{x \in \mathbb{R}^n : \left\|x - x^*\right\| < \varepsilon\right\}.$$

When $n = 1$, $N(x^*, \varepsilon)$ is the interval $(x^* - \varepsilon,\ x^* + \varepsilon)$; when $n = 2$, we have the circle (deprived of the circumference, i.e. of the border) of center x^* and radius ε; when $n = 3$, we have the sphere (or ball) of center x^* and radius ε (sphere deprived from the "spherical surface"). A neighborhood of radius ε of x^* is also denoted by $N_\varepsilon(x^*)$, $B_\varepsilon(x^*)$, etc. When the radius is not important, the notations $N(x^*)$, $B(x^*)$, etc., are used. Obviously, it is possible to introduce also "closed" neighborhoods, by means of a weak inequality.

On this occasion (we have just mentioned "closed neighborhoods"), let us recall some basic notions of ordinary topology in $\mathbb{R}^n$. Given a set $A \subset \mathbb{R}^n$ and a point (vector) $x^0 \in \mathbb{R}^n$, we say that x^0 is *interior* to A if, not only $x^0 \in A$, but furthermore there exists a neighborhood $N(x^0, \varepsilon)$ which entirely belongs to A, i.e. $N(x^0, \varepsilon) \subset A$. We say that x^0 is *exterior* to A if, not only $x^0 \notin A$, but furthermore there exists a neighborhood $N(x^0, \varepsilon)$ entirely disjoint from A, i.e. $A \cap N(x^0, \varepsilon) = \varnothing$.

Given a set $A \subset \mathbb{R}^n$, the points that are neither interior, nor exterior (with respect to A), are said *boundary points.* A point $x^0 \in \mathbb{R}^n$ is therefore a boundary point for A, if *every* its neighborhood contains a point of A and a point of A^c, being A^c the *complementary set* of A with respect to $\mathbb{R}^n$: $A^c = \mathbb{R}^n \setminus A$.

Note that boundary points can belong to A or also not. The set of all interior points of A is denoted by $\text{int}(A)$ or also by $\overset{\circ}{A}$ and it is called the *interior* of A. The set of all boundary points of A is denoted by ∂A and it is called the *boundary* of A.

The point $x^0 \in \mathbb{R}^n$ is an *accumulation point* for A if *all* its neighborhoods contain at least one point $x \neq x^0$, with $x \in A$. Note that in this case every neighborhood of x^0 necessarily contains infinite points of A. The point $x^0 \in A$ is said to be *isolated* if there exists a neighborhood $N(x^0)$ which contains no point of A, i.e. $N(x^0)$ contains only the point x^0.

Example 1.2.9 Given the set

$$A = [3, 10) \cup \{15\}$$

it results:

- 5 is interior to A and is also an accumulation point for A.
- 3, 10, 15 are boundary points for A; the points 3 and 10 are also accumulation points for A.
- The point 0 is exterior to A.
- The point 15 is an isolated point for A.

Obviously every interior point of a set $A \subset \mathbb{R}^n$ is an accumulation point for A; if a point is exterior to A it cannot be a boundary point, nor an accumulation point for A. Interior and isolated points of A necessarily belong to A. If a point is a boundary point or an accumulation point for A it can belong as well as not belong to A. Surely if a point is exterior to A it does not belong to A.

The set of all accumulation points of A is also called the *derived set* of A and usually denoted by A'. The *closure* of A is the set, denoted by $\bar{A}$ or also by $\text{cl}(A)$, made of all interior and boundary points of A, i.e. $\bar{A} = \text{int}(A) \cup \partial A$. It results also $\bar{A} = A \cup A'$.

Example 1.2.10 With reference to the set of Example 1.2.9, it holds:

- $\text{int}(A) = (3, 10)$.
- $\partial A = \{3, 10, 15\}$.
- $A' = [3, 10]$.
- $\bar{A} = [3, 10] \cup \{15\}$.

Those sets of $\mathbb{R}^n$ made only of interior points are called *open sets* (hence $A \subset \mathbb{R}^n$ is open if and only if $A = \text{int}(A)$). Therefore no boundary point belongs to an open set. The set A is *closed* if all its boundary points belong to$A : \partial A \subset A$. Equivalently: if its closure coincides with $A : A = \text{cl}(A) = \partial A \cup A$. The empty set $\varnothing$ and the whole space $\mathbb{R}^n$ are considered both closed and open. There are also sets which are neither closed nor open, for example the set A of Examples 1.2.9 and 1.2.10. Moreover, it can be shown that A is closed if and only if it contains all its accumulation points: $A' \subset A$. It can also be proved that A is closed (respectively: open) if and only if its complementary set A^c is open (respectively: closed). The union of open sets is an open set; the intersection of a *finite number* of open sets is an open set. The intersection of closed sets is a closed set; the union of a *finite number* of closed sets is a closed set.

A set $A \subset \mathbb{R}^n$ for which there exists a neighborhood of the origin $N([0], \varepsilon)$ which contains the said set A, i.e. $A \subset N([0], \varepsilon)$, is said to be *bounded;* if $A \subset \mathbb{R}^n$ is both closed and bounded is also called a *compact set.*

Closed and compact sets of $\mathbb{R}^n$ can also be characterized by means of convergent sequences of vectors of $\mathbb{R}^n$. Let $\{x^k\}$ be a sequence of vectors in $\mathbb{R}^n$. We say that $\{x^k\}$ *converges* to x^0 for $k \longrightarrow +\infty$ (or that it admits the limit x^0 for $k \longrightarrow +\infty$) if for every neighborhood $N(x^0)$ there exists a number $k^* = k^*(N)$ such that for $k > k^*$ it holds $x^k \in N(x^0)$. We write

$$\lim_{k \longrightarrow +\infty} x^k = x^0$$

or also

$$x^k \longrightarrow x^0 \text{ (for } k \longrightarrow +\infty).$$

We can reformulate the above definition of convergence as follows: we say that $\{x^k\}$ *converges* to x^0 for $k \longrightarrow +\infty$ if, for every $\varepsilon > 0$ there exists a number $k^* = k^*(\varepsilon)$ such that

$$\left\| x^k - x^0 \right\| < \varepsilon, \text{ for } k > k^*.$$

It can be proved that the convergence of $\{x^k\}$ to x^0 is equivalent to the convergence component by component (the first component $\{x_1^k\}$ converges to x_1^0, the second component $\{x_2^k\}$ converges to x_2^0, etc.).

It can be proved that a set $A \subset \mathbb{R}^n$ is *closed* if and only if any converging sequence $\{x^k\} \subset A$ converges to a point $x^0 \in A$:

$$A \text{ closed } \iff \left\{ (x^k \in A,\ x^k \longrightarrow x^0) \Longrightarrow x^0 \in A \right\}.$$

Let $\{x^k\}$ be a sequence in $\mathbb{R}^n$. Consider a strictly increasing sequence $k_1 < k_2 < k_3 < \ldots$ of natural numbers, and let $y^j = x^{k_j}$ for $j = 1, 2, \ldots$. The sequence $\{y^j\}$ is called a *subsequence* of $\{x^k\}$ and is often denoted by $\{x^{k_j}\}$. All terms of the subsequence $\{x^{k_j}\}$ are present in the original sequence $\{x^k\}$.

An important feature of compact sets is that any sequence of vectors $\{x^k\}$ defined on a compact set of $\mathbb{R}^n$ must contain a *subsequence* that actually converges to some vector of the said set. This result is known also as the *Bolzano-Weierstrass Theorem.* Hence:

- Any sequence of vectors contained in a closed and bounded subset of $\mathbb{R}^n$ has a convergent subsequence.

The Bolzano-Weierstrass Theorem has a converse:

- Let A be a subset of $\mathbb{R}^n$ with the property that any sequence in A has a convergent subsequence with limit in A. Then A is closed and bounded.

Hence we have, for a subset A of $\mathbb{R}^n$, two characterizations of compactness:

(1) A is closed and bounded;
(2) Every sequence in A has a convergent subsequence (to some vector of A).

1.3 Basic Topics on Matrices, Determinants and Linear Systems

A *matrix A of order* (m, n) is defined to be an array of numbers ("elements") arranged into m rows and n columns. It is written as follows

$$A = \left[a_{ij}\right], \quad i = 1, \ldots, m; \; j = 1, \ldots, n,$$

or in an extended form,

$$A = \begin{bmatrix} a_{11} & a_{12} & \cdots & a_{1n} \\ a_{21} & a_{22} & \cdots & a_{2n} \\ \vdots & \vdots & \vdots & \vdots \\ a_{m1} & a_{m2} & \cdots & a_{mn} \end{bmatrix}.$$

This notion should be well known from previous courses of Mathematics, hence we shall give here only some hints on this subject. First we note that the set of all matrices of order (m, n), with a_{ij} real or also complex, is a *linear space* (of dimension $m \cdot n$).

If $m = n$, we speak of *square matrices.* For these matrices the elements $a_{11}, a_{22}, \ldots, a_{nn}$, are the elements of the *main diagonal.*

The *sum* of two (or more) matrices is defined only for matrices of the same order. If A and B are two matrices, both of order (m, n), the sum $A + B = C$ is defined as

$$C = \left[c_{ij}\right] = \left[a_{ij} + b_{ij}\right], \; i = 1, \ldots, m; \; j = 1, \ldots, n.$$

For example

$$\begin{bmatrix} 1 & 0 & 2 \\ -1 & 2 & 3 \end{bmatrix} + \begin{bmatrix} 0 & 4 & -1 \\ 3 & 0 & 1 \end{bmatrix} = \begin{bmatrix} 1 & 4 & 1 \\ 2 & 2 & 4 \end{bmatrix}.$$

The *multiplication of a matrix* A of order (m, n) by a scalar $\alpha \in \mathbb{R}$ is defined as

$$\alpha A = B = \left[\alpha a_{ij}\right], \; i = 1, \ldots, m; \; j = 1, \ldots, n.$$

For example

$$2\begin{bmatrix} 1 & 6 \\ 2 & -1 \\ 5 & 0 \end{bmatrix} = \begin{bmatrix} 2 & 12 \\ 4 & -2 \\ 10 & 0 \end{bmatrix}.$$

On the ground of what previously asserted, the following properties hold ([0] here denotes the *zero matrix*, i.e. the matrix with all zero elements).

- $A + B = B + A$, i.e. the commutative property holds.
- $(A + B) + C = A + (B + C)$, i.e. the associative property holds.
- $[0] + A = A + [0] = A$.
- $A + (-1)A = [0]$.
- $\alpha(A + B) = \alpha A + \alpha B$.
- $(\alpha + \beta)A = \alpha A + \beta A$.
- $(\alpha\beta)A = \alpha(\beta A)$.
- $1A = A$.
- $0A = [0]$.

Similarly to what made for vectors of $\mathbb{R}^n$, it is possible to compare two matrices, with real elements, $A = [a_{ij}]$ and $B = [b_{ij}]$ *of the same order* (m, n) :

- $A = B$ if (and only if) $a_{ij} = b_{ij}, \forall i, j$. ($A \neq B$ in the opposite case).
- $A \geqq B$, if $a_{ij} \geqq b_{ij}, \forall i, j$.
- $A \geq B$, if $A \geqq B$, but $A \neq B$. In other words, we have $a_{ij} \geqq b_{ij}, \forall i, j$, but there exists at least a pair $a_{ij},\ b_{ij}$, such that $a_{ij} > b_{ij}$.
- $A > B$, if $a_{ij} > b_{ij}, \forall i, j$.

If we denote by [0] the *zero matrix* (of order (m, n)), we say:

- *A non negative,* if $A \geqq [0]$;
- *A semipositive,* if $A \geq [0]$;
- *A positive,* if $A > [0]$.

The meaning of A nonpositive, A seminegative and A negative should be now clear.

We shall denote by $A_i,\ i = 1, \ldots, m$, the i-th *row vector* of A and by $A^j,\ j = 1, \ldots, n$, the j-th *column vector*. This allows to consider a real matrix A of order (m, n) as an ordered set of m rows vectors (of $\mathbb{R}^n$), or as an ordered set of n columns vectors (of $\mathbb{R}^m$). Hence, it is possible to write

$$A = \begin{bmatrix} A_1 \\ A_2 \\ \vdots \\ A_m \end{bmatrix} \quad \text{or } A = \left[A^1, A^2, \ldots, A^n\right].$$

When one refers without distinction, to rows or columns of a matrix A, it is customary to speak also of the *lines* of A.

Unless otherwise specified, we shall consider *real* matrices. The *transpose* of a matrix A of order (m, n) is a matrix formed from A by interchanging rows and columns. The transpose is usually denoted by $A^\top$ (sometimes also by A'). Therefore, if $A = \left[a_{ij}\right],\ i = 1, \ldots, m;\ j = 1, \ldots, n$, we have $A^\top = \left[a_{ji}\right],\ j = 1, \ldots, n;\ i = 1, \ldots, m$. Obviously we have $(A^\top)^\top = A$ ("involution property"). We recall that if A is *square* (i.e. $m = n$), then A is *symmetric* if $A = A^\top$, i.e. if $a_{ij} = a_{ji},\ \forall i, j$. It is *skew-symmetric* if $A = -A^\top$, i.e. $a_{ij} = -a_{ji}, \forall i, j$ and hence $a_{ii} = 0, \forall i$. For example, the matrix

$$A = \begin{bmatrix} 0 & -2 & -3 \\ 2 & 0 & 1 \\ 3 & -1 & 0 \end{bmatrix}$$

is skew-symmetric.

Always by considering square matrices, a *diagonal matrix* D is a square matrix with elements outside the *main diagonal* $a_{11}, a_{22}, \ldots, a_{nn}$, which are all zero; these elements are also called "off-diagonal elements" or "extra-diagonal elements".

$$D = \begin{bmatrix} \lambda_1 & 0 & \cdots & 0 \\ 0 & \lambda_2 & \cdots & 0 \\ \vdots & \vdots & \ddots & \vdots \\ 0 & 0 & \cdots & \lambda_n \end{bmatrix}.$$

We write also $D = \text{diag}(\lambda_1, \lambda_2, \ldots, \lambda_n)$. If we introduce the so-called *Kronecker symbol*

$$\delta_{ij} = \begin{cases} 1, & \text{if } i = j \\ 0, & \text{if } i \neq j, \end{cases}$$

it results $D = \left[\delta_{ij}\lambda_i\right],\ i, j = 1, \ldots, n$. If in the diagonal matrix D we have $\lambda_1 = \lambda_2 = \cdots = \lambda_n$, we speak also of a *scalar matrix* and if $\lambda_1 = \lambda_2 = \cdots = \lambda_n = 1$, we speak of the *identity matrix,* denoted by I :

$$I = \left[\delta_{ij}\right],\ i, j = 1, \ldots, n, \text{ i.e. } I = \begin{bmatrix} 1 & 0 & \cdots & 0 \\ 0 & 1 & \cdots & 0 \\ \vdots & \vdots & \ddots & \vdots \\ 0 & 0 & \cdots & 1 \end{bmatrix}.$$

A square matrix $A = \left[a_{ij}\right]$ is said to be *upper triangular* if $a_{ij} = 0$ for all $i > j$ (i.e. all entries below the main diagonal are zero), it is said to be *lower triangular* if $a_{ij} = 0$ for all $i < j$. If we make no distinction, we speak of "triangular matrices".

Always with reference to square matrices, a *permutation matrix*, usually denoted by P or Π, is a matrix obtained by interchanging two or more rows (or columns) of the identity matrix I.

Example 1.3.1 From

$$I = \begin{bmatrix} 1\,0\,0 \\ 0\,1\,0 \\ 0\,0\,1 \end{bmatrix}$$

it is possible to obtain

$$P = \begin{bmatrix} 0\,0\,1 \\ 0\,1\,0 \\ 1\,0\,0 \end{bmatrix}, \text{ or also } P = \begin{bmatrix} 1\,0\,0 \\ 0\,0\,1 \\ 0\,1\,0 \end{bmatrix}, \text{ or also}$$

$$P = \begin{bmatrix} 0\,1\,0 \\ 1\,0\,0 \\ 0\,0\,1 \end{bmatrix}, \text{ etc.}$$

Now we recall the definition of *multiplication (or product)* between two matrices. This operation not always is possible. Let us consider two matrices $A = \left[a_{ij}\right]$, of order (m, p) and $B = \left[b_{ij}\right]$, of order (p, n). Note that the *number of columns of A is equal to the number of rows of B*. Then we say that A and B are *conformable for the product* to yield the product $C = AB$ (in this order!), defined as

$$C = \left[c_{ij}\right] = A_i B^j = \sum_{k=1}^{p} a_{ik} b_{kj}, \; i = 1, \ldots, m; \; j = 1, \ldots, n.$$

In other words, the generic element c_{ij} of the product AB is given by the *scalar product* between the i-the row of A and the j-th column of B. It turns out that C has order (m, n). A practical rule to check if the product AB (in this order!) is defined, is to write the order of B nearby the order of A:

$$(m, p) \;\; (p, n).$$

If the two "internal numbers" coincide, then A and B are conformable for the product AB. In this case the two "external numbers" give the order of $C = AB$, i.e. (m, n).

In the above case, we also say that A is the *pre-multiplier matrix* and B is the *post-multiplier matrix*. For example, we have, with

$$A = \begin{bmatrix} 1 & 2 \\ 2 & -1 \\ 3 & 6 \\ 5 & -2 \end{bmatrix}; \quad B = \begin{bmatrix} 1 & -1 & 1 \\ 2 & 0 & 3 \end{bmatrix},$$

$$C = AB = \begin{bmatrix} A_1B^1 & A_1B^2 & A_1B^3 \\ A_2B^1 & A_2B^2 & A_2B^3 \\ A_3B^1 & A_3B^2 & A_3B^3 \\ A_4B^1 & A_4B^2 & A_4B^3 \end{bmatrix} = \begin{bmatrix} 5 & -1 & 7 \\ 0 & -2 & -1 \\ 15 & -3 & 21 \\ 1 & -5 & -1 \end{bmatrix}.$$

We recall that it may be possible to compute the product AB, *but not* the product BA. If A is of order (m, n) and B is of order (n, m), then it is possible to define both AB and BA (the same is true obviously if A and B are square of the same order). In general $AB \neq BA$, even if A and B are square of the same order n. In general the product of two matrices AB and BA, when defined, is *not commutative* (and also the product cancellation law does not hold). For example, if

$$A = \begin{bmatrix} 0 & 1 \\ 0 & 0 \end{bmatrix}; \quad B = \begin{bmatrix} 1 & 0 \\ 2 & 1 \end{bmatrix},$$

we have

$$AB = \begin{bmatrix} 2 & 1 \\ 0 & 0 \end{bmatrix} \neq BA = \begin{bmatrix} 0 & 1 \\ 0 & 2 \end{bmatrix}.$$

If

$$A = \begin{bmatrix} 1 & 1 \\ 1 & 1 \end{bmatrix}; \quad B = \begin{bmatrix} 1 & 1 \\ -1 & -1 \end{bmatrix},$$

we have

$$AB = \begin{bmatrix} 0 & 0 \\ 0 & 0 \end{bmatrix}.$$

Two square matrices A and B of the same order n, *commute* if $AB = BA$. For example, we have $AI = IA = A$.

The transpose of the product of two matrices (if this operation is defined) A and B is given by the product of the two transposes but in a *reverse order:*

$$(AB)^\top = B^\top A^\top.$$

Note that, if AB is defined, surely also $B^\top A^\top$ is defined. Hence, the product of a matrix A by its transpose $A^\top$ is always defined. In general, however

$$AA^\top \neq A^\top A.$$

Moreover, $AA^\top$ and $A^\top A$ are always *symmetric* matrices: indeed, from $B = AA^\top$ we have

$$B^\top = (AA^\top)^\top = AA^\top = B.$$

Similarly for $A^\top A$. More generally, we have (if the products are defined)

$$(ABC)^\top = C^\top B^\top A^\top,$$

and so on.

If we define the *trace* of a square matrix A, of order n, denoted by $\mathrm{tr}(A)$, the sum of the elements of its main diagonal, i.e.

$$\mathrm{tr}(A) = \sum_{i=1}^{n} a_{ii},$$

we have $\mathrm{tr}(AA^\top) = \mathrm{tr}(A^\top A)$.

A square matrix such that

$$AA^\top = A^\top A$$

is said to be *normal.* If

$$AA^\top = A^\top A = I$$

ten A is said to be *orthogonal* (for example the permutation matrices P). Hence A is orthogonal if $A^\top = A^{-1}$, being A^{-1} the *inverse matrix* of A, i.e. that matrix (if existing) such that $AA^{-1} = A^{-1}A = I$. We shall revert on this important concept, after the introduction of the notion of "determinant". For example, the matrix

$$A = \begin{bmatrix} 2 & 1 \\ -1 & 2 \end{bmatrix}$$

is normal, whereas the matrix

$$A = \begin{bmatrix} \frac{1}{\sqrt{2}} & 0 & -\frac{1}{\sqrt{2}} \\ 0 & 1 & 0 \\ \frac{1}{\sqrt{2}} & 0 & \frac{1}{\sqrt{2}} \end{bmatrix}$$

is orthogonal, as well as the matrices

$$B = \begin{bmatrix} \cos\alpha & -\sin\alpha \\ \sin\alpha & \cos\alpha \end{bmatrix}; \quad C = \begin{bmatrix} \cos\alpha & \sin\alpha \\ \sin\alpha & -\cos\alpha \end{bmatrix}.$$

The transpose of an orthogonal matrix is orthogonal. The product of orthogonal matrices of the same order is an orthogonal matrix. Furthermore, it can be proved that, given a square matrix A, the following conditions are equivalent:

(1) A is orthogonal;
(2) The column vectors of A are orthonormal;
(3) The row vectors of A are orthonormal.

Given the square matrix A, we call *n-power of A* (n positive integer), the product of n factors, all equal to A :

$$(A)^n = \underbrace{AAA\ldots A}_{n \text{ factors}}.$$

Obviously, $(I)^n = I$, $(A)^1 = A$; with A different from the zero matrix, we put $(A)^0 = I$, just as it is done for the scalar case. Note however, that for matrices the rules of the algebra of scalars usually do not hold: For example, we have in general, $(A + B)^2 \neq (A)^2 + 2AB + (B)^2$, being in general $AB \neq BA$.

Now we recall the important notion of *determinant* of a *square* matrix A. First, we wish to develop some properties of permutations of natural numbers. The sequence of the first (positive) natural numbers

$$1, 2, 3, 4, \ldots, n$$

is said to be the *natural sequence* of the first n natural numbers. Every reordering of the said sequence is a *permutation.* For example, if we have the natural sequence

$$1, 2, 3, 4, 5$$

the reorderings

$$1, 2, 3, 5, 4; \quad 5, 1, 2, 3, 4$$

are two permutations.

Every permutation contains a certain number of *inversions:* the number of inversions of a permutation is the number of the pairs (of non necessarily adjacent elements) where the second number is less than the first number. For example, given the natural sequence $1, 2, 3, 4, 5$, the permutation $1, 5, 2, 3, 4$ contains *three inversions:* $(5, 2)$, $(5, 3)$, $(5, 4)$.

A permutation is said to be of *even order,* when the number of inversions contained is even; it is said to be of *odd order,* when the number of inversions is odd. For extension, the natural sequence $1, 2, 3, \ldots, n$, called also the *natural permutation,* contains zero inversions and is therefore considered of even order. As it is well known, given the first positive n natural numbers, we have $n(n-1)(n-2)\ldots 2 \cdot 1 = n!$ permutations of the same.

The *determinant* of a *square matrix* A, of order n, denoted by $\det(A)$ or also by $|A|$, is a number, given by the formula

$$|A| = \det(A) = \sum (\pm) a_{ip} a_{2q} a_{3r}, \ldots a_{ns},$$

where the sum is taken over all permutations of the second subscripts $p, q, r, \ldots, s$ (column indices). Every addendum has the sign $(+)$ if the related permutation is of even order; it has the sign $(-)$, if the related permutation is of odd order.

For example,

$$\begin{vmatrix} a_{11} & a_{12} \\ a_{21} & a_{22} \end{vmatrix} = a_{11}a_{22} - a_{21}a_{12}.$$

$$\begin{vmatrix} a_{11} & a_{12} & a_{13} \\ a_{21} & a_{22} & a_{23} \\ a_{31} & a_{32} & a_{33} \end{vmatrix} = a_{11}a_{22}a_{23} - a_{12}a_{21}a_{33} + a_{12}a_{23}a_{31} - a_{13}a_{22}a_{31}$$

$$+ a_{13}a_{21}a_{32} - a_{11}a_{23}a_{32}.$$

For the case $n = 3$ (and only for this case!) the well known *Sarrus Rule* applies. Write out the first two columns nearby the third column, so to obtain the array (of three rows and five columns)

$$\begin{array}{ccccccccc} a_{11} & & a_{12} & & a_{13} & & a_{11} & & a_{12} \\ & \searrow & & \nearrow\searrow & & \nearrow\searrow & & \nearrow & \\ a_{21} & & a_{22} & & a_{23} & & a_{21} & & a_{22} \\ & \nearrow & & \nearrow\searrow & & \nearrow\searrow & & \searrow & \\ a_{31} & & a_{32} & & a_{33} & & a_{31} & & a_{32} \end{array}$$

Then add the product of the elements of the main diagonal and of the two other "descending" diagonals ($\searrow$) and *subtract* the product of the elements of the "secondary diagonal" (i.e. a_{31}, a_{22}, a_{13}) and of the two other "ascending" diagonals ($\nearrow$) :

$a_{11}a_{22}a_{23} + a_{12}a_{23}a_{31} + a_{13}a_{21}a_{32} - a_{31}a_{22}a_{13} - a_{32}a_{23}a_{11} - a_{33}a_{21}a_{12}.$

The main properties of determinants are:

(1) $|A| = |A^{\top}|$.

(2) If we interchange two rows or two columns of A, we have, by denoting with $\bar{A}$ the new matrix,

$$\left|\bar{A}\right| = |A| .$$

(3) The determinant of a square matrix where two rows (or two columns) are equal or proportional is zero. If a row or a column is the zero vector, then the determinant is zero

(4) By multiplying a row or a column of A by a scalar α, then if we denote by B the new matrix so obtained, we have

$$|B| = \alpha\, |A| .$$

Hence, the determinant of αA (A of order n) is

$$|\alpha A| = \alpha^n\, |A| .$$

Note that in general $|\alpha A| \neq \alpha\, |A|$, unless $\alpha = 1$ or $\alpha = 0$ or $\alpha = -1$ with n even.

(5) If we add to the i-th row of A (or to the i-th column) any row (or column) real vector, we obtain a matrix B, where $|B|$ is given by the sum of $|A|$ and the determinant of the matrix obtained from A by substituting that row (or column) with the said vector. For example, if we build

$$B = \left[A^1 + v;\, A^2;\, A^3\right],$$

then we have

$$|B| = |A| + \left|v;\, A^2;\, A^3\right| .$$

In other words, $|A|$ is a *linear function* of the lines of A (see Sect. 1.4).

(6) If a row (or a column) is a linear combination of other rows (or of other columns), then $|A| = 0$. Indeed, if, for example, given a matrix of order 3, where the third column is a linear combination of the first and the second column, we have

$$\begin{vmatrix} a_{11} & a_{12} & (\lambda_1 a_{11} + \lambda_2 a_{12}) \\ a_{21} & a_{22} & (\lambda_1 a_{21} + \lambda_2 a_{22}) \\ a_{31} & a_{32} & (\lambda_1 a_{31} + \lambda_2 a_{32}) \end{vmatrix} = \begin{vmatrix} a_{11} & a_{12} & \lambda_1 a_{11} \\ a_{21} & a_{22} & \lambda_1 a_{21} \\ a_{31} & a_{32} & \lambda_1 a_{31} \end{vmatrix} + \begin{vmatrix} a_{11} & a_{12} & \lambda_2 a_{12} \\ a_{21} & a_{22} & \lambda_2 a_{22} \\ a_{31} & a_{32} & \lambda_2 a_{32} \end{vmatrix}$$

$$= \lambda_1 \begin{vmatrix} a_{11} & a_{12} & a_{11} \\ a_{21} & a_{22} & a_{21} \\ a_{31} & a_{32} & a_{31} \end{vmatrix} + \lambda_2 \begin{vmatrix} a_{11} & a_{12} & a_{12} \\ a_{21} & a_{22} & a_{22} \\ a_{31} & a_{32} & a_{32} \end{vmatrix} = \lambda_1 0 + \lambda_2 0 = 0.$$

This can be obviously extended to the case where the order of the matrix is n. We shall see (Theorem 1.3.3) that also the vice-versa holds. In other words, the determinant is a sort of algorithm to test wether n vectors of $\mathbb{R}^n$ are linearly dependent or linearly independent.

(7) If we add to a row (or column) of A a linear combination of the other rows (or columns) of A, then for the new matrix B obtained, we have

$$|B| = |A| .$$

(8) If A and B are two square matrices of the same order n, we have (*Theorem of Binet-Cauchy*):

$$|AB| = |A| \cdot |B| .$$

We have

$$|ABC| = |A| \cdot |B| \cdot |C| ,$$

and so on. Note that in general

$$|A + B| \neq |A| + |B| .$$

(9) The determinant of a *diagonal* matrix or of a *triangular* matrix is given by the product of the elements of the main diagonal. Hence $|I| = 1$.

If $|A| = 0$, then A is called a *singular matrix* (or also a *noninvertible matrix*); if $|A| \neq 0$, then A is called a *nonsingular matrix* (also an *invertible matrix).* Always under the assumption that A is square, we recall that the *complementary minor* of a_{ij}, denoted by A_{ij}, is the determinant of the square matrix $\bar{A}$, obtained from A by deleting the i-th row and the j-th column. The *cofactor* of a_{ij}, denoted by M_{ij}, is given by

$$(-1)^{i+j} A_{ij} .$$

We recall the *first theorem of Laplace* and the *second theorem of Laplace.*

Theorem 1.3.2

(1) ***(First Theorem of Laplace).*** *The determinant of A is given by the sum of the products of the elements of any line of A by the corresponding cofactors, i.e.* $|A| = \sum_{i=1}^{n} a_{ij} A_{ij}$ *or also* $|A| = \sum_{j=1}^{n} a_{ij} A_{ij}$.

(2) ***(Second Theorem of Laplace).*** *The sum of the products of the elements of a line of A by the corresponding cofactors of the elements of a parallel line, is always zero, i.e.* $\sum_{i=1}^{n} a_{ij} A_{ik} = 0$*, with* $j \neq k$*; or also* $\sum_{j=1}^{n} a_{ij} A_{kj} = 0$*, with* $i \neq k$.

Now we recall the following important theorem, which justifies, in a certain sense, the notion, somewhat abstruse, of determinant of a square matrix.

Theorem 1.3.3 *The lines of a square matrix A are linearly independent if and only if* $|A| \neq 0$; *hence the lines of A are linearly dependent if and only if* $|A| = 0$.

Proof If the lines of A are linearly dependent, it is possible to get one of them as a linear combination of the remaining lines, and by Property 6 of the determinants, we have $|A| = 0$. Now, vice-versa, let us suppose that $|A| = 0$. If $A = [0]$, i.e. the lines are all zero, then the lines are linearly dependent and there is nothing else to prove, If this is not the case, there exists a square submatrix B, with $|B| \neq 0$, of maximum order $r \geqq 1$, $r \leqq n - 1$. Without loss of generality, let us suppose that B is given by the first r rows and the first r columns of A. Thus we have

$$B = \begin{bmatrix} a_{11} & a_{12} & \cdots & a_{1r} \\ a_{21} & a_{22} & \cdots & a_{2r} \\ \vdots & \vdots & \cdots & \vdots \\ a_{r1} & a_{r2} & \cdots & a_{rr} \end{bmatrix}, \quad |B| \neq 0.$$

Let us build the bordered matrix, of order $(r + 1)$

$$M = \begin{bmatrix} a_{11} & \cdots & a_{1r} & a_{1q} \\ \vdots & \cdots & \vdots & \vdots \\ a_{r1} & \cdots & a_{rr} & a_{rq} \\ a_{p1} & \cdots & a_{pr} & a_{pq} \end{bmatrix},$$

with $|M| = 0$ by construction, where the elements a_{pj} and a_{iq} are taken respectively from any row and any column not included in B. Let $k_1, k_2, \ldots, k_{r+1} = |B|$ be the respective cofactors of the elements $a_{1q}, a_{2q}, \ldots, a_{rq}, a_{pq}$ of the last column of M. Then we have

$$k_1 a_{1i} + k_2 a_{2i} + \cdots + k_r a_{ri} + k_{r+1} a_{pi} = 0, \; i = 1, 2, \ldots, r,$$

and by hypothesis we have

$$k_1 a_{1q} + k_2 a_{2q} + \cdots + k_r a_{rq} + k_{r+1} a_{pq} = |M| = 0.$$

Now, let the last column of M be replaced by another of the remaining columns, say the column numbered s, not appearing in B. The cofactors of the elements of this column are precisely the k's obtained above, so that

$$k_1 a_{1s} + k_2 a_{2s} + \cdots + k_r a_{rs} + k_{r+1} a_{ps} = 0.$$

Thus

$$k_1 a_{it} + k_2 a_{2t} + \cdots + k_r a_{rt} + k_{r+1} a_{pt} = 0, \;\; t = 1, 2, \ldots, n,$$

and hence

$$k_1 A_1 + k_2 A_2 + \cdots + k_r A_r + k_{r+1} A_p = [0] .$$

Since $k_{r+1} \neq 0$, A_p is a linear combination of the r vectors $A_1, A_2, \ldots, A_r$. But A_p was any one of the $(n-r)$ vectors $A_{r+1}, A_{r+2}, \ldots, A_n$; hence each of these can be expressed as a linear combination of $A_1, A_2, \ldots, A_r$. Hence $A_1, A_2, \ldots, A_n$ are linearly dependent. A similar proof can be performed by considering the columns of A. □

We recall again that, given $n \in \mathbb{N}_+$ (set of positive natural numbers), the *factorial of n* is the number

$$n! = n(n-1)(n-2)\ldots 3 \cdot 2 \cdot 1.$$

By definition we have $0! = 1$. The *binomial coefficient* is the number, denoted by $\binom{n}{k}$ (read: "n over k")

$$\binom{n}{k} = \frac{n!}{k!(n-k)!} .$$

Now we give the following definitions.

Definition 1.3.4 Let be given A of order (m, n); a *minor of order t* of A is any determinant of the square submatrix of A, of order t, formed by t rows and t columns of A (the choice of the columns is independent from the choice of the rows).

Obviously we have $1 \leqq t \leqq \min(m, n)$. It can be proved that, from A of order (m, n) it is possible to obtain

$$\binom{m}{t}\binom{n}{t}$$

minors of order t. For example, if A is of order $(3, 4)$, it is possible to obtain:

- 4 minors of order 3, i.e.

$$\left[A^1; A^2; A^3\right]; \quad \left[A^1; A^2; A^4\right]; \quad \left[A^1; A^3; A^4\right]; \quad \left[A^2; A^3; A^4\right].$$

- 18 minors of order 2.
- 12 minors of order 1.

We take the opportunity to recall that the *rank* of a matrix A, of order (m, n), is the *maximum order of its nonzero minors*. This number, denoted by $\operatorname{rank}(A)$, $r(A)$

or simply by r (when there is no possibility of confusion), coincides therefore with the maximum number of linearly independent rows (or equivalently: columns) of the matrix A. We have therefore:

$$A \text{ of order } (m, n) \Longrightarrow 0 \leqq \text{rank}(A) \leqq \min(m, n),$$

being $\text{rank}(A) = 0$ if and only if $A = [0]$. A basic property of the rank is

$$\text{rank}(A) = \text{rank}(A^\top).$$

If it holds $\text{rank}(A) = \min(m, n)$ we say that A has *full rank* or *maximum rank*. Other important properties of the rank of a matrix are:

- If for the matrices A and B the product AB exists, it holds

$$\text{rank}(AB) \leqq \min\{\text{rank}(A), \text{rank}(B)\}.$$

- If A, B and C are matrices such that AB and CA exist, B and C are square and $|B| \neq 0$, $|C| \neq 0$, it holds

$$\text{rank}(AB) = \text{rank}(CA) = \text{rank}(A).$$

A useful method to compute the rank of a matrix A of order (m, n) is given by the *Kronecker algorithm:*

(a) Consider a square submatrix of A, of order k. We call this submatrix “pivotal matrix”.
(b) Consider then the $(m-k)(n-k)$ minors of order $(k+1)$, obtained by bordering the pivotal matrix by a row and a column of A, chosen in all possible ways among the rows and columns which do not belong to the pivotal matrix.
(c) If all minors of order $(k+1)$, obtained from the previous point b), are zero, then we conclude that $\text{rank}(A) = k$. If one of the said minors of order $(k+1)$ is different from zero, it will be the new pivotal minor and the above procedure begins from this pivotal minor.
(d) If it is no more possible to obtain a bordered minor from a pivotal minor, obviously we have that $\text{rank}(A) = \min(m, n)$.

If, for example, we have to compute the rank of

$$A = \begin{bmatrix} 1 & 2 & 1 & 0 \\ 4 & 2 & -2 & 3 \\ 0 & 1 & 1 & 4 \end{bmatrix}$$

with the algorithm of Kronecker, we can choose, as first pivotal minor

$$\begin{vmatrix} 1 & 2 \\ 4 & 2 \end{vmatrix} = -6 \neq 0.$$

Then, we have to consider the two following minors, obtained by bordering the above minor with the third row and the third column of A and with the third row and the fourth column of A. We have

$$\begin{vmatrix} 1 & 2 & 1 \\ 4 & 2 & -2 \\ 0 & 1 & 1 \end{vmatrix} = 0, \text{ but } \begin{vmatrix} 1 & 2 & 0 \\ 4 & 2 & 3 \\ 0 & 1 & 4 \end{vmatrix} = -31.$$

We deduce that $\text{rank}(A) = 3$.

Definition 1.3.5 Given A square of order n, a *principal minor of order* t is any determinant of the square submatrix of order t formed by t rows of A and the *corresponding* t columns.

It can be proved that from A, of order n, it is possible to get $\binom{n}{t}$ principal minors of order t. For example, from

$$A = \begin{bmatrix} a_{11} & a_{12} & a_{13} \\ a_{21} & a_{22} & a_{23} \\ a_{31} & a_{32} & a_{33} \end{bmatrix}$$

it is possible to obtain:

- 1 principal minor of order 3, i.e. $|A|$.
- $\binom{3}{2} = 3$ principal minors of order 2, i.e.

$$\begin{vmatrix} a_{11} & a_{12} \\ a_{21} & a_{22} \end{vmatrix}, \quad \begin{vmatrix} a_{11} & a_{13} \\ a_{31} & a_{33} \end{vmatrix}, \quad \begin{vmatrix} a_{22} & a_{23} \\ a_{32} & a_{33} \end{vmatrix}.$$

- $\binom{3}{1} = 3$ principal minor of order 1 : $|a_{11}|$, $|a_{22}|$, $|a_{33}|$.

The sum t_k of all $\binom{n}{k}$ principal minors of order k is called the *trace of order k of* A. The trace of order 1, i.e.

$$t_1 = a_{11} + a_{22} + \cdots + a_{nn},$$

is said simply the *trace* of A, usually denoted by $\text{tr}(A)$. Obviously $t_n = |A|$.

Definition 1.3.6 Given A square of order n, the *North-West principal minors* or *leading principal minors* or *consecutive principal minors* of A are all those determinants formed by the *first* t rows and the *first* t columns. The North-West principal minors of A are therefore given by the following sequence of determinants:

$$|a_{11}|, \quad \begin{vmatrix} a_{11} & a_{12} \\ a_{21} & a_{22} \end{vmatrix}, \quad \begin{vmatrix} a_{11} & a_{12} & a_{13} \\ a_{21} & a_{22} & a_{23} \\ a_{31} & a_{32} & a_{33} \end{vmatrix}, \ldots, |A|.$$

We have already previously recalled the notion of *inverse* (if it exists!) of a square matrix A : it is the matrix, denoted by A^{-1}, such that

$$AA^{-1} = A^{-1}A = I.$$

Hence, if A admits its inverse A^{-1}, then A and A^{-1} *commute.* The following properties hold:

(1) Being

$$|A| \cdot \left|A^{-1}\right| = \left|A^{-1}\right| \cdot |A| = |I| = 1,$$

we deduce that A admits the inverse A^{-1} if and only if $|A| \neq 0$, i.e. if and only if A is *nonsingular* (also: "invertible").

(2) The inverse A^{-1} (if existing) is unique.

(3) If A is a *diagonal matrix,* with $a_{ii} \neq 0,\ i = 1, \ldots, n$, then also its inverse A^{-1} is a diagonal matrix, given by

$$A^{-1} = \begin{bmatrix} \frac{1}{a_{11}} & 0 & \cdots & 0 \\ 0 & \frac{1}{a_{22}} & \cdots & 0 \\ \vdots & \vdots & \ddots & \vdots \\ 0 & 0 & \cdots & \frac{1}{a_{nn}} \end{bmatrix}.$$

Therefore $I^{-1} = I$.

(4) If A^{-1} exists, then it holds $(A^{-1})^{-1} = A$ (involution property), $(A^{\top})^{-1} = (A^{-1})^{\top}$ and $\left|A^{-1}\right| = \frac{1}{|A|}$.

(5) If A and B (square of the same order) are both nonsingular, then

$$(AB)^{-1} = B^{-1}A^{-1}.$$

Determinants are also useful to compute the inverse of A : assume $|A| \neq 0$ and compute the *adjoint* of A, denoted by A^{+}, which is the *transpose* of the matrix A^{*}, obtained from A by replacing each element a_{ij} of A by its cofactor A_{ij}, i.e.

$$A^{+} = (A^{*})^{\top},$$

where

$$A^* = \begin{bmatrix} A_{11} & A_{12} & \cdots & A_{1n} \\ A_{21} & A_{22} & \cdots & A_{2n} \\ \vdots & \vdots & \ddots & \vdots \\ A_{n1} & A_{n2} & \cdots & A_{nn} \end{bmatrix}.$$

We have the following basic property

$$AA^+ = A^+A = |A|\, I,$$

from which

$$A^{-1} = \frac{1}{|A|}A^+.$$

Obviously, there are other methods, more efficient from a computational point of view, to compute the inverse of a square matrix.

Finally, we recall that if $AB = [0]$, A and B square of the same order and $|A| \neq 0$, it holds that $B = [0]$. If $BA = [0]$, with $|A| \neq 0$, it holds that $B = [0]$. Indeed, if, for example, $|A| \neq 0$ and $AB = [0]$, we have

$$A^{-1}AB = A^{-1}[0] = [0],$$

i.e.

$$IB = [0] \iff B = [0].$$

When A^{-1} exists, we put, for each $k > 0$ ($k \in \mathbb{N}$):

$$A^{-k} = A^{-1}A^{-1} \ldots A^{-1} \quad (k \text{ times}).$$

We recall now shortly the main issues concerning *linear systems.*As it should be known from previous courses of mathematics, we intend to refer to "systems of algebraic linear equations", i.e. a system of m equations in the unknowns $x_1, x_2, \ldots, x_n$, of the type

$$\begin{cases} a_{11}x_1 + a_{12}x_2 + \cdots + a_{1n}x_n = b_1 \\ a_{21}x_1 + a_{22}x_2 + \cdots + a_{2n}x_n = b_2 \\ \ldots\ldots\ldots\ldots\ldots\ldots\ldots\ldots \\ a_{m1}x_1 + a_{m2}x_2 + \cdots + a_{mn}x_n = b_m, \end{cases}$$

which can be rewritten in the matrix form

$$Ax = b,$$

where A of order (m, n) is the *matrix of the coefficients,* $x \in \mathbb{R}^n$ is the *vector of the unknowns* and $b \in \mathbb{R}^m$ is the *constant terms vector* or also the *right-hand side (column) vector.* If $b = [0]$ we speak of "homogeneous systems" ; if $b \neq [0]$, of "non homogeneous systems". The usual questions related to the above system are:

- To establish the conditions such that the system admits solutions, i.e. it is *consistent.*
- If the system is consistent, to establish how many solutions it admits.
- In case the system is consistent, to find its solutions.

We note that we can rewrite the above system in the form

$$x_1 A^1 + x_2 A^2 + \cdots + x_n A^n = b,$$

relation which puts into evidence that the constant terms vector b is generated by the linear combination of the columns of A : the coefficients of the combinations are just the components of the vector of the unknowns. Hence, the system $Ax = b$ admits solutions if and only if b can be expressed as a linear combination of the columns of A. This is equivalent to the following basic result:

Theorem of Rouché-Capelli A necessary and sufficient condition such that in $Ax = b$ the vector b is given by a linear combination of the columns of A, i.e. such that $Ax = b$ is consistent, is that $\text{rank}(A) = \text{rank}(A \mid b)$, where $(A \mid b)$ is the so-called *augmented (or complete or bordered) matrix* of the system, i.e. that matrix, of order $(m, n + 1)$, obtained from A by putting near the last column of A, the constant terms vector b.

If it results, moreover, that $\text{rank}(A) = \text{rank}(A \mid b) = n$, the solution is *unique;* if $\text{rank}(A) = \text{rank}(A \mid b) = r < n$, the system is *indeterminate,* i.e. it admits infinitely many solutions, more precisely it is also said that there are ∞^{n-r} solutions or that the system admits $n - r$ "degrees of freedom". We point out that if a linear system admits two distinct solutions, then it admits infinite solutions; in other words, if a linear system is consistent, then either it admits one solution or it admits infinite solutions. Indeed, if x^1 and x^2 are two distinct solutions, we have by definition $Ax^1 = b$ and $Ax^2 = b$. Let t be any real number and consider the infinite vectors $x(t) = tx^1 + (1-t)x^2$. All these infinite vectors are solutions of our system: indeed, we have $Ax(t) = A\left[tx^1 + (1 - t)x^2\right] = tAx^1 + Ax^2 - tAx^2 = tb + b - tb = b$.

Summing up, if we have the linear system

$$Ax = b$$

with A of order (m, n), $x \in \mathbb{R}^n$, $b \in \mathbb{R}^m$, then the following cases may occur:

(a) The system is *inconsistent, i.e. it admits no solutions,* i. e. the theorem of Rouché-Capelli does not hold, i.e.

$$\text{rank}(A \mid b) \neq \text{rank}(A).$$

More precisely: $\text{rank}(A) = r$ and $\text{rank}(A \mid b) = r + 1$.

(b) The system is *consistent,* i.e. it admits solutions, i.e.

$$\text{rank}(A \mid b) = \text{rank}(A) = r.$$

Note that the theorem of Rouché-Capelli is *automatically verified* if $\text{rank}(A) = m$ (in this case we speak also of "normal systems") or also if $b = [0]$ ("homogeneous systems").

- (b_1) If (and only if) $\text{rank}(A) = \text{rank}(A \mid b) = n$, the solution is *unique;* this is, for example, the case of the so-called *Cramerian systems,* i.e. those square systems where $|A| \neq 0$. We recall that a linear system $Ax = b$, with n equations and n unknowns, where $|A| \neq 0$, is called a *Cramerian system* (from the name of the Swiss mathematician G. Cramer). For these systems the well known *Cramer's Rule* holds: Let be given a Cramerian system $Ax = b$; to obtain the solution x_k, $k = 1, \ldots, n$, we divide by $|A|$ the determinant of the matrix formed from A by replacing the k-th column of A with the vector b. In other words,

$$x_1 = \frac{1}{|A|} \begin{vmatrix} b_1 & a_{12} & \cdots & a_{1n} \\ b_2 & a_{22} & \cdots & a_{2n} \\ \vdots & \vdots & \ldots & \vdots \\ b_n & a_{2n} & \cdots & a_{nn} \end{vmatrix}, \ldots, x_n = \frac{1}{|A|} \begin{vmatrix} a_{11} & a_{12} & \cdots & b_1 \\ a_{21} & a_{22} & \cdots & b_2 \\ \vdots & \vdots & \ldots & \vdots \\ a_{n1} & a_{n2} & \cdots & b_n \end{vmatrix}.$$

  This is equivalent to compute the solution

$$x = A^{-1}b.$$

- (b_2) If $\text{rank}(A) = \text{rank}(A \mid b) = r < n$, then there exist $n - r$ variables that can be chosen freely, whereas the remaining r variables are uniquely determined by the choice of these $n - r$ free variables. The system has $n - r$ *degrees of freedom.* It is also customary to say that the system admits ∞^{n-r} solutions (this is the case, for example, of "normal systems, i.e. when $\text{rank}(A) = m < n$).

We wish to recall some basic properties of *homogeneous linear systems,* i.e. the systems

$$Ax = [0]$$

which will be the starting point of the theory of eigenvalues and eigenvectors (Sect. 1.6). As already observed, the Theorem of Rouché-Capelli here holds "by default": indeed, these systems always admit the zero solution (also called "trivial solution") $\bar{x} = [0]$. On the grounds of what previously observed, we can assert that:

- If and only if $\text{rank}(A) = n$, the zero solution $\bar{x} = [0]$ is the *unique solution* of the system $Ax = [0]$.
- If and only if $\text{rank}(A) < n$, there are, besides the zero solution, other solutions $\bar{x} \neq [0]$, also called "non trivial solutions".
- If $Ax = [0]$ is a *square system,* it will admit nonzero solutions if and only if $|A| = 0$.

Indeed, we have, by definition,

$$Ax = [0] \iff x_1 A^1 + x_2 A^2 + \cdots + x_n A^n = [0]$$

and it holds $x_1 = x_2 = \cdots = x_n = 0$, *unique choice* of the coefficients of the linear combination of the columns of A, i.e. unique choice of the solutions of the system, if and only if the columns $A^1, A^2 \dots, A^n$ are *linearly independent,* i.e. if and only if $\text{rank}(A) = n$.

The following result is quite important. Some authors (for example [22]) calls this result "The Fundamental Theorem of Linear Algebra".

Theorem 1.3.7 *Let be A of order (m, n) and* $\text{rank}(A) = r$. *The set of solutions X^0 of $Ax = [0]$ is a linear subspace of $\mathbb{R}^n$, of dimension $(n-r)$, i.e.* $\dim(X^0) = n - r$.

Hence, if $r = n$, the dimension of X^0 is zero, i.e. X^0 contains only the zero vector (as already said in previous considerations). If $r = 0$, the dimension of X^0 is n, i.e. X^0 coincides with the whole $\mathbb{R}^n$: indeed, in this case $A = [0]$ and obviously $[0]\,x = [0]$, for any $x \in \mathbb{R}^n$.

Proof of Theorem 1.3.7 Proving that X^0 is a linear subspace of $\mathbb{R}^n$ is easy. Let be $X^0 = \{x : Ax = [0]\}$; from $x^1 \in X^0,\ x^2 \in X^0, \lambda \in \mathbb{R}$ it follows:

(*a*) $A(x^1 + x^2) = Ax^1 + Ax^2 = [0] + [0] = [0]$, i.e. $x^1 + x^2 \in X^0$;
(*b*) $A(\lambda x^1) = \lambda A x^1 = \lambda\,[0] = [0]\,,\ \forall \lambda \in \mathbb{R}$, i.e. $\lambda x^1 \in X^0,\ \forall \lambda \in \mathbb{R}$.

We have already observed that if $\text{rank}(A) = n$, the system admits the zero solution only, hence $\dim(X^0) = n - n = 0$; if $\text{rank}(A) = 0$, then $A = [0]$ and $\dim(X^0) = n - 0 = n$. In the other cases we put $\text{rank}(A) = r,\ 1 \leqq r \leqq n - 1$. We perform the proof by means of some steps.

(1) The system is equivalent to a system of r independent equations in the same n unknowns:

$$\bar{A}x = [0]\,,$$

with $\bar{A}$ of order $(r, n),\ [0] \in \mathbb{R}^r$ and $\text{rank}(\bar{A}) = r$.

(2) As $\text{rank}(\bar{A}) = r$, there are r linearly independent columns in $\bar{A}$, and the remaining columns are given by a linear combination of these r columns. Without loss of generality, we can suppose that the linearly independent columns of $\bar{A}$ are the first columns $\bar{A}^1, \ldots, \bar{A}^r$. Let us denote by $A(1)$ the *nonsingular* matrix formed by the first r columns and by $A(2)$ the matrix formed by the remaining $n-r$ columns:

$$\bar{A} = (A(1), A(2));$$

the corresponding vector of the unknowns will be

$$x = \begin{bmatrix} x^1 \\ x^2 \end{bmatrix}, \; x^1 \in \mathbb{R}^r, \; x^2 \in \mathbb{R}^{n-r}.$$

We can therefore rewrite the system in the following form

$$A(1)x^1 + A(2)x^2 = [0] \in \mathbb{R}^r$$

i.e.

$$A(1)x^1 = -A(2)x^2.$$

(3) As $A(1)$ is nonsingular, we have

$$x^1 = -[A(1)]^{-1} A(2)x^2.$$

The solutions of the systems are all vectors $x \in \mathbb{R}^n$ given by the expression

$$x = \begin{pmatrix} -[A(1)]^{-1} A(2)x^2 \\ x^2 \end{pmatrix}$$

with x^2 any vector of $\mathbb{R}^{n-r}$.

(4) Let us consider now the standard vectors $e^i \in \mathbb{R}^{n-r}$, $i = 1, \ldots, n-r$. We observe that the vectors

$$\left\{ x(i) = \begin{pmatrix} -[A(1)]^{-1} A(2)e^i \\ e^i \end{pmatrix} \right\}, \; i = 1, \ldots, n-r,$$

are a spanning set (or set of generators) of X^0. It is simple to verify that these vectors are also linearly independent, hence they are a *basis* for X^0. We can therefore conclude that X^0 has a dimension coinciding with the number of the vectors $x(i)$, i.e. $n-r$. □

We have already observed that the number $(n-r)$ is also said the "number of the freedom degrees" of the system $Ax = [0]$.

Now we recall the so-called "superposition principle" we shall find again in the chapters treating dynamical systems.

Theorem 1.3.8 (Superposition Principle) *The general solution of the system $Ax = b$ can be obtained by summing to the general solution of the associated homogeneous system, $Ax = [0]$, any particular solution of the system $Ax = b$, i.e. any vector x^* which solves the system $Ax = b$.*

Proof Let be $\hat{X} = \{x : Ax = b\}$ and let be $\hat{x} \in \hat{X}$ any given solution (particular solution) of $Ax = b$. Let x^0 be any vector of $X^0 = \{x : Ax = [0]\}$. If we consider $x = \hat{x} + x^0$, it is quite immediate to note that x solves $Ax = b$. Indeed, $Ax = A(\hat{x} + x^0) = A\hat{x} + Ax^0 = b + [0] = b$. □

1.4 Linear Functions

One of the subjects of modern mathematics is the study of functions whose domain and codomain are linear spaces, i.e. $f : X \longrightarrow Y$, with X and Y (general) linear spaces. These functions are also called *applications, transformations, maps, etc.* We shall treat the simplest examples of the said functions, i.e. the *linear functions* or *linear transformations* or *linear maps* from $\mathbb{R}^n$ to $\mathbb{R}^m$. A function $f : \mathbb{R}^n \longrightarrow \mathbb{R}^m$ (hence a *vector function of several variables:* see Chap. 2) is called a *linear function* if the following two properties are satisfied:

(1) $f(x^1 + x^2) = f(x^1) + f(x^2)$, $\forall x^1, x^2 \in \mathbb{R}^n$ ("additive property").
(2) $f(\alpha x) = \alpha f(x)$, $\forall x \in \mathbb{R}^n$, $\forall \alpha \in \mathbb{R}$ (property of "homogeneity of degree 1").

The first property simply says that the image of the sum is equal to the sum of the images. If f is a *production function,* the second property says that if we double, triple, etc. the inputs, also the outputs come out to be doubled, tripled, etc. In Economic Analysis this property is also called "constant returns to scale".

Any linear function $f : \mathbb{R}^n \longrightarrow \mathbb{R}^m$ can be characterized by an appropriate matrix. The following result is known also as the "representation theorem of linear functions".

Theorem 1.4.1 (Representation Theorem of Linear Functions) *All and only the linear functions of the type $f : \mathbb{R}^n \longrightarrow \mathbb{R}^m$ are identified by a matrix A, of order (m, n) and it holds*

$$f(x) = Ax, \quad \forall x \in \mathbb{R}^n,$$

where A is the matrix whose columns are the image, according to f, of the standard vectors e^i, $i = 1, \ldots, n$:

$$A = \left[f(e^1), f(e^2), \ldots, f(e^n) \right].$$

Proof We put $A^1 = f(e^1), A^2 = f(e^2), \dots, A^n = f(e^n)$ and observe that

$$x = \sum_{i=1}^{n} x_i e^i, \ \forall x \in \mathbb{R}^n.$$

Then we apply properties (1) and (2) above and obtain

$$f(x) = f(\sum_{i=1}^{n} x_i e^i) = \sum_{i=1}^{n} f(x_i e^i) = \sum_{i=1}^{n} x_i f(e^i) = \sum_{i=1}^{n} x_i A^i = Ax.$$

Now we prove that A is unique. Let us absurdly suppose that there exists a matrix $B = \left[B^1, B^2, \dots, B^n\right]$, $B \neq A$, such that $f(x) = Bx$. Then, if $x = e^i$, $i = 1, \dots, n$, from $f(x) = Ax$ we get $f(e^i) = A^i$, whereas from $f(x) = Bx$ we get $f(e^i) = B^i$, whence $A = B$. Moreover, easily we get that if $f(x) = Ax$, then f is linear; indeed, $A(x + y) = Ax + Ay$ and $A(\alpha x) = \alpha(Ax)$. Therefore the theorem is proved. □

Example 1.4.2 Let be given $f : \mathbb{R}^2 \longrightarrow \mathbb{R}^3$, with $f(x) = [3x_1 - 2x_2;\ x_1]^\top$. This function is linear, as it is quite immediate to verify. Then we have $f(e^1) = [3, 1]^\top$; $f(e^2) = [-2, 0]^\top$, hence

$$A = \begin{bmatrix} 3 & -2 \\ 1 & 0 \end{bmatrix}.$$

Indeed, let us compute the product

$$Ax = \begin{bmatrix} 3 & -2 \\ 1 & 0 \end{bmatrix} \begin{bmatrix} x_1 \\ x_2 \end{bmatrix} = \begin{bmatrix} 3x_1 - 2x_2 \\ x_1 \end{bmatrix}.$$

Example 1.4.3 Check if the function $f : \mathbb{R}^3 \longrightarrow \mathbb{R}^2$, with $f(x) = [x_1 + 2;\ x_2 + x_3]^\top$ is linear or not.

Let be $x = [x_1, x_2, x_3]^\top$; $y = [y_1, y_2, y_3]^\top$. We have

$$f(x) = [x_1 + 2;\ x_2 + x_3]^\top;\ f(y) = [y_1 + 2;\ y_2 + y_3]^\top.$$

$$f(x + y) = \begin{bmatrix} x_1 + y_1 + 2 \\ x_2 + y_2 + x_3 + y_3 \end{bmatrix} \neq f(x) + f(y) = \begin{bmatrix} x_1 + y_1 + 4 \\ x_2 + y_2 + x_3 + y_3 \end{bmatrix}.$$

Linear functions justify also the peculiar definition somewhat "unusual" of the product between two conformable matrices. Let be given, for instance, two linear functions $f : \mathbb{R}^n \longrightarrow \mathbb{R}^m$ and $g : \mathbb{R}^m \longrightarrow \mathbb{R}^p$; for these functions the following representation, in terms of matrices, holds:

$$f(x) = Ax, \quad \forall x \in \mathbb{R}^n$$

$$g(y) = By, \quad \forall y \in \mathbb{R}^m,$$

being A of order (m, n) and B of order (p, m). It is quite easy to see that the composite function $z = g[f(x)]$ is linear and that it holds

$$z(x) = g[f(x)] = BAx,$$

where BA is given by the usual product between conformable matrices.

Given $y = f(x) = Ax$, the set of all $y \in \mathbb{R}^m$ which are the image of a vector $x \in \mathbb{R}^n$, according to f, is said the *range* or also the *codomain* of f and usually denoted by $\mathrm{Im}(f)$. Hence this set is the linear subspace of $\mathbb{R}^m$ generated by the columns of A (indeed we have $Ax = x_1A^1 + x_2A^2 + \cdots + x_nA^n$, and by varying the coefficients $x_1, x_2, \ldots, x_n$, we obtain the linear subspace generated by the columns of A). Hence it results

$$\dim(\mathrm{Im}(f)) = \mathrm{rank}(A) = r.$$

Any vector $x \in \mathbb{R}^n$ such that $Ax = y$, is the *inverse image* of y according to the linear function represented by A. The set of all inverse images of the zero vector $[0]$, i.e. the set of solutions of the homogeneous system $Ax = [0]$, is called the *kernel* or also the *nullspace* of f (or of A) and denoted by $\ker(f)$, $\ker(A)$, $N(f)$, $N(A)$:

$$\ker(f) = \ker(A) = \left\{x \in \mathbb{R}^n : Ax = [0]\right\}.$$

We have already remarked that:

- $\ker(A)$ is always a nonempty set (it always contains at least the zero vector $[0]$);
- $\ker(A)$ is a linear subspace of $\mathbb{R}^n$, of dimension $n - \mathrm{rank}(A)$. See Theorem 1.3.7. Hence, with $\mathrm{rank}(A) = r$,

$$\dim(\ker(A)) = n - r.$$

The dimension of the kernel of the linear function f (or of the matrix A associated to f) is called also the *nullity* of f (of A), or also, as previously said, the *number of the freedom degrees* of f. We can therefore state the following result, important but immediate, as soon as we remind that $\dim(\mathrm{Im}(f)) = \mathrm{rank}(A) = r$.

Theorem 1.4.4 *Let be given the linear function $f : \mathbb{R}^n \longrightarrow \mathbb{R}^m$; it results*

$$\dim(\mathrm{Im}(f)) + \dim(\ker(f)) = n.$$

Sometimes the functions of the type

$$f(x) = Ax + b,$$

with A matrix of order (m, n) and $b \in \mathbb{R}^m$ are called "linear affine".

We recall that a function $f : A \longrightarrow B$ is called

- *one-to-one,* if for any $a_1, a_2 \in A,\ a_1 \neq a_2$, we have $f(a_1) \neq f(a_2)$.
- *onto* (B), if $f(A) = B$.
- *bijective* or *invertible,* if f is both one-to-one and onto, i.e. there is a one-to-one correspondence between A and B and vice-versa.

Therefore, we deduce the following facts.

- A linear function $y = Ax$ is one-to-one if and only if $\ker(A) = \{[0]\}$, i.e. if and only if $\text{rank}(A) = r = n$. Indeed, in this case the system $Ax = y$ admits, for each $y \in \mathbb{R}^m$ a unique solution: it cannot hold $Ax^1 = \bar{y}$ and $Ax^2 = \bar{y}$, with $x^1 \neq x^2$ and $\bar{y} \in \text{Im}(f)$.
- A linear function $y = Ax$ is onto if and only if $\text{rank}(A) = m$, i.e. if and only if the linear system $Ax = y$ is "normal" and hence it admits solutions for *every* $y \in \mathbb{R}^m$.
- A linear function $y = Ax$ is bijective if and only if $\text{rank}(A) = m = n$, i.e. if and only if the system $Ax = b$ is *Cramerian.*

Finally, we present some exercises on linear functions and their properties.

Exercise 1.4.5 Let be given the linear function $f : \mathbb{R}^3 \longrightarrow \mathbb{R}^3$ such that $f(e^1) = [1, 0, 1]^\top$; $f(e^2) = [0, 2, 2]^\top$; $f(e^3) = [1, -1, 0]^\top$.

(i) Build the matrix A which represents f.
(ii) valuate $\dim(\text{Im}(f))$ and $\dim(\ker(f))$.
(iii) Find all vectors of $\ker(f)$ and a basis of $\ker(f)$.
(iv) Find the image, according to f, of the vector $x^* = [4, -2, -4]^\top$.

Solution

(i)

$$A = \begin{bmatrix} 1 & 0 & 1 \\ 0 & 2 & -1 \\ 1 & 2 & 0 \end{bmatrix}.$$

(ii) We have $|A| = 0$, $\text{rank}(A) = 2$ and hence $\dim(\text{Im}(f)) = 2$, $\dim(\ker(f)) = n - \text{rank}(A) = 3 - 2 = 1$.
(iii) We have to solve the system $Ax = [0]$. As $\text{rank}(A) = 2 < n = 3$, we have ∞^1 solutions:

$$\bar{x} = \begin{bmatrix} -\bar{x}_3 \\ \frac{1}{2}\bar{x}_3 \\ \bar{x}_3 \end{bmatrix}, \quad \bar{x}_3 \in \mathbb{R}.$$

A basis of $\ker(f)$ is, e.g., with $\bar{x}_3 = 1$,

$$\left[-1, \frac{1}{2}, 1\right]^\top .$$

(iv) We note that $x^* = [4, -2, -4]^\top \in \ker(f)$: if we put in $\bar{x}$, $\bar{x}_3 = -4$, we obtain the said vector x^*, therefore $Ax^* = [0]$, i.e. the image of x^*, according to f, is the zero vector $[0]$.

Exercise 1.4.6 Let be $f : \mathbb{R}^3 \longrightarrow \mathbb{R}^3$ a linear function such that the inverse images of the vectors $e^1, e^2, e^3 \in \mathbb{R}^3$, are, respectively, the vectors

$$w^1 = [0, 1, 0]^\top ; \; w^2 = [2, 0, 1]^\top ; \; w^3 = [1, 0, 1]^\top .$$

(a) Build the matrix A associated to f.
(b) Consider all linear combinations of the vectors

$$a^1 = [1, 1, -2]^\top ; \; a^2 = [1, 1, 1]^\top .$$

Compute the set of the images, according to f, of all the above linear combinations.

Solution
(a) We have to find the matrix A such that: $Aw^1 = e^1$; $Aw^2 = e^2$; $Aw^3 = e^3$, i.e.

$$A \begin{bmatrix} 0 \\ 1 \\ 0 \end{bmatrix} = \begin{bmatrix} 1 \\ 0 \\ 0 \end{bmatrix} ; \; A \begin{bmatrix} 2 \\ 0 \\ 1 \end{bmatrix} = \begin{bmatrix} 0 \\ 1 \\ 0 \end{bmatrix} ; \; A \begin{bmatrix} 1 \\ 0 \\ 1 \end{bmatrix} = \begin{bmatrix} 0 \\ 0 \\ 1 \end{bmatrix},$$

i.e., with $W = \left[w^1, w^2, w^3\right]$,

$$AW = I.$$

Being $|W| = -1$, we get

$$A = W^{-1}.$$

It results then

$$W^{-1} = \begin{bmatrix} 0 & 1 & 0 \\ 0 & 0 & -1 \\ -1 & 0 & 2 \end{bmatrix} = A.$$

(b) The linear combination of the vectors a^1 and a^2 are all vectors of the form

$$x = \lambda_1 \begin{bmatrix} 1 \\ 1 \\ -2 \end{bmatrix} + \lambda_2 \begin{bmatrix} 1 \\ 1 \\ 1 \end{bmatrix} = \begin{bmatrix} \lambda_1 + \lambda_2 \\ \lambda_1 + \lambda_2 \\ -2\lambda_1 + \lambda_2 \end{bmatrix}, \ \lambda_1, \lambda_2 \in \mathbb{R}.$$

We find all images of $f(x) = Ax$ by means of the expression

$$Ax = \begin{bmatrix} 0 & 1 & 0 \\ 0 & 0 & -1 \\ -1 & 0 & 2 \end{bmatrix} \begin{bmatrix} \lambda_1 + \lambda_2 \\ \lambda_1 + \lambda_2 \\ -2\lambda_1 + \lambda_2 \end{bmatrix} = \begin{bmatrix} \lambda_1 + \lambda_2 \\ 3\lambda_1 \\ -5\lambda_1 + \lambda_2 \end{bmatrix}, \ \lambda_1, \lambda_2 \in \mathbb{R}.$$

Hence we have a linear space of dimension 2 generated as follows:

$$\lambda_1 \begin{bmatrix} 1 \\ 3 \\ -5 \end{bmatrix} + \lambda_2 \begin{bmatrix} 1 \\ 0 \\ 1 \end{bmatrix}.$$

Exercise 1.4.7 Let be given the linear function $f : \mathbb{R}^3 \longrightarrow \mathbb{R}^4$ and the following vectors of $\ker(f)$:

$$z^1 = \begin{bmatrix} 1 \\ 0 \\ 2 \end{bmatrix}; \ z^2 = \begin{bmatrix} -2 \\ 1 \\ -1 \end{bmatrix}; \ z^3 = \begin{bmatrix} -1 \\ 2 \\ 4 \end{bmatrix}.$$

Let $[0] \neq b \in \mathbb{R}^4$ be a vector such that

$$f \begin{bmatrix} 1 \\ 0 \\ 1 \end{bmatrix} = b.$$

(1) Evaluate the rank of the matrix A associated with f.
(2) Decide if the vector

$$\hat{x} = \begin{bmatrix} 0 \\ 1 \\ 3 \end{bmatrix}$$

is or not a solution of the equation $f(x) = b$.

Solution

(1) We first build the following matrix M :

$$M = \begin{bmatrix} 1 & -2 & -1 \\ 0 & 1 & 2 \\ 2 & -1 & 4 \end{bmatrix}.$$

We have $|M| = 0$ and $\text{rank}(M) = 2$. Hence we can say that $\text{rank}(\ker(f)) = n_1 \geqq 2$. As the system

$$f \begin{bmatrix} 1 \\ 0 \\ 1 \end{bmatrix} = b \neq [0]$$

has a solution, we deduce that A cannot be the zero matrix and hence $\text{rank}(A) = n_2 \geqq 1$. Being $f : \mathbb{R}^3 \longrightarrow \mathbb{R}^4$, it must hold

$$n_1 + n_2 = 3.$$

Hence $\text{rank}(A) = 1$ and $\text{rank}(\ker(A)) = 2$.

(2) We know, from the previous point, that $\ker(f)$ is given, for example, by

$$\lambda_1 \begin{bmatrix} 1 \\ 0 \\ 2 \end{bmatrix} + \lambda_2 \begin{bmatrix} -2 \\ 1 \\ -1 \end{bmatrix}.$$

We have a particular solution of the nonhomogeneous system: the vector

$$\begin{bmatrix} 1 \\ 0 \\ 1 \end{bmatrix}.$$

By the superposition principle we must check if the system

$$\lambda_1 \begin{bmatrix} 1 \\ 0 \\ 2 \end{bmatrix} + \lambda_2 \begin{bmatrix} -2 \\ 1 \\ -1 \end{bmatrix} + \begin{bmatrix} 1 \\ 0 \\ 1 \end{bmatrix} = \begin{bmatrix} 0 \\ 1 \\ 3 \end{bmatrix}$$

is or not consistent. We have

$$\lambda_1 \begin{bmatrix} 1 \\ 0 \\ 2 \end{bmatrix} + \lambda_2 \begin{bmatrix} -2 \\ 1 \\ -1 \end{bmatrix} = \begin{bmatrix} -1 \\ 1 \\ 2 \end{bmatrix}.$$

Applying the Theorem of Rouché-Capelli, we see that the same is *not* verified. We deduce that

$$\hat{x} = \begin{bmatrix} 0 \\ 1 \\ 3 \end{bmatrix}$$

is *not* a solution of $f(x) = b$.

1.5 Introduction to Complex Numbers

In the present section we introduce, without pretensions of completeness and avoiding axiomatic approaches, the so-called *complex numbers.* It is well known that, if we want to solve the equation

$$ax = b, \quad a, b \in \mathbb{N} \text{ or also } a, b \in \mathbb{Z}, a \neq 0,$$

we have to introduce the set of *rational numbers* $\mathbb{Q}$. If we want to solve the equation

$$x^2 = 2,$$

we have to introduce the set of *real numbers* $\mathbb{R}$. Indeed, as it should be known, there exists no rational number of the form $\frac{p}{q}$, $q \neq 0$, such that

$$\left(\frac{p}{q}\right)^2 = 2.$$

If we want to solve, by means of real numbers, the simple equation

$$x^2 = -1,$$

we cannot find a solution of the same. While, for instance, the equation $x^2 = 1$ has two solutions in $\mathbb{R}$, the above equation is totally missing of solutions in $\mathbb{R}$. A sort of "par condicio" leads us to find a remedy to this inconvenient.

We call *imaginary unit,* denoted by the letter i, the number that by definition, satisfies the relation

$$i^2 = -1.$$

It is called *Complex Analysis* that topic of Mathematical Analysis which treats *complex numbers* and *complex variables,* i.e. those numbers (or variables) formed by two components, a *real component* and an *imaginary component.* A *complex number,* in its algebraic form, is written as

$$z = a + bi$$

where $a, b \in \mathbb{R}$ and i is the imaginary unit, previously introduced. The number a is called *the real part* and is usually denoted by $a = \text{Re}(z)$. The number b is called *the coefficient of the imaginary unit,* and is usually denoted by $b = \text{Im}(z)$.

If $a = 0$ we obtain the so-called *pure imaginary numbers;* obviously if $b = 0$, we obtain the *real numbers.* Then real numbers are a subset of complex numbers. We can therefore write

$$z = \text{Re}(z) + \text{Im}(z)i.$$

We note that $z = 0$ if and only if $a = 0,\ b = 0$. Moreover, $a + bi = c + di$ if and only if $a = c,\ b = d$. Obviously, it is equivalent to write ib instead of bi. In the set of complex numbers it is not possible to define *order relations* of the type $\geqq,\ \leqq,\ >,\ <$, as in the set of real numbers.

The complex number $\bar{z}$ is called the *conjugate* of z if it has the same real part of z and the coefficient of the imaginary part is of sign *opposed to the one of* z:

$$z = a + bi \Longrightarrow \bar{z} = a - bi; \quad z = a - bi \Longrightarrow \bar{z} = a + bi.$$

It results $\overline{(\bar{z})} = z$ and the pair $a \pm bi$ is also called a *conjugate pair.*

Now we are able to solve, for instance, the equation $x^2 + 9 = 0$. We find two roots (pure imaginary numbers) $x_{1,2} = \pm 3i$. If, for instance, we have the equation $x^2 - 2x + 65 = 0$, being $\Delta = 4 - 260$, we find the two solutions (complex conjugate) $1 \pm 8i$.

The set of complex numbers is usually denoted by $\mathbb{C}$ and hence $\mathbb{C}^n$ is used to denote the set of vectors with n complex components. The sum and difference of complex numbers is defined in the same way as in the algebra of real quantities:

$(a_1 + b_1 i) + (a_2 + b_2 i) = (a_1 + a_2) + (b_1 + b_2)i;$

$(a_1 + b_1 i) - (a_2 + b_2 i) = (a_1 - a_2) + (b_1 - b_2)i.$

Hence, for instance, $3i + 5i = 8i$; $(6 - 4i) - (3 + i) = 3 - 5i$; $7i - 8i = -i$.

It results $z + \bar{z} = a + bi + a - bi = 2a \in \mathbb{R}$.

The product between complex numbers is built with the usual rules of the basic algebra, with the care that $i^2 = -1$:

$$z_1 z_2 = (a + bi)(c + di) = ac + adi + bci + bdi^2 = (ac - bd) + (ad + bc)i.$$

For example, $(5 - 3i)(2 + 4i) = 22 + 14i$.

It results $z\bar{z} = (a + bi)(a - bi) = a^2 - b^2 i^2 = a^2 + b^2 \in \mathbb{R}$.

Now we consider the consecutive *powers* of the imaginary unit i. We have:

$$i \cdot i = i^2 = -1;\; i^3 = i^2 \cdot i = -1 \cdot i = -i;\; i^4 = (i^2)(i^2) = 1;$$
$$i^5 = i^4 \cdot i = i \text{ (i.e. } i^1);\; i^6 = (i^4)(i^2) = -1 \text{ (i.e. } i^2).$$

We deduce that, in general, it holds

$$i^m = i^{m+4k}, \quad k \geqq 0, \; k \text{ integer.}$$

In other words, the powers of i are the same, with *period equal to* 4: for instance, $i^{237} = i^{(4 \cdot 59)+1} = i^1 = i$.

Now let us consider the quotient between two complex numbers (with a nonzero denominator!). The simplest way to introduce this operation is to multiply numerator and denominator for the conjugate of the denominator:

$$\frac{a+bi}{c+di} = \frac{(a+bi)(c-di)}{(c+di)(c-di)} = \frac{ac+bd}{c^2+d^2} + \frac{bc-ad}{c^2+d^2} i.$$

Example 1.5.1

$$\frac{3-2i}{5+4i} = \frac{(3-2i)(5-4i)}{(5+4i)(5-4i)} = \frac{15-12i-10i+8i^2}{25+16} =$$
$$= \frac{7-22i}{41} = \frac{7}{41} - \frac{22}{41} i.$$

The *modulus* or *absolute value* of the complex number $z = a \pm bi$ is the nonnegative (real) quantity $\sqrt{a^2+b^2}$.

Summing up, we have the following Euler-Venn diagram (see Fig. 1.1).

It is quite convenient to give a geometric representation of a complex number $z = a + bi$ on a plane partitioned by two orthogonal oriented straight lines, a plane that in the present context is called the *complex plane* or *Gauss-Argand plane:* the abscissas axis, said *real axis,* gathers the real numbers (i.e. $b = 0$); the ordinate

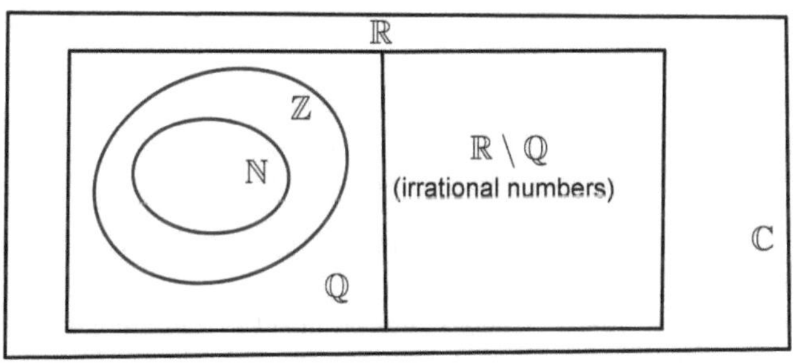

Fig. 1.1 Euler-Venn diagram of different classes of numbers

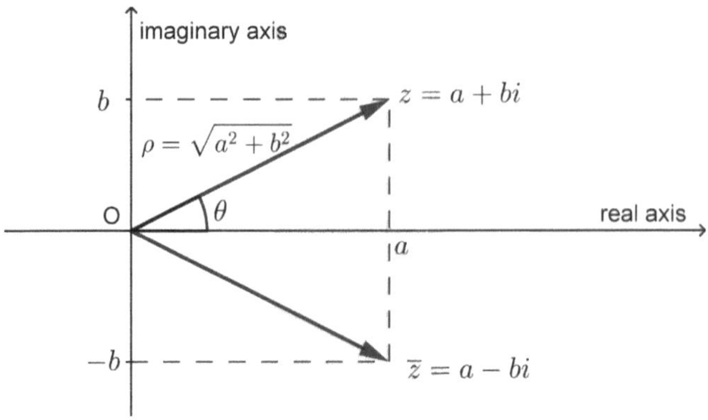

Fig. 1.2 Geometric representation of a complex number $z = a + bi$ and its conjugate $\overline{z} = a - bi$. ρ =modulus of z, θ =argument of z

axis, said *imaginary axis,* gathers the pure imaginary numbers ($a = 0$). Obviously there is also the possibility to represent a complex number, not only by the point of coordinates (a, b), but also by means of an arrow (just as for the bidimensional real vectors), with an extremum on the origin and the other extremum (the point of the arrow) on the point (a, b). The conjugate complex numbers will lie in a symmetric position with respect to the real axis. The lenght of the "arrow" which locates the complex number $z = a + bi$ is given by the modulus of z : $\sqrt{a^2 + b^2}$.

Now we give a short overview of the *trigonometric form of complex numbers.* We make reference to Fig. 1.2. We denote by ρ the modulus of $z = a + bi$ and we denote by θ the measure (in radians) of the angle that the arrow forms with the real axis (the measure has its sign, according to the usual rules of Trigonometry). The measure θ is said *argument* of the complex number z. The argument of z is denoted by $\arg(z)$, i.e. $\arg(z) = \theta$. Obviously, the argument is determined up to a number of complete "tours" of the trigonometric circumference, i.e. up to multipliers of 2π. More precisely, if θ is the argument of the complex number z, also any other angle $\theta' = \theta + 2k\pi$, with k any integer, ($k \in \mathbb{Z}$) is also an argument of the same complex number. When one does not want to take into consideration this indeterminacy of the argument, it is possible to impose some restrictions, of the type $\theta \in [0, 2\pi)$, or $\theta \in (-\pi, \pi]$, etc. The last written argument, i.e. $\theta \in (-\pi, \pi]$ is usually called the "principal argument".

Then it results $a = \rho \cos\theta;\ b = \rho \sin\theta$, hence

$$z = a + bi = \rho\cos\theta + i\rho\sin\theta =$$
$$= \rho(\cos\theta + i\sin\theta),\quad \theta \in [0, 2\pi)\,.$$

This formula is the so-called *trigonometric form* of the complex number $z = a + bi$.

For instance, if $z = 1 + \sqrt{3}i$, we have $a = 1,\ b = \sqrt{3},\ \rho = \sqrt{1+3} = 2,\ \cos\theta = \frac{a}{\rho} = \frac{1}{2},\ \sin\theta = \frac{b}{\rho} = \frac{\sqrt{3}}{2}$, from which $\theta = \frac{\pi}{3}$ and hence

$$1 + \sqrt{3}i = 2(\cos\frac{\pi}{3} + i\sin\frac{\pi}{3}).$$

If $z = 2 + 2i$, we have $a = 2,\ b = 2,\ \rho = 2\sqrt{2},\ \theta = \frac{\pi}{4}$ (represent z on the Argand-Gauss plane), hence

$$z = 2\sqrt{2}(\cos\frac{\pi}{4} + i\sin\frac{\pi}{4}).$$

We can observe that:

- A *positive* real number has argument $\theta = 0$.
- A *negative* real number has argument $\theta = \pi$.
- A pure imaginary number with a *positive* coefficient has argument $\theta = \pi/2$.
- A pure imaginary number with a *negative* coefficient has argument $\theta = \frac{3}{2}\pi$.

Let be $z = \rho(\cos\theta + i\sin\theta)$ and $w = r(\cos\psi + i\sin\psi)$ two complex numbers. Then:

(1) $zw = \rho r(\cos(\theta+\psi) + i\sin(\theta+\psi))$;
(2) If $w \neq 0$, then $\frac{z}{w} = \frac{\rho}{r}(\cos(\theta-\psi) + i\sin(\theta-\psi))$.

The rules of the product of two complex numbers obviously can be extended to any number of factors and, when the factors are all equal, we obtain the so-called *De Moivre formula* for the integer powers of a complex number:

$$z^n = [\rho(\cos\theta + i\sin\theta)]^n = \rho^n[\cos(n\theta) + i\sin(n\theta)].$$

If $z \neq 0$, the previous formula holds for $n \in \mathbb{Z}$.

Finally, we give some hints on the so-called *exponential form of complex numbers*. We recall the "Mac Laurin series expansion" for a real function which possesses derivatives of any order on a neighborhood of the origin. It results

$$f(x) = f(0) + f'(0)x + \frac{f''(0)}{2}x^2 + \dots + \frac{f^{(n)}(0)}{n!}x^n + \dots = \sum_{k=0}^{\infty}\frac{1}{k!}f^{(k)}(0)x^k,$$

with the convention that in the summation $f^{(0)} \equiv f$, and under appropriate conditions, verified by all elementary functions. Now we apply this expansion to the elementary functions $f(x) = \cos x,\ f(x) = \sin x,\ f(x) = e^x$.

$$\begin{aligned}\cos x &= \cos(0) - x\sin(0) - \frac{x^2}{2!}\cos(0) + \frac{x^3}{3!}\sin(0) + \dots\\ &= 1 - \frac{x^2}{2!} + \frac{x^4}{4!} - \frac{x^6}{6!} + \dots\end{aligned}$$

$$\sin x = x - \frac{x^3}{3!} + \frac{x^5}{5!} - \frac{x^7}{7!} + \dots$$

$$e^x = 1 + x + \frac{x^2}{2!} + \frac{x^3}{3!} + \frac{x^4}{4!} + \dots = \sum_{k=0}^{\infty} \frac{x^k}{k!}.$$

Hence we have formally

$$\begin{aligned} z &= \rho\,[\cos\theta + i\,\sin\theta] \\ &= \rho\left[(1 - \frac{\theta^2}{2!} + \frac{\theta^4}{4!} - \frac{\theta^6}{6!} + \dots) + i(\theta - \frac{\theta^3}{3!} + \frac{\theta^5}{5!} - \frac{\theta^7}{7!} + \dots)\right] \\ &= \rho\left[1 + i\theta - \frac{\theta^2}{2!} - i\frac{\theta^3}{3!} + \frac{\theta^4}{4!} + \frac{i\theta^5}{5!} - \frac{\theta^6}{6!} - \frac{i\theta^7}{7!} + \dots\right] \\ &= \rho\left[1 + i\theta + \frac{(i\theta)^2}{2!} + \frac{(i\theta)^3}{3!} + \frac{(i\theta)^4}{4!} + \frac{(i\theta)^5}{5!} + \frac{(i\theta)^6}{6!} + \frac{(i\theta)^7}{7!} + \dots\right]. \end{aligned}$$

It is therefore possible to write

$$z = \rho e^{i\theta}. \tag{1.1}$$

Relation (1.1) is said the *exponential form* of the complex number z. We remark that it holds

$$e^{i\theta} = \cos\theta + i\,\sin\theta;$$
$$e^{-i\theta} = \cos\theta - i\,\sin\theta.$$

By summing both members of the last two relations, we obtain

$$e^{i\theta} + e^{-i\theta} = 2\cos\theta$$

from which

$$\cos\theta = \frac{e^{i\theta} + e^{-i\theta}}{2}. \tag{1.2}$$

By subtracting member to member, we obtain

$$e^{i\theta} - e^{-i\theta} = 2i\,\sin\theta$$

from which

$$\sin\theta = \frac{e^{i\theta} - e^{-i\theta}}{2i}. \tag{1.3}$$

Relations (1.2) and (1.3) are the so-called *Euler formulas* (for the functions $\sin\theta$ and $\cos\theta$). We note that in particular we obtain

$$e^{i\pi} = \cos\pi + i\sin\pi$$

and, being $\cos\pi = -1$ and $\sin\pi = 0$, we obtain $e^{i\pi} = -1$, i.e.

$$e^{i\pi} + 1 = 0,$$

a wonderful and astonishing relation between the numbers $i, \pi, e, 0, 1$.

Another curiosity discovered by Euler is the fact that the number i^i is real. Indeed, from $z = \rho e^{i\theta}$ we get $i = e^{i\frac{\pi}{2}}$; hence $(i)^i = (e^{i\frac{\pi}{2}})^i = e^{i^2\frac{\pi}{2}} = e^{-\frac{\pi}{2}} \simeq 0,208$.

We conclude this section by noting that obviously it is possible to introduce *complex vectors*, as elements of the space $\mathbb{C}^n$, complex space with n dimensions:

$$x = [a_1 + ib_1,\ a_2 + ib_2, \ldots, a_n + ib_n]^\top.$$

We have to note that in $\mathbb{C}^n$ is not possible to introduce order relations of the type $\geqq, \leqq, >, <$, etc. The *conjugate vector of* x is the vector

$$\bar{x} = [a_1 - ib_1,\ a_2 - ib_2, \ldots, a_n - ib_n]^\top.$$

The *scalar product* between two complex vectors x and y of $\mathbb{C}^n$ is defined as follows:

$$x^\top y = \sum_{k=1}^{n} x_k \bar{y}_k.$$

We note that the above definition collapses to the usual one (for real vectors) when x and y are vectors of $\mathbb{R}^n$. The *norm* of the complex vector $x \in \mathbb{C}^n$ is defined as

$$\|x\| = \sqrt{\sum_{k=1}^{n} x_k \bar{x}_k} = \sqrt{\sum_{k=1}^{n} (a_k + ib_k)(a_k - ib_k)} = \sqrt{\sum_{k=1}^{n} \left[(a_k)^2 + (b_k)^2\right]}.$$

The vector $x \in \mathbb{C}^n$ is said to be *normalized* if $\|x\| = 1$. A complex vector $x \neq [0]$ can always be normalized by considering its associated vector

$$v = \frac{1}{\|x\|} x,$$

which results to be normalized.

Also matrices can have complex elements. The *transpose* of a complex matrix is called *Hermitian transpose* (from the French mathematician C. Hermite, 1822–1901), it is denoted by A^H and is defined as the usual transpose of the *conjugate matrix* $\bar{A}$ of A, whose elements are the conjugate of the elements of A. Hence it results

$$A^H = (\bar{A})^\top.$$

For instance, with

$$A = \begin{bmatrix} 1+i & 1 & 0 \\ -i & 2 & 3-2i \end{bmatrix},$$

it results

$$\bar{A} = \begin{bmatrix} 1-i & 1 & 0 \\ i & 2 & 3+2i \end{bmatrix},$$

$$A^H = \begin{bmatrix} 1-i & i \\ 1 & 2 \\ 0 & 3+2i \end{bmatrix}.$$

A square matrix A, with complex elements, is said to be *Hermitian* if

$$A = A^H.$$

A square matrix A, with complex elements, is said to be *unitary* if

$$AA^H = A^H A = I.$$

Do not make confusion with the identity matrix I. For instance

$$A = \begin{bmatrix} 3 & i & 3 \\ -i & 2 & 1+i \\ 3 & 1-i & 1 \end{bmatrix}$$

is an Hermitian matrix. The matrix

$$A = \frac{1}{2}\begin{bmatrix} 1+i & -1+i \\ 1+i & 1-i \end{bmatrix}$$

is an unitary matrix.

1.6 Eigenvalues and Eigenvectors of Square Matrices

Historically, the concepts of eigenvalue and eigenvector of square matrices arose in a rather curious way, within the studies of Celestial Mechanics: mathematicians and astronomers were trying to compute the perturbations of the orbit of Uranus, perturbations perhaps due to the presence of an unknown planet. The computations generated a particular equation, subsequently called "characteristic equation" and also, in Mathematical Physics, "secular equation", strictly connected to the elements of a square matrix. So, it was discovered Neptune, before it was possible to see it through telescopes!

From a strict mathematical and formal point of view, the problem can be stated as follows: let be given a square (real) matrix A, of order n; we have to find a vector $x \neq [0]$, such that the linear transformation Ax produces a multiple λx of the said vector x:

$$Ax = \lambda x, \; x \neq [0], \quad \lambda \in \mathbb{C}.$$

Therefore, even if we start from a real matrix A, the scalar λ may be complex! This will be clear after the next lines. Let us rewrite the last relation in the form

$$Ax - \lambda x = [0],$$

that is

$$Ax - \lambda I x = [0],$$

that is

$$(A - \lambda I)x = [0].$$

Therefore, we have to find the *non trivial solutions* of a square homogeneous linear system (in the vector x), where the coefficient matrix is $(A - \lambda I)$, matrix which may have complex elements, as $\lambda \in \mathbb{C}$. This fact must not worry, as linear systems with a complex matrix enjoy of the same properties and theorems of linear systems with a real coefficient matrix. Hence, in order to obtain $x \neq [0]$, it must be $\operatorname{rank}(A - \lambda I) < n$, i.e. $|A - \lambda I| = 0$, i.e.

$$\begin{vmatrix} (a_{11} - \lambda) & a_{12} & \cdots & a_{1n} \\ a_{21} & (a_{22} - \lambda) & \cdots & a_{2n} \\ \vdots & \vdots & \ddots & \vdots \\ a_{n1} & a_{n2} & \cdots & (a_{nn} - \lambda) \end{vmatrix} = 0. \tag{1.4}$$

Owing to the classical definition of determinant of a square matrix, relation (1.4) is an *algebraic equation of degree n* in λ, which is said the *characteristic equation of*

A. It is obvious that it is equivalent to write $|\lambda I - A| = 0$ instead of $|A - \lambda I| = 0$. It can be proved that (1.4) has the following expression

$$\lambda^n + c_1\lambda^{n-1} + c_2\lambda^{n-2} + \cdots + c_{n-2}\lambda^2 + c_{n-1}\lambda + c_n = 0$$

where the generic real coefficient c_k is given by $c_k = (-1)^k t_k$, and t_k is the *trace of order k* of A (see Sect. 1.3):

$$t_k = \sum_{i=1}^{\binom{n}{k}} \Delta_i(k),$$

and $\Delta_i(k)$ is the i-th principal minor of order k of A.

Now we recall the so-called *Fundamental Theorem of Algebra,* called also, inappropriately, *D'Alembert Theorem*; "inappropriately" as it was for the first time proved in a sound way by C. F. Gauss, in his degree thesis!

Theorem 1.6.1 *Every algebraic equation, with complex coefficients, and of degree* n :

$$a_0x^n + a_1x^{n-1} + \cdots + a_n = 0$$

admits n solutions (or roots), real or complex, $\alpha_1, \alpha_2, \ldots, \alpha_n$, *counted with their respective multiplicity.*

In other words: any polynomial

$$P(x) = a_0x^n + a_1x^{n-1} + \cdots + a_n$$

can be "factorized" as follows

$$P(x) = a_0(x - \alpha_1)(x - \alpha_2)\ldots(x - \alpha_n). \tag{1.5}$$

If in (1.5) the root α_i appears k times, then we say that α_i has *multiplicity* equal to k. If $k = 2$, we say also that α_i is a *double root.* If $k = 3$, we say also that α_i is a *triple root.* If $k = 1$ we say also that α_i is a *simple root.*

Theorem 1.6.2 *If all coefficients* $a_0, a_1, \ldots, a_n$ *are real and if* $P(x) = 0$ *has a complex root* $a + bi$, *then* $P(x)$ *has also the conjugate complex root* $a - bi$.

Corollary 1.6.3 *If the degree of* $P(x)$ *is odd and the coefficients* $a_0, a_1, \ldots, a_n$ *are real, then* $P(x) = 0$ *has at least one real root.*

Hence the characteristic equation $|A - \lambda I| = 0$ has n roots, real or complex, counted with their multiplicity order, say $\lambda_1, \lambda_2, \ldots \lambda_n$. These roots are called *eigenvalues* or *characteristic values* or *latent roots* of the matrix A. These roots may be simple or multiple. Moreover, as we started from a real matrix, if there exists a complex eigenvalue $\lambda = a + bi$, there will be also its conjugate eigenvalue $\bar{\lambda} = a - bi$. The set of all eigenvalue of A is called the "spectrum" of A. It is important to note that A and $A^\top$ have the same spectrum (the spectrum is the result of a determinant!).

Theorem 1.6.4

(1) $\lambda = 0$ *is an eigenvalue of* A *if and only if* $|A| = 0$ *(this fact is trivial:* $|A - 0I| = 0 \iff |A| = 0$*).*
(2) If A *is a diagonal matrix or even a triangular matrix, then its eigenvalues are given by the elements of the main diagonal.*
(3) If $\lambda_1, \lambda_2, \dots, \lambda_n$ *are the elements of the spectrum of* A*, then:*

   (a) $\lambda_1\lambda_2 \dots \lambda_n = |A|$.
   (b) $\lambda_1 + \lambda_2 + \cdots + \lambda_n = \sum_{i=1}^n a_{ii} = t_1 = \mathrm{tr}(A)$*, the trace of* A.

(4) If A *is invertible, the eigenvalues of* A^{-1} *are the reciprocals of the eigenvalues of* A*, but the eigenvectors are the same: indeed from* $Ax = \lambda x$ *we get* $x = \lambda A^{-1}x$*, from which* $A^{-1}x = \frac{1}{\lambda}x$.
(5) If λ *is an eigenvalue of* A*, then* λ^p *is an eigenvalue of* $(A)^p$*, with* p *positive integer.*

Example 1.6.5 Compute the eigenvalues of:

$$(1) \quad A = \begin{bmatrix} 4 & -1 \\ 2 & 1 \end{bmatrix}; \quad (2) \quad A = \begin{bmatrix} 1 & -4 \\ 1 & 1 \end{bmatrix}.$$

(1) It results

$$|A - \lambda I| = \begin{vmatrix} 4-\lambda & -1 \\ 2 & 1-\lambda \end{vmatrix}.$$

We have $(4 - \lambda)(1 - \lambda) + 2 = 0$, i.e. $\lambda^2 - 5\lambda + 6 = 0$, from which $\lambda_1 = 2,\ \lambda_2 = 3$. Note that $\lambda_1\lambda_2 = 6 = |A|;\ \lambda_1 + \lambda_2 = 5 = \mathrm{tr}(A)$.
(2) It results

$$|A - \lambda I| = \begin{vmatrix} 1-\lambda & -4 \\ 1 & 1-\lambda \end{vmatrix}.$$

We have therefore $(1 - \lambda)^2 + 4 = 0$, i.e. $\lambda^2 - 2\lambda + 5 - 0$, from which $\lambda_1 = 1 + 2i;\ \lambda_2 = 1 - 2i$.

If and only if the scalar λ is an eigenvalue of A, then the system

$$Ax = \lambda x, \quad \lambda \in \mathbb{C},$$

has a non trivial solution $x \neq [0]$. The vectors $x^i \neq [0]$ such that $Ax^i = \lambda_i x^i$, $i = 1, \ldots, n$, are the *eigenvectors* or *characteristic vectors of A associated with the eigenvalue* λ_i. Note that, being $|A - \lambda I| = 0$, the system $(A - \lambda I)x = [0]$ has infinitely many eigenvectors x^i, associated with λ_i, $i = 1, \ldots, n$.

Example 1.6.6 Let be given the matrix

$$A = \begin{bmatrix} 2 & -2 & 3 \\ 1 & 1 & 1 \\ 1 & 3 & -1 \end{bmatrix}.$$

Its characteristic equation is

$$-\lambda^3 + 2\lambda^2 + 5\lambda - 6 = 0$$

which admits three (distinct) real roots: $\lambda_1 = 1$, $\lambda_2 = -2$, $\lambda_3 = 3$. Let us find, for instance, the eigenvectors (there will be infinitely many eigenvectors!) associated with $\lambda_3 = 3$. We have to solve the system $(A - 3I)x^3 = [0]$, $x^3 \neq [0]$. It results

$$B = A - 3I = \begin{bmatrix} -1 & -2 & 3 \\ 1 & -2 & 1 \\ 1 & 3 & -4 \end{bmatrix}.$$

It results $\operatorname{rank}(B) = 2$. We choose, for instance, the first two equations (the third equation is a linear combination of the first two equations) and we solve the system

$$\begin{cases} -x_1 - 2x_2 = -\bar{x}_3 \\ x_1 - 2x_2 = -\bar{x}_3. \end{cases}$$

It results $\bar{x}_1 = \bar{x}_3$; $\bar{x}_2 = \bar{x}_3$, and hence all vectors of $\mathbb{R}^3$ of the type

$$\bar{x}^3 = \begin{bmatrix} \alpha \\ \alpha \\ \alpha \end{bmatrix}, \quad \alpha \neq 0,$$

are eigenvectors of A associated with $\lambda_3 = 3$.

We have already remarked, in the previous example, that, being $\operatorname{rank}(A - \lambda I) < n$, we find infinitely many eigenvectors associated with each eigenvalue: more precisely, ∞^{n-r} eigenvectors, being r the rank of $(A - \lambda I)$.

If A (square and real) has a *complex eigenvector* λ_k, its associated eigenvector x^k is complex too (if x^k would be real we would have that the real quantity Ax^k is equal to the complex quantity $\lambda_k x^k$). Furthermore, the matrix admits the *conjugate eigenvalue* $\bar{\lambda}_k$ and it is always possible to choose $\bar{x}^k$ as the conjugate of the complex eigenvector x^k.

Example 1.6.7 Let be given the matrix

$$A = \begin{bmatrix} 0 & 0 & 0 \\ 1 & 0 & -1 \\ 0 & 1 & 0 \end{bmatrix}.$$

It results $|A - \lambda I| = -(\lambda^3 + \lambda)$, and hence $\lambda_1 = 0$, $\lambda_2 = i$, $\lambda_3 = -i$. The (real) eigenvectors associated with λ_1 are of the type $x^1 = \alpha\,[1, 0, 1]^\top$, $\alpha \in \mathbb{R}$, $\alpha \neq 0$. The eigenvectors associated with λ_2 are of the type $x^2 = \alpha\,[0, 1, -i]^\top$, α any, $\alpha \neq 0$. The eigenvectors associated with λ_3 are of the type $x^3 = \alpha\,[0, 1, i]^\top$, α any, $\alpha \neq 0$. Note that x^2 and x^3 are conjugate (by choosing the same α for both).

The eigenvectors introduced until now are column vectors x which therefore post-multiply A; they are also said *column eigenvectors* or *right-hand eigenvectors.* It is obviously possible to introduce *row eigenvectors* or *left-hand eigenvectors,* which are those row vectors $y^\top \neq [0]$ solutions of

$$y^\top A = \lambda y^\top .$$

This system obviously can be rewritten in the form

$$A^\top y = \lambda y$$

from which we deduce that if A is *symmetric*, right-hand eigenvectors coincide with left-hand eigenvectors (apart from typographic aspects). The following property is less trivial.

Theorem 1.6.8 *Let be $y^\top$ a row eigenvector of A, associated with the eigenvalue λ and let be x a column eigenvector of A associated with the eigenvalue μ, with $\mu \neq \lambda$. Then y and x are orthogonal vectors: $y^\top x = 0$.*

Proof By assumption $y^\top A = \lambda y^\top$ and $Ax = \mu x$. Hence $y^\top Ax = y^\top(Ax) = y^\top(\mu x) = \mu(y^\top x)$ and also $y^\top Ax = (y^\top A)x = (\lambda y^\top)x$. Therefore $\mu(y^\top x) = \lambda(y^\top x)$, i.e. $(\mu - \lambda)(y^\top x) = 0$ and being $\mu \neq \lambda$, it holds $y^\top x = 0$. □

Let be λ^* an eigenvalue of A and let be X^* the set of all eigenvectors of A associated with λ^*. For what previously said, the set $X^* \cup \{[0]\}$ is a linear subspace of $\mathbb{C}^n$, of dimension $n - \text{rank}(A - \lambda^* I)$, subspace called *eigenspace* or *invariant space* associated with λ^*; the eigenspace is obviously the kernel of $(A - \lambda^* I)$.

Definition 1.6.9 Let λ_i be an eigenvalue of the matrix (real and square) A of order n.

- The *geometric multiplicity* of λ_i, denoted by

$$m_i^G,$$

is the dimension of the eigenspace associated with λ_i, i.e.

$$m_i^G = \dim(\ker(A - \lambda_i I)) = n - \text{rank}(A - \lambda_i I),$$

that is the maximum number of linearly independent eigenvectors associated with λ_i (this number cannot be zero).

- The *algebraic multiplicity* of λ_i, denoted by

$$m_i^A,$$

is the order of multiplicity of λ_i, as a root of the characteristic equation of A.

Obviously, the sum of the algebraic multiplicities of all eigenvalues of A is equal to n. There exists an important relation between algebraic multiplicity and geometric multiplicity, given by the following result.

Theorem 1.6.10 *It holds*

$$1 \leqq m_i^G \leqq m_i^A, \quad \forall i.$$

This theorem will be proved in the next pages, after Theorem 1.6.21.

It may happen that, for some index i,

$$m_i^G < m_i^A.$$

In this case we say that A is *defective* or that A presents a *defect of multiplicity.*

Example 1.6.11 Let be

$$A = \begin{bmatrix} 1 & 0 \\ 1 & 1 \end{bmatrix}.$$

It results $|A - \lambda I| = (1 - \lambda)^2$, hence $\lambda_{1,2} = 1$. Therefore $m^A = 2$. Then it holds $m^G = 2 - \text{rank}(A - 1I) = 2 - \text{rank}(A - I)$ and being

$$A - I = \begin{bmatrix} 0 & 0 \\ 1 & 0 \end{bmatrix},$$

it results $m^G = 2 - 1 = 1$.

If it results $m_i^G = m_i^A$, that is if $n - \text{rank}(A - \lambda_i I) = m_i^A$, then the eigenvalue λ_i is said to be *regular* and if all eigenvalues are regular, the matrix A has no defect of multiplicity. The defect of multiplicity shows, roughly speaking, a loss of dimension. Taking into account that the sum of all algebraic multiplicities gives n, if we want that the number of linearly independent eigenvectors of A is just n, it is not difficult to realize the validity of the following result.

Theorem 1.6.12 *The square matrix A of order n has n linearly independent eigenvectors $x^1, x^2, \ldots, x^n$ if and only if every eigenvalue of A is regular, that is*

$$m_i^G = m_i^A, \ \forall i,$$

that is

$$n - \text{rank}(A - \lambda_i I) = m_i^A, \quad \forall i.$$

We observe that if λ_i is a *simple* root of the characteristic equation, i.e. if $m_i^A = 1$, as $1 \leqq m_i^G \leqq m_i^A = 1$, we have obviously $m_i^G = 1 = m_i^A$, hence the following result holds true.

Corollary 1.6.13 *If the n eigenvalues $\lambda_1, \ \lambda_2, \ldots, \lambda_n$ are all distinct roots, then A has n linearly independent eigenvectors.*

We now collect in the next theorem the main spectral properties of *symmetric matrices.*

Theorem 1.6.14 *Let A be a symmetric matrix, of order n. The following properties hold.*

(1) The eigenvalues of A are all real.
(2) A has n eigenvectors which can be always chosen two by two orthogonal.
(3) A admits n linearly independent eigenvectors (this is a consequence of the previous property and Theorem 1.2.7.
(4) Columns eigenvectors and row eigenvectors (associated to the same eigenvalue) coincide.

Now we give some notions on the so-called *similarity transformations* between two square matrices (of the same order). Analytic and algorithmic issues lead to the research of "transformations" of a square matrix A into another matrix B, of the same order of A, and with the same spectrum of A, but more suitable from a computational point of view, than the original matrix A. We recall again that the problem of computing the eigenvalues of a square matrix is a problem connected with the research of the roots of an algebraic equation of degree n, a problem surely not trivial, for $n > 4$. We begin with the following basic definition.

Definition 1.6.15 Two matrices A and B, square of the same order n, are said to be *similar* if there exists a nonsingular square matrix S, of order n, such that

$$B = S^{-1}AS$$

that is

$$A = SBS^{-1}.$$

The operation described in the above definition, which transforms A into B or B into A, is called *similarity transformation.* The matrix S is also called *transition matrix.* From the previous definition it follows that:

(a) Every square matrix is similar to itself ("reflexive property").
(b) If A is similar to B, then B is similar to A (symmetric property").
(c) If A and B are two similar matrices and B and C are similar, then also A and C are similar ("transitive property").

In other words, similarity transformations are equivalence relations.

Theorem 1.6.16 *The matrices A and $B = S^{-1}AS$ have the same spectrum.*

Proof We have to prove that

$$|A - \lambda I| = |B - \lambda I| .$$

It results

$$\begin{aligned} |B - \lambda I| &= \left|S^{-1}AS - \lambda I\right| = \left|S^{-1}AS - \lambda S^{-1}IS\right| = \left|S^{-1}(AS - \lambda IS)\right| \\ &= \left|S^{-1}(A - \lambda I)S\right| = \text{(Binet-Cauchy Theorem)} \left|S^{-1}\right| |A - \lambda I| \, |S| \\ &= \frac{1}{|S|} |A - \lambda I| \, |S| = |A - \lambda I| . \end{aligned}$$

□

Therefore, two similar matrices A and $S^{-1}AS$ have the same eigenvalues (the vice-versa does not hold), but they have not the same eigenvectors and hence in general the two eigenspaces do not coincide. However, if $\bar{x}$ is an eigenvector of A associated with the eigenvalue λ, then $\bar{z} = S^{-1}\bar{x}$ is an eigenvector of B, associated with the same eigenvalue λ. Indeed, if $A\bar{x} = \lambda\bar{x}$, then $(S^{-1}AS)S^{-1}\bar{x} = \lambda(S^{-1}\bar{x})$, or $(S^{-1}AS)\bar{z} = \lambda\bar{z}$, i.e. $\bar{z}$ is an eigenvector of B associated with λ. If the transformed matrix B is a diagonal matrix, its eigenvalues appear as the elements of its main diagonal. Then it is natural to find (if there exists!) an invertible matrix V which "diagonalizes" A, that is, such that

$$V^{-1}AV = D,$$

with D a diagonal matrix. We speak then of *diagonalization* of A and in this case therefore

$$V^{-1}AV = \begin{bmatrix} \lambda_1 & 0 & \cdots & 0 \\ 0 & \lambda_2 & \cdots & 0 \\ \vdots & \vdots & \cdots & \vdots \\ 0 & 0 & \cdots & \lambda_n \end{bmatrix},$$

being $\lambda_1, \lambda_2, \ldots, \lambda_n$ the eigenvalues of A. The matrix V which performs the diagonalization is also called *polar matrix or modal matrix of* A.

Theorem 1.6.17 *A square matrix A of order n is diagonalizable if and only if it admits n linearly independent eigenvectors* $x^1, x^2, \ldots, x^n$*, that is if and only if each its eigenvalue is regular (algebraic multiplicity = geometric multiplicity), that is if and only if the following equality holds true for every eigenvalue:*

$$m_i^A = n - \text{rank}(A - \lambda_i I).$$

Proof Let $x^1, x^2, \ldots, x^n$ be n linearly independent eigenvectors of A, associated to the respective eigenvalues $\lambda_1, \lambda_2, \ldots, \lambda_n$. The n relations $Ax^i = \lambda_i x^i$ can be put in the matrix relation $AV = VD$, where V is the matrix obtained by putting together the n column eigenvectors of A and D is the diagonal matrix having the eigenvalues of A on its main diagonal. Being, by assumption, $|V| \neq 0$, we obtain at once $A = VDV^{-1}$, that is $D = V^{-1}AV$, that is the result that A is similar to a diagonal matrix.

Vice-versa, if the equality $A = VDV^{-1}$ holds, with D a diagonal matrix, we obtain $AV = VD$, that is the relations $Ax^i = \lambda_i x^i$, $i = 1, \ldots, n$. The column of V are therefore the linearly independent eigenvectors of A, being $|V| \neq 0$. □

We again point out that the polar matrix V is formed by the n linearly independent column eigenvectors of A.

Corollary 1.6.18 *If the square matrix A, of order n, has all distinct eigenvalues, then A is diagonalizable.*

Another result is concerned with *symmetric matrices.*

Theorem 1.6.19 *Let A be symmetric. Then*

(1) A is always diagonalizable (even if A has multiple eigenvalues).
(2) It is always possible to diagonalize A by means of an orthogonal matrix:

$$D = V^\top AV.$$

We omit the proof. We remark that (1) of the previous theorem is not really a "novelty", on the grounds of Theorem 1.6.14. Point (2) is called also "Spectral Theorem" or "Theorem of the principal axes" and has some interesting applications (for example in the theory of *quadratic forms;* see the next section). When V is orthogonal, then it is formed by the normalized and orthogonal eigenvectors of the symmetric matrix A. In other words, the columns of the polar matrix V are formed by *orthonormal vectors.* Another class of (square) matrices always diagonalizable is the class of *normal matrices* ($AA^\top = A^\top A$).

Example 1.6.20 The symmetric matrix

$$A = \begin{bmatrix} 1 & 2 \\ 2 & 1 \end{bmatrix}$$

admits two real and distinct eigenvalues $\lambda_1 = -1$ and $\lambda_2 = 3$. The two associated eigenvectors are

$$x^1 = k \begin{bmatrix} 1 \\ -1 \end{bmatrix}; \quad x^2 = h \begin{bmatrix} 1 \\ 1 \end{bmatrix}, \quad k, h \neq 0.$$

Note that, with $k = h$, x^1 and x^2 are orthogonal. Let us choose them as normalized vectors, i.e. we put

$$h = k = \frac{1}{\sqrt{2}} = \frac{\sqrt{2}}{2}$$

and we obtain the polar matrix V:

$$V = \begin{bmatrix} \frac{\sqrt{2}}{2} & \frac{\sqrt{2}}{2} \\ -\frac{\sqrt{2}}{2} & \frac{\sqrt{2}}{2} \end{bmatrix}.$$

This matrix is orthogonal and diagonalizes A, i.e. $A = VDV^\top$ (check!).

If A is diagonalizable, then its powers are easily obtained. Indeed, from $A = VDV^{-1}$ we have

$$(A)^k = (VDV^{-1})^k = \underbrace{(VDV^{-1})(VDV^{-1})\dots(VDV^{-1})}_{k \text{ times}}$$

$$= VD(V^{-1}V)D(V^{-1}V)\dots(V^{-1}V)DV^{-1} = VDD\dots DV^{-1} = V(D)^k V^{-1}.$$

The advantage is in the fact that $(D)^k$ is the power of a diagonal matrix, which is a diagonal matrix formed by the powers of the elements of its main diagonal:

$$(D)^k = \begin{bmatrix} \lambda_1^k & 0 & \cdots & 0 \\ 0 & \lambda_2^k & \cdots & 0 \\ \vdots & \vdots & \ddots & 0 \\ 0 & 0 & \cdots & \lambda_n^k \end{bmatrix}.$$

We have seen that the diagonalization of a square matrix A is not always possible. However, it holds always the following result, known as *Theorem of Schur,* which ensures that *any* square matrix can be "triangularized".

Theorem 1.6.21 (Theorem of Schur) *Any square matrix is similar to a triangular matrix T, lower or upper:*

$$T = S^{-1}AS.$$

On the grounds of the above theorem the eigenvalues of A are the elements of the main diagonal of T. If A (real square matrix) has all real eigenvalues, it is possible to choose, as a transition matrix S, an orthogonal matrix. The triangular matrix T is also called *Schur canonical form.* The product

$$A = STS^{-1}$$

is also known as *Schur decomposition of* A. The theorem of Schur allows also to give an easy proof that $\det A = \lambda_1\lambda_2 \dots \lambda_n$. Indeed, A and T have the same eigenvalues and, being T a triangular matrix, it holds $\det T = \lambda_1\lambda_2 \dots \lambda_n$. Furthermore, as similar matrices also have the same trace, we have $\mathrm{tr}(A) = \mathrm{tr}(T) = \lambda_1 + \lambda_2 + \cdots + \lambda_n$.

Now we are able to give a simple proof of the property

$$1 \leqq m_i^G \leqq m_i^A, \text{ for every index } i,$$

i.e. the thesis of Theorem 1.6.10.

Proof of Theorem 1.6.10. Let us suppose that $m^A(\lambda) = k$. The Theorem of Schur assures the existence of a transition matrix S such that

$$S^{-1}AS = \begin{bmatrix} T_{11} & T_{12} \\ [0] & T_{22} \end{bmatrix},$$

where T_{11} is an upper triangular matrix of order k, and whose diagonal elements are equal to λ. The matrix T_{22} is an upper triangular matrix of order $(n-k)$, whose spectrum *does not contain* λ. Hence $T_{22} - \lambda I$ is nonsingular and we have (the rank of a matrix remains unchanged if the matrix is premultiplied or postmultiplied by a nonsingular square matrix):

$$\begin{aligned} \mathrm{rank}(A - \lambda I) = \mathrm{rank}(S^{-1}(A - \lambda I)S) = \mathrm{rank} \begin{pmatrix} T_{11} - \lambda I & T_{12} \\ [0] & T_{22} - \lambda I \end{pmatrix} \\ \geqq \mathrm{rank}(T_{22} - \lambda I) = n - k. \end{aligned}$$

The inequality follows from the fact that the rank of a matrix is greater or equal to the rank of any its submatrix. Therefore,

$$\mathrm{rank}(A - \lambda I) \geqq n - k$$

that is

$$k \geqq n - \text{rank}(A - \lambda I), \quad \text{that is } k \geqq m^G(\lambda).$$

□

The nice result of Schur (Theorem 1.6.21) can be "refined" by means of the so-called *Jordan canonical form*. In other words, it is always possible to obtain a similarity transformation for a square matrix A, such that A is transformed in an "almost diagonal form":

$$S^{-1}AS = J,$$

where J is a "diagonal block matrix", said *Jordan matrix or Jordan canonical form or Jordan normal form*, having the following properties:

(i) The elements of the main diagonal of J are the eigenvalues of A and the multiple eigenvalues appear on the main diagonal in a consecutive position.
(ii) The elements of the diagonal above the main diagonal are either zero or one. In any case the said elements are 1 if they are near an eigenvalue which has algebraic multiplicity greater than one.
(iii) All other elements of J are zero.

For instance, with $n = 4$, the matrix J could be the following one:

$$J = \begin{bmatrix} \lambda_1 & 1 & 0 & 0 \\ 0 & \lambda_1 & 0 & 0 \\ 0 & 0 & \lambda_2 & 1 \\ 0 & 0 & 0 & \lambda_2 \end{bmatrix}.$$

With reference to this form, the two submatrices

$$\begin{bmatrix} \lambda_1 & 1 \\ 0 & \lambda_1 \end{bmatrix}; \quad \begin{bmatrix} \lambda_2 & 1 \\ 0 & \lambda_2 \end{bmatrix}$$

are called "Jordan blocks". When A is diagonalizable, the matrix J "collapses" to the diagonal matrix D which has on its diagonal the eigenvalues of A. The Jordan canonical form can always be written as a sum of two matrices

$$J = D + N,$$

where D is the diagonal matrix of the eigenvalues of A, with their multiplicity, and N is a particular matrix whose elements are all zero, except the elements above the main diagonal, which are either zero or the number 1. Moreover, it must be pointed out that the consecutive powers of N have the property that, in each multiplication, the diagonal which contains the numbers 1 is shifted of one place upwards. For example, in the case of a matrix of order 4, if

$$N = \begin{bmatrix} 0 & 1 & 0 & 0 \\ 0 & 0 & 1 & 0 \\ 0 & 0 & 0 & 1 \\ 0 & 0 & 0 & 0 \end{bmatrix},$$

we have

$$(N)^2 = \begin{bmatrix} 0 & 0 & 1 & 0 \\ 0 & 0 & 0 & 1 \\ 0 & 0 & 0 & 0 \\ 0 & 0 & 0 & 0 \end{bmatrix}, \quad (N)^3 = \begin{bmatrix} 0 & 0 & 0 & 1 \\ 0 & 0 & 0 & 0 \\ 0 & 0 & 0 & 0 \\ 0 & 0 & 0 & 0 \end{bmatrix}, \quad (N)^4 = [0] \, .$$

Hence, if n is the order of N, the power $(N)^n$ is given by the zero matrix $[0]$. In other words, N is a so-called *nilpotent matrix (of index n), that is* $(N)^n = [0]$, but $(N)^{n-1} \neq [0]$.

Further insights on the rather deep theory of Jordan canonical form are beyond the aims of the present book and we refer the reader to the references.

Another important result is the following one.

Theorem 1.6.22 (Theorem of Cayley-Hamilton) *Let A be a square matrix of order n and its characteristic equation $|A - \lambda I| = 0$ is given by the expression*

$$a_0\lambda^n + a_1\lambda^{n-1} + \cdots + a_{n-1}\lambda + a_n = 0.$$

Then, it holds

$$a_0(A)^n + a_1(A)^{n-1} + \cdots + a_{n-1}A + a_n I = [0] \, .$$

In other words, as it is customary to say, every square matrix A verifies (in a matrix sense!) its characteristic equation.

Proof The proof for the general case is quite long and tedious. We give the simple proof for the case where A is diagonalizable. Let be $A = VDV^{-1}$, with D diagonal matrix containing the eigenvalues of A as its diagonal elements. We have therefore

$$a_0(A)^n + a_1(A)^{n-1} + a_2(A)^{n-2} + \cdots + a_n I$$

$$= \sum_{i=0}^{n} a_i (VDV^{-1})^{n-i} = \sum_{i=0}^{n} a_i V(D)^{n-i} V^{-1} = V\left(\sum_{i=0}^{n} a_i (D)^{n-i}\right)V^{-1}.$$

The expression between brackets is

$$\sum_{i=0}^{n} a_i \begin{bmatrix} \lambda_1^{n-i} & 0 & \cdots & 0 \\ 0 & \lambda_2^{n-i} & \cdots & 0 \\ \vdots & \vdots & \ddots & \vdots \\ 0 & 0 & \cdots & \lambda_n^{n-i} \end{bmatrix}$$

$$= \begin{bmatrix} \sum_{i=0}^{n} a_i \lambda_1^{n-i} & 0 & \cdots & 0 \\ 0 & \sum_{i=0}^{n} a_i \lambda_2^{n-i} & \cdots & 0 \\ \vdots & \vdots & \ddots & \vdots \\ 0 & 0 & \cdots & \sum_{i=0}^{n} a_i \lambda_n^{n-i} \end{bmatrix}$$

$$= \begin{bmatrix} p_n(\lambda_1) & 0 & \cdots & 0 \\ 0 & p_n(\lambda_2) & \cdots & 0 \\ \vdots & \vdots & \ddots & \vdots \\ 0 & 0 & \cdots & p_n(\lambda_n) \end{bmatrix},$$

where $p_n(\lambda)$ is the characteristic polynomial of A. Being $\lambda_1, \lambda_2, \ldots, \lambda_n$ the eigenvalues of A, it will hold $p_n(\lambda_1) = 0,\ p_n(\lambda_2) = 0, \ldots, p_n(\lambda_n) = 0$, hence the above matrix is the zero matrix. It follows

$$\sum_{i=0}^{n} a_i (VDV^{-1})^{n-i} = V(\sum_{i=0}^{n} a_i\, [0])V^{-1} = V\,[0]\,V^{-1} = [0]\,.$$

□

Example 1.6.23 Consider the matrix

$$A = \begin{bmatrix} 4 & -1 \\ 2 & 1 \end{bmatrix}.$$

We have $|A - \lambda I| = \lambda^2 - 5\lambda + 6$. Then it holds

$$\left(\begin{bmatrix} 4 & -1 \\ 2 & 1 \end{bmatrix}\right)^2 - 5\begin{bmatrix} 4 & -1 \\ 2 & 1 \end{bmatrix} + 6\begin{bmatrix} 1 & 0 \\ 0 & 1 \end{bmatrix} = \begin{bmatrix} 0 & 0 \\ 0 & 0 \end{bmatrix}.$$

One of the many applications of the Theorem of Cayley-Hamilton is a method to compute the inverse of a nonsingular matrix. If the square matrix A of order n admits inverse, from

$$a_0(A)^n + a_1(A)^{n-1} + \cdots + a_{n-1}A + a_n I = [0]$$

we have, by multiplying both members by A^{-1},

$$a_0(A)^{n-1} + a_1(A)^{n-2} + \cdots + a_{n-1}I + a_n A^{-1} = [0]$$

from which

$$A^{-1} = \frac{1}{a_n}(-a_0(A)^{n-1} - a_1(A)^{n-2} - \cdots - a_{n-1}I).$$

Now we give some hints on *matrix series.* Let us consider the spectrum of a real square matrix A; as previously seen, there may exist also complex eigenvalues. By $|\lambda|$ we denote the modulus of the eigenvalue λ; obviously the modulus coincides with the absolute value if λ is real.

Definition 1.6.24 Let A be a real square matrix of order n; the nonnegative number

$$\rho_A = \max_{1 \leqq i \leqq n} \{|\lambda_i|\}$$

is called the *spectral radius of* A.

Example 1.6.25 Let

$$A = \begin{bmatrix} 2 & 0 & 1 \\ 2 & 1 & 1 \\ 0 & -1 & 0 \end{bmatrix}.$$

We have $|A - \lambda I| = -\lambda^3 + 3\lambda^2 - 3\lambda$. It results $\lambda_1 = 0,\ \lambda_2 = \frac{1}{2}(3+\sqrt{3}\,i),\ \lambda_3 = \frac{1}{2}(3 - \sqrt{3}\,i)$. Then

$$\left|\frac{3}{2} + \frac{\sqrt{3}}{2}i\right| = \left|\frac{3}{2} - \frac{\sqrt{3}}{2}i\right| = \sqrt{\frac{9}{4} + \frac{3}{4}} = \sqrt{3} = \rho_A.$$

It should be known, from the basic courses of Mathematics, the notion of limit of a (numerical) sequence. It is natural to introduce the notion of limit of a *sequence of matrices* (with real elements). Consider a sequence of matrices, all of the same order (m, n) :

$$B_{(1)},\ B_{(2)}, \ldots,\ B_{(k)},\ B_{(k+1)}, \ldots$$

We say that this sequence *converges* to the matrix B and write

$$\lim_{k \longrightarrow +\infty} B_{(k)} = B = \left[b_{ij}\right]$$

whenever

$$\lim_{k \longrightarrow +\infty} b_{ij(k)} = b_{ij}, \quad \forall i, j.$$

Consider now a square matrix of order n. Put $(A)^0 = I$; the expression

$$\sum_{k=0}^{+\infty} (A)^k = I + A + (A)^2 + (A)^3 + \cdots + \ldots \tag{1.6}$$

is a *power series of matrices.* If we put

$$S_{(k)} = I + A + (A)^2 + (A)^3 + \cdots + (A)^k,$$

and if the sequence $\left\{S_{(k)}\right\}$ converges to the matrix S, that is

$$\lim_{k \longrightarrow +\infty} S_{(k)} = S,$$

we say that the series (1.6) *converges to* S and write

$$\sum_{k=0}^{+\infty} (A)^k = S.$$

In particular, if

$$\lim_{k \longrightarrow \infty} (A)^k = [0]$$

we say that A is "small" or also that A is "convergent". Hence A is convergent if all elements $(a_{ij})^k$ of $(A)^k$ tend to zero, as $k \longrightarrow +\infty$. We have the following basic result.

Theorem 1.6.26 *Let be given the square matrix A of order n. Then A is convergent if and only if $\rho_A < 1$, where ρ_A is the spectral radius of A. More generally, if $\mu > 0$, then μA is convergent if and only if $\rho_A < \frac{1}{\mu}$.*

Proof Let J be the Jordan canonical form of A, obtained by a similarity transformation of the type VAV^{-1}. It follows that $(\mu J)^n = V(\mu A)^n V^{-1}$ and hence $V^{-1}(\mu J)^n V = (\mu A)^n$. From this it appears immediately that the conditions of convergence of $(\mu A)^n$ coincide with the conditions of convergence of $(\mu J)^n$. First, let us suppose that A is *diagonalizable,* i.e. $J = D$. In this case if (and only if) $\rho_A < \frac{1}{\mu}$, all the elements of the main diagonal of D are in absolute value less than one and hence their k-th powers converge to zero, as $k \longrightarrow +\infty$. It follows

$$\lim_{k \longrightarrow +\infty} (\mu D)^k = [0]$$

and hence also

$$\lim_{k \longrightarrow +\infty} (\mu A)^k = [0] .$$

If A is *not diagonalizable,* it is always possible to write the k-th power of the corresponding Jordan form of A as a binomial expansion formula, because D and N commute: $DN = ND$. We have

$$(\mu J)^k = (\mu D + \mu N)^k = (\mu D)^k + k(\mu D)^{k-1}\mu N + \binom{k}{2}(\mu D)^{k-2}(\mu N)^2$$

$$+ \ldots + \binom{k}{k-1} k\mu D(\mu N)^{k-1} + (\mu N)^k.$$

If n is the order of A, the previous addenda are zero for $k \geqq n$. Therefore we can write

$$(\mu J)^k = \sum_{i=0}^{n} \frac{k!}{(k-i)!i!} \mu^k (D)^{k-i} (N)^i .$$

But being n a (given) finite number, the conditions of convergence of (μJ), for $k \longrightarrow +\infty$, coincide with the conditions of convergence of (μD). It follows that under the given assumptions, $\lim_{k \longrightarrow +\infty} (\mu D)^k = [0]$ and also

$$\lim_{k \longrightarrow +\infty} (\mu J)^k = [0] .$$

Hence, also when A is not diagonalizable, we have

$$\lim_{k \longrightarrow +\infty} (\mu A)^k = [0] .$$

□

Now we can give an iterative method for computing the inverse matrix $(I - \mu A)^{-1}$ and the inverse matrix $(\lambda I - A)^{-1}$, with μ and λ positive numbers.

Let be given $\mu > 0$ and a square matrix A, of order n. We consider the following expression

$$\sum_{i=0}^{k} (\mu A)^i = (\mu A)^0 + \mu A + (\mu A)^2 + \cdots + (\mu A)^k .$$

Now, we multiply both members of the above expression by $(I - \mu A)$:

$$(I - \mu A)\sum_{i=0}^{k}(\mu A)^i = (I - \mu A)\left[(I + \mu A + (\mu A)^2 + \cdots + (\mu A)^k\right]$$

$$= I + \mu A + (\mu A)^2 + \cdots + (\mu A)^k - \mu A - (\mu A)^2 - \cdots - (\mu A)^k - (\mu A)^{k+1}$$

$$= I - (\mu A)^{k+1}.$$

We know from Theorem 1.6.26 that if and only if $\rho_A < \frac{1}{\mu}$, it holds

$$\lim_{k \longrightarrow +\infty} (\mu A)^{k+1} = [0]\,.$$

Hence the series

$$(I - \mu A)\left[I + \mu A + (\mu A)^2 + (\mu A)^3 + \dots\right]$$

is convergent and its sum is the identity matrix I:

$$(I - \mu A)\,[\dots] = I.$$

It follows that the series in square brackets is nothing but the inverse of the matrix $(I - \mu A)$. Hence, when $\rho_A < \frac{1}{\mu}$, $\mu > 0$, we have

$$(I - \mu A)^{-1} = I + \mu A + (\mu A)^2 + (\mu A)^3 + \dots$$

i.e.

$$(I - \mu A)^{-1} = \sum_{k=0}^{\infty} \mu^k (A)^k.$$

Now we put $\lambda = \frac{1}{\mu}$, $\mu > 0$. Then

$$(I - \mu A) = \frac{1}{\lambda}(\lambda I - A)$$

and hence

$$(I - \mu A)^{-1} = \left[\frac{1}{\lambda}(\lambda I - A)\right]^{-1} = \lambda(\lambda I - A)^{-1}.$$

If (and only if) $\lambda > \rho_A$, it holds

$$(\lambda I - A)^{-1} = \frac{1}{\lambda}\left[I + \frac{1}{\lambda}A + (\frac{1}{\lambda}A)^2 + (\frac{1}{\lambda}A)^3 + \dots\right]$$

i.e.

$$(\lambda I - A)^{-1} = \frac{1}{\lambda}\sum_{k=0}^{\infty}\frac{1}{\lambda^k}(A)^k.$$

The above power series (of matrices) are called "power series of C. Neumann". We shall see an application of these series when, in Sect. 1.9, we shall treat some linear economic models, due to W. Leontief and to P. Sraffa.

We conclude the present section with some exercises on the last subjects treated.

Example 1.6.27

(1) Given the matrix

$$A = \begin{bmatrix} 1 & 5 & 5 \\ 0 & 2 & 1 \\ 0 & 0 & \alpha \end{bmatrix}$$

find the values of $\alpha \in \mathbb{R}$ such that A is diagonalizable.

Solution Being A a triangular matrix, its eigenvalues are $\lambda_1 = 1, \lambda_2 = 2, \lambda_3 = \alpha$. Therefore:

(a) If $\alpha \neq 1$, $\alpha \neq 2$, we have three distinct eigenvalues and hence A is diagonalizable.

(b) If $\alpha = 1$, the eigenvalue $\lambda = 1$ has algebraic multiplicity $m^A_{\lambda=1} = 2$. Let us compute $\text{rank}(A - 1I) = \text{rank}(A - I)$. We have

$$A - I = \begin{bmatrix} 0 & 5 & 5 \\ 0 & 1 & 1 \\ 0 & 0 & 0 \end{bmatrix};$$

hence $\text{rank}(A - I) = 1$, and $m^G_{\lambda=1} = 3 - 1 = 2$. Therefore, for $\alpha = 1$, the given matrix is diagonalizable.

(c) If $\alpha = 2$, we have $m^A_{\lambda=2} = 2$ and

$$A - 2I = \begin{bmatrix} -1 & 5 & 5 \\ 0 & 0 & 1 \\ 0 & 0 & 0 \end{bmatrix}.$$

Hence $\operatorname{rank}(A-2I)=2$ and $m^G_{\lambda=2}=3-2=1\neq m^A_{\lambda=2}$. Therefore for $\alpha=2$, the given matrix is *not* diagonalizable.

(2) Diagonalize the following symmetric matrix

$$A=\begin{bmatrix}2 & \sqrt{2}\\ \sqrt{2} & 1\end{bmatrix}$$

by means of an orthogonal transformation.

Solution We have

$$\begin{vmatrix}2-\lambda & \sqrt{2}\\ \sqrt{2} & 1-\lambda\end{vmatrix}=\lambda^2-3\lambda,$$

hence $\lambda_1=0,\ \lambda_2=3$.

Eigenvectors associated to $\lambda_1=0$. We have to solve the system

$$\begin{cases}2x_1+\sqrt{2}x_2=0\\ \sqrt{2}x_1+x_2=0.\end{cases}$$

We have

$$x^1=\begin{bmatrix}-\frac{1}{\sqrt{2}}x_2\\ x_2\end{bmatrix},\quad x_2\in\mathbb{R},\, x_2\neq 0.$$

We have then $\|x^1\|=\sqrt{(x^1)^\top x^1}=\sqrt{\frac{3}{2}}\,|x_2|$. Now we normalize x^1:

$$\begin{bmatrix}\dfrac{-\frac{1}{\sqrt{2}}x}{\sqrt{\frac{3}{2}}|x_2|}\\ \dfrac{x_2}{\sqrt{\frac{3}{2}}|x_2|}\end{bmatrix}=\begin{bmatrix}\frac{-1}{\sqrt{3}}\\ \sqrt{\frac{2}{3}}\end{bmatrix}\ \text{or also}\ \begin{bmatrix}\frac{1}{\sqrt{3}}\\ -\sqrt{\frac{2}{3}}\end{bmatrix}.$$

Eigenvectors associated to $\lambda_2=3$. We have to solve the system

$$\begin{cases}2x_1+\sqrt{2}x_2=3x_1\\ \sqrt{2}x_1+x_2=3x_2.\end{cases}$$

We find

$$x^2=\begin{bmatrix}\sqrt{2}x_2\\ x_2\end{bmatrix},\quad x_2\in\mathbb{R},\, x_2\neq 0.$$

Now we normalize x^2, obtaining

$$x^2 = \begin{bmatrix} \sqrt{\frac{2}{3}} \\ \frac{1}{\sqrt{3}} \end{bmatrix} \text{ or also } x^2 = \begin{bmatrix} -\sqrt{\frac{2}{3}} \\ -\frac{1}{\sqrt{3}} \end{bmatrix}.$$

The normalized vectors x^1 and x^2 are orthogonal; for instance

$$\begin{bmatrix} -\frac{1}{\sqrt{3}}, \sqrt{\frac{2}{3}} \end{bmatrix} \begin{bmatrix} \sqrt{\frac{2}{3}} \\ \frac{1}{\sqrt{3}} \end{bmatrix} = -\frac{\sqrt{2}}{3} + \frac{\sqrt{2}}{3} = 0.$$

Therefore, we have, for instance,

$$V = \begin{bmatrix} -\frac{1}{\sqrt{3}} & \sqrt{\frac{2}{3}} \\ \sqrt{\frac{2}{3}} & \frac{1}{\sqrt{3}} \end{bmatrix}.$$

It results $V^\top = V^{-1}$ (check!). Finally it results

$$V^\top AV = \begin{bmatrix} 0 & 0 \\ 0 & 3 \end{bmatrix} = \begin{bmatrix} \lambda_1 & 0 \\ 0 & \lambda_2 \end{bmatrix}.$$

(3) Establish if the matrix

$$A = \begin{bmatrix} 1 & 1 & 0 \\ 2 & 2 & 0 \\ 0 & 0 & 0 \end{bmatrix}$$

is or not diagonalizable. If A is diagonalizable, find a polar matrix S which diagonalizes A.

Solution We have

$$\begin{vmatrix} 1-\lambda & 1 & 0 \\ 2 & 2-\lambda & 0 \\ 0 & 0 & -\lambda \end{vmatrix} = \lambda^2(\lambda - 3).$$

Hence $\lambda_1 = \lambda_2 = 0$; $\lambda_3 = 3$. We have $\text{rank}(A - 0I) = \text{rank}(A) = 1$. Therefore $m^A_{\lambda=0} = 2 = 3 - m^G_{\lambda=0} = 3 - 1 = 2$. Hence A is diagonalizable. Now we find the polar matrix S.

Eigenvectors associated to $\lambda = 0$:

$$\begin{bmatrix} x_1 \\ -x_1 \\ x_3 \end{bmatrix}; \text{ for instance } \begin{bmatrix} 1 \\ -1 \\ 0 \end{bmatrix}, \begin{bmatrix} 0 \\ 0 \\ 1 \end{bmatrix}, \text{ etc.}$$

Eigenvectors associated to $\lambda = 3$:

$$\begin{bmatrix} x_1 \\ 2x_1 \\ 0 \end{bmatrix}, \quad x_1 \neq 0; \quad \text{for instance } \begin{bmatrix} 1 \\ 2 \\ 0 \end{bmatrix}, \text{ etc.}$$

Let us verify that the three chosen vectors are linearly independent:

$$\begin{vmatrix} 1 & 0 & 1 \\ -1 & 0 & 2 \\ 0 & 1 & 0 \end{vmatrix} = -3 \neq 0.$$

Hence we have, for instance,

$$S = \begin{bmatrix} 1 & 0 & 1 \\ -1 & 0 & 2 \\ 0 & 1 & 0 \end{bmatrix}.$$

1.7 Quadratic Forms

In the present section we deal with a particular class of real functions of vectors $x \in \mathbb{R}^n$, called *quadratic forms;* these functions have several applications in many sectors of pure and applied mathematics.

Definition 1.7.1 Let be $A = \left[a_{ij}\right]$ a real square and *symmetric* matrix of order n. We call *quadratic form* associated to A the following function of several real variables $x_1, x_2, \ldots, x_n$:

$$Q(x) = x^\top Ax.$$

The above formulation can be rewritten as

$$Q(x) = \sum_{i=1}^{n}\sum_{j=1}^{n} a_{ij}x_i x_j,$$

where $a_{ij} = a_{ji}$, $\forall i \neq j$. In a more extended form we have also

$$\begin{aligned}Q(x) = a_{11}x_1x_1 + a_{12}x_1x_2 + a_{13}x_1x_3 + \cdots + a_{1n}x_1x_n \\ +a_{21}x_2x_1 + a_{22}x_2x_2 + a_{23}x_2x_3 + \cdots + a_{2n}x_2x_n \\ +\cdots\cdots\cdots\cdots\cdots\cdots\cdots\cdots\cdots\cdots\cdots\cdots \\ +a_{n1}x_nx_1 + a_{n2}x_nx_2 + a_{n3}x_nx_3 + \cdots + a_{nn}x_nx_n.\end{aligned}$$

Without loss of generality, it is always convenient to suppose A symmetric. Indeed, if A were not symmetric, being $Q(x)$ a scalar quantity (and hence a matrix of order 1, whose transpose coincides with the same quantity), and being $(x^\top Ax)^\top = x^\top A^\top x$, we have at once

$$\begin{aligned}Q(x) &= \frac{1}{2}(x^\top Ax + x^\top Ax) = \frac{1}{2}(x^\top Ax + x^\top A^\top x) \\ &= x^\top \left\{\frac{1}{2}(A + A^\top)\right\} x.\end{aligned}$$

Therefore we can associate with the given $Q(x)$ the matrix $\frac{1}{2}(A + A^\top)$, which is symmetric. The fact to be able to choose a symmetric matrix, associated with a given quadratic form, without changing the values of the same, is rather convenient for the further developments.

Example 1.7.2 Let be given

$$A = \begin{bmatrix} 8 & 5 & 2 \\ 5 & 1 & 0 \\ 2 & 0 & 3 \end{bmatrix}.$$

It results

$$\begin{aligned}Q(x) &= 8(x_1)^2 + (x_2)^2 + 3(x_3)^2 + 5x_1x_2 + 2x_1x_3 + 5x_2x_1 + 2x_3x_1 \\ &= 8(x_1)^2 + (x_2)^2 + 3(x_3)^2 + 10x_1x_2 + 4x_1x_3.\end{aligned}$$

We are interested in studying the "sign" of a quadratic form. First of all we observe that it always results $Q([0]) = 0$. Then, we adopt the following classification of quadratic forms.

Definition 1.7.3 Let be given $Q(x) = x^\top Ax$ (A symmetric and of order n).

(1) $Q(x)$ is *definite* if it assumes always the same sign for every $x \neq [0]$. More precisely: $Q(x)$ is *positive definite* if $Q(x) > 0, \forall x \in \mathbb{R}^n, x \neq [0]$; $Q(x)$ is *negative definite* if $Q(x) < 0, \forall x \in \mathbb{R}^n, x \neq [0]$.
(2) $Q(x)$ is *positive semidefinite* if $Q(x) \geqq [0], \forall x \in \mathbb{R}^n$; $Q(x)$ is *negative semidefinite* if $Q(x) \leqq [0], \forall x \in \mathbb{R}^n$.
(3) $Q(x)$ is *indefinite* if there exists a pair of vectors x^1 and $x^2 \in \mathbb{R}^n$ such that it results $Q(x^1)Q(x^2) < 0$.

Before going on, let us see some easy examples.

Example 1.7.4

- $Q(x) = (x_1)^2 + (x_2)^2$ is obviously positive definite. Here the symmetric matrix associated with the form is the identity matrix I of order 2.
- $Q(x) = -(x_1)^2 - (x_2)^2$ is negative definite, being $A = -I$ (of order 2).
- $Q(x) = (x_1)^2 - 2x_1x_2 + (x_2)^2 = (x_1 - x_2)^2$ is positive semidefinite: $Q(x) \geqq 0, \ \forall x \in \mathbb{R}^2$ and $Q(x) = 0$ on the straight line $x_2 = x_1$.
- $Q(x) = (x_1)^2 - (x_2)^2$ is indefinite.

It should be clear that $Q(x)$ or the associated symmetric matrix A of the form, belong to one of the types described in Definition 1.7.3, according to the structure of A. Indeed, we speak also of (symmetric) matrices positive definite, negative definite, positive semidefinite, negative semidefinite, indefinite if the associated quadratic form belongs to the said types. Do not make confusion with the classes of positive matrices, semipositive matrices, nonnegative matrices, etc., which are not necessarily square nor symmetric (and have other completely different definitions).

The cases examined in Example 1.7.4 are very easy and it is easy to decide the type of the related quadratic form. But in general things may be not so trivial. However, we have some criteria for determining the *sign* of a quadratic form (or of the associated symmetric matrix A). We shall see two criteria: the first one is based on the eigenvalues of A (eigenvalues which are all *real,* being A symmetric); the second one is based on the principal minors and on the North-West principal minors of A and it is perhaps more easy to work with.

Theorem 1.7.5 (First Criterion for Determining the Sign of a Quadratic Form) *Let A be square of order n and symmetric and let be $Q(x) = x^\top Ax$.*

(1) $Q(x)$ is positive definite (negative definite) if and only if every eigenvalue of A is positive (negative).
(2) $Q(x)$ is positive semidefinite (negative semidefinite) if and only if every eigenvalue of A is nonnegative (non positive).
(3) $Q(x)$ is indefinite if and only if there exists a pair of eigenvalues of A, say λ_1 and λ_2, with $\lambda_1\lambda_2 < 0$.

Proof Being A symmetric, thanks to Theorem 1.6.19 ("Spectral Theorem"), there exists always an orthogonal matrix V (i.e. it holds $V^\top = V^{-1}$) such that

$$D = V^\top AV.$$

Let be $Q(x) = x^\top Ax$; we put $x = Vy$. Being V invertible, the linear transformation $x = Vy$ produces a bijective correspondence between the vectors

x and y; hence, if x varies on $\mathbb{R}^n$, $Q(x)$ assumes the same values assumed by $Q(y)$, when y varies on $\mathbb{R}^n$ (and vice-versa). Therefore It results

$$Q(y) = (Vy)^\top A(Vy) = y^\top V^\top AVy = y^\top Dy.$$

Hence $y^\top Dy$ has the same sign of $x^\top Ax$, but it is more easy to study than $x^\top Ax$, being

$$y^\top Dy = \sum_{i=1}^{n} \lambda_i (y_i)^2.$$

This expression proves the thesis of the theorem. □

The previous results are also known as the "diagonalization of a quadratic form" or also as the "reduction of a quadratic form to its canonical form".

Example 1.7.6 Let be given the symmetric matrix

$$A = \begin{bmatrix} 4 & 2 \\ 2 & 7 \end{bmatrix}.$$

It results $|A - \lambda I| = \lambda^2 - 11\lambda + 24$; $\lambda_1 = 8$, $\lambda_2 = 3$. $Q(x)$ is therefore positive definite. Let us check the validity of our assertion. It results

$$Q(x) = [x_1, x_2] \begin{bmatrix} 4 & 2 \\ 2 & 7 \end{bmatrix} \begin{bmatrix} x_1 \\ x_2 \end{bmatrix}$$

$$= 4(x_1)^2 + 4x_1x_2 + 7(x_2)^2 = 4(x_1)^2 + 4x_1x_2 + (x_2)^2 + 6(x_2)^2$$

$$= (2x_1 + x_2)^2 + 6(x_2)^2.$$

It must be observed that when it is not easy to compute the eigenvalues of A, it is always possible to exploit other procedures which transform $Q(x)$ into a *sum of squares,* where the coefficients of the said sum are not necessarily the eigenvalues of A. There are, at least theoretically, infinitely many ways to operate such a transformation. One of the most useful methods is due to J. L. Lagrange, a method we do not describe here. It must be pointed out that, even if the coefficients of the sum of squares can vary, according to the type of transformation chosen, it holds the following basic result, known also as the *inertia law of quadratic forms.*

Theorem 1.7.7 *In any transformation of a quadratic form $Q(x)$ into a sum of squares, the number of positive coefficients of the sum, of the negative coefficients and of the zero coefficients, does not depend from the transformation chosen.*

From the previous result, we obtain that the sign of a quadratic form can be obtained by means of *any* reduction of the same to a sum of squares. The triplet

$\{n_1, n_2, n_3\}$, where n_1 is the number of positive eigenvalues of the symmetric matrix A, n_2 is the number of negative eigenvalues and n_3 is the number of zero eigenvalues (all roots must be counted with their multiplicity), is called the *inertia* of the quadratic form (or of the associated symmetric matrix A).

We point out that if A is symmetric and invertible, then also A^{-1} is symmetric, i.e. $(A^{-1})^\top = A^{-1}$. Indeed, we have $(A^{-1})^\top = (A^\top)^{-1} = A^{-1}$. Hence, if A is symmetric and positive definite (negative definite), its inverse A^{-1} is positive definite (negative definite): recall that the eigenvalues of the inverse matrix are the reciprocals of the eigenvalues of the matrix in question. Note also that, given the symmetric and invertible matrix A, its inverse cannot be semidefinite (but not definite).

We also point out that the criterion of Theorem 1.7.5 relies on the knowledge of the *sign* of the eigenvalues of A and is therefore a qualitative result. Sometimes it can be useful a particular version of the classical *Descartes Rule:*

- Let be $P(x) = a_n x^n + a_{n-1} x^{n-1} + \cdots + a_0$ a polynomial with *real coefficients,* of degree n and with all *real roots.* Then the number of the *positive* roots of $P(x)$, counted with their respective multiplicities, is equal to the number of the *sign variations* in the sequence of the coefficients of $P(x)$.

We recall the concept of the number of the sign variations in the sequence of the coefficients of an algebraic equation. In order to obtain the said number, we write all coefficients of the polynomial, for example in a decreasing order with respect the powers of the variable x, including a_0 but *neglecting the zero coefficients.* Then we consider all pairs of the contiguous numbers of the sequence obtained. If in one of these pairs the signs of the numbers are *different,* we say that we have a *sign variation.* For instance, if

$$P(x) = x^7 + 3x^5 - 5x^4 - 8x^2 + 7x + 2,$$

the sequence of the coefficients is

$$1, 3, -5, -8, 7, 2$$

and hence the number of sign variations is 2.

Therefore, if we have the information that all roots are all real (and this is the case of the characteristic equation of a symmetric matrix!), it is possible to know, without making computations, the number of positive roots and the number of negative roots. It is then immediate to know the number of zero roots. For instance, if we have the symmetric matrix

$$A = \begin{bmatrix} 1 & 0 & -4 \\ 0 & 1 & 3 \\ -4 & 3 & 1 \end{bmatrix},$$

we have

$$P(\lambda) = -\lambda^3 + 3\lambda^2 + 22\lambda - 24.$$

We have the sequence: $-1, 3, 22, -24$. We have two variations (and one permanence), hence we have 2 positive roots and one negative root. The related quadratic form is therefore indefinite.

The second criterion for determining the sign of a quadratic form relies on the principal minors of A and on the North-West (or leading) principal minors of A. This criterion is also known as the *Sylvester criterion* (from the English mathematician J. Sylvester).

Theorem 1.7.8 (Second Criterion for Determining the Sign of a Quadratic Form) *Let A be symmetric of order n. Then $Q(x) = x^\top Ax$ is:*

(i) Positive definite if and only if all its North-West principal minors are positive, i.e.

$$a_{11} > 0; \quad \begin{vmatrix} a_{11} & a_{12} \\ a_{21} & a_{22} \end{vmatrix} > 0; \quad \begin{vmatrix} a_{11} & a_{12} & a_{13} \\ a_{21} & a_{22} & a_{23} \\ a_{31} & a_{32} & a_{33} \end{vmatrix} > 0; \ldots; |A| > 0.$$

(ii) Negative definite if and only if the elements of the sequence of point (i) alternate in sign, with the first element negative (i.e. $a_{11} < 0$).

(iii) Positive semidefinite if and only if all $\binom{n}{k}$ principal minors of A ($1 \leqq k \leqq n$) are nonnegative.

(iv) Negative semidefinite if and only if the principal minors of A of even order are nonnegative, and the principal minors of odd order are non positive.

(v) Indefinite in all other cases.

Proof We prove only the first two points (i) and (ii), following [22]. We put forward the following remark. Let us suppose that A is a symmetric matrix of order n and Q a nonsingular matrix of order n. Then $Q^\top AQ$ is symmetric:

$$(Q^\top AQ)^\top = Q^\top A^\top Q = Q^\top AQ.$$

Let us suppose, for instance, that $Q^\top AQ$ is *positive definite.* Let $x \neq [0]$ be an arbitrary vector of $\mathbb{R}^n$; being Q nonsingular, there exists $y \neq [0]$ such that $x = Qy$. Hence

$$x^\top Ax = (Qy)^\top A(Qy) = y^\top Q^\top AQy = y^\top(Q^\top AQ)y > 0, \ \forall y \neq [0],$$

as $Q^\top AQ$ is positive definite, so A is positive definite.

Now let us suppose that A is *positive definite* and let $z \neq [0]$ an arbitrary vector of $\mathbb{R}^n$. It results $x = Qz \neq [0]$, as Q is nonsingular. Therefore

$$0 < (Qz)^\top A(Qz) = z^\top Q^\top AQz = z^\top (Q^\top AQ)z$$

and hence $Q^\top AQ$ is positive definite.

Let us denote by Γ_k, $k = 1, \dots, n$, the North-West principal submatrices of A:

$$\Gamma_1 = [a_{11}], \; \Gamma_2 = \begin{bmatrix} a_{11} & a_{12} \\ a_{21} & a_{22} \end{bmatrix}, \dots, \Gamma_n = A.$$

If A is positive definite, its eigenvalues are all positive and therefore

$$\det(\Gamma_n) = \det(A) = \lambda_1 \lambda_2 \dots \lambda_n > 0.$$

Moreover, if A is positive definite, it is also positive definite the quadratic form associated with Γ_k, $k = 1, \dots, n$, as, with $x = [x_1, \dots, x_k, 0, \dots, 0]^\top$ it results

$$x^\top Ax > 0 \Longrightarrow [x_1, \dots, x_k]\, \Gamma_k \begin{bmatrix} x_1 \\ \vdots \\ x_k \end{bmatrix} > 0.$$

Hence it must result $\det(\Gamma_k) > 0$, $k = 1, \dots, n$, that is A has all its North-West principal minors positive.

In order to prove the sufficiency part of the theorem, we use the method of induction, showing that if Γ_k is positive definite and $\det(\Gamma_{k+1}) > 0$, then Γ_{k+1} is positive definite.

The result is trivially true for $k = 1$: it is obvious that $a_{11} > 0$ and $\det(\Gamma_2) > 0$ imply that $\lambda_1 > 0$ and $\lambda_2 > 0$, and hence Γ_2 is positive definite. Then let be Γ_k positive definite for a generic k. We consider the matrix Γ_{k+1} built in the following way:

$$\Gamma_{k+1} = \begin{bmatrix} \Gamma_k & v \\ v^\top & \alpha \end{bmatrix}$$

where

$$v = \begin{bmatrix} a_{1,k+1} \\ \vdots \\ a_{k,k+1} \end{bmatrix}, \quad \alpha = a_{k+1,k+1}.$$

Let us consider the matrix B, square and of order $k + 1$, defined as

$$B = \begin{bmatrix} I_k & w \\ [0]^\top & 1 \end{bmatrix}$$

where I_k is the identity matrix of order k, $[0] \in \mathbb{R}^k$ and $w = -\Gamma_k^{-1} v$ (the matrix Γ_k is positive definite by assumption and hence it is invertible!). Being $\det(B) = 1$ (B is an upper triangular matrix with elements on its main diagonal all equal to 1), the linear transformation $y = Bx$ has its range coinciding with the whole $\mathbb{R}^{k+1}$, therefore we can write the quadratic form associated with Γ_{k+1} as

$$y^\top \Gamma_{k+1} y = x^\top B^\top \Gamma_{k+1} B x = x^\top C x,$$

where

$$C = B^\top \Gamma_{k+1} B = \begin{bmatrix} I_k & [0] \\ v^\top & \alpha \end{bmatrix} \begin{bmatrix} \Gamma_k & v \\ [0]^\top & 1 \end{bmatrix} = \begin{bmatrix} \Gamma_k & [0] \\ [0]^\top & \beta \end{bmatrix}$$

where $\beta = \alpha + v^\top v$. The sign of the quadratic form associated to Γ_{k+1} coincides with the sign of the quadratic form associated to C, for what remarked at the beginning of this proof. In order to establish this sign, we find the eigenvalues of C. The matrix

$$C - \lambda I_{k+1} = \begin{bmatrix} \Gamma_k - \lambda I_k & [0] \\ [0]^\top & \beta - \lambda \end{bmatrix}$$

has the following determinant (use the first Theorem of Laplace, considering the last row):

$$\det(C - \lambda I_{k+1}) = (\beta - \lambda) \det(\Gamma_k - \lambda I_k).$$

Therefore the matrix C has k eigenvalues ρ_s equal to the eigenvalues λ_s of the matrix Γ_k, with $s = 1, \ldots, k$ and a last eigenvalue

$$\rho_{k+1} = \beta.$$

But then the first k eigenvalues of C are positive, as they coincide with the eigenvalues of Γ_k, positive definite by assumption. We have only to establish the sign of the last eigenvalue. We have

$$\det(C) = \beta \det(\Gamma_k)$$

and also

$$\det(C) = \det(B^\top \Gamma_{k+1} B) = \det(B^\top)\det(\Gamma_{k+1})\det(B) = \det(\Gamma_{k+1})$$

by the Theorem of Binet-Cauchy and being $\det(B) = 1$. Taking into account the two previous relations, together with the induction assumptions, we get

$$\beta \det(\Gamma_k) = \det(\Gamma_{k+1}) = \beta > 0.$$

Being C positive definite, also Γ_{k+1} is positive definite and the proof is finished.
The proof of the point (ii) follows at once, by remarking that A is negative definite if and only if $-A$ is positive definite and that, for the properties of determinants, it results $\det(-\Gamma_k) = (-1)^k \det(\Gamma_k)$. □

Theorem 1.7.9 *A symmetric matrix A positive semidefinite (negative semidefinite) is positive definite (resp. negative definite) if and only if* $\det(A) \neq 0$.

This last result (already anticipated) says that a symmetric matrix semidefinite, but not definite, must be singular, i.e. $\det(A) = 0$, that is it must have at least one eigenvalue equal to zero, which is established also by the first criterion for the determination of the sign of a quadratic form.

Example 1.7.10 We take again into consideration the matrix of Example 1.7.6. We have

$$4 > 0, \quad \begin{vmatrix} 4 & 2 \\ 2 & 7 \end{vmatrix} = 24 > 0,$$

which confirms that the related quadratic form is positive definite.
Now let be $Q(x) = 3(x_1)^2 + 2(x_2)^2 + 10(x_3)^2 + 2x_1x_2 - 8x_2x_3$. It results

$$A = \begin{bmatrix} 3 & 1 & 0 \\ 1 & 2 & -4 \\ 0 & -4 & 10 \end{bmatrix}.$$

Therefore

$$a_{11} = 3 > 0; \quad \begin{vmatrix} 3 & 1 \\ 1 & 2 \end{vmatrix} = 5 > 0; \quad |A| = 2 > 0.$$

The quadratic form here considered is therefore positive definite.

We remark that the above second criterion (Sylvester criterion) requires, for the "semidefinite case", the inspection of the sign of *all principal minors* of the matrix and not only the sign of the North-west principal minors (as for the "definite case"). Consider, for instance, the matrix

$$A = \begin{bmatrix} 0 & 0 \\ 0 & -1 \end{bmatrix}.$$

If we take into consideration the two North-West principal minors, which are both zero, we could conclude that the related quadratic form is both positive semidefinite and negative semidefinite (which is wrong). In fact A is negative semidefinite, as it has a principal minor of order one which is negative ($a_{22} = -1$).

Example 1.7.11 Let us consider the quadratic form

$$Q(x) = (x_1)^2 - 4ax_1x_2 + 4(x_2)^2 + 3a^2(x_3)^2, \quad a \neq 0.$$

We have (we denote by Δ_k the North-West principal minor of order k):

$$A = \begin{bmatrix} 1 & -2a & 0 \\ -2a & 4 & 0 \\ 0 & 0 & 3a^2 \end{bmatrix},$$

$\Delta_2 = 4 - 4a^2 = 4(1 - a^2)$. Hence Δ_2 is negative for $a \in (-\infty, -1) \cup (1, +\infty)$, is positive for $a \in (-1, 1)$ and is equal to zero at $a = -1$ and $a = 1$.

Then we have $\Delta_3 = \det(A) = 12a^2 - 12a^4 = 12a^2(1 - a^2)$. We conclude that:

- If $a \in (-1, 1)$ then $Q(x)$ is positive definite.
- If $a = 1$ or if $a = -1$, then $Q(x)$ is positive semidefinite.
- If $a \in (-\infty, -1) \cup (1, +\infty)$, then $Q(x)$ is indefinite.

Example 1.7.12 Let us consider the symmetric matrix

$$A = \begin{bmatrix} 1 & 2 & 3 \\ 2 & 0 & 1 \\ 3 & 1 & 1 \end{bmatrix}.$$

We have

$$\Delta_1 = a_{11} = 1; \quad \Delta_2 = \begin{vmatrix} 1 & 2 \\ 2 & 0 \end{vmatrix} = -4.$$

It is useless to compute $\Delta_3 = \det(A)$, as being $\Delta_1 > 0,\ \Delta_2 < 0$, we can deduce that A is indefinite.

The quadratic forms examined previously are also said "unconstrained quadratic forms", as the vector x of the independent variables is "free" to move over the whole $\mathbb{R}^n$. However, in many applications it is interesting to know the sign of a quadratic form when x varies over a proper subset of $\mathbb{R}^n$. In this case we speak of "constrained quadratic forms". Obviously the problem makes sense when the quadratic form in question is not definite: it is obvious that if a quadratic form is definite over the whole $\mathbb{R}^n$, a fortiori it will be definite over a proper subset of $\mathbb{R}^n$. Usually the problem of the constrained quadratic forms is the following one (a sound treatment is given by Debreu [6]).

Let be given $Q(x) = x^\top Ax$, with A a symmetric matrix of order n; we want to study the sign of $Q(x)$, when $x \neq [0]$ is the solution of the linear homogeneous system

$$Bx = [0],$$

with B of order (m, n) and $\text{rank}(B) = m$. We build the following bordered matrix

$$M = \begin{bmatrix} [0] & B \\ B^\top & A \end{bmatrix},$$

which is obviously square of order $(m + n)$. The following result holds true.

Theorem 1.7.13 *Let be* $\text{rank}(B) = m$ *and, without loss of generality, suppose that the first m columns of B are linearly independent. Then, a necessary and sufficient condition so that $Q(x) = x^\top Ax$ is positive definite on the set of nonzero solutions of the system $Bx = [0]$, is that the North-West principal minors of M, of order $2m + 1, \ldots, m + n$, have the sign of $(-1)^m$:*

$$(-1)^m \Delta_h > 0, \quad \forall h = 2m + 1, \ldots, m + n,$$

where Δ_h here denotes the North-West principal minor of M of order h.

A necessary and sufficient condition so that $Q(x) = x^\top Ax$ is negative definite on the set of nonzero solutions of the system $Bx = [0]$, is that the North-West principal minors of M, of order $2m+1, \ldots, m+n$, alternate in sign, beginning with the sign of $(-1)^{m+1}$.

Corollary 1.7.14 *If we have one constraint, of the type $b^\top x = 0$, with $b_1 \neq 0$ (i.e. $m = 1$ in the matrix B), the conditions expressed by Theorem 1.7.13 become:*

(a) $Q(x)$ is positive definite on the set of nonzero solutions of $b^\top x = 0$ if and only if

$$\Delta_3 = \begin{vmatrix} 0 & b_1 & b_2 \\ b_1 & a_{11} & a_{12} \\ b_2 & a_{21} & a_{22} \end{vmatrix} < 0, \ldots, |M| < 0.$$

(b) *$Q(x)$ is negative definite on the set of nonzero solutions of $b^\top x = 0$ if and only if*

$$\Delta_3 = \begin{vmatrix} 0 & b_1 & b_2 \\ b_1 & a_{11} & a_{12} \\ b_2 & a_{21} & a_{22} \end{vmatrix} > 0, \quad \Delta_4 = \begin{vmatrix} 0 & b_1 & b_2 & b_3 \\ b_1 & a_{11} & a_{12} & a_{13} \\ b_2 & a_{21} & a_{22} & a_{23} \\ b_3 & a_{31} & a_{32} & a_{33} \end{vmatrix} < 0, \quad etc.$$

For $n = 2$, $m = 1$, the results of Corollary 1.7.14 can be justified as follows. Consider the quadratic form

$$Q(x) = a_{11}x_1^2 + 2a_{12}x_1x_2 + a_{22}x_2^2,$$

subject to: $b_1x_1 + b_2x_2 = 0$, with $b_1 \neq 0$. Solving the constraint with respect to x_1, we have $x_1 = (-b_2x_2)/b_1$. Substituting into the quadratic form we have

$$\begin{aligned} Q(x) &= a_{11}\left(-\frac{b_2x_2}{b_1}\right)^2 + 2a_{12}\left(-\frac{b_2x_2}{b_1}\right)x_2 + a_{22}x_2 \\ &= \frac{1}{b_1^2}(a_{11}b_2^2 - 2a_{12}b_1b_2 + a_{22}b_1^2)x_2^2. \end{aligned}$$

Hence $Q(x)$ is, e.g., positive definite under the above constraint, if and only if

$$a_{11}b_2^2 - 2a_{12}b_1b_2 + a_{22}b_1^2 > 0.$$

The reader is invited to verify that the above inequality is equivalent to:

$$\begin{vmatrix} 0 & b_1 & b_2 \\ b_1 & a_{11} & a_{12} \\ b_2 & a_{12} & a_{22} \end{vmatrix} < 0.$$

(Use the Sarrus Rule).

Example 1.7.15 Evaluate the sign of $Q(x) = (x_1)^2 - 4(x_2)^2 - (x_3)^2 + 2x_1x_2$ over the set of the nonzero solutions of the constraint

$$2x_1 + x_2 + x_3 = 0.$$

It results

$$A = \begin{bmatrix} 1 & 1 & 0 \\ 1 & -4 & 0 \\ 0 & 0 & -1 \end{bmatrix},$$

and this matrix is indefinite over $\mathbb{R}^3$. Then we have $m = 1$, $b^\top = [2, 1, 1]$. Therefore

$$\Delta_3 = \begin{vmatrix} 0 & 2 & 1 \\ 2 & 1 & 1 \\ 1 & 1 & -4 \end{vmatrix} = 19 > 0,$$

$$\Delta_4 = \begin{vmatrix} 0 & 2 & 1 & 1 \\ 2 & 1 & 1 & 0 \\ 1 & 1 & -4 & 0 \\ 1 & 0 & 0 & -1 \end{vmatrix} = |M| = -14 < 0.$$

Hence the constrained quadratic form is negative definite on the constraint considered.

Example 1.7.16 Evaluate the sign of the quadratic form $Q(x) = (x_1)^2 - 4x_1x_2 + 4(x_2)^2 - 5(x_3)^2$ on the set of nonzero solutions of the system

$$\begin{cases} x_1 - x_3 = 0 \\ 2x_1 + x_2 = 0. \end{cases}$$

We have

$$B = \begin{bmatrix} 1 & 0 & -1 \\ 2 & 1 & 0 \end{bmatrix}$$

$$M = \begin{bmatrix} 0 & 0 & 1 & 0 & -1 \\ 0 & 0 & 2 & 1 & 0 \\ 1 & 2 & 1 & -2 & 0 \\ 0 & 1 & -2 & 4 & 0 \\ -1 & 0 & 0 & 0 & -5 \end{bmatrix}.$$

Being $2m + 1 = 5$, we have only to check the sign of the determinant of M; we have $|M| = 20 > 0$. Hence the quadratic form is positive definite on the nonzero solutions of the system, as $|M|$ has the sign of $(-1)^m = (-1)^2 > 0$.

Example 1.7.17 Evaluate the sign of $Q(x) = x^\top Ax$, where

$$A = \begin{bmatrix} 1 & -1 & -1 \\ 0 & 1 & -1 \\ 0 & 0 & 1 \end{bmatrix}.$$

Then evaluate the sign of the quadratic form $x^\top Ax$ under the constraint $x_1 + 2x_3 = 0$.

Being A not symmetric, first we consider the transformation

$$C = \frac{1}{2}(A + A^\top),$$

as $x^\top Cx$ has the same values of $x^\top Ax$, but C is symmetric. We have

$$C = \begin{bmatrix} 1 & -\frac{1}{2} & -\frac{1}{2} \\ -\frac{1}{2} & 1 & -\frac{1}{2} \\ -\frac{1}{2} & -\frac{1}{2} & 1 \end{bmatrix}.$$

Then:

$$\begin{vmatrix} 1 & -\frac{1}{2} \\ -\frac{1}{2} & 1 \end{vmatrix} = \frac{3}{4} > 0, \quad |C| = 0.$$

Therefore the quadratic form cannot be definite positive. The principal minors of order one are all positive. The other principal minors of order 2 are:

$$\begin{vmatrix} 1 & -\frac{1}{2} \\ -\frac{1}{2} & 1 \end{vmatrix} = \frac{3}{4} > 0, \quad \begin{vmatrix} 1 & -\frac{1}{2} \\ -\frac{1}{2} & 1 \end{vmatrix} = \frac{3}{4} > 0.$$

Hence the quadratic form $x^\top Cx = x^\top Ax$ is positive semidefinite.
The constraint is given by $b^\top x = 0$, with $b^\top = [1, 0, 2]$. We form the matrix

$$M = \begin{bmatrix} 0 & 1 & 0 & 2 \\ 1 & 1 & -\frac{1}{2} & -\frac{1}{2} \\ 0 & -\frac{1}{2} & 1 & -\frac{1}{2} \\ 2 & -\frac{1}{2} & -\frac{1}{2} & 1 \end{bmatrix}.$$

We have (with the usual notations) $\Delta_3 = -1$, $\Delta_4 = \det(M) = -\frac{27}{4}$. Hence the quadratic form is positive definite on the given constraint.

We have an alternative different condition to evaluate the sign of a constrained quadratic form.

Theorem 1.7.18 *A necessary and sufficient condition so that a quadratic form $Q(x) = x^\top Ax$, with A symmetric of order n, is positive definite on the set of nonzero solutions of the linear homogeneous system $Bx = [0]$, with B of order (m, n) and* $\text{rank}(B) = m$, *is that all roots of the equation in λ*

$$\det \begin{bmatrix} A - \lambda I & B^\top \\ B & [0] \end{bmatrix} = 0$$

are positive. The form is negative definite on the same set if and only if all the above roots are negative.

Remark 1.7.19 We point out that the assumption in Theorem 1.7.13 that the rank of B is given by the *first* m columns of B (assumption which is always possible to obtain by means of suitable renumeration of the variables), allows to give necessary and sufficient conditions to evaluate the sign of a constrained quadratic form. Without this assumption, the conditions of Theorem 1.7.13 and of Corollary 1.7.14 become only *sufficient.* Take into consideration the following simple example. It is quite obvious that the quadratic form $Q(x) = (x_1)^2 + (x_2)^2 - (x_3)^2$ is positive definite on the constraint $x_3 = 0$ (here we have therefore $b^\top = [0, 0, 1]$ and the above assumptions are not verified). However it results

$$\Delta_3 = \begin{vmatrix} 0 & 0 & 0 \\ 0 & 1 & 0 \\ 0 & 0 & 1 \end{vmatrix} = 0.$$

Note that if we use the criterion of Theorem 1.7.18, we have

$$\det \begin{bmatrix} 1-\lambda & 0 & 0 & 0 \\ 0 & 1-\lambda & 0 & 0 \\ 0 & 0 & 1+\lambda & 1 \\ 0 & 0 & 1 & 0 \end{bmatrix} = -(1-\lambda)^2 = 0,$$

which has two coincident positive roots $\lambda_1 = \lambda_2 = 1$, hence the constrained quadratic form is positive definite.

1.8 Theorems of Perron and Frobenius

In the present section we shall be concerned with some results particularly important in the analysis of some linear economic models. We begin with the following definition.

Definition 1.8.1 Let A be a square matrix of order n. Then A is said to be *decomposable or reducible* (some authors use also the term "*non connected*") whenever it is possible, by means of permutations of the rows of A and of the *corresponding* columns, to obtain one of the following "block triangular" forms:

$$A = \begin{bmatrix} A_{11} & A_{12} \\ [0] & A_{22} \end{bmatrix} \text{ or, equivalently, } A = \begin{bmatrix} A_{11} & [0] \\ A_{21} & A_{22} \end{bmatrix},$$

with A_{11} *square matrix* (and hence also A_{22} square matrix). If this operation is not possible, (after performing all possible permutations of the rows and of the corresponding columns!), A is called *indecomposable or irreducible* (or also "connected").

It is easy to see that if we obtain one of the above triangular forms, then we obtain the other one and vice-versa. If A is decomposable (respectively: indecomposable), so it is also its transpose $A^\top$. The square zero matrix [0] is obviously decomposable (it is already decomposed!). Every diagonal matrix is decomposable, as well as every triangular matrix. On the other hand, if a square matrix does not contain zero elements, it is obviously indecomposable. An indecomposable matrix cannot contain a row or a column of all zeros. However, the presence of zero elements is only a necessary condition for A to be decomposable, not a sufficient one.

Example 1.8.2 The matrix

$$A = \begin{bmatrix} 0 & 1 & 4 \\ 0 & 5 & 0 \\ 3 & 9 & 1 \end{bmatrix}$$

is decomposable; this is seen by interchanging the second and the third row and, subsequently, the second and the third column. We obtain the form

$$A = \begin{bmatrix} 0 & 4 & 1 \\ 3 & 1 & 9 \\ 0 & 0 & 5 \end{bmatrix},$$

which is in the first form of Definition 1.8.1. The matrix

$$A = \begin{bmatrix} 2 & 0 & 1 & 1 \\ 0 & 6 & 0 & 5 \\ 3 & 1 & 4 & 0 \\ 0 & 7 & 0 & 8 \end{bmatrix}$$

is decomposable: by interchanging the second and the third row and subsequently the second and the third column we obtain the form

$$A = \begin{bmatrix} 2 & 1 & 0 & 1 \\ 3 & 4 & 1 & 0 \\ 0 & 0 & 6 & 5 \\ 0 & 0 & 7 & 8 \end{bmatrix}.$$

The matrix

$$A = \begin{bmatrix} 0 & 1 \\ 1 & 0 \end{bmatrix}$$

is indecomposable, even if it has two zero elements on the total of 4 elements! Similarly, the matrix

$$A = \begin{bmatrix} 0 & 0 & -2 \\ 1 & 0 & 0 \\ 0 & -1 & 0 \end{bmatrix}$$

is indecomposable, even if it has only three nonzero elements, on the total of 9 elements.

Remark 1.8.3

(*a*) The permutations of the rows and of the corresponding columns which may put into evidence the decomposability of a square matrix, can be formalized by means of a suitable *permutation matrix* P. Definition 1.8.1 can therefore be reformulated as follows: A is decomposable if there exists a permutation matrix P, such that PAP^{-1} (or equivalently, $PAP^{\top}$, being P and orthogonal matrix) is in one of the two "block triangular forms" evidenced in Definition 1.8.1. For instance, with reference to the second matrix of Example 1.8.2, we have

$$P = \begin{bmatrix} 1 & 0 & 0 & 0 \\ 0 & 0 & 1 & 0 \\ 0 & 1 & 0 & 0 \\ 0 & 0 & 0 & 1 \end{bmatrix}.$$

(*b*) It is possible to show that a characterization of a decomposable matrix, equivalent to the one given above, is the following one: A square of order n is decomposable if and only if it is possible to find a partition $\{N_1, N_2\}$ of the set $N = \{1, 2, \ldots, n\}$ such that $\{i \in N_1,\ j \in N_2\} \Longrightarrow a_{ij} = 0$.

(*c*) It is possible to show that A is indecomposable if and only if, for any choice in N of the indices i and j, it is always possible to find in N some indices $h_1, h_2, \ldots, h_r, h_s$, such that

$$a_{ih_1} a_{h_1 h_2} \ldots a_{h_r h_s} a_{h_s j} \neq 0,$$

that, is, there exists a *chain* which *connects* i with j. This has (see further) an interesting economic interpretation.

(*d*) In Definition 1.8.1 the submatrices A_{11} and/or A_{22} may be in turn decomposable. We obtain in this case a block upper triangular form or a block lower triangular form. For instance, we can obtain the following form

$$A = \begin{bmatrix} A_{11} & A_{12} & \cdots & A_{1k} \\ [0] & A_{22} & \cdots & A_{2k} \\ \vdots & \vdots & \cdots & \vdots \\ [0] & [0] & \cdots & A_{kk} \end{bmatrix}, \tag{1.7}$$

where every submatrix A_{ii} is square and indecomposable. If the matrix A is "fully decomposable" , this means that we obtain, by means of permutations of the rows and the corresponding columns, a block diagonal form of the type

$$A = \begin{bmatrix} A_{11} & [0] & \cdots & [0] \\ [0] & A_{22} & \cdots & [0] \\ \vdots & \vdots & \ddots & \vdots \\ [0] & [0] & \cdots & A_{kk} \end{bmatrix},$$

where every submatrix A_{ii} is square and indecomposable. Further generalizations are possible: see, e.g., the basic books by F. R. Gantmacher, quoted in the References.

(e) The fact that A is decomposable allows to "simplify" the structure of the linear system $Ax = b$. Indeed, in this case we have

$$\begin{bmatrix} A_{11} & A_{12} \\ [0] & A_{22} \end{bmatrix} \begin{bmatrix} x^1 \\ x^2 \end{bmatrix} = \begin{bmatrix} b^1 \\ b^2 \end{bmatrix},$$

with A_{11} of order k and A_{22} of order $(n - k)$. Therefore the system is decomposed into the form

$$\begin{cases} A_{11}x^1 + A_{12}x^2 = b^1 \\ \qquad A_{22}x^2 = b^2. \end{cases}$$

The second system is "independent" from the first one, whereas the first system depends from the second one. Hence, the components of x^2 are independent from the ones of x^1, and the components of x^1 depend from the components of x^2. If we have, furthermore, $A_{12} = [0]$ (that is A is "fully decomposable"), then x^1 and x^2 would be completely independent. On the contrary, if A is indecomposable, there is no independence relation between x^1 and x^2, nor in one sense (of x^2 from x^1), nor in the other one (of x^1 from x^2).

(f) Recognizing the decomposability or indecomposability of a square matrix is not a trivial problem. Being $n!$ the number of permutations of n, it is clear that it is hard to find all possible permutation matrices P, if the order n of the matrix is relevant. We observe that the property of decomposability (or indecomposability) of a square matrix is a *qualitative* property, not a quantitative property. It matters that suitable elements a_{ij} are zero or different from zero. A second remark concerns the fact that the decomposability or the indecomposability of a square matrix *do not depend* from the diagonal elements a_{ii} : in $PAP^\top$, where P is a permutation matrix, the diagonal elements of A change of position but they remain on the main diagonal, hence they are not relevant for the assessment of decomposability or indecomposability. Finally we note that we can associate to A another square matrix $M = [m_{ij}]$, of order n, whose elements are defined as follows:

$$m_{ii} = 0, \; i = 1, \ldots, n;$$

$$m_{ij} = \begin{cases} 1, \text{ if } a_{ij} \neq 0, \\ 0, \text{ if } a_{ij} = 0, \end{cases} \quad i \neq j.$$

A is decomposable (resp. indecomposable) if and only if M is decomposable (resp. indecomposable). M is called the *adjacency matrix* of an "oriented graph G, with n vertices" . *Graph Theory* is an important branch of Operations Research, theory that however will not be treated in the present book. See, for basic facts on graph theory [4] and [8]. One of the standard reference books on graph theory is [14]. See also [3] for relationships with nonnegative matrices.

It can be proved that M is indecomposable if and only if G is "strongly connected" . There are efficient algorithms to test the above fact and hence, thanks to Graph Theory, we have efficient algorithms to test the decomposability or indecomposability of a given square matrix A, also of high order n.

There are several important results concerning eigenvalues and eigenvectors of nonnegative square matrices. Historically, these results are mainly due to two German mathematicians: O. Perron (1880–1975) and G. F. Frobenius (1849–1917). The related theorems have a "strong version" , concerning the case of indecomposable semipositive matrices and a "weak version" , where indecomposability is not assumed.

Theorem 1.8.4 (Theorem of Perron-Frobenius, "Strong Version") *Let be $A \geq [0]$, square of order n and indecomposable.*

(1) There is an eigenvalue, denoted by λ^* or $\lambda^*(A)$ or also dom(A), *said dominant eigenvalue or Frobenius eigenvalue, such that:*

(α) $\lambda^* > 0$ (λ^* is therefore a positive real number).

(β) λ^* is a "dominant" eigenvalue, in the sense that it holds $\lambda^* \geqq |\lambda|$, being λ any other eigenvalue of A (recall that we can have also complex eigenvalues, even if A is a real matrix!). In other words, the dominant eigenvalue is the spectral radius of A.

(γ) λ^* is a simple root of the characteristic equation of A, i.e. of $|A - \lambda I| = 0$.

(2) Every eigenvector associated with λ^* has all nonzero components and with the same sign. The dominant root λ^* is the unique eigenvalue to have this property. Hence, for instance, the problem

$$\begin{cases} Ax = \lambda x \\ x > [0] \end{cases}$$

(where obviously A satisfies the assumptions of the present theorem) has a solution if and only if $\lambda = \lambda^*$. *Furthermore, the positive eigenvector* x^*, *associated with* λ^*, *is unique up to a positive scalar. That is, all positive eigenvectors are of the form* αx^*, $\alpha > 0$.

(3) *The matrix* $[\mu I - A]^{-1}$, *where* $\mu \in \mathbb{R}$, *exists and is positive if and only if* $\mu > \lambda^*$.

(4) λ^* *is a strictly increasing function of the elements of* A, *in the sense that it holds*

$$B \geq A \Longrightarrow \lambda^*(B) > \lambda^*(A).$$

Corollary 1.8.5 (Criterion of Brauer-Solow) *Let* m *be the minimum of the sums of the elements of the columns of* A *and let* M *be the maximum of the said sums, that is*

$$m = \min\left\{e^\top A^1, \ldots, e^\top A^n\right\},$$

$$M = \max\left\{e^\top A^1, \ldots, e^\top A^n\right\},$$

where $e^\top = [1, 1, \ldots, 1]$. *Then it holds*

$$m \leqq \lambda^*(A) \leqq M.$$

The sign of equality holds only in the case when $m = M$ *(and hence in this case we have* $\lambda^* = m = M$*). Otherwise it holds* $m < \lambda^*(A) < M$.

We point out that, being $|A| = \left|A^\top\right|$ and being the eigenvalues the roots of a determinant, we can assert that Corollary 1.8.5 holds also if m and M are defined with reference to the sums of the rows of A.

Example 1.8.6 Compute the Frobenius eigenvalue of

$$A = \begin{bmatrix} 3 & 2 & 1 \\ 1 & 3 & 2 \\ 1 & 1 & 4 \end{bmatrix}.$$

Being A indecomposable and all its row sums equal to 6, we conclude at once that $\lambda^* = 6$.

In case at point 1,β of Theorem 1.8.4 it holds $\lambda^* > |\lambda|$, being λ any other eigenvalue of A, then A is said to be a *primitive matrix or acyclic matrix.* These matrices are particularly important in some issues of the theory of stochastic processes, above all in the theory of "Markov chains" ; for some basic notions see, e.g., De Giuli et al. [8]. It can be proved that a sufficient conditions for $A \geq [0]$, indecomposable, to be primitive is that at least one element a_{ii} of its main diagonal is positive: $a_{ii} > 0$ (see [23]). Therefore any positive square matrix is necessarily primitive.

We do not give the proof of all assertions of Theorem 1.8.4, but only the proof of the more relevant results. There are many proofs available in the mathematical literature, but some of them are not completely correct (see the proof in [21]). We follow essentially the proof by Debreu and Herstein [7], quite elegant and short, but which needs some preliminary results.

Lemma 1.8.7 *Let $f : \mathbb{R}^n \longrightarrow \mathbb{R}^n$ be continuous on a compact (i.e. closed and bounded) convex set $S \subset \mathbb{R}^n$ to itself (i.e. $f : S \longrightarrow S$). Then f has a fixed point, i.e. there exists $x^* \in S$ such that $f(x^*) = x^*$.*

The previous result is nothing but the *Brouwer fixed point theorem,* an apparently simple, however deep result of Topology. For the notions of continuous function, see Chap. 2 and for the notion of convex set, see Chap. 3.

Consider the set

$$S = \left\{ x \geq [0] : \sum_{i=1}^{n} x_i = 1 \right\}.$$

The above set is the so-called *fundamental simplex of* $\mathbb{R}^n$; it is a convex and compact set.

Lemma 1.8.8 *Let $A \geq [0]$, square of order n and indecomposable, and let be $x \geq [0]$. Then $Ax \geq [0]$.*

Proof Surely $Ax \geqq [0]$. Suppose $Ax = [0]$; then A must have a column of zeros, which contradicts the indecomposability assumption. □

We now prove the following assertions of Theorem 1.8.4.

- (1) Let be $A \geq [0]$, square of order n and indecomposable. Then A has a positive eigenvalue $\lambda^*(A)$ and a corresponding positive eigenvector.

Proof Consider the map $f : S \longrightarrow \mathbb{R}^n$, where S is the fundamental simplex of $\mathbb{R}^n$ and where the i-th component of f is given by

$$f_i(x) = \frac{\sum_{j=1}^{n} a_{ij} x_j}{\sum_i \sum_j a_{ij} x_j}, \quad i = 1, \ldots, n.$$

Note that $f(x) = \frac{1}{e^\top Ax} Ax$, $Ax \geq [0]$ for all $x \geq [0]$ by Lemma 1.8.8 and so $e^\top Ax > 0$, f is well defined and $f(x) \geq [0]$ for all $x \geq [0]$. As $\sum_{i=1}^{n} f_i(x) = 1$, we have $f : S \longrightarrow S$ and f is continuous on S. By Lemma 1.8.7 there exists $x^* \in S$ such that $f(x^*) = x^*$. Hence $(e^\top Ax^*)x^* = Ax^*$ and $e^\top Ax^* > 0$. Put $\lambda^*(A) = e^\top Ax^* > 0$. Suppose $x^* \not> [0]$. Consider the equation

$$Ax^* = \lambda^*(A)x^*$$

or

$$PAP^{-1}Px^* = \lambda^*(A)Px^*$$

where

$$Px^* = \begin{bmatrix} \bar{x} \\ [0] \end{bmatrix}, \quad \bar{x} > [0]$$

or

$$\begin{bmatrix} A_{11} & A_{12} \\ A_{21} & A_{22} \end{bmatrix} \begin{bmatrix} \bar{x} \\ [0] \end{bmatrix} = \lambda^*(A) \begin{bmatrix} \bar{x} \\ [0] \end{bmatrix}.$$

So, $A_{21}\bar{x} = [0]$. As $\bar{x} > [0]$, it follows that $A_{21} = [0]$. This contradicts the indecomposability assumption on A. Hence $x^* > [0]$. □

Remark 1.8.9 The previous proof has been conducted with reference to the right-hand eigenvector x^*. Obviously, the same proof could be repeated with reference to left-hand eigenvectors. Now let $B \geqq [0]$ be a square matrix of order n, and let α be an eigenvalue (maybe complex) of B. Let $p^\top$ be the associated left-hand eigenvector (maybe complex). We have

$$p^\top B = \alpha p^\top,$$

i.e.

$$\sum_{i=1}^{n} p_i b_{ij} = \alpha p_j, \quad j = 1, \ldots, n.$$

If we apply the triangular inequality we can write

$$\sum_{i=1}^{n} |p_i| b_{ij} \geqq |\alpha| \left|p_j\right|, \quad j = 1, \ldots, n. \tag{1.8}$$

If we denote by $(p^*)^\top$ the vector $[|p_1|, \ldots, |p_n|]$, relation (1.8) can be written as

$$(p^*)^\top B \geqq |\alpha| (p^*)^\top. \tag{1.9}$$

We are now ready to prove the following part of the Perron-Frobenius Theorem 1.8.4 (for semipositive indecomposable square matrices).

- (2) If α is an eigenvalue of $A \geq [0]$, A indecomposable of order n, then it holds $|\alpha| \leqq \lambda^*(A)$. In other words, $\lambda^*(A)$ is the *dominant eigenvalue or spectral radius* of A.

Proof Let us consider a square matrix B, of order n, such that

$$[0] \leq B \leqq A. \tag{1.10}$$

Let α be an eigenvalue of B and let $p^\top$ be the associated left-hand eigenvector. Moreover, let $\lambda^*(A)$ and x^* be, respectively, the Frobenius eigenvalue of A and its associated Frobenius eigenvector. By (1.9) and (1.10) we can write

$$|\alpha|\, p^\top \leqq p^\top B \leqq p^\top A. \tag{1.11}$$

Now, we multiply both members of relation (1.11) by $x^* > [0]$:

$$|\alpha|\, p^\top x^* \leqq p^\top A x^*$$

and hence

$$|\alpha|\, p^\top x^* \leqq \lambda^*(A) p^\top x^* :$$

But, since $x^* > [0]$, we get

$$|\alpha| \leqq \lambda^*(A).$$

In particular, with $B = A$, we obtain the thesis. □

- (3) Let α be any real eigenvalue of $A \geq [0]$, indecomposable of order n, with $\alpha \neq \lambda^*(A)$. Then, if x is the associated eigenvector corresponding to α, x has at least a negative component.

Proof By definition, we have

$$Ax = \alpha x. \tag{1.12}$$

We know from the previous proposition that $|\alpha| \leqq \lambda^*(A)$, and hence in our case we have

$$\alpha < \lambda^*(A). \tag{1.13}$$

Let us consider two cases:

(*i*) $\alpha < 0$. Then (1.12) cannot be satisfied if $x > [0]$, hence x must contain at least a negative component.

(ii) $\alpha \geqq 0$. We have, with $\bar{p} > [0]$

$$(\bar{p})^\top A = \lambda^*(A)(\bar{p})^\top. \tag{1.14}$$

Now, we post-multiply both members of relation (1.14) by x and pre-multiply both members of relation (1.12) by $\bar{p}$; then we make the subtraction between the members of the relations. We get

$$(\lambda^*(A) - \alpha)(\bar{p})^\top x = 0.$$

By relation (1.14) we have $(\bar{p})^\top x = 0$, and since $\bar{p} > [0]$, we conclude that x also in this case must contain at least a negative element. □

In order to prove the next result we need a further lemma.

Lemma 1.8.10 (Lemma of Wielandt) *Let* $A \geq [0]$, *square of order* n *and indecomposable. Then* $(I + A)^{n-1} > [0]$.

Proof We prove that for any $x \geq [0]$, it holds $(I + A)^{n-1}x > [0]$, and hence $(I + A)^{n-1} > [0]$. Consider a vector $y \geq [0]$ and write

$$z = (I + A)y = y + Ay. \tag{1.15}$$

Since $A \geq [0]$, $Ay \geqq [0]$ and z has at least as many nonzero (and hence positive) elements as y. If y is not already positive, we prove that z has at least one more nonzero element than y.

It is clear that the matrix properties used in the statement of the lemma are invariant under permutation transformations. So, if it is assumed that z has no more positive elements than y, we can write (without loss of generality)

$$y = \begin{bmatrix} u \\ [0] \end{bmatrix}, \quad z = \begin{bmatrix} v \\ [0] \end{bmatrix}$$

where $u > [0]$ and, since $z = y + Ay$, $v > [0]$. If we partition A accordingly,

$$A = \begin{bmatrix} A_{11} & A_{12} \\ A_{21} & A_{22} \end{bmatrix}$$

we have, by (1.15), $A_{21}u = [0]$.

Since $u > [0]$ and $A_{21} \geqq [0]$, this implies that $A_{21} = [0]$, contradicting the assumption that A is indecomposable. By repeating the argument $n - 1$ times we find that, for an arbitrary $y \geq [0]$,

$$(I + A)^{n-1}y > [0].$$

Putting $y = e^i$, $i = 1, \ldots, n$, we see that this implies $(I + A)^{n-1} > [0]$ (hence $(I + A)^n > [0]$, $(I + A)^{n+1} > [0]$, etc.). □

- (4) If we consider the *positive* real number $\mu = \frac{1}{\rho}$, we have, with $A \geq [0]$ square of order n and indecomposable,

$$(\mu I - A)^{-1} > [0]$$

$$(I - \rho A)^{-1} > [0]$$

if and only if $\mu > \lambda^*(A)$, and hence $\rho < \frac{1}{\lambda^*(A)}$.

Proof We know (see Sect. 1.6) that, under our assumptions, it holds

$$(I - \rho A)^{-1} = I + \rho A + (\rho A)^2 + (\rho A)^3 + \cdots = \sum_{k=0}^{\infty} \rho^k (A)^k \tag{1.16}$$

and

$$(\mu I - A)^{-1} = \frac{1}{\mu}\left[I + \frac{1}{\mu}A + (\frac{1}{\mu}A)^2 + (\frac{1}{\mu}A)^3 + \ldots\right] = \frac{1}{\mu}\sum_{k=0}^{\infty}\frac{1}{\mu^k}(A)^k. \tag{1.17}$$

We note that the first n addenda in the right-hand side of the two above relations are, leaving out of consideration the positive coefficients, the same addenda of the binomial expansion

$$(I + A)^n = I + nA + \binom{n}{2}(A)^2 + \cdots + \binom{n}{n-1}(A)^{n-1} + (A)^n.$$

We know, from Lemma 1.8.10 (Wielandt Lemma) that $(I+A)^n > [0]$. Moreover, all coefficients of the above binomial expansion are positive. It follows that the sum of the first n addenda of the two series (1.16) and (1.17), is a positive matrix. A fortiori the sum of the whole series will be positive. Hence, we have

$$(I - \rho A)^{-1} > [0]; \;\; (\mu I - A)^{-1} > [0].$$

□

- (5) The *Brauer-Solow corollary* can be proved as follows.

Proof We have, by definition,

$$Ax^* = \lambda^*(A)x^*,$$

i.e.

$$\sum_{j=1}^{n} a_{ij}x_j^* = \lambda^*(A)x_i^*, \; i = 1, \ldots, n.$$

Summing over i we get

$$\sum_{i=1}^{n}\sum_{j=1}^{n} a_{ij}x_j^* = \sum_{j=1}^{n}\left(x_j^*\sum_{i=1}^{n} a_{ij}\right) = \sum_{j=1}^{n} x_j^* e^\top A^j = \lambda^*(A)\sum_{j=1}^{n} x_j^*.$$

We have immediately

$$\lambda^*(A) = \frac{\sum_{j=1}^{n} x_j^* e^\top A^j}{\sum_{j=1}^{n} x_j^*} \tag{1.18}$$

so that $\lambda^*(A)$ is a nonnegative weighted average of the column sums of A, the weights being the various x_j^*. It now immediately follows that

$$\max\left\{e^\top A^j\right\} \geqq \lambda^*(A) \geqq \min\left\{e^\top A^j\right\}. \tag{1.19}$$

Now, suppose that all sums are equal, but that one of the equalities in (1.19) holds. Then, all column sums not equal to $\lambda^*(A)$ must have a zero weight in (1.18), that is the corresponding x_j^* must be zero, but this is in contrast with the assumption of indecomposability of A. □

If we drop the assumption of indecomposability of A, the Theorem of Perron-Frobenius assumes a weaker version.

Theorem 1.8.11 (Theorem of Perron-Frobenius, "Weak Version") *Let be* $A \geqq [0]$, *square of order* n.

(1) The characteristic equation of A has a root, denoted by λ^* *or* $\lambda^*(A)$ *or* dom(A), *such that:*

 (a) $\lambda^* \geqq 0$.
 (b) $\lambda^* \geqq |\lambda|$, *being* λ *any other eigenvalue of* A.
 (c) λ^* *is not necessarily a simple root of* $|A - \lambda I| = 0$.

(2) It is possible to associate with λ^* *an eigenvector* $x^* \geq [0]$ *or* $x^* \leq [0]$.
(3) If and only if $\mu > \lambda^*$ $(\mu \in \mathbb{R})$, *it holds* $[\mu I - A]^{-1} \geq [0]$.
(4) λ^* *is an increasing function of the elements of* A :

$$B \geq A \Longrightarrow \lambda^*(B) \geqq \lambda^*(A).$$

The "Brauer-Solow" criterion has the version:

$$m \leqq \lambda^*(A) \leqq M,$$

without any other consideration.

Remark 1.8.12

(1) We note that, given a square matrix $A \geq [0]$, its Frobenius eigenvector is its spectral radius, which is in this case a real nonnegative number (or even positive): it is that real eigenvalue which coincides with the spectral radius. For instance, if a matrix $A \geq [0]$ has the following spectrum: $\lambda_1 = 5,\ \lambda_2 = 4,\ \lambda_3 = 2,\ \lambda_4 = 3 + 4i,\ \lambda_5 = 3 - 4i$, it results $\rho_A = 5$ (as $\lambda_1 = 5$ and $|3 + 4i| = |3 - 4i| = \sqrt{9 + 16} = 5$) and its Frobenius eigenvalue is obviously $\lambda^* = 5$, given by the real eigenvalue λ_1).

(2) With A decomposable and nonzero, it may be $\lambda^* = 0$. Consider, for instance,

$$A = \begin{bmatrix} 0 & 1 \\ 0 & 0 \end{bmatrix}.$$

We have $\lambda^* = 0$ (double root). It is possible to associate with λ^* eigenvector $x^* \geq [0]$, for instance we have

$$\begin{bmatrix} 0 & 1 \\ 0 & 0 \end{bmatrix} \begin{bmatrix} 1 \\ 0 \end{bmatrix} = 0 \begin{bmatrix} 1 \\ 0 \end{bmatrix}.$$

However, it is not possible to associate eigenvectors $x^* > [0]$ with the same Frobenius eigenvalue. Moreover, if A is decomposable, $\lambda^*(A)$ is not the unique eigenvalue for which it is possible to associate semipositive (or seminegative) eigenvectors. Consider, e.g., the matrix

$$A = \begin{bmatrix} 3 & 0 \\ 0 & 2 \end{bmatrix}$$

for which it is possible to associate eigenvectors $x \geq [0]$ both to $\lambda_1 = \lambda^* = 3$, and to $\lambda_2 = 2$. Furthermore, for the semipositive vectors, say x^* and x^{**}, associated with $\lambda^*(A)$, we may have $x^{**} \neq \alpha x^*,\ \alpha > 0$.

(3) It is possible to obtain $x^* > [0]$ also when A is decomposable? Beyond trivial cases, such as a diagonal matrix of the form αI, with $\alpha > 0$, on the grounds of what previously asserted, the answer is: in general it is not possible. However, under additional assumptions the answer is affirmative. Let A be decomposed in the form

$$A = \begin{bmatrix} A_{11} & A_{12} \\ [0] & A_{22} \end{bmatrix},$$

with $A_{12} \neq [0]$. If we have $\lambda^*(A_{22}) > \lambda^*(A_{11})$, then it is possible to associate to $\lambda^*(A_{22})$, which is the Frobenius eigenvalue of A, an eigenvector $x^* > [0]$.
Let us consider the matrix

$$A = \begin{bmatrix} 0 & 1 \\ 0 & 1 \end{bmatrix}.$$

We have $\lambda^*(A) = 1,\ \lambda^*(a_{11}) = 0,\ \lambda^*(a_{22}) = 1$. Any vector

$$x^* = \begin{bmatrix} \alpha \\ \alpha \end{bmatrix},\quad \alpha > 0,$$

is a Frobenius eigenvector:

$$\begin{bmatrix} 0 & 1 \\ 0 & 1 \end{bmatrix}\begin{bmatrix} \alpha \\ \alpha \end{bmatrix} = 1\begin{bmatrix} \alpha \\ \alpha \end{bmatrix}.$$

The above result is a simplified version of a more general result, due to F. R. Gantmacher (see the References).

(4) We remark that if A is decomposed into a "block upper triangular form" (1.7) or in a similar "block lower triangular form", λ is an eigenvalue of A if and only if λ is an eigenvalue of at least one of the diagonal blocks of A, in other words the spectrum of A is given by the union of the spectra of the various diagonal blocks.

The following result may be considered a generalization of the well known (in Economic Analysis!) "Hawkins-Simon conditions", useful in the study of several linear economic models, such as, for example, the Leontief models (see the next section).

Theorem 1.8.13 *Let be given* $A \geqq [0]$, *square of order* n. *The North-West principal minors of the matrix* $[\mu I - A]$, $\mu \in \mathbb{R}$, *are positive if and only if* $\mu > \lambda^*(A)$.

From the previous theorem we conclude that, given A square and nonnegative, the matrix $[\mu I - A]^{-1}$ exists and is semipositive (i.e. with semipositive lines) or even positive, if A is indecomposable, if and only if the North-West principal minors of the matrix $[\mu I - A]$, are positive, which is more "operative" than computing the Frobenius eigenvalue of A; (see [7]).

1.9 Introduction to the Multi-Sectoral Models of Leontief and Sraffa

Among the various economic, financial, probabilistic, etc. applications of the results surveyed in the previous section, we shall be concerned with some basic considerations on two well known linear economic models: a model due to the Russian economist (but naturalized as American) W. Leontief (1905–1999), Nobel prize for economics in 1973, and a model due to the Italian economist Piero Sraffa (1896–1983), who worked for the most part of his life at the University of Cambridge (U.K.).

We begin with one of the simplest models proposed by W. Leontief, models that in economic literature are known as "input-output models", "multifactorial economic models", etc. Let us consider a production system formed by n industries or economic activities or production sectors, each of them producing only one type of goods: the production processes are "simple", i.e. there is not joint production in the system under consideration. Every good is produced by employing as production factors (or inputs) the n goods produced by the economic system and no good enters in the production process for more than one period (in other words, "capital goods" are not taken into consideration).

Let $x_i > 0,\ i = 1, \ldots, n$, be the quantity of the i-th good obtained by the i-th industry and let $q_{ij} \geqq 0$ be the quantity of the i-th good necessary, as an input, to produce the total quantity x_j of the j-th good ($i, j = 1, \ldots, n$). The net output of the i-th industry, i.e. the quantity

$$x_i - \sum_{j=1}^{n} q_{ij}, \quad i = 1, \ldots, n,$$

is available for the market consumption (it is an "exogeneous" quantity) or "final consumption". Therefore we have the following "equilibrium conditions" or "balance conditions":

$$x_i = \sum_{j=1}^{n} q_{ij} + c_i, \quad i = 1, \ldots, n,$$

where $c_i \geqq 0,\ i = 1, \ldots, n$, represents the net output of the i-th industry, that is the consume of the market of the i-th good. Now we make the following basic assumption: we suppose that for each good the quantity of the i-th good necessary as an input, to produce the j-th good, is in a constant ratio with respect to the total produced quantity of the j-th good. In other words

$$q_{ij} = a_{ij} x_j, \quad i, j = 1, \ldots, n.$$

In the above relation the elements (nonnegative and constant) a_{ij} express the quantity of the i-th good necessary to produce *one unity* of the j-th good. The numbers $a_{ij},\ i, j = 1, \ldots, n$, are also called "input-output coefficients" or "technological coefficients" . This assumption, in other words, states that all production processes are *linear,* or with *constant returns to scale* or *with fixed production coefficients.* Let us assemble the coefficients a_{ij} into a square nonnegative matrix, of order n,

$$A = \left[a_{ij}\right], \ \ i, j = 1, \ldots, n,$$

which is usually called *input-output matrix, Leontief matrix, technical coefficients matrix, etc.* In this matrix each row $A_i,\ i = 1, \ldots, n$, describes the use, the destination of the i-th good for the n production activities, under the assumption that the production processes are activated at unitary production intensities. It is convenient to assume

$$A_i \geq [0]\,, \ i = 1, \ldots, n,$$

in other words, every good has at least one destination in the economic system under consideration. Every column $A^j,\ j = 1, \ldots, n$, describes, in a certain sense, the technological feature of the various production processes. Also for the columns it is convenient to assume

$$A^j \geq [0]\,, \ \ j = 1, \ldots, n,$$

in other words, every industry needs at least one input; the economist (and Nobel prize for economics) T. C. Koopmans says that the "Land of Cockaigne" must be excluded. If we denote by c the vector of the "basket of consumption goods" and assume $c \geq [0]$, we must have the following equilibrium conditions:

$$\sum_{j=1}^{n} a_{ij}x_j + c_i = x_i, \ i = 1, \ldots, n,$$

that is

$$Ax + c = x,$$

where A and c are given (as "exogenous data"), $A \geq [0]$, $c \geq [0]$, and $x > [0]$ represents the vector of the global quantities of the various goods to be produced in order to meet the equilibrium conditions. This is, roughly speaking, the so-called "open Leontief system" , i.e. an economic system whose final consumption demand has to be met (in the so-called "closed Leontief system" all inputs into production are produced and all outputs exist merely to serve as inputs).

We have to find this vector $x > [0]$. If $[I - A]$ is invertible, the unique solution of our problem is obviously given by

$$x = [I - A]^{-1} c.$$

This is the mathematical solution, but not necessarily the "economic solution" : it may result that x has not all positive components, even if A has all semipositive lines and c is a semipositive vector. The following counterexample is taken from Nicola (1976).

$$A = \begin{bmatrix} 2 & 0.4 & 0.3 \\ 0.5 & 0 & 0.2 \\ 0.1 & 0.5 & 0 \end{bmatrix}; \quad c = \begin{bmatrix} 2 \\ 4 \\ 0 \end{bmatrix}.$$

It holds

$$[I - A]^{-1} = \frac{1}{1.213} \begin{bmatrix} -0.9 & -0.55 & -0.38 \\ -0.52 & 1.03 & 0.05 \\ -0.35 & 0.46 & 1.2 \end{bmatrix},$$

$$[I - A]^{-1} c = x = \begin{bmatrix} -3.297661 \\ 2.53916 \\ 0.94763 \end{bmatrix}.$$

Hence the total quantity of the first good that must be produced is negative, which obviously makes no sense. If the matrix $[I - A]^{-1}$ would be semipositive or even positive, the said anomaly of course would not occur. If we make the further assumption that A is *indecomposable* and recall the point (3) of Theorem 1.8.4 (Perron-Frobenius theorem in its "strong version"), we can conclude that $[I - A]^{-1} > [0]$ if and only if

$$\lambda^*(A) < 1.$$

A Leontief matrix $A \geq [0]$ for which $\lambda^*(A) < 1$ is said, in the economic literature, *productive matrix* and the related condition is said "productivity condition" . The same terminology is also used for the matrix $(I - A)$. Therefore this condition therefore ensures that

$$[I - A]^{-1} c = x > [0].$$

Furthermore, we have $x > [0]$ for *any choice* of the vector $c \geq [0]$. It is also obvious that if $c > [0]$ we can drop the assumption of indecomposability of A, as the point 3) of Theorem 1.8.11 of the previous section assures that $[I - A]^{-1} \geq [0]$

if and only if $\lambda^*(A) < 1$ and, being $[I - A]^{-1}$ with all semipositive lines, again we obtain $x > [0]$.

The productivity condition $\lambda^*(A) < 1$ is however not too meaningful from an economic point of view. There are other equivalent productivity conditions, more interesting for economic purposes or for computational purposes. The related theory is known in Linear Algebra as the "Theory of M-matrices" or also "Theory of K-matrices"; see, e.g., [3] and [8].

Theorem 1.9.1 *Let be given $A \geq [0]$, square, of order n. Then the following conditions are equivalent.*

(i) A is productive, i.e. $\lambda^*(A) < 1$.
(ii) For any $c \geq [0]$ there exists $x \geq [0]$ such that $[I - A]x = c$.
(iii) There exists $x > [0]$ such that $[I - A]x > [0]$.
(iv) There exists $x \geq [0]$ such that $[I - A]x > [0]$.
(v) There exists $c > [0]$ such that the system $[I - A]x = c$ admits a solution $x \geq [0]$.
(vi) $[I - A]^{-1}$ exists and it holds $[I - A]^{-1} \geq [0]$.
(vii) The series $\sum_{k=0}^{\infty}(A)^k$ converges and it holds $\sum_{k=0}^{\infty}(A)^k = [I - A]^{-1} \geq [0]$.
(viii) It holds $\lim_{k \longrightarrow \infty} (A)^k = [0]$, i.e. A is "small" or "convergent".
(ix) The principal minors of $(I - A)$ are all positive.
(x) The North-West principal minors of $(I - A)$ are all positive (this is the so-called "Hawkins-Simon condition"; see, e.g., [19, 24]).

We can remark that in the previous theorem conditions (ii), (iii), (iv) and $v)$ have a non trivial economic meaning: A is productive if and only if there exists a production vector $x \geq [0]$ (or also $x > [0]$) such that the economic system in question is able to produce a positive net output vector.

For example, if there exists $x > [0]$ such that $(I - A)x > [0]$, then $\lambda^*(A) < 1$. Indeed, if $\lambda^*(A) \geqq 0$ is the Frobenius root of A and $(x^*)^\top \geq [0]$ is the corresponding semipositive row eigenvector, we have $(x^*)^\top A = \lambda^*(A)(x^*)^\top$ and also $(x^*)^\top Ax = \lambda^*(A)(x^*)^\top x$, with $x > [0]$ satisfying the above assumptions. But from the same assumptions we get $(x^*)^\top x > (x^*)^\top Ax$, i.e. $(x^*)^\top x > \lambda^*(A)(x^*)^\top x$, but being $(x^*)^\top x > 0$, it will hold $\lambda^*(A) < 1$. From this result we have, thanks to the weak version of the Perron-Frobenius Theorem, $(I - A)^{-1} \geq [0]$. This last result implies in turn that there exists $x > [0]$ such that $x > Ax$: choose any $y > [0]$, and let $x = (I - A)^{-1}y$. Then, $(I - A)^{-1} \geq [0]$ implies $x > [0]$ and $(I - A)x = y > [0]$ implies $x > Ax$.

In case $A \geq [0]$, of order n, is *indecomposable,* some results of the previous theorem can be reformulated as follows.

Theorem 1.9.2 *Let $A \geq [0]$, square of order n, be indecomposable. Then the following conditions are equivalent:*

(a) There exists $x > [0]$ such that $[I - A]x \geq [0]$.
(b) There exists $x \geq [0]$ such that $[I - A]x \geq [0]$.
(c) $[I - A]^{-1}$ exists and it holds $[I - A]^{-1} > [0]$.

Also the celebrated "Hawkins-Simon condition" is not too much meaningful from an economic point of view, but it is more "operative" than the other conditions of Theorems 1.9.1 and 1.9.2. If A is indecomposable, by means of the "Brauer-Solow criterion" (Corollary 1.8.5), we have that a sufficient condition to have $\lambda^*(A) < 1$ is that all row (or column) sums of A are less than or equal to one, and that at least one sum is strictly less than one.

Again we make the assumption that the Leontief matrix A is productive: $\lambda^*(A) < 1$. By (vii) of Theorem 1.9.1 (see also what said after Theorem 1.6.26) we have ("C. Neumann series")

$$[0] \leq [I - A]^{-1} = I + A + (A)^2 + (A)^3 + \ldots$$

Hence, being $x = [I - A]^{-1} c$, we have

$$x = Ic + Ac + (A)^2 c + (A)^3 c + \ldots$$

that is

$$x = c + Ac + A(Ac) + A(A)^2 c + \ldots$$

Also this matrix series is susceptible of an economic interpretation: the total production vector x is given by the sum of the net output vector c (basket of market consumes) plus the vector Ac, which represents the quantity of inputs necessary to produce the said vector c, plus the vector $A(Ac)$, which represents the quantity of inputs necessary to produce the vector Ac, and so on. Every addendum of the said series can be interpreted as one of the various "rounds" of the production process, in order to obtain the gross production vector x, rounds ordered backward.

Finally, with reference to the indecomposability or decomposability assumption on a technological Leontief matrix A, by recalling also what said at the points b) and c) of Remark 1.8.3, we can assert that the assumption of indecomposability of A is equivalent to the fact that all goods enter *directly or indirectly* in the production of *all* goods of the system (in other words, any good enters directly or indirectly in the production of itself and in the production of the other $(n - 1)$ goods of the system). In Economic Analysis it is said that in this case the system is "perfectly integrated" , or also, following the terminology of P. Sraffa, that all goods of the systems are "basic" (Sraffa speaks of "basic commodities"). If, vice-versa, A is decomposable, this means that the economic system splits into at least two blocks which do not interact between them in the two directions. For example, we may have a first block of goods which enter directly and indirectly in all the n goods produced by the system, and a second block of goods which enter only in the production of the goods of this second block (they are the so-called "non-basic commodities" in Sraffa's terminology).

We give now some hints on one of the various economic models considered by P. Sraffa in his famous book "Production of Commodities by Means of Commodities" (1960), a rather difficult book, paradoxically owing to the decision of the author to not use mathematical tools in treating the related economic models. The book of Sraffa is divided into two parts: in the first one the author considers *simple production models* (i.e. each industry produces only one type of commodity, just as in the Leontief model previously considered) and in the second one the author is concerned with *joint production models* (and hence here it is possible to consider the presence of fixed capital, beyond the "circulating capital"). This second part entails some rather complex mathematical issues, such as a Perron-Frobenius theory for pair of nonnegative matrices, and we shall not be concerned with these types of models. From a formal-mathematical point of view, the simple production models of Sraffa may be viewed as a "side" on prices of an input-output model or in general of a linear economic model treating quantities. Sraffa, however introduces also an explicit quantity of *labour,* used in each one of the n industries; furthermore, as in the Leontief model, also "intermediate commodities" produced by the system, are used in the production of the n commodities (the title of the book just refers to this kind of economic situation). Even if Sraffa does not mention in an explicit way an assumption of "constant returns to scale" , we shall consider a technological matrix A, square, semipositive and of order n, whose elements have the same meaning than the ones of the Leontief model. We assume the two conditions, previously commented,

$$A_i \geq [0]\,,\ \ i = 1, \ldots, n;$$
$$A^j \geq [0]\,,\ \ j = 1, \ldots, n.$$

Thanks to the "Brauer-Solow criterion" (Corollary 1.8.5), these conditions assure that $\lambda^*(A) > 0$. Then we consider a positive row vector of the direct labour requirements:

$$\ell^\top = [\ell_1, \ell_2, \ldots, \ell_n] > [0]\,.$$

The total available quantity of labour $\ell_1 + \ell_2 + \cdots + \ell_n$ will be denoted by L. The labour force is rewarded with a wage rate $w \geqq 0$, equal for all the n industries.

The n inputs of commodities, necessary for the production and which come from the previous production cycle, are anticipated by the entrepreneurs at the beginning of the production period, The value of all these commodities is the so-called "circulating capital" . The entrepreneurs receive profits, proportional in every sector, to the circulating capital. This "proportionality factor" is called "profit rate" and denoted by $r\ :\ r \geqq 0$. We suppose that all gross productions are *unitary* and moreover we suppose that $L = 1$, so that w assumes the role of *unitary wage.* It is possible to give an economic justification of the fact that both r and w are equal in all sectors, by assuming a "perfect competition" . The problem we wish to study is the following one.

With A and ℓ given, find:

- The *equilibrium prices* of the n commodities produced by the system, i.e. a vector $p > [0]$ which satisfies n equilibrium conditions (balance economic-financial conditions, to be specified).
- The corresponding unitary wage rate w.
- The corresponding profit rate r.

The previous equilibrium conditions entail that in each sector it must hold:

gross return=value of the circulating capital+total wages+total profits.

As each total production is unitary by assumption, we have, with reference to the j-th commodity ($j = 1, \ldots, n$):

$$p_j = \sum_{i=1}^{n} p_i a_{ij} + w\ell_j^\top + r \sum_{i=1}^{n} p_i a_{ij}.$$

The above relation must hold for $j = 1, \ldots, n$, hence we can write

$$p^\top = p^\top A + w\ell^\top + r(p^\top A),$$

that is

$$p^\top A + rp^\top A + w\ell^\top = p^\top,$$

that is

$$(1+r)p^\top A + w\ell^\top = p^\top. \tag{1.20}$$

Therefore, with A and ℓ data of the problem, we have to find a triplet (p, r, w), with $p > [0]$, $r \geqq 0$ and $w \geqq 0$, which solves (1.20). Note that the problem is "over determined": the system contains n equations in $(n+2)$ unknowns (p, r, w). We must therefore fix one of the two unknowns and then treat the other one as a parameter; this is the so-called *normalization problem.* For instance, we can put $w = 1$, or put equal to one a certain price (e.g. $p_1 = 1$), or (we shall follow this way) put equal to 1 the value of the *total net product,* value computed at the equilibrium prices. From a formal point of view, this normalization is expressed as follows. If we denote by $e^\top = [1, 1, \ldots, 1]$ the gross productions vector, then the net productions vector, assumed to be semipositive, is given by

$$y = e - Ae = [I - A]\, e \geq [0]\,.$$

Hence the vector $p^\top y = p^\top(e - Ae) = p^\top e - p^\top Ae$, represents the value of the total net product, value that we put equal to one:

$$p^\top y = 1.$$

A benefit of this normalization condition is that now w not only measures the wage rate, but also a distribution quote. For example, if $w = 0.35$, this means that 35% of the value of the total net production is cashed by the workers, whereas the 65% of the same values is cashed by the entrepreneurs. Furthermore, $(p^\top Ae)$ measures the total circulating capital anticipated by the entrepreneurs and $(rp^\top Ae)$ measures the total profits perceived by the entrepreneurs class. It is rather obvious (but we shall prove this fact) that if the value of the net production is a "cake" to be divided between the wage-earners and the capitalists, the sum of the two respective slices reassembles the whole cake:

$$1 = p^\top y = w + r(p^\top Ae). \tag{1.21}$$

We prove relation (1.21) by starting from the system

$$(1+r)p^\top A + w\ell^\top = p^\top,$$

where $p > [0]$, $r \geqq 0$, $w \geqq 0$. We have

$$p^\top A + rp^\top A + w\ell^\top = p^\top,$$

that is

$$p^\top - p^\top A = w\ell^\top + rp^\top A.$$

Let us multiply both members of the last relation by the summation vector $e = [1, 1, \dots, 1]^\top$:

$$(p^\top - p^\top A)e = (w\ell^\top)e + (rp^\top A)e.$$

But $(p^\top - p^\top A)e = p^\top(I - A)e = p^\top y = 1$, whereas $(w\ell^\top)e = w(\ell^\top)e = wL = w1 = w$. Hence $1 = w + r(p^\top Ae)$ and the game is done.

We continue to assume $y \geq [0]$, where $y = e - Ae$, so that it is possible to choose the normalization condition $p^\top y = 1$; we are looking for the triplets (p, r, w) representing the equilibrium solutions of the system

$$(1+r)p^\top A + w\ell^\top = p^\top, \tag{1.22}$$

$$p > [0], \; r \geqq 0, \; w \geqq 0, \tag{1.23}$$

$$p^\top y = 1. \tag{1.24}$$

Let us suppose the existence of solutions with $w > 0$ (we shall consider only this case; the case $w = 0$ characterizes a "slave economy" , perhaps interesting from a mathematical point of view, but luckily not too common at present). We rewrite (1.22) as

$$p^\top [I - (1+r)A] = w\ell^\top . \tag{1.25}$$

We have

$$[I - (1+r)A] = (1+r)\left[\frac{1}{1+r}I - A\right]$$

and

$$[I - (1+r)A]^{-1} \geq [0] \iff \left[\frac{1}{1+r}I - A\right]^{-1} \geq [0] .$$

By the Theorem of Perron-Frobenius ("weak version") the above last inequality holds if and only if

$$\frac{1}{1+r} > \lambda^*(A) \iff 1 > (1+r)\lambda^*(A) \iff r < \frac{1}{\lambda^*(A)} - 1.$$

The last inequality makes sense as $\lambda^*(A) > 0$, thanks to the assumptions on the sign of the lines of A. The productivity of A, i.e. $\lambda^*(A) < 1$, assured by the fact that $(I - A)e \geq [0]$, implies that the quantity $\frac{1}{\lambda^*(A)} - 1$ is positive and hence the interval $\left[0, \frac{1}{\lambda^*(A)} - 1\right)$ does not shrink to a single point or to the empty set. Hence, by choosing r in the interval

$$\left[0, \frac{1}{\lambda^*(A)} - 1\right),$$

the matrix $[I - (1+r)A]$ admits inverse, and this inverse has all semipositive lines; therefore we have

$$p^\top = w\ell^\top [I - (1+r)A]^{-1} , \tag{1.26}$$

with $p > [0]$. The quantity

$$\left(\frac{1}{\lambda^*(A)} - 1\right) \equiv r^*$$

is called by Sraffa "maximum profit rate" and denoted by this author, by R. Now we use the normalization condition (1.24) in order to deduce from (1.26) the following relation

$$1 = p^\top y = w\ell^\top [I - (1+r)A]^{-1} y,$$

relation which gives the wage rate w:

$$w = \frac{1}{\ell^\top [I - (1+r)A]^{-1} y} > 0, \quad r \in \left[0, r^*\right).$$

This formula states also that w is, in the interval $\left[0, \frac{1}{\lambda^*(A)} - 1\right)$, a continuous and differentiable function of the profit variable r. In Economic Analysis it is also said that the said relation represents the *wage-profit frontier* for the model in consideration. If we compute the derivative of w with respect to r, we obtain

$$\frac{dw}{dr} = \frac{-\ell^\top [I - (1+r)A]^{-2} y}{\left\{\ell^\top [I - (1+r)A]^{-1} y\right\}^2}.$$

The second member of this equality is always negative, which is rather obvious, as the profit rate and the wage rate are two "antagonistic" variables.

If we choose $r = 0$ on the wage-profit frontier, then we have that $w = 1$ is the maximum wage rate (from (1.21) with $r = 0$ we get $w = 1$): with this choice the entrepreneurs have no profit and the whole value of the net production goes to the workers. From (1.26) we get in the said case

$$p^\top = \ell^\top [I - A]^{-1}.$$

Being A productive ($\lambda^*(A) < 1$), we can write the following series expansion

$$[I - A]^{-1} = I + A + (A)^2 + (A)^3 + \dots$$

and hence we obtain

$$p^\top = \ell^\top + \ell^\top A + \ell^\top (A)^2 + \ell^\top (A)^3 + \dots$$

The expression in the right-hand side of the previous relation has the following economic meaning: the elements of ℓ are, as previously already said, the quantities of direct labour requirements to obtain a unit of every commodity; $\ell^\top A$ is therefore the vector of the quantities of labour used to produce the production factors necessary to obtain a unit of the various commodities; $\ell^\top (A)^2$ is the vector of the quantities of labour used to produce the production factors necessary to produce the production factors in order to produce a unity of every commodity, and so on.

We can therefore assert that the above series represents the total quantity of labour, direct and indirect, used in the unitary production of the n commodities of the system. In other words, under the choice of $r = 0$, the equilibrium prices are equal to the values of the labour directly and indirectly "incorporated" in the commodities. This is the kernel of the mathematical formulation of the so-called *labour-value theory,* supported, for instance, by D. Ricardo and mainly by K. Marx.

Besides the works quoted along the chapter, we refer to the reader interested in the topics of this chapter to the references [1, 2, 5, 9–13, 15, 16, 18, 25–27].

References

1. G. Abraham-Frois, E. Berrebi, *Theory of Value, Prices and Accumulation* (Cambridge University Press, Cambridge, 1979)
2. R. Bellman, *Dynamic Programming* (Princeton University Press, Princeton, 1960)
3. A. Berman, R. Plemmons, *Nonnegative Matrices in the Social Sciences* (SIAM, Philadelphia, 1994)
4. M. Bertocchi, S. Stefani, G. Zambruno, *Matematica per l'Economia e la Finanza* (McGraw-Hill, Milano, 1995)
5. Y. Chabrillac, J.P. Crouzeix, Definiteness and semidefiniteness of quadratic forms revisited. Linear Algebra Appl. **63**, 283–292 (1984)
6. G. Debreu, Definite and semidefinite quadratic forms. Econometrica **20**, 295–300 (1952)
7. G. Debreu, I.N. Herstein, Nonnegative square matrices. Econometrica **21**, 597–607 (1953)
8. M.E. De Giuli, G. Giorgi, M.A. Maggi, U. Magnani, *Matematica per l'Economia e la Finanza* (Zanichelli, Bologna, 2008)
9. V.N. Faddeeva, *Computational Methods of Linear Algebra* (Dover Publications, New York, 1959)
10. W. Farebrother, Necessary and sufficient conditions for a quadratic form to be positive whenever a set of homogeneous linear constraints is satisfied. Linear Algebra Appl. **16**, 39–42 (1977)
11. G. Giorgi, *Elementi di Algebra Lineare* (Giappichelli, Torino, 1998)
12. G. Giorgi, Various proofs of the Sylvester criterion for quadratic forms. J. Math. Res. **9**(4), 168–184 (2017)
13. G. Hadley, *Linear Algebra* (Addison-Wesley, Reading, 1969)
14. F. Harary, *Graph Theory* (Addison-Wesley, Reading, 1969)
15. H.D. Kurz, S. Salvadori, *Theory of Production. A Long-Period Analysis* (Cambridge University Press, Cambridge, 1995)
16. K. Lancaster, *Mathematical Economics* (Macmillan, London, 1968)
17. C.D. Meyer, *Matrix Analysis and Applied Linear Algebra* (SIAM, Philadelphia, 2000)
18. L. Mirsky, *An Introduction to Linear Algebra* (Dover, Mineola, 1990)
19. H. Nikaido, *Convex Structures and Economic Theory* (Academic Press, New York, 1968)
20. B. Noble, J.W. Daniel, *Applied Linear Algebra* (Prentice Hall, Upper Saddle River, 1998)
21. L.L. Pasinetti, *Lectures on the Theory of Production* (Macmillan, London, 1977)
22. C.P. Simon, L. Blume, *Mathematics for Economists* (W. W. Norton, New York, 1994)
23. R. Solow, On the structure of linear models. Econometrica **20**, 29–46 (1952)
24. A. Takayama, *Mathematical Economics* (Cambridge University Press, Cambridge, 1985)
25. R. Varga, Matrix Iterative Analysis (Prentice-Hall, Englewood Cliffs, 1962)
26. J.E. Woods, *Mathematical Economics* (Longman, London, 1978)
27. J.E. Woods, *The Production of Commodities. An Introduction to Sraffa* (Macmillan, London, 1990)

Chapter 2
Introduction to Functions of Several Real Variables

2.1 Functions of Several Real Variables: Limits and Continuity

From previous courses of Mathematics the reader has learnt the concept of "function" ("map" , "correspondence"), that is, a relation which associates with each element of a given set A one and only one element of another set B:

$$f : A \longrightarrow B.$$

Usually, the following classification is adopted:

- *Real functions of one real variable,* if $A \subset \mathbb{R}$, $B \subset \mathbb{R}$;
- *Real functions of several real variables or scalar functions of real vectors,* if $A \subset \mathbb{R}^n$, $B \subset \mathbb{R}$;
- *Vector functions of one real variable,* if $A \subset \mathbb{R}$, $B \subset \mathbb{R}^m$;
- *Vector functions of real vectors or vector functions of several real variables (or applications or transformations),* if $A \subset \mathbb{R}^n$, $B \subset \mathbb{R}^m$.

It is mainly in economic and financial sciences that the link "cause-effect" cannot be expressed by a function $f : \mathbb{R} \longrightarrow \mathbb{R}$; indeed, it is usual in these fields that an effect is generated by a multiplicity of causes or also that a multiplicity of causes produces a multiplicity of effects. From this, the opportunity to be concerned with functions of the type

$$f : \mathbb{R}^n \longrightarrow \mathbb{R}^m,$$

G. Giorgi et al., *Lectures on Mathematics for Economic and Financial Analysis*,
https://doi.org/10.1007/978-3-031-83339-7_2

that is, in an extended form,

$$\begin{bmatrix} y_1 \\ y_2 \\ \vdots \\ y_m \end{bmatrix} = \begin{bmatrix} f_1(x_1, x_2, \ldots x_n) \\ f_2(x_1, x_2, \ldots, x_n) \\ \vdots \\ f_m(x_1, x_2, \ldots x_n) \end{bmatrix}.$$

In other words, these functions associate with vectors of $\mathbb{R}^n$ other vectors of $\mathbb{R}^m$. By varying n and m we obtain the cases previously described.

We shall call *domain* of the function f the set $A \subset \mathbb{R}^n$, set denoted also by $\text{dom}(f)$; the set $B = f(A) \subset \mathbb{R}^m$ is also called *codomain or set of the images* of f and is denoted also by $\text{Im}(f)$. An example of functions from $\mathbb{R}^n$ to $\mathbb{R}^m$, is given by *linear functions,* a type of functions we have already met in the previous chapter, Sect. 1.4:

$$f(x) = Ax,$$

with A real matrix of order (m, n). If $\text{dom}(f) = \mathbb{R}^n$ and $m = 1$, we have the important example of *quadratic forms,* also these ones already met in Sect. 1.7 of the previous chapter:

$$f(x) = x^\top Ax,$$

where A is a real symmetric matrix of order n.

The determination of the domain of a function $f : \mathbb{R}^n \longrightarrow \mathbb{R}^m$ may present some difficulties. However, if $n = 2$ and $m = 1$, often the problem is simplified, as in this case the domain is obviously a subset of $\mathbb{R}^2$ and we can try to represent the said set on the Cartesian plane.

Definition 2.1.1 Given $f : A \subset \mathbb{R}^n \longrightarrow \mathbb{R}^m$, we call *graph* of f the set

$$G_f = \{(x, y) : x \in A, \ y = f(x)\}.$$

If $n = 2, \ m = 1$, we shall have in general a *surface* of $\mathbb{R}^3$. There are at present several mathematical softwares which visualize on the desktop of a computer the graphs of real functions of two real variables. It is also possible, always in the case of $n = 2, \ m = 1$, to use the "device" of *level lines or level curves or level sets,* we have encountered on various geographical maps (where these lines were called "contour lines"). By this methodology we represent on the Cartesian plane xOy the geometrical locus of the points where $f(x, y)$ assumes a certain "level" h:

$$\text{lev}_{=h} f = \left\{(x, y) \in \mathbb{R}^2 : f(x, y) = h\right\}.$$

In other words: given $z = f(x, y)$, the level curve or level line equal to h is the orthogonal projection on the plane xOy of the sets of points of the surface generated by f and having value $z = h$. By varying h, we obtain the various level curves which usually (with a bit of imagination) give a good representation of the real function of two variables in question. Again: these level curves may also be interpreted as the orthogonal projection on the plane xOy of the section of the surface generated by f, section made through a plane parallel to the plane xOy and of equation $z = h$.

For instance, given $f(x, y) = x^2 + y^2$, whose graph in $\mathbb{R}^3$ is a so-called "paraboloid" (see Fig. 2.1), generates level curves given by the equation $x^2 + y^2 = h$, $h \geqq 0$. These level curves are therefore concentric circumferences, with center at the origin O and radius $\sqrt{h}$ (see Fig. 2.2).

If we consider the economic example of a production function of the type "Cobb-Douglas" :

$$z = kx^{\alpha}y^{\beta}, \quad k, \alpha, \beta > 0,$$

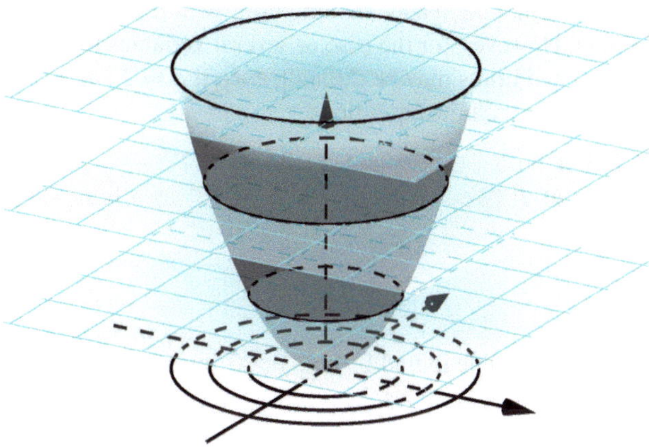

Fig. 2.1 Graph of $f(x, y) = x^2 + y^2$ and several sections

Fig. 2.2 Level curves

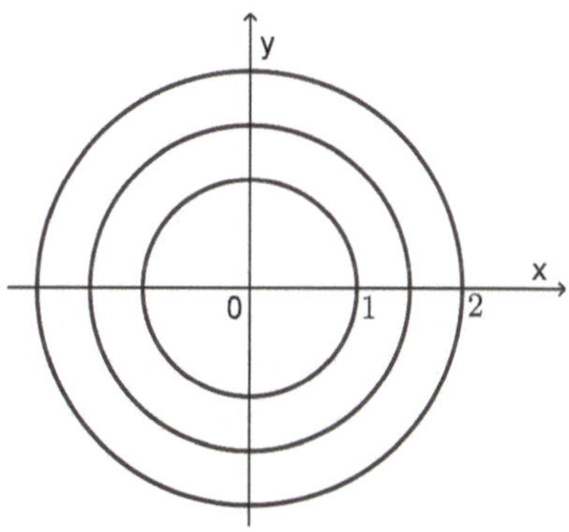

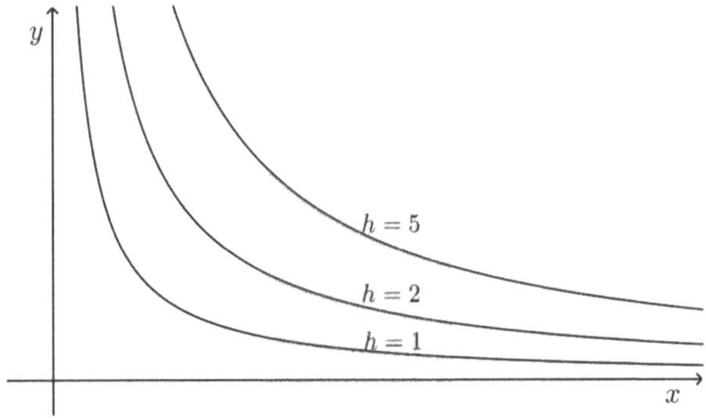

Fig. 2.3 Level curves (isoquants)

the same admits the following level curves, that in the economic literature are called *isoquants.* See Fig. 2.3.

Obviously, also for functions $f : \mathbb{R}^n \longrightarrow \mathbb{R}^m$ it is possible to introduce the notion of *limit,* notion that the reader should know, with reference to real functions of one real variable. However, for the case of several variables the determination of the values of the related limits may be difficult, even if $n = 2,\ m = 1$. We shall not be concerned with these technical questions.

Definition 2.1.2 Let be $T \subset \mathbb{R}^n$ and let be $f : T \longrightarrow \mathbb{R}^m$. Let x^0 be an accumulation point of T. We say that f has $L \in \mathbb{R}^m$ as a limit when x tends to x^0, and we write

$$\lim_{x \longrightarrow x^0} f(x) = L$$

if for every $\varepsilon > 0$ it is possible to find $\delta = \delta(\varepsilon) > 0$, such that for every $x \in T,\ x \neq x^0$, such that $\left\| x - x^0 \right\| < \delta$, it holds

$$\| f(x) - L \| < \varepsilon.$$

Intuitively, this means that *anyway* the vector x approaches to the point x^0, the images of f belong to the neighborhood of L of radius ε.

It is shown rather easily that Definition 2.1.2 (where f is a vector function) is equivalent to impose that for each component of f it holds

$$\lim_{x \longrightarrow x^0} f_i(x) = L_i \in \mathbb{R},\ i = 1, \ldots, m.$$

Moreover, it can be proved that Definition 2.1.2 is equivalent to the so-called "sequential definition". We recall the notion of convergence for a sequence $\left\{x^k\right\}$ of vectors of $\mathbb{R}^n$. We say that $\left\{x^k\right\}$ is *convergent* to a vector $x^0 \in \mathbb{R}^n$ and we write

$$\lim_{k \longrightarrow +\infty} x^k = x^0$$

Fig. 2.4 Paths to point x^0

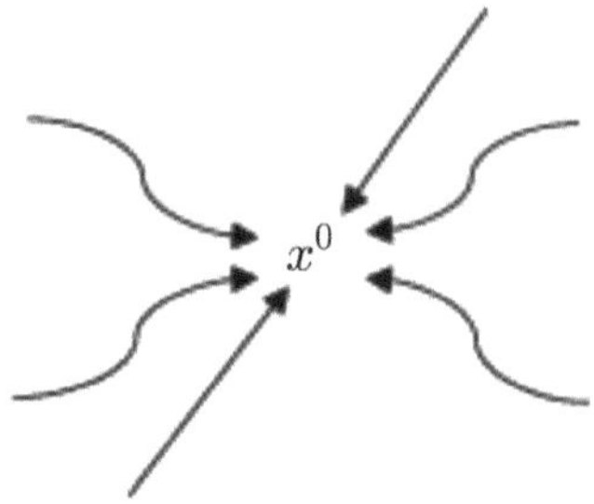

or also

$$x^k \longrightarrow x^0 \text{ for } k \longrightarrow +\infty,$$

if the sequences of the n components of x^k converge to the respective components of x^0. Then we can give the following result.

Theorem 2.1.3 *Let be given $f : T \longrightarrow \mathbb{R}^m$ and x^0, accumulation point for T; then we have*

$$\lim_{x \longrightarrow x^0} f(x) = L \in \mathbb{R}^m$$

if and only if for every sequence $\{x^k\}$ of vectors of T, with $x^k \neq x^0$, and converging to x^0, the corresponding sequence of the images $\{f(x^k)\}$ converges to L.

As previously remarked, the computations of limits for functions of several variables is not a trivial problem, even with $n = 2$, $m = 1$. The difficulties come from the fact that the vector x must be closer and closer to x^0, in all possible ways, not only with all possible rectilinear paths, but also with all possible curvilinear paths. See Fig. 2.4.

Therefore, if we find two "directions" along which the approach to x^0 produces two different values, we can surely conclude that the related limit does not exist. It is more easy to prove the non existence of a limit, rather than its existence.

Example 2.1.4 Decide about the existence of the limit

$$\lim_{(x,y) \longrightarrow (0,0)} f(x, y) = \frac{x}{y}.$$

Let us choose the approaches of the type $y = x$; we obtain

$$\lim_{(x,y) \longrightarrow (0,0)} \frac{x}{x} = 1.$$

Now we choose the approaches to the origin of $\mathbb{R}^2$ along the axis of the ordinates ($x = 0$); we obtain

$$\lim_{(x,y)\longrightarrow(0,0)} \frac{0}{y} = 0.$$

We conclude that the proposed limit does not exist.

Remark 2.1.5 In the case of real functions of several real variables it makes no sense to speak of right-hand limit and left-hand limit, as for functions of one real variable, where the approach to x^0 is made on the real axis and hence the said approach may be only from the right side, from the left side or from both directions. For functions of several variables, as already remarked, the approach to x^0 is to be made by means of *any trajectory,* contained obviously on $\mathrm{dom}(f)$. If $\mathrm{dom}(f)$ is not bounded, for functions of several variables it makes no sense to speak of limits for $x \longrightarrow +\infty$ or for $x \longrightarrow -\infty$. Vice-versa it has a meaning the problem of the existence of the limit of f for $x \longrightarrow \infty$ (with no sign!), when $\mathrm{dom}(f)$ is not bounded. In this case we say that $f : T \subset \mathbb{R}^n \longrightarrow \mathbb{R}^m$ has the limit $L \in \mathbb{R}^m$ for $x \longrightarrow \infty$ and we write

$$\lim_{x\longrightarrow\infty} f(x) = L$$

or also

$$\lim_{\|x\|\longrightarrow+\infty} f(x) = L$$

if and only if for *every* neighborhood of L of radius ε, there exists $\delta > 0$ such that for every $x \in T$, with $\|x\| > \delta$, it holds $\|f(x) - L\| < \varepsilon$.

Also the notion of continuity for functions $f : \mathbb{R}^n \longrightarrow \mathbb{R}^m$ does not present particular novelties.

Definition 2.1.6 Let be given $f : T \subset \mathbb{R}^n \longrightarrow \mathbb{R}^m$ and let $x^0 \in T$ be an accumulation point for T. If

$$\lim_{x\longrightarrow x^0} f(x) = f(x^0)$$

then f is continuous at x^0. If f is continuous at every point of T, then f is continuous on T and we write in this case $f \in C^0(T)$.

It is easy to verify that the following classes of functions are formed by continuous functions.

- The quadratic forms $Q(x) = x^\top Ax$, with A symmetric matrix of order n.
- The quadratic functions $f(x) = x^\top Ax + a^\top x + \beta$, with A symmetric matrix of order n.

- The elementary functions (exponential, power, logarithmic, trigonometric) in their respective domains.
- The functions obtained by composition of elementary functions, in their respective domains.

Similarly to what happens for the case $n = 1$, $m = 1$, we have the important Theorem of Weierstrass (for scalar functions):

Theorem 2.1.7 *Let be $f : T \subset \mathbb{R}^n \longrightarrow \mathbb{R}$ be continuous on the compact set T (i.e. T is closed and bounded). Then f admits global maximum and global minimum over T :*

$$\exists x^* \in T : f(x^*) \geqq f(x), \; \forall x \in T;$$
$$\exists x^{**} \in T : f(x^{**}) \leqq f(x), \; \forall x \in T.$$

We shall revert further on this result in Chap. 4. Also for functions $f : \mathbb{R}^n \longrightarrow \mathbb{R}^m$ a result on the continuity of composed functions holds true.

Theorem 2.1.8 *Let be $y = f(z)$, $z \in Z \subset \mathbb{R}^p$, $y \in \mathbb{R}^m$ and $z = g(x)$, $x \in T \subset \mathbb{R}^n$; let us consider the composite function (we suppose that the same is defined) $f[g(x)]$. If $f(x)$ is continuous on Z and $g(x)$ is continuous on T, then also $f[g(x)]$ is continuous on T.*

2.2 Partial Derivatives: Differentiability

We give now a quick overview of the main questions concerning the existence of partial derivatives for scalar functions of several real variables. Let be $f : T \subset \mathbb{R}^n \longrightarrow \mathbb{R}$ and let be $x^0 \in \text{int}(T)$. We define the following *partial difference quotient* with respect to the i-th component of the vector x^0 (hence the other $(n-1)$ components remain fixed):

$$\frac{f(x_1^0, \ldots, x_i^0 + h, \ldots, x_n^0) - f(x_1^0, \ldots, x_n^0)}{h}.$$

If the following limit exists *finite*

$$\lim_{h \longrightarrow 0} \frac{f(x_1^0, \ldots, x_i^0 + h, \ldots, x_n^0) - f(x_1^0, \ldots, x_n^0)}{h} = L \in \mathbb{R},$$

then f admits *a partial derivative* at x^0, with respect to x_i and L is the *partial derivative* of f at x^0, with respect to x_i, denoted by

$$\frac{\partial f}{\partial x_i}(x^0).$$

When all n partial derivatives of f at $x^0 \in \mathrm{int}(T)$ exist, then f is said to be *derivable* at x^0 and the vector of all the said partial derivatives at x^0 is called the *gradient* of f at x^0, usually denoted by $\nabla f(x^0)$ (this symbol is also called "nabla"):

$$\nabla f(x^0)^\top = \left[\frac{\partial f}{\partial x_1}(x^0), \frac{\partial f}{\partial x_2}(x^0), \ldots, \frac{\partial f}{\partial x_n}(x^0)\right].$$

For instance, if $f(x) = a^\top x,\ a, x \in \mathbb{R}^n$, we have $\nabla f(x) = a$; if $f(x) = Q(x) = x^\top Ax$, we have $\nabla Q(x) = 2Ax$.

Again:

(1) If $f(x_1, x_2, x_3) = \log(x_1 + 3x_2x_3)$; $x^0 = [0, 2, 1]^\top$, we have

$$\frac{\partial f}{\partial x_1}(x) = \frac{1}{x_1 + 3x_2x_3}; \ \frac{\partial f}{\partial x_2}(x) = \frac{3x_3}{x_1 + 3x_2x_3}; \ \frac{\partial f}{\partial x_3}(x) = \frac{3x_2}{x_1 + 3x_2x_3}.$$

$$\nabla f(x^0) = \left[\frac{1}{6}, \frac{1}{2}, 1\right]^\top.$$

(2) The Euclidean norm $\|x\|$ of $\mathbb{R}^n$ is a function which admits partial derivatives everywhere in $\mathbb{R}^n$, *except the origin.* Indeed:

$$f(x) = \|x\| = \sqrt{\sum_{i=1}^{n}(x_i)^2} = \left[(x_1)^2 + (x_2)^2 + \cdots + (x_n)^2\right]^{\frac{1}{2}};$$

$$\frac{\partial f}{\partial x_i}(x) = \frac{1}{2}\left[\sum_{i=1}^{n}(x_i)^2\right]^{-\frac{1}{2}} 2x_i = \frac{x_i}{\|x\|}$$

from which

$$\nabla f(x) = \frac{1}{\|x\|}\left[x_1, x_2, \ldots, x_n\right]^\top, \ x \neq [0].$$

Let now be given the *vector* function $f : T \subset \mathbb{R}^n \longrightarrow \mathbb{R}^m$; if f admits, with reference to its m components, all $m \cdot n$ partial derivatives, we can build the *Jacobian matrix* (from the German mathematician C. G. J. Jacobi), usually denoted by $Jf(x)$ or also $\nabla f(x)$, of order (m, n):

$$Jf(x) = \nabla f(x) = \begin{bmatrix} \nabla f_1(x)^\top \\ \nabla f_2(x)^\top \\ \vdots \\ \nabla f_m(x)^\top \end{bmatrix}.$$

We have to note that some authors define the Jacobian matrix as that matrix whose *columns* are the gradients of the components f_i of f. In this case the Jacobian matrix is of order (n, m) and obviously its transpose coincides with the Jacobian matrix defined above.

Now we come back to scalar functions, which admits all partial derivatives on an open set $T \subset \mathbb{R}^n$. The gradient $\nabla f(x)$ is therefore a vector function of several real variables:

$$\nabla f(x) : T \subset \mathbb{R}^n \longrightarrow \mathbb{R}^n.$$

If $\nabla f(x)$ admits all its partial derivatives, with respect to all its components and with respect to all its variables, then we speak of *second-order partial derivatives* or *partial derivatives of the second order.*

These derivatives can be:

- "Pure second-order derivatives" , when the derivation operation is made with respect to the same variable considered in the first-order derivative. Following the Jacobi notation these derivatives are denoted by

$$\frac{\partial^2 f}{\partial x^2}(x) \text{ which represents the operation } \frac{\partial^2 f}{\partial x_i \partial x_i}(x).$$

- "Mixed second-order derivatives" , when the derivation operation is made with respect to a variable *different* from the one considered in the first-order derivative. The Jacobi notation is then

$$\frac{\partial^2 f}{\partial x_i \partial x_j}(x) \text{ (with } i \neq j).$$

If we have the function $f : T \subset \mathbb{R}^n \longrightarrow \mathbb{R}$ which admits all its second-order partial derivatives, the square matrix, of order n, formed by the said n^2 second-order partial derivatives, is called *Hessian matrix,* from the name of the German mathematician L. O. Hesse, and denoted usually by

$$Hf(x) \text{ or also } \nabla^2 f(x).$$

We may say also that the Hessian matrix is the Jacobian matrix of the vector function $\nabla f(x)$.

Example 2.2.1 Let $f : \mathbb{R}^2 \longrightarrow \mathbb{R}$ be given by $f(x_1, x_2) = (x_1)^3 x_2 + 3(x_2)^3$. We have

$$\nabla f(x) = \left[3(x_1)^2 x_2;\ \ (x_1)^3 + 9(x_2)^2\right]^\top.$$

$$Hf(x) = \begin{bmatrix} 6x_1x_2 & 3(x_1)^2 \\ 3(x_1)^2 & 18x_2 \end{bmatrix}.$$

Note that in the previous example $Hf(x)$ is symmetric. This is not fortuitous. If, given the function $f : T \subset \mathbb{R}^n \longrightarrow \mathbb{R}$, all its second-order partial derivatives are continuous on T, we say that f is "of class C^2 on T" and write $f \in C^2(T)$. Under these assumptions we have the important *Theorem of Schwarz* (called also sometimes Theorem of Young), concerned with the symmetry of the second-order partial derivatives.

Theorem 2.2.2 (Theorem of Schwarz) *If $f : A \subset \mathbb{R}^n \longrightarrow \mathbb{R}$, with A open set, and if $f \in C^2(A)$, then for every $x \in A$, it holds, for every $i \neq j$,*

$$\frac{\partial^2 f}{\partial x_i \partial x_j}(x) = \frac{\partial^2 f}{\partial x_j \partial x_i}(x).$$

That is, the Hessian matrix is symmetric.

Remark 2.2.3 If in the previous theorem we do not respect the assumptions, it is quite likely that the thesis does not work (but this happens in all theorems!). In other words, the Hessian matrix can be non symmetric. Consider, e.g., the function

$$f(x, y) = \begin{cases} xy\frac{x^2-y^2}{x^2+y^2} & \text{for } (x, y) \neq [0] \\ 0 & \text{for } (x, y) = [0] . \end{cases}$$

This function admits at the origin of $\mathbb{R}^2$ all its four second-order partial derivatives, which however are not continuous. It is seen that it holds

$$\frac{\partial^2 f}{\partial x \partial y}(0, 0) = 1,$$

$$\frac{\partial^2 f}{\partial y \partial x}(0, 0) = -1.$$

Hence the Hessian matrix is not symmetric. The Theorem of Schwarz can be given also in a weaker form, by requiring that only the second-order mixed partial derivatives are continuous.

Let us now briefly recall the concept of differentiable real functions of one real variable, $f : \mathbb{R} \longrightarrow \mathbb{R}$. Let us consider a function $f : T \subset \mathbb{R} \longrightarrow \mathbb{R}$ and let be $x_0 \in \text{int}(T)$. Let f admit the first-order derivative at x^0. Then the *differential* of f at x_0 is given by $f'(x_0)h$ and it holds

$$\Delta y_0 = f(x_0 + h) - f(x_0) = f'(x_0)h + \beta(h),$$

being

$$\lim_{h \longrightarrow 0} \beta(h) = 0;$$

$$\lim_{h \longrightarrow 0} \frac{\beta(h)}{h} = 0.$$

We may also write

$$f(x_0 + h) - f(x_0) = f'(x_0)h + o(h), \quad \text{for } h \longrightarrow 0,$$

where the symbol $o(h)$ (read: "small o of h"), said also "small o of Landau" (from the name of the German mathematician E. Landau), for $h \longrightarrow 0$, stands for "a function of h which is infinitesimal of order higher than h, for $h \longrightarrow 0$" . Hence, for instance, $o(1)$, for $x \longrightarrow x_0$, denotes an infinitesimal function for $x \longrightarrow x_0$.

From the previous definition of differential it is clear that:

1. If f is differentiable at x_0, then f is continuous at x_0.
2. If f admits first-order derivative at x_0, then f is also differentiable at x_0, and also the vice-versa holds true; indeed, from

$$f(x_0 + h) - f(x_0) = f'(x_0)h + o(h),$$

by dividing both members by h ($h \neq 0$) and taking the limit for $h \longrightarrow 0$, we obtain what just asserted. Hence, for functions $f : \mathbb{R} \longrightarrow \mathbb{R}$ the class of functions which admit the first-order derivative coincides with the class of differentiable functions.

The previous two properties are alas lost for functions $f : \mathbb{R}^n \longrightarrow \mathbb{R}$, with $n > 1$. We remark at once that the existence of the gradient $\nabla f(x^0)$ *does not imply* that f is continuous at x^0.

Example 2.2.4

(a) Let be

$$f(x, y) = \begin{cases} 0, & \text{if } xy \neq 0; \\ 1, & \text{if } xy = 0. \end{cases}$$

It holds

$$\frac{\partial f}{\partial x}(0,0) = \lim_{t \longrightarrow 0} \frac{f(t,0) - f(0,0)}{t} = \lim_{t \longrightarrow 0} \frac{1-1}{t} = 0;$$
$$\frac{\partial f}{\partial y}(0,0) = \lim_{t \longrightarrow 0} \frac{f(0,t) - f(0,0)}{t} = \lim_{t \longrightarrow 0} \frac{1-1}{t} = 0.$$

Obviously, the function is discontinuous at $(0, 0)$.

(b) Let be

$$f(x, y) = \begin{cases} \frac{xy}{x^2+y^2} & \text{for } (x, y) \neq (0, 0); \\ 0 & \text{for } (x, y) = (0, 0). \end{cases}$$

This function is not continuous at the origin. Indeed, if we approach the origin along the bisecting line of equation $y = x$, we have

$$\lim_{(x,y)\longrightarrow(0,0)} f(x, y) = \lim_{(x,y)\longrightarrow(0,0)} \frac{x^2}{2x^2} = \frac{1}{2}.$$

However, $f(0, 0) = 0$. Yet the function admits at the origin both its partial derivatives and it holds

$$\frac{\partial f}{\partial x}(0, 0) = \frac{\partial f}{\partial y}(0, 0) = 0.$$

The previous considerations justify the introduction of the following basic definition.

Definition 2.2.5 Let be $f : T \subset \mathbb{R}^n \longrightarrow \mathbb{R}$ and let be $x^0 \in \text{int}(T)$; f is said to be *differentiable at* x^0 if there exists a vector $a \in \mathbb{R}^n$, depending only from x^0 (and from f) such that, for $h \longrightarrow [0]$:

$$f(x^0 + h) - f(x^0) = a^\top h + o(\|h\|). \tag{2.1}$$

The "linear part" $a^\top h$ is also called "first-order differential" of f at x^0 and it is denoted by $df(x^0) : df(x^0) = a^\top h$. Some classical books of mathematical analysis calls the quantity $df(x^0)$ "total differential" . On the grounds of Definition 2.2.5 we can assert that if $f : T \subset \mathbb{R}^n \longrightarrow \mathbb{R}$ is differentiable at x^0, the variation of the function when the vector of independent variables from x^0 becomes $x^0 + h$, is $df(x^0)$, with an approximation error which is an infinitesimal quantity, for $h \longrightarrow [0]$, of order greater than the (Euclidean) norm of h. The following theorem justifies, in a certain sense, Definition 2.2.5.

Theorem 2.2.6 *Let $f : T \subset \mathbb{R}^n \longrightarrow \mathbb{R}$ be differentiable at $x^0 \in \text{int}(T)$. Then:*

1. *f is continuous at x^0.*
2. *f admits all first-order partial derivatives at x^0 and in* (2.1) *it holds $a = \nabla f(x^0)$.*

Proof

1. The thesis is immediate. From (2.1), taking the limit for $h \longrightarrow [0]$ of both members, we have

$$\lim_{h\longrightarrow[0]} \left(f(x^0 + h) - f(x^0) \right) = 0,$$

that is, by putting $x = x^0 + h$,

$$\lim_{x\longrightarrow x^0} f(x) = f(x^0).$$

2. As the notion of differentiable function does not depend from the variation vector h, but it depends only form the point x^0, let us consider again relation (2.1) and let us choose $h = te^j,\ j = 1, \ldots, n$. We have therefore

$$f(x^0 + te^j) - f(x^0) = a^\top te^j + o(\left\| te^j \right\|),\ \ j = 1, \ldots, n.$$

That is (recall that $\left\| e^j \right\| = 1,\ \forall j = 1, \ldots, n$)

$$f(x^0 + te^j) - f(x^0) = ta^\top e^j + o(|t|),\ \ j = 1, \ldots, n.$$

By dividing both members of the last relation by t we get

$$\frac{f(x^0 + te^j) - f(x^0)}{t} = a_j + \frac{o(|t|)}{t},\ \ j = 1, \ldots, n.$$

Taking the limit for $t \longrightarrow 0$ we obtain at last (being $o(t)/t = o(1)$):

$$\frac{\partial f}{\partial x_j}(x^0) = a_j,\ \ j = 1, \ldots, n.$$

Therefore $\nabla f(x^0) = a$.

□

Hence, the differentiability of f assures the continuity of f, which is not assured by the existence of all partial derivatives. Moreover, the differentiability of f not only assures that in (2.1) we have $\nabla f(x^0) = a$, but also that if there exists $a \in \mathbb{R}^n$ such that (2.1) holds, also $\nabla f(x^0)$ exists and these two quantities are equal. It may indeed happen that $\nabla f(x^0)$ exists, but that $a \in \mathbb{R}^n$ satisfying (2.1) does not exist. From these considerations we have the conclusion that the differentiability property of a function $f : \mathbb{R}^n \longrightarrow \mathbb{R}$ (with $n > 1$) is a "stronger" requirement than the simple existence of the gradient.

The differentiability at a point x^0 is, in a sense, a "regularity property" of the function in a multi-dimensional neighborhood of x^0. This property assures that the function has a "good behaviour" in all points "near" to x^0, according to all possible approaching paths (in other words, the same considerations made on the existence of limits for functions of several variables hold also here). The existence of the first-order partial derivatives requires, on the contrary, a "good behaviour" only with respect to those directions parallel to the various axes: this good behaviour may occur, without that the same occurs with respect to all other (infinite, rectilinear and not) directions. In the case of real functions of one real variable, differentiability and existence of first-order derivatives are two coincident properties, as the related neighbourhoods are uni-dimensional: given a point x_0 on the real axis, we can approach this point by moving from its right hand or from its left hand or from both the said directions, but there are no other possibilities.

Fig. 2.5 Relationships among differentiability, class $\mathcal{C}^1$, continuity and partial derivative

$$\begin{array}{ccc} f \text{ is of class } \mathcal{C}^1 & & \\ \Downarrow \quad \not\Uparrow & & \\ f \text{ is differentiable} & \Rightarrow & f \text{ admits partial derivative} \\ \Downarrow & \not\Uparrow & \\ & f \text{ is continuous} & \end{array}$$

A useful sufficient condition for the differentiability of $f : \mathbb{R}^n \longrightarrow \mathbb{R}$ (with $n > 1$) is given by the following basic result (we do not prove). We recall that a function $f : A \subset \mathbb{R}^n \longrightarrow \mathbb{R}$ is of *class* $\mathcal{C}^1$ *on* A or *continuously differentiable on* A (we write also $f \in \mathcal{C}^1(A)$) if f is continuous over A, together with all its first-order partial derivatives.

Theorem 2.2.7 *Let* $f : A \subset \mathbb{R}^n \longrightarrow \mathbb{R}$ *be of class* $\mathcal{C}^1$ *on the open set* A. *Then* f *is differentiable on* A. *If* f *admits all its first-order partial derivatives in a neighbourhood of* $x^0 \in \text{int}(A)$ *and they are continuous at* x^0, *then* f *is differentiable at* x^0.

The previous theorem is deeper than what it may appear at a first hasty reading: we have repeated that the existence of all first-order partial derivatives of a function $f : \mathbb{R}^n \longrightarrow \mathbb{R}$ (with $n > 1$) is not sufficient to obtain a "good behaviour" at a point of its domain. Indeed, Theorem 2.2.6 (point 2) says that if $a \in \mathbb{R}^n$ exists, then $a = \nabla f(x^0)$, but if f is not differentiable at x^0, then $\nabla f(x^0)$ may exist, without being $\nabla f(x^0) = a$. However, if the first-order partial derivatives are continuous, then the information the same derivatives yield is more pregnant, as in this case they assure differentiability, that is a "good behaviour" (at least locally). The situations can be summarized as follows ($x \in \mathbb{R}^n$, $n > 1$, see Fig. 2.5).

The notion of differentiability can be extended without particular difficulties to vector functions $f : \mathbb{R}^n \longrightarrow \mathbb{R}^m$.

Definition 2.2.8 Let be $f : T \subset \mathbb{R}^n \longrightarrow \mathbb{R}^m$ and let $x^0 \in \text{int}(T)$. We say that f is differentiable at x^0 if there exists a matrix M, of order (m, n), depending only from the point x^0, such that, for $h \longrightarrow [0]$ it holds

$$f(x^0 + h) - f(x^0) = M \cdot h + \underline{o}(\|h\|).$$

In the previous definition the symbol $\underline{o}(\|h\|)$ denotes a vector of m "small o's of Landau" . The matrix M (which is unique) is nothing but the Jacobian matrix of f evaluated at x^0 : $M = Jf(x^0)$.

The notion of differentiability allows also to obtain the notion of *tangent hyperplane* to the surface generated by a differentiable function. This notion generalizes to the case of $f : \mathbb{R}^n \longrightarrow \mathbb{R}$ ($n > 1$) what seen in previous courses of Mathematics for what concerns the existence of a tangent straight line to the diagram of a function $f : \mathbb{R} \longrightarrow \mathbb{R}$. It can be shown that the equation of the tangent hyperplane to the point $(x^0, f(x^0))$ of a function $f : \mathbb{R}^n \longrightarrow \mathbb{R}$ $(n > 1)$, with f *differentiable at* x^0, is

$$z = f(x^0) + \nabla f(x^0)^\top (x - x^0).$$

Notice the "parallelism" with the equation of the tangent straight line.

Example 2.2.9 Find the equation of the tangent hyperplane to the point $([1, 1]^\top, f(1, 1))$ of the function

$$z = e^{\sqrt{xy}}.$$

The function is of C^1 class, and hence differentiable, in a neighbourhood of $(1, 1)$. The function is defined on the first and third quadrant of the Cartesian plane. We have

$$\frac{\partial f}{\partial x} = e^{\sqrt{xy}} \frac{y}{2\sqrt{xy}}; \quad \frac{\partial f}{\partial y} = e^{\sqrt{xy}} \frac{x}{2\sqrt{xy}}.$$

Hence

$$z = e + \left[\frac{e}{2}, \frac{e}{2}\right] \begin{bmatrix} x - 1 \\ y - 1 \end{bmatrix},$$

that is

$$z = e + \frac{e}{2}x - \frac{e}{2} + \frac{e}{2}y - \frac{e}{2},$$

that is

$$z = \frac{e}{2}(x + y).$$

It is possible to introduce the notion of *twice differentiability* for functions $f : \mathbb{R}^n \longrightarrow \mathbb{R}$ and also the notion of *k-differentiability* $(k > 2)$, but for these notions we refer the reader to a good textbook of Mathematical Analysis.

2.3 Differentiability of Composite Functions

We begin by taking into consideration the following rather elementary cases.

- Let $f(x, y)$ be differentiable on the open set $A \subset \mathbb{R}^2$ and let $x = g(t)$, $y = h(t)$, with $t \in [a, b]$. Let us suppose that $g(t)$ and $h(t)$ are differentiable (i.e. they admit first-order derivative, being functions of one variable!) on $[a, b]$ and that

$[x = g(t);\ y = h(t)] \in A$ for $t \in [a, b]$. Then the composite function

$$z = f\,[g(t), h(t)]$$

is differentiable with respect to t and it holds ("chain rule")

$$z'(t) = \frac{\partial f\,[g(t), h(t)]}{\partial x} g'(t) + \frac{\partial f\,[g(t), h(t)]}{\partial y} h'(t),$$

where the partial derivatives are to be computed for $x = g(t)$ and $y = h(t)$.

Example 2.3.1 Let be $f(x, y) = x^2 - y^2 + 3x + 4y$, function which is differentiable on the whole $\mathbb{R}^2$ and let be $x = 2t;\ \ y = t^2$.

(a) It holds, by applying the above formula,

$$\frac{\partial f}{\partial x} = 2x + 3, \text{ that is, with } x = 2t, \ \frac{\partial f}{\partial x} = 4t + 3.$$

$$\frac{\partial f}{\partial y} = -2y + 4, \text{ that is, with } y = t^2, \ \frac{\partial f}{\partial y} = -2t^2 + 4.$$

Then we have $g'(t) = 2,\ h'(t) = 2t$. Therefore it holds

$$z'(t) = (4t + 3)2 + (-2t^2 + 4)2t = -4t^3 + 16t + 6.$$

(b) By making the substitution we get directly: $z(t) = f\,[g(t), h(t)] = (2t)^2 - (t^2)^2 + 3(2t) + 4t^2 = -t^4 + 8t^2 + 6t$. Hence we have

$$z'(t) = -4t^3 + 16t + 6,$$

which confirms the validity of the "chain rule".

- Let us consider a particular case of the previous one, that is a function $z = f(x, y)$ where the variable y depends from the variable x, i.e. $y = \varphi(x)$. We assume the hypotheses of the previous case. We have therefore $z = [x, \varphi(x)] = F(x)$. It holds, on the grounds of the previous result,

$$F'(x) = \frac{\partial f}{\partial x} + \frac{\partial f}{\partial y}\varphi'(x).$$

Example 2.3.2 Let be $f(x, y) = xy + ye^x$, which is differentiable on the whole $\mathbb{R}^2$. Let then be $y = \varphi(x) = \log x$, differentiable for $x > 0$. It holds $\varphi\prime(x) = \frac{1}{x}$.

(*a*) We apply the formula of the "chain rule":

$$\frac{\partial f}{\partial x} = y + ye^x, \text{ that is, with } y = \log x, \ \frac{\partial f}{\partial x} = \log x + e^x \log x.$$

$$\frac{\partial f}{\partial y} = x + e^x.$$

Hence

$$F'(x) = \log x + e^x \log x + (x + e^x)\frac{1}{x} = (\log x)(1 + e^x) + \frac{e^x}{x} + 1.$$

(*b*) By making the substitution we get directly $F(x) = x \log x + e^x \log x$, and hence

$$\begin{aligned} F'(x) &= \log x + x \cdot \frac{1}{x} + e^x \log x + e^x \cdot \frac{1}{x} = \log x + 1 + e^x \log x + \frac{e^x}{x} \\ &= (\log x)(1 + e^x) + \frac{e^x}{x} + 1. \end{aligned}$$

- A case a bit more general than the previous ones is given by the following situation. Let $x_1(t),\ x_2(t),\ \ldots,\ x_n(t)$ be differentiable functions on an interval $I \subset \mathbb{R}$ and let us suppose that, with t varying on I, the point $P[x_1(t), x_2(t), \ldots, x_n(t)]$ belongs to an open set $A \subset \mathbb{R}^n$, where it is defined the function $f(x_1, x_2, \ldots, x_n)$, differentiable on A. Then we can consider the composite function $F(t) = f[x(t)]$, which is differentiable on I and for which it holds

$$F'(t) = \nabla f[x(t)]^\top \cdot x'(t),$$

being

$$x'(t) = \begin{bmatrix} x_1'(t) \\ x_2'(t) \\ \vdots \\ x_n'(t) \end{bmatrix}.$$

- For completeness let us consider also the general case of composition of two vector functions. Let be given $f : \mathbb{R}^p \longrightarrow \mathbb{R}^m$ and $g : \mathbb{R}^n \longrightarrow \mathbb{R}^p$ and suppose that there exists the composite function $z(t) = f[g(t)]$, being $z : \mathbb{R}^n \longrightarrow \mathbb{R}^m$. If g is differentiable at t^0, with Jacobian matrix $Jg(t^0)$, and f is differentiable at $y^0 = g(t^0)$, with Jacobian matrix $Jf(y^0) = Jf\left[g(t^0)\right]$, then $z(t) = f\left[g(t)\right]$ is differentiable at t^0 and its Jacobian matrix evaluated at t^0 is given by

$$Jz(t^0) = Jf\left[g(t^0)\right] \cdot Jg(t^0).$$

We explain this last case by an example.

Example 2.3.3

$$g(t) = \begin{bmatrix} t_1 + (t_2)^2 \\ \cos(t_1 + t_2) \\ t_1 - (t_2)^2 \end{bmatrix},$$

$$f(y) = \begin{bmatrix} y_1 + y_2 y_3 \\ \sin(y_1 y_3) \end{bmatrix}.$$

The Jacobian matrices of g and f are, respectively,

$$Jg(t) = \begin{bmatrix} 1 & 2t_2 \\ -\sin(t_1 + t_2) & -\sin(t_1 + t_2) \\ 1 & -2t_2 \end{bmatrix},$$

$$Jf(y) = \begin{bmatrix} 1 & y_3 & y_2 \\ y_3 \cos(y_1 y_3) & 0 & y_1 \cos(y_1 y_3) \end{bmatrix}.$$

Let us consider the origin of the space $\mathbb{R}^2$ to which vectors t belong. It holds

$$y^0 = g(0, 0) = \begin{bmatrix} 0 \\ 1 \\ 0 \end{bmatrix};$$

$$Jg(0, 0) = \begin{bmatrix} 1 & 0 \\ 0 & 0 \\ 1 & 0 \end{bmatrix}.$$

We have then

$$Jf\left[g(0, 0)\right] = Jf(0, 1, 0) = \begin{bmatrix} 1 & 0 & 1 \\ 0 & 0 & 0 \end{bmatrix}$$

and hence the Jacobian matrix of the composite function z, evaluated at $[0] \in \mathbb{R}^2$ is

$$Jz(0,0) = \begin{bmatrix} 1 & 0 & 1 \\ 0 & 0 & 0 \end{bmatrix} \begin{bmatrix} 1 & 0 \\ 0 & 0 \\ 1 & 0 \end{bmatrix} = \begin{bmatrix} 2 & 0 \\ 0 & 0 \end{bmatrix}.$$

The willing reader can make the direct substitution and check the coincidence of the two results obtained.

2.4 Directional Derivatives

We have seen that partial derivatives are defined as limits of difference quotients evaluated with reference to directions parallel to the axes. We could compute the said limits with reference to "oblique" directions and consider the difference quotients of the variations generated by a function $f : \mathbb{R}^n \longrightarrow \mathbb{R}$ when the initial point x^0 moves along any direction, i.e. along a straight line, not necessarily parallel to the axes. This entails the fact that all n independent variables are involved in this operation. A *direction* in $\mathbb{R}^n$ is determined by a nonzero vector $v \in \mathbb{R}^n$ that for convenience we can assume to be *normalized:* $\|v\| = 1$. We recall that normalized vectors are called also *versors.* We recall the so-called *parametric equation* of a straight line in $\mathbb{R}^n$ passing through the point x^0 and having the direction v:

$$x = x^0 + tv, \quad t \in \mathbb{R}.$$

This line is represented in Fig. 2.6.

Now we define the *difference quotient of* $f : \mathbb{R}^n \longrightarrow \mathbb{R}$ *at the point* x^0 *and in the direction* $v \in \mathbb{R}^n$ the following expression

$$\frac{f(x^0 + tv) - f(x^0)}{t}, \quad t \in \mathbb{R}, t \neq 0, \quad x^0 + tv \in \text{dom}(f).$$

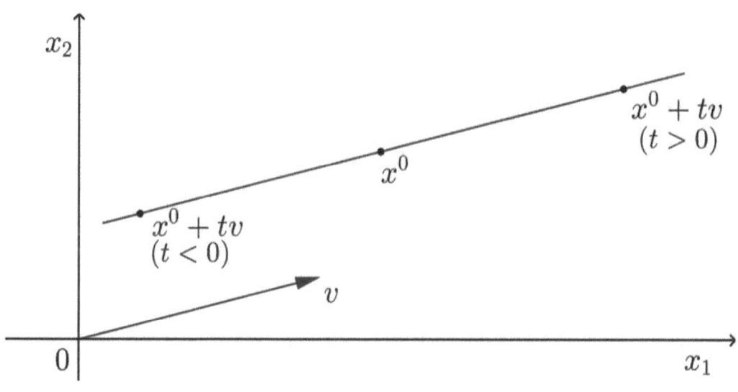

Fig. 2.6 Straight line through x^0 and direction v

If the following limit exists finite

$$\lim_{t \longrightarrow 0} \frac{f(x^0 + tv) - f(x^0)}{t} \tag{2.2}$$

we say that f is *directionally differentiable at* x^0 *in the direction* v or that f admits a *directional derivative at* x^0 *in the direction* v and call the said limit the *directional derivative of* f *at* x^0 *in the direction* v, directional derivative denoted by

$$D_v f(x^0) \text{ or } Df(x^0, v).$$

Obviously, with $v = e^j$, $j = 1, \dots, n$, relation (2.2) gives the partial derivative of f with respect to the j-th variable:

$$Df(x^0, e^j) = \frac{\partial f}{\partial x_j}(x^0).$$

If we put $\phi(t) = f(x^0 + tv)$, i.e. $\phi(t)$ is the *restriction of* f to the straight line passing through x^0 and with direction v, it is quite immediate to verify that the difference quotient

$$\frac{\phi(t) - \phi(0)}{t}$$

coincides with the previous written difference quotient of $f : \mathbb{R}^n \longrightarrow \mathbb{R}$ at the point x^0 and in the direction $v \in \mathbb{R}^n$. Hence the directional derivative defined by (2.2) coincides with the usual derivative of ϕ evaluated at zero: $\phi'(0)$.

Example 2.4.1

(1) Let be $f(x, y) = x^2 + y^2 + xy + x$; let be $x^0 = (1, -1)$ and let v be the versor $(1/\sqrt{2}, 1/\sqrt{2})$. We have

$$f(x^0 + tv) = \phi(t) = 2 + \frac{t}{\sqrt{2}} + \frac{3}{2}t^2.$$

Hence $D_v f(x^0) = \phi'(0) = \frac{1}{\sqrt{2}}$.

(2) Let be $f(x, y, z) = e^{x+y+z}$ and let $v = (\frac{1}{\sqrt{3}}, \frac{1}{\sqrt{3}}, \frac{1}{\sqrt{3}})$. Let be $x^0 = (0, 0, 0)$. We have $\phi(t) = e^{\sqrt{3}t}$ and hence $D_v f(x^0) = \phi'(0) = \sqrt{3}$.

If the following limit exists finite

$$\lim_{t \longrightarrow 0^+} \frac{f(x^0 + tv) - f(x^0)}{t} = D_v^+ f(x^0) = D^+ f(x^0, v),$$

we speak of *right-hand sided directional derivative* or also of *right-hand sided radial derivative* of f at x^0 in the direction v; similarly, the *left-hand-sided directional derivative of* f at x^0 in the direction v is defined by

$$\lim_{t \longrightarrow 0^-} \frac{f(x^0 + tv) - f(x^0)}{t} = D_v^- f(x^0) = D^- f(x^0, v).$$

For $v = 0$, both $D^+ f(x^0, v)$ and $D^- f(x^0, v)$ are assumed to be equal to zero. It can be shown that

$$- D^+ f(x^0, -v) = D^- f(x^0, v)$$

so that we can limit ourselves to consider only $D^+ f(x^0, v)$. Indeed, this is done usually in the so-called "Nonsmooth Analysis", that is that part of Mathematical Analysis where the related functions are not necessarily differentiable, and also in the modern Convex Analysis. Obviously, if $D^+ f(x^0, v) = D^- f(x^0, v)$, then there exists the (bilateral) directional derivative of f, as defined in (2.2). However, the two one-sided directional derivatives could both exist and be different.

Theorem 2.4.2 *Let be $f : T \subset \mathbb{R}^n \longrightarrow \mathbb{R}$ be differentiable at $x^0 \in \text{int}(T)$. Then f is directionally differentiable at x^0 in each direction v and it holds*

$$Df(x^0, v) = \nabla f(x^0)^\top v,$$

that is every directional derivative is given by the linear combination of the n partial derivatives, with coefficients given by the components of the direction v.

Proof Being f differentiable at x^0, we have

$$f(x^0 + h) - f(x^0) = \nabla f(x^0)^\top h + o(\|h\|), \quad \text{for } h \longrightarrow [0] .$$

By choosing $h = tv$, we find (being $\|v\| = 1$):

$$f(x^0 + tv) - f(x^0) = \nabla f(x^0)^\top tv + o(|t|).$$

By dividing both members by t we obtain

$$\frac{f(x^0 + tv) - f(x^0)}{t} = \nabla f(x^0)^\top v + o(1).$$

Taking the limit for $t \longrightarrow 0$ of both members of the last relation we obtain the thesis. □

Example 2.4.3 Let us apply Theorem 2.4.2 to the differentiable function $f(x, y) = x^2 + y^2 + xy + x$ of point (1) of Example 2.4.1. We have

$$\frac{\partial f}{\partial x} = 2x + y + 1; \quad \frac{\partial f}{\partial y} = 2y + x.$$

$$\nabla f(1, -1) = \begin{bmatrix} 2 \\ -1 \end{bmatrix}.$$

Therefore we have

$$Df((1, -1), v) = [2, -1] \begin{bmatrix} \frac{1}{\sqrt{2}} \\ \frac{1}{\sqrt{2}} \end{bmatrix} = \frac{2}{\sqrt{2}} - \frac{1}{\sqrt{2}} = \frac{1}{\sqrt{2}}.$$

We observe that Theorem 2.4.2 is not reversible, that is if f admits at x^0 directional derivatives in any directions, then f may be not differentiable at x^0, nor continuous at x^0 (the reasons are the ones already observed: in defining directional derivatives we take into considerations only "rectilinear" directions, and this is not sufficient for a "regular" behaviour of a function of $n > 1$ independent variables, in a neighborhood of a point).

Example 2.4.4 Consider the function

$$f(x_1, x_2) = \begin{cases} \frac{(x_1)^2}{x_2} & \text{if } x_2 \neq 0 \\ 0 & \text{if } x_2 = 0. \end{cases}$$

It is quite easy to verify that at $x^0 = (0, 0)$ the function admits all directional derivatives in any direction (and hence it admits also all n partial derivatives), but it is not continuous at x^0, as, for instance $f(x_1, 0) = 0$, whereas $f(x_1, (x_1)^2) = 1$, if $x_1 \neq 0$. Obviously f is not differentiable at x^0.

Figure 2.7 summarizes the situations.

The formula $Df(x^0, v) = \nabla f(x^0)^\top v$, which holds if f is differentiable at x^0, allows also to detect the directions of maximal and minimal increase of a differentiable function at a point x^0, whenever its gradient is different from zero

$\exists \nabla f(x)$ $\qquad$ f continuous

$\Uparrow$ $\qquad$ $\Uparrow$

f is differentiable

$\Downarrow$ $\qquad$ $\Downarrow$

f class C^1 $\qquad$ f directionally diff. in all directions

Fig. 2.7 Several notions related to the differentiability

at x^0. We recall that the angle θ between two nonzero vector x and y of $\mathbb{R}^n$ is measured by

$$\cos\theta = \frac{x^\top y}{\|x\| \cdot \|y\|},$$

so for the angle between two nonzero vectors $\nabla f(x^0)$ and v, we have

$$Df(x^0, v) = \nabla f(x^0)^\top v = \left\|\nabla f(x^0)\right\| \cdot \|v\| \cos\theta.$$

Remember that $\cos\theta \leqq 1$ for all θ and $\cos 0 = 1$. So, when $\|v\| = 1$, it follows that at points where $\nabla f(x^0) \neq [0]$, the number $Df(x^0, v)$ is largest when $\theta = 0$, i.e. when v points in the same direction as $\nabla f(x^0)$, while $Df(x^0, v)$ is smallest when $\theta = \pi$ (and hence $\cos\theta = -1$), i.e. when v points in the opposite direction of $\nabla f(x^0)$. Moreover, it follows that the length of $\nabla f(x^0)$ equals the magnitude of the maximum directional derivative. In other words $\left\|\nabla f(x^0)\right\|$ measures how fast the function increases in the direction of maximal increase. Similarly, we conclude that the direction of minimal increase corresponds to $-\nabla f(x^0)$.

2.5 Homogeneous Functions

In the present section we are concerned with a class of functions which assume a particular importance in Economic Theory: *homogeneous functions.* We recall that in elementary algebra a polynomial of several variables is said to be *homogeneous of k degree* if all its monomials have the same degree k. For instance, $P(x, y, z) = x^2 + 3y^2 - xz$ is homogeneous of degree 2. With reference to the said trinomial we have also, with $\lambda \in \mathbb{R}$, $P(\lambda x, \lambda y, \lambda z) = \lambda^2 P(x, y, z)$. Indeed, we have $(\lambda x)^2 + 3(\lambda y)^2 - \lambda x \lambda z = \lambda^2(x^2 + 3y^2 - xz)$.

This justifies what follows. First of all we give the following definition.

Definition 2.5.1 A set $C \subset \mathbb{R}^n$ is said to be a *cone* (with vertex at the origin) when

$$x \in C \Longrightarrow \lambda x \in C, \quad \forall \lambda > 0.$$

Hence, if C is a cone and $x \in C$, also all vectors λx, with $\lambda > 0$, will belong to C, that is C contains all the half-lines (or rays) starting from the origin and passing through any point x of C (see Fig. 2.8). For example, in $\mathbb{R}^2$ the first quadrant of the Cartesian plane is a cone, as well as the union of the first quadrant and the third quadrant. Obviously, the whole $\mathbb{R}^2$ is a cone. Similarly, with reference to $\mathbb{R}^n$. Again: the set

$$C = \left\{(x, y) \in \mathbb{R}^2 : x = 0 \cup y = 0\right\}$$

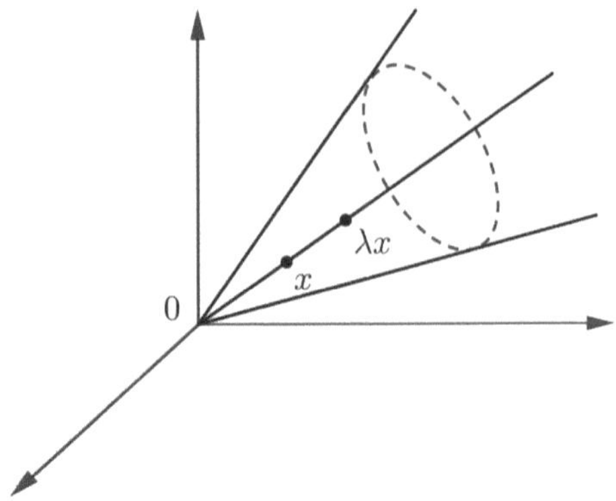

Fig. 2.8 A cone

is a cone. Any linear subspace of $\mathbb{R}^n$ is a cone, therefore the following sets

$$C = \left\{x \in \mathbb{R}^n : Ax = [0]\right\}, \quad C = \left\{x \in \mathbb{R}^n : p^\top x \geqq 0, \quad p \neq [0]\right\},$$

with A matrix of order (m, n), p and x vectors of $\mathbb{R}^n$, are cones.

Definition 2.5.2 Let be $f : C \subset \mathbb{R}^n \longrightarrow \mathbb{R}$, where C is a cone. This function is said *(positively) homogeneous of degree* $\alpha \in \mathbb{R}$ if

$$f(\lambda x) = \lambda^\alpha f(x), \quad \forall \lambda > 0, \quad \forall x \in C. \tag{2.3}$$

Therefore: f is homogeneous of degree α, when by multiplying by $\lambda > 0$ every independent variable, the value of the function results multiplied by λ^α. Note that f must be defined on a cone, otherwise the first member of (2.3) makes no sense, as the related equality must hold for any $\lambda > 0$ and for any $x \in C$. If it results, in the said relation, $\alpha = 1$, we speak also of *linearly homogeneous functions;* in any case α is the *homogeneity degree* of the function.

Example 2.5.3

(1) Every polynomial of the type $f(x) = a_1x_1 + a_2x_2 + \cdots + a_nx_n$ (that is any *linear function* $f : \mathbb{R}^n \longrightarrow \mathbb{R}$) is homogeneous of degree 1 (i.e. $f(\lambda x) = \lambda f(x)$):

$$a_1\lambda x_1 + a_2\lambda x_2 + \cdots + a_n\lambda x_n = \lambda(a_1x_1 + a_2x_2 + \cdots + a_nx_n).$$

(2) Every quadratic form $f(x) = x^\top Ax$ (with A symmetric matrix of order n) is homogeneous of degree 2 (i.e. $f(\lambda x) = \lambda^2 f(x)$):

$$(\lambda x)^\top A(\lambda x) = \lambda^2 x^\top Ax.$$

(3) The Cobb-Douglas production function of the type

$$f(x_1, x_2) = bx_1^h x_2^k, \quad x_1, x_2 > 0, \; b, h, k > 0,$$

is homogeneous of degree $\alpha = h + k$. Indeed:

$$f(\lambda x_1, x_2) = b(\lambda x_1)^h (\lambda x_2)^k = b\lambda^h x_1^h \lambda^k x_2^k = \lambda^{h+k} b x_1^h x_2^k = \lambda^{h+k} f(x_1, x_2).$$

Usually, in Economic Analysis, it is taken into consideration the Cobb-Douglas function of the type $f(K, L) = bK^\alpha L^{1-\alpha}$ (where K denotes the capital and L denotes the labour). Hence, this function is homogeneous of degree 1.

Remark 2.5.4

(*a*) A homogeneous function is not necessarily continuous, as shown by the following example: $f(x_1, x_2) = x_1 x_2$, if $x_1 - x_2 \leqq 0$ and $f(x_1, x_2) = 2x_1 x_2$ otherwise. This function is homogeneous of degree one, but it is not continuous at $(1, 1)$.

(*b*) The homogeneity degree is unique (except the trivial case of a function identically equal to zero).

(*c*) From relation (2.3) it follows

$$\frac{f(\lambda x)}{\lambda^\alpha} = f(x).$$

Hence, f is homogeneous if and only if the ratio $f(\lambda x)/\lambda^\alpha$ is *independent* from λ, for all x.

(*d*) If the origin belongs to $\mathrm{dom}(f)$, as relation (2.3) must hold for every $\lambda > 0$, if $\alpha \neq 0$, necessarily we have $f([0]) = 0$.

(*e*) If f represents a production function from $\mathbb{R}_+^n$ to $\mathbb{R}_+$, then if $\alpha = 1$ we have the so-called *constant returns to scale;* if $\alpha > 1$ and $\lambda > 1$, we have the so-called *increasing returns to scale;* if $\alpha < 1$ and $\lambda > 1$ we have the so-called *decreasing returns to scale.*

If in (2.3) we have $\alpha = 0$, that is $f(\lambda x) = f(x), \; \forall \lambda > 0, \; \forall x \in C$, this means that the function assumes constant values on every ray coming out from the origin (with the exclusion of the origin). In economic analysis usually demand functions are assumed to be homogeneous of degree zero. For example, if the demand of a certain good is a function of the income and of the prices of all goods exchanged in a given market, the fact that this function is homogeneous of degree zero means that the same is not influenced by the same variations of the income and of the prices: all these quantities vary in the same proportion and hence the final demand remains the same, i.e. it has no variation. In Economics it is also said that in this case we have no phenomenon of "money illusion". An example of a homogeneous function of degree zero is

$$f(x, y) = \frac{(x - y)^2}{xy}$$

(note that this function is not defined at the origin of $\mathbb{R}^2$). We have

$$f(\lambda x, \lambda y) = \frac{(\lambda x - \lambda y)^2}{\lambda x \lambda y} = \frac{\lambda^2 (x - y)^2}{\lambda^2 xy} = \frac{(x - y)^2}{xy} = f(x, y).$$

Theorem 2.5.5 *Let $f : C \subset \mathbb{R}^n \longrightarrow \mathbb{R}$ be homogeneous of degree α and endowed with all first-order partial derivatives on the open cone C. Then its partial derivatives are homogeneous functions of degree $\alpha - 1$:*

$$\frac{\partial f(\lambda x)}{\partial x_i} = \lambda^{\alpha-1} \frac{\partial f(x)}{\partial x_i}, \quad i = 1, \ldots, n.$$

Proof It results

$$\begin{aligned} \frac{\partial f(\lambda x)}{\partial x_i} &= \lim_{h \longrightarrow 0} \frac{f(\lambda x + \lambda h e^i) - f(\lambda x)}{\lambda h} \\ &= \lim_{h \longrightarrow 0} \frac{\lambda^\alpha f(x + h e^i) - \lambda^\alpha f(x)}{\lambda h} \\ &= \lambda^{\alpha-1} \lim_{h \longrightarrow 0} \frac{f(x + h e^i) - f(x)}{h} = \lambda^{\alpha-1} \frac{\partial f(x)}{\partial x_i}. \end{aligned}$$

□

Example 2.5.6 Let be given $f(x_1, x_2) = (x_1)^4 + 2x_1(x_2)^3 - 5(x_2)^4$. Verify that this function is homogeneous of degree $\alpha = 4$ and verify that its partial derivatives are homogeneous of degree $\alpha = 3$.

The function is a trinomial whose monomials are all homogeneous of degree $\alpha = 4$, hence f is homogeneous of the same degree. Its partial derivatives are, respectively,

$$\frac{\partial f}{\partial x_1} = 4(x_1)^3 + 2(x_2)^3; \quad \frac{\partial f}{\partial x_2} = 6x_1(x_2)^2 - 20(x_2)^3.$$

These partial derivatives are in turn two binomials whose monomials are of third degree. Hence they are both homogeneous functions of degree $\alpha = 3$.

If $f : \mathbb{R}^n \longrightarrow \mathbb{R}$ has a (one-sided) directional derivative at x^0, say $D^+ f(x^0, v)$, then this directional derivative is positively homogeneous of degree 1, with respect to the direction v, i.e. it results

$$D^+ f(x^0, hv) = h D^+ f(x^0, v), \quad \forall h > 0.$$

Indeed, we have

$$\begin{aligned} D^+ f(x^0, hv) &= \lim_{t \longrightarrow 0^+} \frac{f(x^0 + thd) - f(x^0)}{t} \\ &= h \lim_{t \longrightarrow 0^+} \frac{f(x^0 + thd) - f(x^0)}{th} \\ &= h D^+ f(x^0, v). \end{aligned}$$

Hence, if f is directionally differentiable at x^0, its variation $f(x^0 + tv) - f(x^0)$ can be approximated in a neighbourhood of x^0 by a positively homogeneous function of degree 1:

$$f(x^0 + tv) - f(x^0) = t D^+ f(x^0, v) + o(t),$$

but in general, not by a *linear* function, as it happens for a differentiable function.

Perhaps the most known result on differentiable homogeneous functions is a result due to L. Euler.

Theorem 2.5.7 (Euler's Theorem) *Let $C \subset \mathbb{R}^n$ be an open cone and $f : C \longrightarrow \mathbb{R}$ be differentiable on C. Then f is homogeneous of degree α if and only if*

$$\alpha f(x) = \nabla f(x)^\top \cdot x, \quad \forall x \in C.$$

Proof We recall that f is homogeneous of degree α if and only if the ratio

$$\rho(\lambda) = \frac{f(\lambda x)}{\lambda^\alpha}$$

is independent from λ (hence it is a constant). Therefore we must have $\rho'(\lambda) = 0, \ \forall \lambda > 0$. Being $\rho(\lambda) = \lambda^{-\alpha} \cdot f(\lambda x)$, we have

$$\begin{aligned} \rho'(\lambda) &= -\alpha \lambda^{-\alpha - 1} f(\lambda x) + \lambda^{-\alpha} \nabla f(\lambda x)^\top \cdot x \\ &= -\frac{\alpha}{\lambda^{\alpha+1}} f(\lambda x) + \frac{1}{\lambda^\alpha} \nabla f(\lambda x)^\top \cdot x \\ &= -\frac{\alpha}{\lambda^{\alpha+1}} f(\lambda x) + \frac{1}{\lambda \lambda^\alpha} \nabla f(\lambda x)^\top \cdot \lambda x \\ &= -\frac{\alpha}{\lambda^{\alpha+1}} f(\lambda x) + \frac{1}{\lambda^{\alpha+1}} \nabla f(\lambda x)^\top \cdot \lambda x. \end{aligned}$$

Hence we have $\rho'(\lambda) = 0 \iff (\alpha f(\lambda x) - \nabla f(\lambda x)^\top \cdot \lambda x) = 0$. That is

$$\alpha f(\lambda x) = \nabla f(\lambda x)^\top \cdot \lambda x. \tag{2.4}$$

Being C a cone, it results $\lambda x \in C \iff x \in C$, hence (2.4) is equivalent to

$$\alpha f(x) = \nabla f(x)^{\top} \cdot \lambda x, \quad \forall x \in C.$$

□

Example 2.5.8 Verify that the function $f(x, y) = 4x^3+5x^2y+y^3$ is homogeneous of degree $\alpha = 3$ on $\mathbb{R}^2$. Then verify the Theorem of Euler.

The proposed function is a trinomial, whose monomials are all of degree 3, hence it is homogeneous of degree $\alpha = 3$. Then we have

$$\frac{\partial f}{\partial x} = 12x^2 + 10xy; \quad \frac{\partial f}{\partial y} = 5x^2 + 3y^2.$$

$$\begin{aligned}\nabla f(x)^{\top} \cdot x &= \begin{bmatrix} 12x^2 + 10xy; & 5x^2 + 3y^2 \end{bmatrix} \begin{bmatrix} x \\ y \end{bmatrix} \\ &= 12x^3 + 15x^2y + 3y^3 = 3(4x^3 + 5x^2y + y^3) \\ &= 3f(x, y).\end{aligned}$$

Note that, if $\alpha = 1$, Euler's theorem simply says that $f(x)$, at every point $x \in C$, is given by the linear combination of its partial derivatives, evaluated at x, with coefficients given by the components of x :

$$f(x) = \sum_{i=1}^{n} \frac{\partial f}{\partial x_i}(x) \cdot x_i.$$

An interesting economic interpretation of the theorem of Euler is the following one, known in Economic Analysis as the *Product Exhaustion Theorem.* Consider a production function of the type $f(K, L)$, $K > 0$ and $L > 0$, differentiable and homogeneous of degree 1. For instance the Cobb-Douglas production function of the type $f(K, L) = \beta K^{\alpha} L^{1-\alpha}$, with $\alpha, \beta > 0$. Then we have (Theorem of Euler)

$$f(K, L) = K\frac{\partial f}{\partial K} + L\frac{\partial f}{\partial L}.$$

This equality, in economic terms, says that the total reward of the two production factors (capital and labour) uses up the total production.

As homogeneous functions assume some importance in Economic Analysis, it is useful and important to "homogeneize" functions originally not homogeneous. The following result is also known as the *Representation Theorem for homogeneous functions.*

Suppose that $f : C \subset \mathbb{R}^n \longrightarrow \mathbb{R}$ is defined on the cone C; we can then consider another function $F(\cdot)$ of $(n+1)$ variables defined as

$$F(x,z) = z^\alpha f\left(\frac{x_1}{z}, \frac{x_2}{z}, \ldots, \frac{x_n}{z}\right).$$

This function is homogeneous of degree α on the cone $C \times \mathbb{R}_+ \subset \mathbb{R}^{n+1}$. Indeed, for $t \in \mathbb{R}_+$ and $(x,z) \in C \times \mathbb{R}_+$, we have

$$F(tx,tz) = (tz)^\alpha f(tx \cdot \frac{1}{tz}) = (tz)^\alpha f(x \cdot \frac{1}{z}) = t^\alpha z^\alpha f(x \cdot \frac{1}{z}) = t^\alpha F(x,z).$$

Also the vice-versa holds. More precisely, if the real function $F(x,z)$, defined on $C \times \mathbb{R}_+$, where C is a cone of $\mathbb{R}^n$, is homogeneous of degree α in $C \times \mathbb{R}_+$ and it results $F(x,1) = f(x),\ \forall x \in C$, then

$$F(x,z) = z^\alpha f(x \cdot \frac{1}{z}),\ \forall (x,z) \in C \times \mathbb{R}_+.$$

The proof is easy. Being $F(\cdot)$ homogeneous of degree α, we have

$$F(x,z) = F(z(x \cdot \frac{1}{z}, 1)) = z^\alpha F(x \cdot \frac{1}{z}, 1) = z^\alpha f(x \cdot \frac{1}{z}).$$

Example 2.5.9 If we consider the function $f(x) = x^k,\ x > 0$, its "homogeinization" of degree 1 is

$$F(x,y) = y\left(\frac{x}{y}\right)^k = x^k y^{1-k}.$$

2.6 Implicit Functions

Since now we have considered functions $f : T \subset \mathbb{R}^n \longrightarrow \mathbb{R}^m$ by means of an expression or representation in an *explicit form:*

$$y = f(x),\ \ x \in T, \tag{2.5}$$

that is, in the usual form which keeps "separated" the independent variables $x_1, x_2, \ldots, x_n$, from the dependent ones $y_1, y_2, \ldots, y_m$. Of course relation (2.5) may be rewritten in the form

$$y - f(x) = [0],\ \ x \in T,$$

where all variables, dependent and independent, are in the first member of the written equality. This is a trivial remark. Less trivial is the question: given an equation (or a relation) of the type

$$g(x, y) = [0], \quad x \in \mathbb{R}^n, \quad y \in \mathbb{R}^m,$$

it is possible to "extract" from this equation a function of the type $y = f(x)$ or of the type $x = h(y)$?

For instance, if we consider the equation

$$ax + by + c = 0, \quad x, y \in \mathbb{R}, \tag{2.6}$$

which obviously is the equation of a straight line in its *general form,* if $b \neq 0$, it is possible to rewrite (2.6) in the equivalent form

$$y = -\frac{a}{b}x - \frac{c}{b},$$

that is, with $m = -\frac{a}{b}$, $q = -\frac{c}{b}$,

$$y = mx + q,$$

which is the equation of a straight line in its *canonical form.*

If we consider, for example, the equation

$$x - xy + e^y = 0, \tag{2.7}$$

we get

$$x = \frac{e^y}{y - 1},$$

but we are not able to obtain from (2.7) a function of the type $y = f(x)$.

Always by considering the simple case of two real variables, x and y, the equation $g(x, y) = 0$ describes generally on the plane xOy all points of the level set of g, with level $k = 0$. If this set is nonempty (as, for instance, by considering the equation $x^2 + y^2 + 1 = 0$, clearly with no real solutions), the question we can ask us is: it is possible, at least locally, to obtain a correspondence between x and y, of the type $y = f(x)$, or of the type $x = h(y)$? Not always this is possible, not only with "global" results, but also by limiting our pretensions to "local" results. If, furthermore, we pretend also uniqueness results, also for this question we could have some disappointments. For instance, let us consider the set of points S in the plane xOy defined as follows

$$S = \left\{(x, y) \in \mathbb{R}^2 : y^2 - x^3 - x^2 = 0\right\}.$$

Fig. 2.9 Curve S defined by the equation $y^2 - x^3 - x^2 = 0$

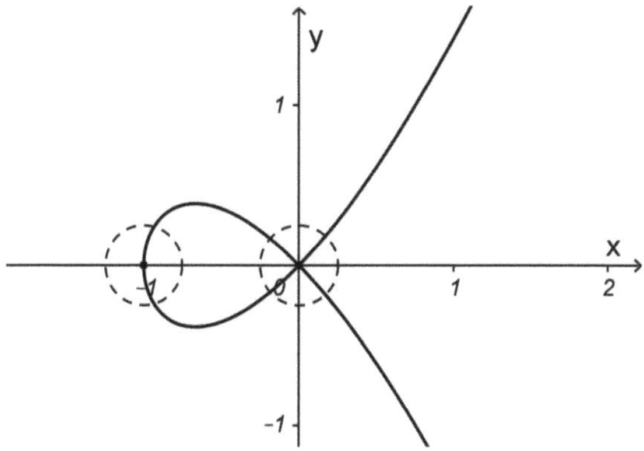

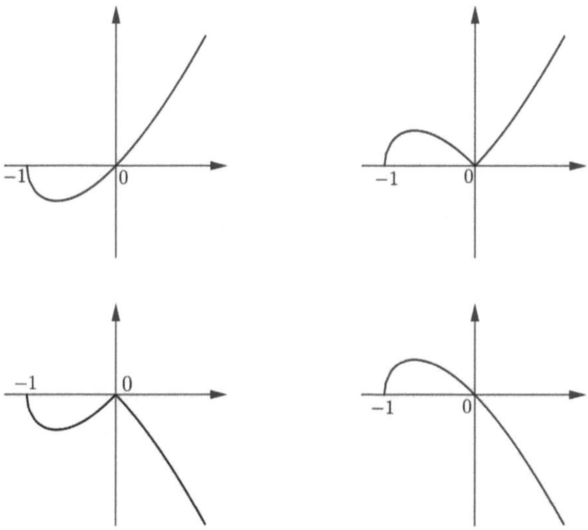

Fig. 2.10 Four continuous functions of the type $y = f(x)$ that satisfy the equation $y^2 - x^3 - x^2 = 0$

This set is drawn in Fig. 2.9.

It is evident that, for values of $x < 1$ the equation $y^2 - x^3 - x^2 = 0$ has no real root. For all other values of x, $x \neq -1$ and $x \neq 0$, the said equation admits two real roots. Hence, if we fix the coordinates of a point $(x_0, y_0) \in S$, $(x_0, y_0) \neq (0, 0)$ and $(x_0, y_0) \neq (-1, 0)$, we can assert that there exists a neighbourhood of (x_0, y_0) which contains a "piece" of the function of the type $y = f(x)$, and that this function is defined in the whole neighbourhood considered. However, if we consider the point $(0, 0)$ or also the point $(-1, 0)$, we see that it is not possible to obtain such a result, even if we consider local results. A fortiori we cannot pretend to obtain "global" results, as the set S "conceals" four continuous functions of the type $y = f(x)$. See Fig. 2.10.

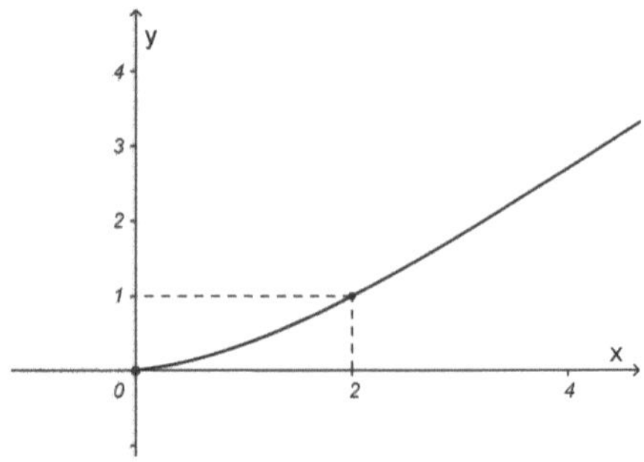

Fig. 2.11 Graph of the curve (2.8)

Furthermore, the set S "conceals" infinitely many (!) non continuous functions, always of the type $y = f(x)$. Now consider the following relation in two variables:

$$x + \log x - y - \log y - 1 - \log 2 = 0. \tag{2.8}$$

From this equation we are not able to obtain $y = f(x)$, nor $x = h(y)$. This *does not mean* that, at least locally, a correspondence of the type $y = f(x)$ or $x = h(y)$ does not exist. It means that it is not possible, by a finite number of algebraic operations, to obtain the *analytic form* of $y = f(x)$ or $x = h(y)$. If we make use of a computer program which plots the graph of a mathematical relation in two variables, we see that the points of the Cartesian plane which satisfy relation (2.8) are represented Fig. 2.11.

It seems the graph of a function of the type $y = f(x)$, continuous on its domain $(0, +\infty)$ and with derivatives of any order; moreover, it is strictly increasing and hence invertible. But we are not able to know the analytic expression of the said function (a similar problem raises also in the theory of integration: from the Fundamental Theorem on Integral Calculus we now that a continuous function admits a primitive (or antiderivative). For example, the function $f(x) = e^{-x^2}$, well known in Statistics and Probability Calculus, is continuous on $(-\infty, +\infty)$, but it is not possible to obtain a primitive with a finite number of algebraic operations).

It is time to take the bull by the horns. For convenience we change the notations used since now and give the following basic definition (we begin with the simplest case).

Definition 2.6.1 Let be given $f : \mathbb{R}^2 \longrightarrow \mathbb{R}$ and consider the equation $f(x, y) = 0$. Let be (x_0, y_0) a point satisfying $f(x_0, y_0) = 0$. We say that the equation $f(x, y) = 0$ *defines implicitly* (at least locally) a function of the type $y = \varphi(x)$ or that the function $y = \varphi(x)$ is *defined implicitly* (locally) by the equation $f(x, y) = 0$, if there exists a neighbourhood $I(x_0)$ such that:

(a) $\varphi(x_0) = y_0$;
(b) $f(x, \varphi(x)) = 0, \ \forall x \in I(x_0)$.

The following result, in its classical form, is due to the Italian mathematician (University of Pisa) Ulisse Dini (1845–1918), and give an answer (at least in a local sense) to the questions we have brought up.

Theorem 2.6.2 (Implicit Function Theorem) *Let be given the equation, in two real variables,* $f(x, y) = 0$ *and consider a point* (x_0, y_0) *such that* $f(x_0, y_0) = 0$. *Furthermore, assume that:*

(1) The function f *is* C^1 *on a neighbourhood* $U(x_0, y_0)$.
(2) It holds $\frac{\partial f}{\partial y}(x_0, y_0) \neq 0$.
Then it holds:

(a) There exists one and only one function of the type $y = \varphi(x)$, *of class* C^1 *and defined on a neighbourhood* $I(x_0)$. *This function satisfies the following properties:*

(1) $f(x, \varphi(x)) = 0, \ \forall x \in I(x_0)$.
(2) $\varphi(x_0) = 0$.

(b) It holds

$$\varphi'(x) = -\frac{\frac{\partial f}{\partial x}(x, \varphi(x))}{\frac{\partial f}{\partial y}(x, \varphi(x))}, \quad \forall x \in I(x_0).$$

Proof Let us suppose that it is, for instance, $\frac{\partial f}{\partial y}(x_0, y_0) > 0$ and observe that, being this partial derivative continuous, by the theorem of sign permanence, we can determine a neighbourhood of (x_0, y_0) of the form

$$R = x_0 - h \leqq x \leqq x_0 + h, \ \ y_0 - k \leqq y \leqq y_0 + k$$

and such that in all its points we have

$$\frac{\partial f}{\partial y}(x, y) > 0. \tag{2.9}$$

Now consider the function $f(x_0, y)$ in the y-variable; this function is *strictly increasing* on the interval $(y_0 - k, \ y_0 + k)$ and as in the point y_0 the same function is equal to zero, we have

$$f(x_0, y_0 - k) < 0; \quad f(x_0, y_0 + k) > 0.$$

From the said relations and by continuity of the two functions $f(x, y_0 - k)$, $f(x, y_0+k)$ and again by the theorem of sign permanence, there exists a positive number $\delta \leqq h$ such that for every x such that $x_0 - \delta \leqq x \leqq x_0 + \delta$, it holds

$$f(x, y_0 - k) < 0, \quad f(x, y_0 + k) > 0. \tag{2.10}$$

Hence we can conclude that, given *any* point x_1 of the interval $(x_0 - \delta,\ x_0 + \delta)$, the function

$$f(x_1, y) \tag{2.11}$$

is, thanks to (2.9), strictly increasing on the whole interval $(y_0 - k, y_0 + k)$ and assumes, by (2.10), a negative value for $y = y_0 - k$ and a positive value for $y = y_0 + k$. But then, thanks to Bolzano's theorem (or theorem of existence of zeros), the function (2.11) is equal to zero *for one and only one value* y_1 of the y-variable, that is it results

$$f(x_1, y_1) = 0.$$

Hence we have been able to build a neighbourhood I of the point x_0, say $(x_0 - \delta,\ x_0 + \delta)$, such that for *every* point x belonging to the said neighbourhood, there exists one and only one value y in the interval $(y_0 - k,\ y_0 + k)$ which, together with x, form a solution (x, y) of the equation $f(x, y) = 0$. It is therefore proved that in the neighbourhood $(x_0 - \delta,\ x_0 + \delta)$ of the point x_0 it is defined a function $\varphi(x)$ satisfying 1) and (2) of point (a) of the theorem.

Now we prove that $\varphi(x)$ is continuous, together with its first-order derivative, in the interval $(x_0 - \delta,\ x_0 + \delta)$. Indeed, if x and $x + \Delta x$ are two points of the said interval, and $\varphi(x)$ and $\varphi(x + \Delta x) = \varphi(x) + \Delta\varphi(x)$ the corresponding values of $\varphi(x)$, we have

$$f(x, \varphi(x)) = 0, \quad f(x + \Delta x, \varphi(x) + \Delta\varphi(x)) = 0. \tag{2.12}$$

On the other hand, by the Mean Value Theorem, we can write

$$f(x + \Delta x, \varphi(x) + \Delta\varphi(x)) - f(x, \varphi(x))$$

$$= \frac{\partial f}{\partial x}(x + \theta\Delta x, \varphi(x) + \theta\Delta\varphi(x))\Delta x + \frac{\partial f}{\partial y}(x + \theta\Delta x, \varphi(x + \theta\Delta\varphi(x))\Delta\varphi(x),$$

with $0 < \theta < 1$. It follows, by (2.12),

$$\frac{\partial f}{\partial x}(x + \theta\Delta x, \varphi(x) + \theta\Delta\varphi(x))\Delta x + \frac{\partial f}{\partial y}(x + \theta\Delta x, \varphi(x + \theta\Delta\varphi(x))\Delta\varphi(x) = 0,$$

that is, being surely $\frac{\partial f}{\partial y}(x + \theta\Delta x, \varphi(x + \theta\Delta\varphi(x)) > 0$, as the point $(x + \theta\Delta x, \varphi(x + \theta\Delta\varphi(x))$ is interior to the rectangle R :

$$\Delta\varphi(x) = -\frac{\frac{\partial f}{\partial x}(x + \theta\Delta x, \varphi(x) + \theta\Delta\varphi(x))}{\frac{\partial f}{\partial y}(x + \theta\Delta x, \varphi(x + \theta\Delta\varphi(x))}\Delta x \tag{2.13}$$

or also

$$\frac{\Delta\varphi(x)}{\Delta x} = -\frac{\frac{\partial f}{\partial x}(x+\theta\Delta x, \varphi(x)+\theta\Delta\varphi(x))}{\frac{\partial f}{\partial y}(x+\theta\Delta x, \varphi(x+\theta\Delta\varphi(x))}.$$

As $\frac{\partial f}{\partial y}(x, y)$ is continuous and positive on the rectangle R, then, as R is closed and bounded, it admits a positive minimum and hence from relation (2.13) we get

$$\lim_{\Delta x \longrightarrow 0} \Delta\varphi(x) = 0,$$

that is

$$\lim_{\Delta x \longrightarrow 0} \varphi(x+\Delta x) = \varphi(x),$$

relation which proves the continuity of $\varphi(x)$. But for $\Delta x \longrightarrow 0$ also $\Delta\varphi(x) \longrightarrow 0$ and by the continuity of $\frac{\partial f}{\partial x}$ and $\frac{\partial f}{\partial y}$, we obtain

$$\lim_{\Delta x \longrightarrow 0} \frac{\Delta\varphi(x)}{\Delta x} = -\frac{\frac{\partial f}{\partial x}(x, \varphi(x))}{\frac{\partial f}{\partial y}(x, \varphi(x))},$$

that is the relation (b) of the theorem. As the second member of the last relation written is a continuous function on I, we have that $\varphi(x)$ is of class C^1 on I. □

Remark 2.6.3

(*a*) In a pragmatic way, we note that, once we have proved that φ is differentiable, its first-order derivative (i.e. the formula (b) of Theorem 2.6.2) is an immediate consequence of the chain rule on differentiability of composed functions (Sect. 2.3). From $f(x, y) = 0$, with $y = \varphi(x)$ and hence $f[x, \varphi(x)] = 0$, by writing this last relation in the form $F(x) = 0$, we obtain

$$F'(x) = \frac{\partial f}{\partial x} + \frac{\partial f}{\partial y}\varphi'(x) = 0$$

from which

$$\varphi'(x) = -\frac{\frac{\partial f}{\partial x}}{\frac{\partial f}{\partial y}}.$$

(*a*) If at a point (x_0, y_0), solution of the equation $f(x, y) = 0$, it holds $\frac{\partial f}{\partial y}(x_0, y_0) = 0$, obviously the assumptions of Dini's Theorem are not respected and it would be a risk to get on the "local" existence of $y = \varphi(x)$. On the other hand, if $\frac{\partial f}{\partial x}(x_0, y_0) \neq 0$ (and also the remaining assumptions of the theorem are verified), then the equation $f(x, y) = 0$ defines a unique function of the type $x = h(y)$, of class C^1 in a suitable neighbourhood of y_0 and such that $h(y_0) = x_0$. If it holds both $\frac{\partial f}{\partial y}(x_0, y_0) = 0$ and $\frac{\partial f}{\partial x}(x_0, y_0) = 0$, nothing can be said on the existence (also in a local sense) and the uniqueness of functions of the type $y = \varphi(x)$ or $x = h(y)$, "concealed" in the relation $f(x, y) = 0$.

Let us consider, for instance, the following examples.

(1) $f(x, y) = x^3 - y^3 = 0$;
(2) $f(x, y) = y^4 - x^4 = 0$;
(3) $f(x, y) = y^4 + x^2 = 0$.

Here $(x_0, y_0) = (0, 0)$. In all three cases we have $\nabla f(0, 0) = [0]$. In the first case the implicit function of the type $y = \varphi(x)$ exists, unique, and we have $y = x$; in the second case we have at least two implicit functions $y = \pm x$, whereas in the third case no implicit function exists.

Remark 2.6.4 Theorem 2.6.2 of Dini yields only *sufficient conditions* for the existence of implicit functions. Let us consider, for instance, the equation $f(x, y) = x - y^3 = 0$. Obviously we have immediately $y = \sqrt[3]{x}$, $\forall x \in \mathbb{R}$. Yet we have $\frac{\partial f}{\partial y} = -3y^2$, which is zero at the origin.

Remark 2.6.5 Again we remark that relation (b) of Theorem 2.6.2 gives the expression of $\varphi'(x)$ in a suitable neighbourhood of x_0, *without any specification on the analytic form of* φ (which could remain unknown!). Furthermore, if we make the assumption that f is of class C^n (i.e. it is continuous, together with all its partial derivatives of order n) in a neighbourhood $U(x_0, y_0)$, then, by assuming the remaining hypotheses of Theorem 2.6.2, we obtain that φ is of class C^n on $I(x_0)$ and hence it will be possible to compute, at least at x_0, also $\varphi''(x_0)$, $\varphi'''(x_0)$, etc.. Therefore, *without knowing* the analytic expression of $y = \varphi(x)$, in this case we can obtain a good approximation of this function, at least in a neighbourhood $I(x_0)$, by means of Taylor's expansion formula. And this fact is another remarkable feature of Dini's Theorem. By curiosity, the expression of $\varphi''(x)$ in a suitable neighbourhood of x_0 (obviously when $\varphi''(x)$ exists) is

$$\varphi''(x) = -\frac{\frac{\partial^2 f}{\partial x^2}\left(\frac{\partial f}{\partial y}\right)^2 - 2\frac{\partial^2 f}{\partial x \partial y}\frac{\partial f}{\partial y}\frac{\partial f}{\partial x} + \frac{\partial^2 f}{\partial y^2}\left(\frac{\partial f}{\partial x}\right)^2}{\left(\frac{\partial f}{\partial y}\right)^3},$$

i.e., taking (b) of Theorem 2.6.2 into account,

$$\varphi''(x) = -\frac{\frac{\partial^2 f}{\partial x^2} + 2\frac{\partial^2 f}{\partial x \partial y}\varphi'(x) + \frac{\partial^2 f}{\partial x^2}\left[\varphi'(x)\right]^2}{\frac{\partial f}{\partial y}}.$$

Summing up, Theorem 2.6.2 answers, at least locally, to the following questions.

(1) If we are not able to find the expressions $y = \varphi(x)$ or $x = h(y)$, "concealed" in the equation $f(x, y) = 0$, it can be possible to assert the existence and the uniqueness (at least locally) of a function of the type $y = \varphi(x)$ or $x = h(y)$, implicitly defined by the equation $f(x, y) = 0$.
(2) It is possible to obtain informations on the local behaviour of $y = \varphi(x)$ or $x = h(y)$, informations given by the sign of $\varphi'(x)$ or $h'(y)$, or also by the sign of the other derivatives of higher order.

There exist in the mathematical literature also "global versions" of Theorem 2.6.2 (and of the other theorems on implicit functions reported in the present section); the reader can see, e.g., the books of [6] and [11]. See the next Theorems 2.6.12, 2.6.13 and 2.6.15.

Example 2.6.6 Again we consider the relation

$$f(x, y) = x + \log x - y - \log y - 1 - \log 2 = 0.$$

(a) Verify that the point $P(2, 1)$ is a zero of the relation and that in a suitable neighbourhood $I(2)$ it is defined a unique function of the type $y = \varphi(x)$.
(b) Compute $\varphi'(2)$.

We note that f is C^1 on its domain. Then it holds

$$f(2, 1) = 2 + \log 2 - 1 - \log 1 - 1 - 2 = 0;$$

$$\frac{\partial f}{\partial y}(2, 1) = \left(-1 - \frac{1}{y}\right)_{(2,1)} = -1 - 1 = -2 \neq 0.$$

Hence there exists a unique function $y = \varphi(x)$ defined in a suitable neighbourhood of 2. Then we have $\frac{\partial f}{\partial x} = 1 + \frac{1}{x}$ which, evaluated in (2, 1) gives $\frac{3}{2}$. Finally we have

$$\varphi'(x) = -\frac{\partial f/\partial x}{\partial f/\partial y}.$$

Therefore

$$\varphi'(2) = -\frac{\frac{3}{2}}{-2} = \frac{3}{4},$$

which informs us that $\varphi(x)$ is strictly increasing at $x_0 = 2$, as already shown by Fig. 2.11.

Example 2.6.7 Show that the equation

$$f(x, y) = 2e^{x+y} + xy^2 + xy - 2 = 0$$

defines implicitly at least in a neighbourhood of the point $(0, 0)$ a unique function of the type $y = \varphi(x)$ and that this function is strictly decreasing at the same point.

We note that f is $\mathcal{C}^1$ on the whole $\mathbb{R}^2$. Then we have $2e^0 + 0 + 0 - 2 = 0$,

$$\frac{\partial f}{\partial x} = 2e^{x+y} + y^2 + y; \quad \frac{\partial f}{\partial x}(0, 0) = 2.$$

$$\frac{\partial f}{\partial y} = 2e^{x+y} + 2xy + x; \quad \frac{\partial f}{\partial y}(0, 0) = 2 \neq 0.$$

$$\varphi'(0) = -\frac{\frac{\partial f}{\partial x}(0, 0)}{\frac{\partial f}{\partial y}(0, 0)} = -\frac{2}{2} = -1 < 0.$$

Therefore φ is strictly decreasing at $x_0 = 0$ in a suitable neighbourhood $I(0)$.

Now we see two theorems which are generalizations of Theorem 2.6.2. The first result is concerned with the case of $x \in \mathbb{R}^n$ and $y \in \mathbb{R}$.

Theorem 2.6.8 *Let be given the equation*

$$f(x_1, x_2, \ldots, x_n, y) = 0,$$

that is $f(x, y) = 0$, *with* $x \in \mathbb{R}^n$ *and* $y \in \mathbb{R}$. *Let* (x^0, y_0) *be a vector of* $\mathbb{R}^{n+1}$ *which satisfies the said equation. Let* f *be of class* $\mathcal{C}^1$ *in a neighbourhood* $U(x^0, y_0)$ *and let be* $\frac{\partial f}{\partial y}(x^0, y_0) \neq 0$. *Then there exists a unique function* $y = \varphi(x)$, *continuously differentiable in a neighbourhood* $I(x^0)$. *Moreover, it holds*

$$\frac{\partial \varphi}{\partial x_i} = -\frac{\frac{\partial f}{\partial x_i}(x, \varphi(x))}{\frac{\partial f}{\partial y}(x, \varphi(x))}, \quad i = 1, \ldots, n, \quad \forall x \in I(x^0).$$

The second result is a version XXL of the theorem on implicit functions, in the sense that it is a version for vector functions of several variables, i.e. for systems of equations.

Theorem 2.6.9 *Let be given the vector equation* $f(x, y) = [0]$, *with* $x \in \mathbb{R}^k$, $y \in \mathbb{R}^m$, *and* $f : \mathbb{R}^{k+m} \longrightarrow \mathbb{R}^m$. *Let* f *be of class* $\mathcal{C}^1$ *on a neighbourhood* $U(x^0, y^0)$, *being* $f(x^0, y^0) = [0]$. *If*

$$\left| J_y f(x^0, y^0) \right| \neq 0,$$

where $J_y f(x, y)$ *denotes the Jacobian matrix of* f, *evaluated with respect to* y, *then there exists a neighbourhood* $V(x^0)$ *where it is defined a unique function of the type*

$y = \varphi(x)$, such that $y^0 = \varphi(x^0)$, $f[x, \varphi(x)] = [0]$, $\forall x \in V(x^0)$. Moreover, φ is $\mathcal{C}^1$ on $V(x^0)$ and it holds, for all $x \in V(x^0)$:

$$J\varphi(x) = -\left[J_y f(x, y)\right]^{-1} \left[J_x f(x, y)\right].$$

Example 2.6.10 Let be given the following system of m equations in $k + m$ variables (hence we have more variables than equations):

$$\begin{cases} x_1 + (x_2)^2 - (x_3)^3 + (y_1)^4 - (y_2)^6 + 1 = 0 \\ x_1 - x_3 + y_2 + 1 = 0. \end{cases}$$

Hence $f : \mathbb{R}^5 \longrightarrow \mathbb{R}^2$ $(k = 3, m = 2)$. Let be $x^0 = [-1, 1, 1]^\top$; $y^0 = [1, 1]^\top$. Let us check that $f(x^0, y^0) = [0]$; indeed

$$\begin{cases} -1 + (1)^2 - (1)^3 + (1)^4 - (1)^6 + 1 = 0 \\ -1 - 1 + 1 + 1 = 0. \end{cases}$$

The function f is $\mathcal{C}^1$. Now we compute the two Jacobian matrices and verify that at (x^0, y^0) the matrix $J_y f(x, y)$ is nonsingular. We have

$$J_y f(x, y) = \begin{bmatrix} 4(y_1)^3 & -6(y_2)^5 \\ 0 & 1 \end{bmatrix}$$

and hence

$$J_y f(x^0, y^0) = \begin{bmatrix} 4 & -6 \\ 0 & 1 \end{bmatrix}.$$

This matrix is nonsingular:

$$\begin{vmatrix} 4 & -6 \\ 0 & 1 \end{vmatrix} = 4.$$

Then it results

$$J_x f(x, y) = \begin{bmatrix} 1 & 2x_2 & -3(x_3)^2 \\ 1 & 0 & -1 \end{bmatrix}$$

and hence

$$J_x f(x^0, y^0) = \begin{bmatrix} 1 & 2 & -3 \\ 1 & 0 & -1 \end{bmatrix}.$$

We have therefore

$$\begin{aligned} J\varphi(x^0) &= -\begin{bmatrix} 4 & -6 \\ 0 & 1 \end{bmatrix}^{-1} \begin{bmatrix} 1 & 2 & -3 \\ 1 & 0 & -1 \end{bmatrix} \\ &= \begin{bmatrix} \frac{1}{4} & \frac{3}{2} \\ 0 & 1 \end{bmatrix} \begin{bmatrix} 1 & 2 & -3 \\ 1 & 0 & -1 \end{bmatrix} \\ &= \begin{bmatrix} -\frac{7}{4} & -\frac{1}{2} & \frac{9}{4} \\ -1 & 0 & 1 \end{bmatrix}. \end{aligned}$$

An important consequence of the implicit function theorem, version XXL, concerns the (local) existence of the inverse of vector functions $f : \mathbb{R}^n \longrightarrow \mathbb{R}^n$, provided that these functions are $\mathcal{C}^1$ with a nonsingular Jacobian matrix, at least at a point x^0 (and hence nonsingular in a suitable neighbourhood of the said point). This refers to a well known property of real functions of one real variable: if f is of class $\mathcal{C}^1$ on a neighborhood $U(x_0) \subset \mathrm{dom}(f)$, and if $f'(x_0) \neq 0$, then f is locally strictly monotone, and hence locally invertible, and we have $(f^{-1})'(y_0) = 1/f'(x_0)$.

Theorem 2.6.11 (Theorem on Local Invertibility of Vector Functions) *Let be given $f : \mathbb{R}^n \longrightarrow \mathbb{R}^n$, a function of class $\mathcal{C}^1$ on a neighbourhood $U(x^0) \subset \mathrm{dom}(f)$. Let be $\left|Jf(x^0)\right| \neq 0$. Then there exists a neighbourhood of $y^0 = f(x^0)$ where $x = h(y) = f^{-1}(y)$. In the said neighbourhood the function $x = f^{-1}(y)$ is $\mathcal{C}^1$ and we have*

$$Jf^{-1}(y^0) = \left[Jf(x^0)\right]^{-1}.$$

It is worthwhile to remark that if the function of Theorem 2.6.11 is linear, that is of the type $f(x) = Ax$, with A square matrix of order n, we refind, under the assumption that A is nonsingular, the "global" result:

$$y = Ax \Longrightarrow x = A^{-1}y,$$

in turn equivalent to the well known Cramer's rule. Indeed, Theorem 2.6.11 can be extended in order to obtain "global results" also for nonlinear functions; these extensions have started just from questions of Economic Analysis, so that the main theorem on the global invertibility of a function $f : \mathbb{R}^n \longrightarrow \mathbb{R}^n$ is due to two mathematical economists: D. Gale and H. Nikaido ("Theorem of Gale-Nikaido"). Obviously, in this case the mathematical questions become rather sophisticated. The reader may consult the book of [9] quoted in the References. We report the following result, due to Gale, Nikaido and Inada, without proof.

Theorem 2.6.12 (Theorem of Gale-Nikaido-Inada) *Let be given the function $f : \mathbb{R}^n \longrightarrow \mathbb{R}^n$, defined and differentiable on the Cartesian product $X = \prod_{i=1}^{n} (a_i, b_i)$, where a_i and b_i, $i = 1, \ldots, n$, may also assume the value, respectively, $-\infty$ and*

$+\infty$. Then, f is globally invertible on X if either one of the following condition (i) or (ii) holds:

(i) The Jacobian matrix $Jf(x)$ has all its principal minors positive over X (in other words, $Jf(x)$ is a so-called P-matrix).
(ii) The Jacobian matrix $Jf(x)$ has continuous elements and all its principal minors are negative on X (in other words, $Jf(x)$ is a so-called N-matrix).

Also implicit function theorems have "global versions". We report a global statement of Theorem 2.6.8 ($x \in \mathbb{R}^n,\ y \in \mathbb{R}$).

Theorem 2.6.13 *Let be given the equation*

$$f(x, y) = 0,$$

with $f : D \subset \mathbb{R}^{n+1} \longrightarrow \mathbb{R}$, defined on the Cartesian product $D = (a_1, b_1) \times (a_2, b_2) \times \cdots \times (a_n, b_n) \times (a_{n+1}, b_{n+1}) \equiv \prod_{s=1}^{n+1} (a_s, b_s)$, where a_s and b_s, $s = 1, \ldots, n$, may also assume the value, respectively, $-\infty$ and $+\infty$. If:

1. *f is continuous on D.*
2. *f is strictly monotone, with respect to y, for all fixed x.*
3. *For every fixed $x = [x_1, \ldots, x_n]^\top$ on $(a, b) \equiv \prod_{s=1}^{n} (a_s, b_s)$, the function f changes sign, when $y \in (a_{n+1}, b_{n+1})$.*

Then, there exists a unique function $y = g(x)$, $g : (a, b) \longrightarrow \mathbb{R}$, such that

$$f\,[x_1, \ldots x_n, g(x_1, \ldots, x_n)] = 0,\ \forall x_s \in (a_s, b_s),\ s = 1, \ldots, n.$$

Moreover, g is continuous on (a, b) and takes values on (a_{n+1}, b_{n+1}).

We consider the following example (recall that, in case of differentiable functions, the related monotonicity can be established by means of the sign of the partial derivatives).

Example 2.6.14 Consider the equation, in three variables,

$$f(x_1, x_2, x_3) = 6x_1 + 2x_2 e^{x_1} - x_3 - e^{x_3} = 0.$$

We have to verify whether the said equation defines implicitly a function of the type $y = g(x_1, x_2)$, with $y = x_3$.

1. f is defined and continuous on the whole $\mathbb{R}^3$.
2. f is differentiable and we have

$$\frac{\partial f}{\partial x_3} = -1 - e^{x_3} < 0, \text{ for every } x_3 \in \mathbb{R},$$

hence f is strictly decreasing with respect to $y = x_3$.

3. In order to check whether f changes sign when $y \in \mathbb{R}$, we can compute the following two limits

$$\lim_{x_3 \longrightarrow -\infty} f(x_1, x_2, x_3) = 6x_1 + 2x_2 e^{x_1} + \infty = +\infty, \text{ for every } x_1 \text{ and } x_2 \text{ (finite);}$$

$$\lim_{x_3 \longrightarrow +\infty} f(x_1, x_2, x_3) = 6x_1 + 2x_2 e^{x_1} - \infty = -\infty, \text{ for every } x_1 \text{ and } x_2 \text{ (finite).}$$

Therefore the said limits have opposite sign and we can deduce (theorem of sign preservation) that f changes sign. Therefore, we conclude that there exists a unique implicit function $g : \mathbb{R}^2 \longrightarrow \mathbb{R}$, continuous and such that

$$f[x_1, x_2, g(x_1, x_2)] = 6x_1 + 2x_2 e^{x_1} - g(x_1, x_2) - e^{g(x_1, x_2)} = 0, \text{ for every } x_1, x_2 \in \mathbb{R}.$$

Notice that we are not able to give x_3 as an explicit function of x_1 and x_2. The reader can verify that the given equation defines implicitly also a function of the type $x_2 = h(x_1, x_3)$, but *not* a function of the type $x_1 = k(x_2, x_3)$.

Theorem 2.6.15 *Let the assumptions of Theorem 2.6.13 be satisfied and, furthermore, let f be differentiable on the Cartesian product $\prod_{i=1}^{n+1} (a_i, b_i)$, with $\frac{\partial f}{\partial y} \neq 0$ everywhere. Then the implicit function $y = g(x)$ is differentiable on the Cartesian product $\prod_{i=1}^{n} (a_i, b_i)$ and it holds*

$$\frac{\partial g}{\partial x_i} = -\frac{\frac{\partial f}{\partial x_i}(x, g(x))}{\frac{\partial f}{\partial y}(x, g(x))}, \quad i = 1, \ldots, n,$$

that is. the thesis of Theorem 2.6.8 holds "in large" (i.e. globally). If, moreover, f is continuously differentiable on $\prod_{i=1}^{n+1} (a_i, b_i)$, then $y = g(x)$ is continuously differentiable on $\prod_{i=1}^{n} (a_i, b_i)$.

One of the classical economic applications of Theorem 2.6.9 concerns the so-called *comparative statics problems,* introduced mainly by P. A. Samuelson (Nobel prize in Economics). Shortly, when economists deal with systems of equations, the variables are often classified *a priori* into two types: *endogenous variables,* which the model tries to determine, and *exogenous variables,* which are determined by forces outside the model. In the formulation of Theorem 2.6.9, $x_1, x_2, \ldots, x_k$ are the exogenous variables, while $y_1, y_2, \ldots, y_m$ are the endogenous variables. The point (x^0, y^0) such that $f(x^0, y^0) = [0]$ is called in this context the "equilibrium solution" or "equilibrium point". In this setting Theorem 2.6.9 gives sufficient conditions for system $f(x, y) = [0]$ to determine the endogenous variables as differentiable functions of the exogenous variables in a neighbourhood of an equilibrium. We refer to the books of [13] and [12] for further insights. Rather good is also the exposition of [10].

Example 2.6.16 Let us consider a financial operation ("cash flow") which at maturities $t_1, t_2, \ldots, t_m$ requires expenditures denoted by $C_1, C_2, \ldots, C_m$, and that at maturities $\tau_1, \tau_2, \ldots, \tau_n$ produces incomes denoted by $R_1, R_2, \ldots, R_n$. It is well known that the *discounted cash flow* is the function $G(x)$ of the discount rate defined as follows

$$G(x) = \sum_{s=1}^{n} R_s(1+x)^{-\tau_s} - \sum_{h=1}^{m} C_h(1+x)^{-t_h}.$$

The *internal rate of return* for the said operation is (if it exists) that positive rate x^* such that $G(x^*) = 0$. There may exist more than one internal rate of returns (and this fact generates serious problems on the financial meaning of the same rates), however several sufficient condition for uniqueness are known. It may be interesting to check that this internal rate of return (i. r. r.) behaves well, in the sense that it points out correctly the variations of the return which gives the financial operation, when one or more amounts vary. If, for instance, the generic income R_s increases, we should expect that also x^* increases. The equation

$$\sum_{s=1}^{n} R_s(1+x^*)^{-\tau_s} - \sum_{h=1}^{m} C_h(1+x^*)^{-t_h} = 0$$

defines the internal rate of return as a function of the cash flows generated by the financial operation we are considering. This equation may be rewritten as

$$g(C_1, C_2, \ldots, C_m, R_1, R_2, \ldots, R_n, x^*) = 0,$$

or, more shortly, $g(C, R, x) = 0$, where C is the vector of the various expenditures and R the vector of the incomes. If we suppose that the assumptions of the Dini implicit function theorem are verified, we obtain

$$\frac{\partial x^*}{\partial R_s} = -\frac{\frac{\partial g}{\partial R_s}(C, R, x^*)}{\frac{\partial g}{\partial x^*}(C, R, x^*)}.$$

It is easy to compute the partial derivative of the numerator: it is given by $(1 + x^*)^{-\tau_s} > 0$, hence in order to have $\frac{\partial x^*}{\partial R_s} > 0$ (as it should be, on the grounds of our expectations), it must be $\frac{\partial g}{\partial x^*}(C, R, x^*) < 0$ (there is the sign "−" in the fraction!). Therefore if the discount rate function intersects the horizontal axes from above to below $\left(\frac{\partial g}{\partial x^*}(C, R, x^*) < 0\right)$, we are in a "normal" situation. Vice-versa, if this function intersects the horizontal axes from below to above, i.e. $\frac{\partial g}{\partial x^*}(C, R, x^*) > 0$, we have that if that income increases, the related internal rate of return x^* decreases!

2.7 Taylor's Formula

We know that if $f : T \subset \mathbb{R}^n \longrightarrow \mathbb{R}$ is differentiable at $x^0 \in \operatorname{int}(T)$, this means that if we remain within a "small" neighbourhood of x^0, the behaviour of f is significantly similar to the behaviour of the linear function

$$P(x) = f(x^0) + \nabla f(x^0)^\top (x - x^0).$$

To be more precise, we have, with $h = x - x^0$,

$$f(x^0 + h) - f(x^0) = \nabla f(x^0)^\top h + o(\|h\|), \quad \text{for } h \longrightarrow [0] .$$

For several questions this (linear) approximation of f around x^0 is not sufficient. In previous courses of mathematics the reader have seen that for functions $f : \mathbb{R} \longrightarrow \mathbb{R}$, which at x_0 possesses all derivatives of the various order, a good approximation is given by a suitable polynomial: the so-called Taylor's polynomial. This approximation, in a neighbourhood of a point x_0, will be more and more accurate by increasing the degree of Taylor's polynomial (that is by increasing the order of the derivatives valuated at the point) and by shrinking the radius of the neighbourhood in question.

With reference to functions of several real variables, we can follow the same type of reasoning and obtain, under suitable assumptions, a "good" approximation of a function $f : T \subset \mathbb{R}^n \longrightarrow \mathbb{R}$ in a neighbourhood of a point $x^0 \in \operatorname{int}(T)$. Even if it is possible to involve in this approximation partial derivatives of order n, we shall consider a Taylor's formula involving only partial derivatives of order one and two. Similarly to Taylor's formula for real function of one real variable, it is possible to formulate the *remainder* in the form given by Peano and in the form given by Lagrange.

We recall that $f : \mathbb{R}^n \longrightarrow \mathbb{R}$ is of *class* C^2 on an open set $A \subset \mathbb{R}^n$ if all its second-order partial derivatives are continuous on A.

Theorem 2.7.1 (Taylor's Formula with Remainder in the Form of Peano) *Let be $f : A \longrightarrow \mathbb{R}$ a function of class C^2 on the open set $A \subset \mathbb{R}^n$ and let be $x^0 \in A$, $U(x^0) \subset A$, with $U(x^0)$ suitable neighbourhood of x^0. Then it holds, for all $x \in U(x^0)$,*

$$f(x) = f(x^0) + \nabla f(x^0)^\top (x - x^0) + \frac{1}{2}(x - x^0)^\top Hf(x^0)(x - x^0) + o(\left\|x - x^0\right\|^2),$$

for $x \longrightarrow x^0$, *that is, with* $h = x - x^0$,

$$f(x^0 + h) = f(x^0) + \nabla f(x^0)^\top h + \frac{1}{2} h^\top Hf(x^0) h + o(\|h\|^2), \text{ for } h \longrightarrow [0].$$

Example 2.7.2 Let be given $f(x_1, x_2) = 3x_1 + 2x_2 + 4x_1x_2 - (x_1)^2 - (x_2)^3$. Consider the point $(0, 0)$ and write the approximation of f around this point by means of Taylor's formula, involving second-order partial derivatives. We have in this case $h = x - 0 = x$. Then

$$\nabla f(x_1, x_2) = \left[3 + 4x_2 - 2x_1;\ \ 2 + 4x_1 - 3(x_2)^2\right]^\top.$$

$$Hf(x_1, x_2) = \begin{bmatrix} -2 & 4 \\ 4 & -6x_2 \end{bmatrix}.$$

Therefore we have $f(0, 0) = 0$; $\nabla f(0, 0) = [3, 2]^\top$;

$$Hf(0, 0) = \begin{bmatrix} -2 & 4 \\ 4 & 0 \end{bmatrix}.$$

Taylor's formula with remainder in the form of Peano is given by

$$f(x_1, x_2) = 0 + [3, 2] \begin{bmatrix} x_1 \\ x_2 \end{bmatrix} + \frac{1}{2} [x_1, x_2] \begin{bmatrix} -2 & 4 \\ 4 & 0 \end{bmatrix} \begin{bmatrix} x_1 \\ x_2 \end{bmatrix} + o(\|x\|^2),$$

that is

$$f(x_1, x_2) = 3x_1 + 2x_2 + \frac{1}{2}(-2(x_1)^2 + 8x_1x_2) + o(\|x\|^2).$$

If we put $h = tv$, with $t \in \mathbb{R}$ and $v \in \mathbb{R}^n$, $\|v\| = 1$, Taylor's formula with remainder in Peano's form can be written as

$$f(x^0 + tv) = f(x^0) + t\nabla f(x^0)^\top v + \frac{t^2}{2} v^\top Hf(x^0) v + o(t^2).$$

Now we give Taylor's formula with remainder in the form of Lagrange (note that in this formula second-order partial derivatives are incorporated in the remainder).

Theorem 2.7.3 (Taylor's Formula with Remainder in the Form of Lagrange) *Consider the function* $f : A \longrightarrow \mathbb{R}$ *and assume that this function satisfies the assumptions given in Theorem 2.7.1. Then, for all* $x \in U(x^0)$, *it holds*

$$f(x) = f(x^0) + \nabla f(x^0)^\top (x - x^0) + \frac{1}{2}(x - x^0) Hf(\bar{x})(x - x^0),$$

where $\bar{x}$ is a point interior to the line segment joining x^0 and x, that is $\bar{x} = x^0 + \theta(x - x^0)$, with $\theta \in (0, 1)$.

We note that if A is a *convex set* (see the next chapter), in this case Taylor's formula holds for every $x \in A$. From Theorem 2.7.3, under the assumptions of once differentiability of f, we obtain the following *Mean Value Theorem* (which is the version for functions $f : \mathbb{R}^n \longrightarrow \mathbb{R}$ of the classical Theorem of Lagrange for $f : \mathbb{R} \longrightarrow \mathbb{R}$).

Theorem 2.7.4 (Mean Value Theorem) *Let $A \subset \mathbb{R}^n$ be an open set and let be $f : A \longrightarrow \mathbb{R}$ differentiable on A; let be $U(x^0) \subset A$. Then, for every $x \in U(x^0)$ we have*

$$f(x) - f(x^0) = \nabla f(\bar{x})^\top (x - x^0),$$

where $\bar{x} = x^0 + \theta(x - x^0)$, with $\theta \in (0, 1)$.

Corollary 2.7.5 *Under the assumptions of Theorem 2.7.4, if $\nabla f(x) = [0]$ for all $x \in U(A)$, then f is constant on $U(A)$.*

Besides the works quoted along the chapter, we refer to the reader interested in the topics of this chapter to the references [1–5, 7, 8].

References

1. T.M. Apostol, *Calculus*, vol. 2 (Blaisdell, Waltham, 1967)
2. R.G. Bartle, *The Elements of Real Analysis* (Wiley, New York, 1976)
3. B.D. Craven, *Functions of Several Variables* (Springer, Dordrecht, 1981)
4. W. Fleming, *Functions of Several Variables*. Undergraduate Texts in Mathematics (Springer, New York, 2012)
5. D. Gale, *The Theory of Linear Economic Models* (McGraw-Hill, New York, 1960)
6. A. Guerraggio, S. Salsa, *Metodi Matematici per l'Economia e le Scienze Sociali* (Giappichelli, Torino, 1997)
7. S. Lang, *Calculus of Several Variables*. Undergraduate Texts in Mathematics (Springer, New York, 2012)
8. A. Mas-Colell, M.D. Whinston, J.R. Green, *Microeconomic Theory* (Oxford University Press, Oxford, 1995)
9. H. Nikaido, *Convex Structures and Economic Theory* (Academic Press, New York, 1968)
10. J. Quirk, R. Saposnik, *Introduction to General Equilibrium Theory and Welfare Economics* (McGraw-Hill, New York, 1968)
11. W. Rudin, *Principles of Mathematical Analysis*, 3rd edn. (McGraw-Hill, New York, 1976)
12. E. Silberberg, W. Suen, *The Structure of Economics: A Mathematical Analysis* (McGraw-Hill, New York, 2000)
13. H.R. Varian, *Microeconomic Analysis* (W. W. Norton, New York, 1992)

Chapter 3
Introduction to Convex Analysis

3.1 Convex Sets

Convex Analysis plays a central role in Optimization and in Mathematical Economics: think, for example, to the games theory of John von Neumann, to linear programming, to the modern theory of general Walrasian economic equilibrium, to nonlinear programming, to the theory of optimal control, etc. We now give the basic definitions of *line segment, hyperplane and half-space* in $\mathbb{R}^n$.

Let x^1, x^2, and p be vectors of $\mathbb{R}^n$, with $p \neq [0]$, and consider the scalars $\lambda, \alpha \in \mathbb{R}$. We call:

• *Line segment* joining x^1 and x^2, the set of vectors $x \in \mathbb{R}^n$ such that

$$x = \lambda x^1 + (1 - \lambda)x^2, \;\; 0 \leqq \lambda \leqq 1,$$

or, which is the same,

$$x = x^2 + \lambda(x^1 - x^2), \;\; 0 \leqq \lambda \leqq 1.$$

More precisely, we speak in this case of *closed line segment,* denoted also by $\left[x^1, x^2\right]$. The set

$$x = \lambda x^1 + (1 - \lambda)x^2, \;\; 0 < \lambda < 1$$

is called *open line segment* (joining x^1 and x^2), denoted also by (x^1, x^2).

• *Hyperplane* (in $\mathbb{R}^n$) the set H of vectors of $\mathbb{R}^n$ defined as follows:

$$H = \left\{x : p^\top x = \alpha\right\} \tag{3.1}$$

G. Giorgi et al., *Lectures on Mathematics for Economic and Financial Analysis*,
https://doi.org/10.1007/978-3-031-83339-7_3

with p a given column vector of $\mathbb{R}^n$ (nonzero!), x column vector of $\mathbb{R}^n$ and α a given scalar. We say that the hyperplane passes through the origin if $[0] \in H$ and then it will hold $\alpha = 0$. Furthermore, for considerations of geometric type, clear when $n = 2$ or $n = 3$, the vector p (as well as any other vector μp, with $\mu \in \mathbb{R}$, $\mu \neq 0$), is said to be *normal* to the hyperplane $p^\top x = \alpha$ and the number $|\alpha| / \|p\|$ measures the *distance* of the hyperplane from the origin. Finally, note that, with p given, by varying α, relation (3.1) describes infinite parallel hyperplanes.

• Let be given $p \neq [0]$ and $\alpha \in \mathbb{R}$; then the following subsets of $\mathbb{R}^n$:

$$X_1 = \left\{x : p^\top x < \alpha\right\};$$

$$X_2 = \left\{x : p^\top x = \alpha\right\};$$

$$X_3 = \left\{x : p^\top x > \alpha\right\},$$

are a partition of $\mathbb{R}^n$ and hence a point $x \in \mathbb{R}^n$ belongs to one and only one of the said subsets. The set X_2, as already said, characterizes the hyperplane of equation $p^\top x = \alpha$, while X_1 and X_2 are called *open half-spaces* of $\mathbb{R}^n$ (associated with the hyperplane $p^\top x = \alpha$) and everyone of the said half-spaces may be considered as the union of infinitely many hyperplanes parallel to the hyperplane X_2, and hence admitting the representation $p^\top x = \beta$, respectively with $\beta < \alpha$ and $\beta > \alpha$. In several applications it is made reference to *closed half-spaces* of $\mathbb{R}^n$, in the sense that, the following sets are considered:

$$H^+ = \left\{x : p^\top x \geqq \alpha\right\};$$

$$H^- = \left\{x : p^\top x \leqq \alpha\right\},$$

being, obviously, $H^+ \cap H^- = H$, $H^+ \cup H^- = \mathbb{R}^n$. Furthermore, it can be proved that H^+ and H^- are indeed closed sets (a closed half-space is closed not only owing to its name!). Therefore the intersection of a finite number of hyperplanes and/or closed half-spaces is a closed set.

We give now the basic definition of a *convex set.*

Definition 3.1.1 A set $X \subset \mathbb{R}^n$ is said to be *convex* if for every pair of elements x^1 and x^2 of X, the line segment joining x^1 and x^2 belongs entirely to X:

$$x = \lambda x^1 + (1 - \lambda)x^2 \in X, \quad \forall x^1, x^2 \in X, \quad \forall \lambda \in [0, 1]. \tag{3.2}$$

By convention we accept that every singleton is convex and also that the empty set $\emptyset$ is convex. In Fig. 3.1 some examples of convex sets and non convex sets in $\mathbb{R}^2$ are given.

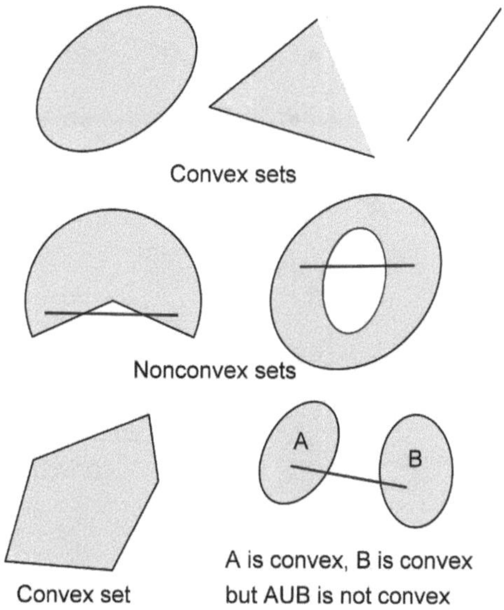

Fig. 3.1 Examples of convex sets and nonconvex sets

The vector x appearing in (3.2) is said to be obtained by *convex combination* of the vectors x^1 and x^2; in fact this operation generates all points of the line segment joining x^1 and x^2. More generally, we say that $x \in \mathbb{R}^n$ is given by the *convex combination* of k vectors of $\mathbb{R}^n$, say $x^1, x^2, \ldots, x^k$, if there exist scalars $\lambda_1, \lambda_2, \ldots, \lambda_k \in \mathbb{R}$, such that

$$x = \sum_{i=1}^{k} \lambda_i x^i, \quad \text{with} \sum_{i=1}^{k} \lambda_i = 1, \quad \lambda_i \geqq 0, \ i = 1, \ldots, k.$$

It is possible to prove that $X \subset \mathbb{R}^n$ is convex if and only if it contains every convex combination of its k ($k \geqq 2$) elements. Definition 3.1.1 perhaps has the advantage of a more direct geometric interpretation.

Example 3.1.2

(a) Every linear subspace of $\mathbb{R}^n$ is a convex set.
(b) *Intervals* are the unique examples of convex sets in $\mathbb{R}$.
(c) The interior and the closure of a convex set are convex sets.
(d) Every hyperplane and every half-space (open or closed) of $\mathbb{R}^n$ are convex sets.
(e) Every neighbourhood (of radius $\delta > 0$) of a point $x^0 \in \mathbb{R}^n$ is a convex set.
(f) The set of solutions of the system $Ax \leqq b$ or $Ax \geqq b$, where A is a matrix of order (m, n), with nonzero rows, $x \in \mathbb{R}^n$ and $b \in \mathbb{R}^m$, is a closed convex set. Indeed, with, for instance, $Ax^1 \leqq b$ and $Ax^2 \leqq b$, we have $A(\lambda x^1 + (1 - \lambda)x^2) = \lambda Ax^1 + (1-\lambda)Ax^2 \leqq \lambda b + (1-\lambda)b = b$. These sets are called *convex polyhedra;* hence a convex polyhedron is a closed convex set. If $b = [0]$, we speak of *convex polyhedral cones.*

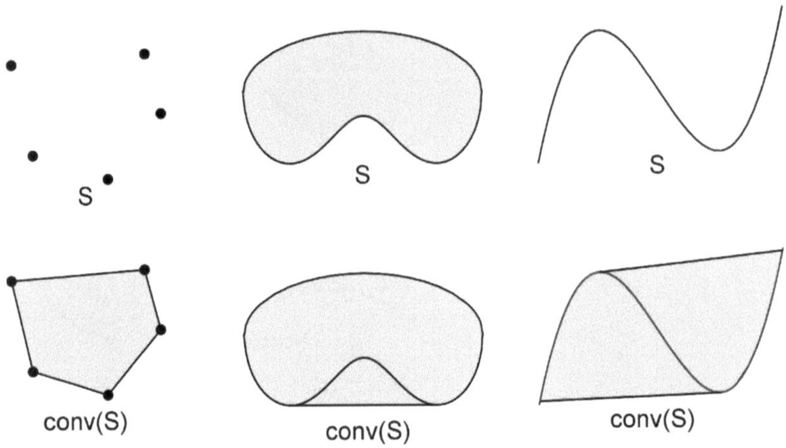

Fig. 3.2 Convex hulls of three sets

(g) The intersection of a finite number of convex sets is a convex set. However, it is not in general a convex set the union of convex sets: consider, e.g., the set $[0, 1] \cup [4, 5]$, which obviously is not a convex set.

Definition 3.1.3 Given a set $S \subset \mathbb{R}^n$, the *convex hull* of S, denoted by $\text{conv}(S)$ or also by $\text{co}(S)$, is the smallest convex set containing S, or, equivalently, the intersection of all convex sets containing S.

We give, in Fig. 3.2, some examples of $\text{conv}(S)$, with $S \subset \mathbb{R}^2$.

Theorem 3.1.4 *The set* $\text{conv}(S)$ *is given by all convex combinations of finitely many vectors of* S*:*

$$\text{conv}(S) = \left\{ \begin{array}{l} x \in \mathbb{R}^n : x = \sum_{i=1}^p \lambda_i x^i, \; \sum_{i=1}^p \lambda_i = 1, \\ \lambda_i \geqq 0, \forall i = 1, \dots, p; \; x^i \in S; \; p \in \mathbb{N}_+. \end{array} \right\}$$

Hence, a point in the convex hull of an arbitrary set $S \subset \mathbb{R}^n$ can be represented as a convex combination of a finite number of points in the set. More precisely, it can be proved (*Theorem of Caratheodory*) that every vector x in the convex hull of a set $S \subset \mathbb{R}^n$ can be represented as a convex combination of. at most, $n + 1$ points of S.

If we have a set S of $\mathbb{R}^n$, formed by a *finite* number of elements (i.e. formed by a finite number of vectors of $\mathbb{R}^n$), then $\text{conv}(S)$ is given by the intersection of a finite number of closed half-spaces and it is in this case a bounded and closed convex polyhedron. Several authors call "*polytope*" a compact convex polyhedron.

We have already encountered in Sect. 2.5 (see Definition 2.5.1), the notion of *cone* (with vertex at the origin), as a set $C \subset \mathbb{R}^n$ such that

$$x \in C \Longrightarrow \lambda x \in C, \;\; \forall \lambda > 0.$$

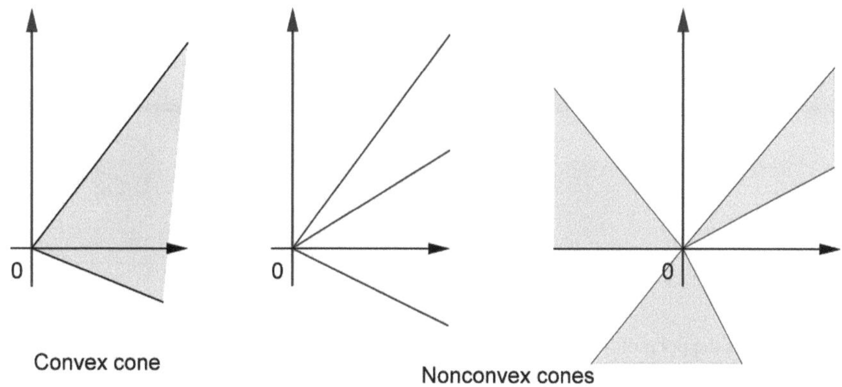

Fig. 3.3 Examples in $\mathbb{R}^2$ of convex and nonconvex cones

Note that the null element $[0]$ does not belong necessarily to the cone C. Some authors require $\lambda \geqq 0$ (obviously in this case $[0] \in C$). A *convex cone* is nothing but a cone which is also convex. It is not difficult to see that a cone $C \subset \mathbb{R}^n$ is convex if and only if, for all $x, y \in C$, it holds $x + y \in C$. Indeed, if C is convex, we have $x + y = 2\left(\frac{1}{2}x + \frac{1}{2}y\right)$, with $\frac{1}{2}x + \frac{1}{2}y \in C$, being C convex; moreover, $2\left(\frac{1}{2}x + \frac{1}{2}y\right) \in C$, being C a cone. Conversely, if the sum of any two elements of the cone C is again an element of C, we have at once $\lambda x + (1 - \lambda)y \in C$, $\forall \lambda \in [0, 1]$, i.e. C is a convex set. See Fig. 3.3.

Example 3.1.5

1. In $\mathbb{R}$ we have that the two half-lines $(-\infty, 0]$ and $[0, +\infty)$ are convex cones.
2. The set of nonnegative vectors of $\mathbb{R}^n$, i.e.

$$\left\{x \in \mathbb{R}^n : x \geqq [0]\right\}$$

   is a convex cone.
3. The sets $\left\{x \in \mathbb{R}^n : Ax \geqq [0]\right\}$; $\left\{x \in \mathbb{R}^n : Ax \leqq [0]\right\}$, where A is a given matrix of order (m, n), are convex cones; they are called, as previously said, *convex polyhedral cones:* these cones are therefore closed and convex. They are also called "finite cones".

The smallest convex cone containing a set $Y \subset \mathbb{R}^n$ is called *convex cone spanned by* Y and is defined in a formal way by the relation

$$C(Y) = \left\{\sum_{i=1}^{k} \lambda_i y^i : \lambda_i \geqq 0,\ y^i \in Y,\ i = 1, \ldots, k,\ k \text{ arbitrary}\right\}.$$

See Fig. 3.4.

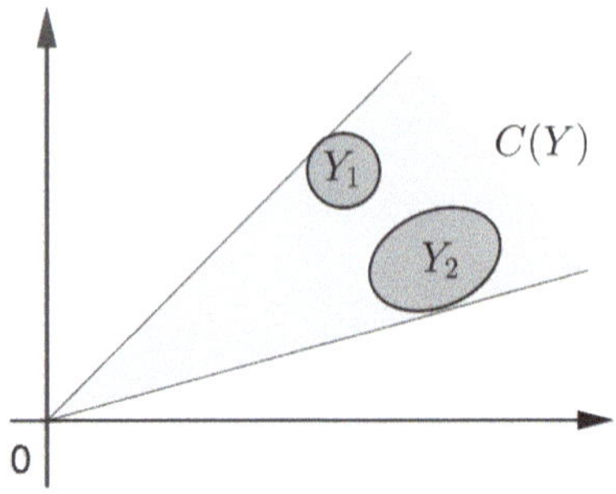

Fig. 3.4 Convex cone spanned by $Y = Y_1 \cup Y_2$

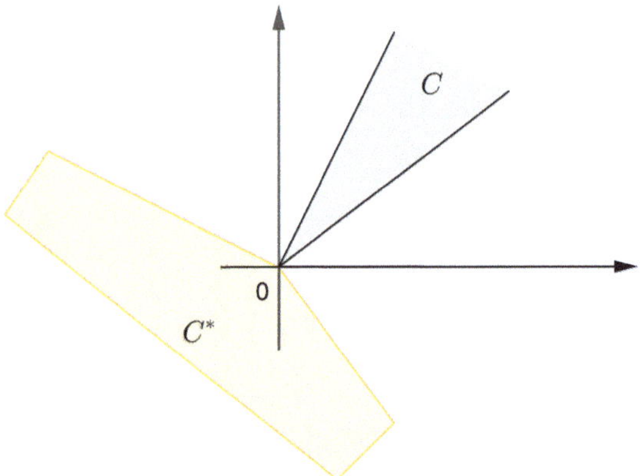

Fig. 3.5 C^* is the polar of the cone C

Definition 3.1.6 Let be given the cone $C \subset \mathbb{R}^n$; the (negative) *polar cone* C^* of the cone C is the set

$$C^* = \left\{ p \in \mathbb{R}^n : p^\top x \leqq 0, \ \ \forall x \in C \right\}.$$

See Fig. 3.5.

The polar cone is a closed and convex cone, with vertex at the origin and with $[0] \in C^*$. Note that these properties hold, independently from the fact that C is closed or not, convex or not. We denote by C^{**} the polar of the polar, called also *bipolar cone* of $C : C^{**} = (C^*)^*$. We have the following non trivial result.

Theorem 3.1.7 (Polarity Theorem) *If $C \subset \mathbb{R}^n$ is a convex cone, then $C = C^{**}$ if and only if C is closed.*

In Example 3.1.5, point 3, we have introduced the notion of polyhedral convex cones, given by the set of solutions $x \in \mathbb{R}^n$ of a system of the type $Bx \leqq [0]$, with B matrix of order (m, n). Note that these cones are closed and convex and hence the Polarity Theorem holds true. We can also equivalently assert that *convex polyhedral cones* (called also *finite cones*) are those convex cones generated by a *finite number* of elements:

$$C(x^1, \ldots, x^m) = \left\{ \sum_{i=1}^{m} \lambda_i x^i : \lambda_i \geqq 0, \ \ i = 1, \ldots, m \right\}.$$

If we consider the vectors x^i which appear in the above expression, as the column vectors of a matrix A of order (n, m), we can rewrite the said expression as

$$C = \left\{ x \in \mathbb{R}^n : A\lambda = x, \ \ \lambda \geqq 0. \right\}$$

In other words, a convex polyhedral cone can be represented in two equivalent ways: as the solutions x of a system $Bx \leqq [0]$, with B matrix of order (m, n), or as all those $x \in \mathbb{R}^n$ such that $\left\{ Ay = x, \ \ y \geqq [0] \right\}$, with A matrix of order (n, m). This is the content of the *Minkowski-Weyl Theorem.*

We conclude this section with the following remarks. It is seen, without particular difficulties, that, given $X \subset \mathbb{R}^n$:

(i) The set X is a *linear space or vector space* if and only if it contains all linear combinations of any two its points x and y : $x, y \in X \Longrightarrow \alpha x + \beta y \in X, \ \forall \alpha, \beta \in \mathbb{R}$.
(ii) The set X is a *convex cone* with vertex at the origin and with $[0] \in X$, if and only if it contains all linear combinations, with nonnegative coefficients, of any two its points x and y : $x, y \in X \Longrightarrow \alpha x + \beta y \in X, \ \forall \alpha, \beta \geqq 0$.
(iii) The set X is a *convex set* if and only if it contains every *convex combination* of any two its points x and y : $x, y \in X \Longrightarrow \alpha x + \beta y \in X, \ \forall \alpha, \beta \geqq 0, \ \alpha + \beta = 1$.

3.2 Convex and Concave Functions

Definition 3.2.1 Let be given $f : X \longrightarrow \mathbb{R}$, with $X \subset \mathbb{R}^n$ convex set; the function f is said to be *convex* on X if

$$f(\lambda x^1 + (1-\lambda)x^2) \leqq \lambda f(x^1) + (1-\lambda) f(x^2), \ \ \forall x^1, x^2 \in X, \ \ \forall \lambda \in [0, 1]. \qquad (3.3)$$

The function f is said to be *concave* on X if (and only if) $-f$ is convex on X, that is, if (3.3) holds with $\geqq$. A function $f : X \longrightarrow \mathbb{R}$ is said to be *strictly convex* on the convex set $X \subset \mathbb{R}^n$ (respectively, *strictly concave* on the convex set $X \subset \mathbb{R}^n$), if (3.3) holds with $<$ (respectively, with $>$), for all $x^1, x^2 \in X$, with $x^1 \neq x^2$, and for all $\lambda \in (0, 1)$.

Remark 3.2.2 The same definitions are valid also for *vector functions* $f : X \longrightarrow \mathbb{R}^m$. Indeed, one realizes at once that a vector function is convex (concave) if and only if all its components $f_1, f_2, , , f_m$, are convex (concave). This fact allows us to consider only real-valued convex (concave) functions.

Remark 3.2.3 Owing to the first member of relation (3.3), a convex (concave) function must be defined on a convex set $X \subset \mathbb{R}^n$. If we have a function $f : \mathbb{R} \longrightarrow \mathbb{R}$, then X will be an *interval.* Furthermore, being f concave if and only if $-f$ is convex, this allows us to "convert" easily to concave functions the properties of convex functions (and vice-versa). Hence not always we shall repeat, with reference to concave functions, what previously asserted for the class of convex functions.

From a geometric point of view, in $\mathbb{R}^2$, the fact that a function is convex (concave) means that the diagram of the function, generated by the points of the line segment joining x^1 and x^2 (that is all points of the type $\lambda x^1 + (1-\lambda)x^2$, $0 \leqq \lambda \leqq 1$) is never above (is never below) the line segment joining the points $(x^1, f(x^1))$, $(x^2, f(x^2))$ (that is the points of the type $(\lambda x_1 + (1-\lambda)x_2, \lambda f(x^1) + (1-\lambda) f(x^2))$, $0 \leqq \lambda \leqq 1$). This fact holds also in $\mathbb{R}^n$, but the geometric interpretation in $\mathbb{R}^2$ should help the reader to grasp the meaning of Definition 3.2.1, without being obliged to retain the same by heart. See Fig. 3.6.

Linear affine functions (that is functions of the type $f(x) = a^\top x + \beta$, with $a \in \mathbb{R}^n$, $\beta \in \mathbb{R}$) are the unique functions to be both convex and concave (but not both strictly convex and strictly concave!). Convex and concave functions are often used in optimization theory and in economic analysis. For instance, the relation which characterizes concave functions

$$f(\lambda x^1 + (1-\lambda)x^2) \geqq \lambda f(x^1) + (1-\lambda) f(x^2), \quad \forall x^1, x^2 \in X, \quad \forall \lambda \in [0, 1]$$

points out some economically acceptable properties, such as:

(i) Diversity is preferable to uniformity: it is better to have a piece of bread and a piece of cheese rather than two pieces of bread or two pieces of cheese.

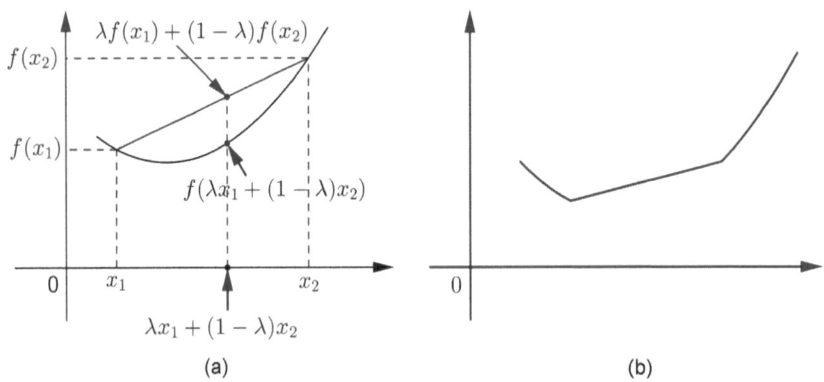

Fig. 3.6 (**a**) Strictly convex function. (**b**) Convex function but not strictly convex

(ii) Certainty is better than uncertainty: it is better to have at once 100 euros than a bet on a winning of either 200 euros or nothing, with probability $\frac{1}{2}$.
(iii) The so-called "law of decreasing returns": the passing from 10 euros to 20 euros is "greater" than the passing from 20 euros to 30 euros.

Examples of convex and concave functions are:

(1) The Euclidean norm in $\mathbb{R}^n$, i.e. $\|x\|$, is a convex function; indeed, we have, from the basic properties of the Euclidean norm,

$$\left\|\lambda x^1 + (1-\lambda)x^2\right\| \leqq \lambda \left\|x^1\right\| + (1-\lambda)\left\|x^2\right\|, \quad \forall \lambda \in [0,1], \quad \forall x^1, x^2 \in \mathbb{R}^n.$$

(2) Any constant function is both convex and concave, but not strictly convex, nor strictly concave.
(3) We have already observed that any linear function $f(x) = a^\top x$, $a \in \mathbb{R}^n$, is both convex and concave. Indeed, we have at once

$$a^\top(\lambda x^1 + (1-\lambda)x^2) = \lambda a^\top x^1 + (1-\lambda)a^\top x^2, \quad \forall x^1, x^2 \in \mathbb{R}^n, \ \forall \lambda \in [0,1].$$

The same property holds for affine functions, i.e. for $f(x) = a^\top x + \alpha$, $\alpha \in \mathbb{R}$. It is possible to prove that this last class of functions is the unique to be both convex and concave (not in the strict sense!).
(4) A quadratic form $Q(x) = x^\top Ax$, A symmetric matrix of order n, is convex if and only if A is positive semidefinite; it is concave if and only if A is negative semidefinite. It is strictly convex (resp. strictly concave) if and only if A is positive definite (resp. negative definite). See Example 3.2.19.

It is possible to prove the following result, known as "Jensen's inequality":

- Let be given $f : X \subset \mathbb{R}^n \longrightarrow \mathbb{R}$, with X a convex set; then f is convex on X if and only if

$$f(\lambda_1 x^1 + \lambda_2 x^2 + \cdots + \lambda_k x^k) \leqq \lambda_1 f(x^1) + \lambda_2 f(x^2) + \cdots + \lambda_k f(x^k),$$

$$\forall x^1, x^2, \ldots, x^k \in X, \ \forall \lambda_i \in [0,1], \ \sum_{i=1}^{k} \lambda_1 = 1, \ k \geqq 2.$$

It is possible to prove (the proof will be given in the next pages) that the Cobb-Douglas production function

$$f(x_1, \ldots, x_n) = \alpha_0 \prod_{i=1}^{n} (x_i)^{\alpha_i},$$

with $x_i > 0$ and $\alpha_0, \alpha_i > 0$ for every $i = 1, \ldots, n$, is concave if and only if $\sum_{i=1}^n \alpha_i \leqq 1$.

The notion of convexity (concavity) of a function of the type $f : \mathbb{R}^n \longrightarrow \mathbb{R}$ is a truly uni-dimensional concept, as it is concerned with its behaviour on every line segment $\lambda x + (1-\lambda)y,\ 0 \leqq \lambda \leqq 1$. We have indeed the following result.

Theorem 3.2.4 *The function $f : X \longrightarrow \mathbb{R}$ is convex (resp. concave) on the convex set $X \subset \mathbb{R}^n$ if and only if all functions (of one variable) of the type*

$$\varphi_{x,h}(t) = f(x+th),\ \text{with } t \in [0,1],\ x \in X \text{ and } x+th \in X,$$

are convex (resp. concave).

Proof

(i) If f is convex, it holds

$$\begin{aligned}\varphi_{x,h}(\lambda t + (1-\lambda)\tau) &= f(x + (\lambda t + (1-\lambda)\tau)h)\\ &= f(\lambda(x+th) + (1-\lambda)(x+\tau h))\\ &\leqq \lambda f(x+th) + (1-\lambda) f(x+\tau h)\\ &= \lambda \varphi_{x,h}(t) + (1-\lambda)\varphi_{x,h}(\tau).\end{aligned}$$

(ii) If all functions φ are convex, it results

$$\begin{aligned}f(\lambda x + (1-\lambda)y) &= f(x + (1-\lambda)(y-x)) = \varphi_{x,y-x}(1-\lambda)\\ &= \varphi_{x,y-x}(\lambda \cdot 0 + (1-\lambda)\cdot 1)\\ &\leqq \lambda\varphi_{x,y-x}(0) + (1-\lambda)\varphi_{x,y-x}(1)\\ &= \lambda f(x) + (1-\lambda) f(y).\end{aligned}$$

□

It is therefore useful to recall the basic properties of convex/concave functions of one real variable, of the type $\varphi : \mathbb{R} \longrightarrow \mathbb{R}$. We have the following classical results.

Theorem 3.2.5 *Let be given the function $\varphi : I \subset \mathbb{R} \longrightarrow \mathbb{R}$, where I is an interval.*

(i) If φ is differentiable on I, then φ is convex (concave) on I if and only if $\varphi'(x)$ is increasing (decreasing) on I.
(ii) If φ admits on I a derivative of the second-order, then it is convex (concave) on I if and only if $\varphi''(x) \geqq 0$ ($\varphi''(x) \leqq 0$), for all $x \in I$.
(iii) If φ is differentiable on I, then φ is strictly convex (strictly concave) on I if $\varphi'(x)$ is strictly increasing (strictly decreasing) on I. Note that here we have only sufficient conditions.
(iv) If φ admits on I a derivative of the second-order, then φ is strictly convex (strictly concave) on I if $\varphi''(x) > 0$ ($\varphi''(x) < 0$), for all $x \in I$. Again note that here we have only sufficient conditions. Indeed, consider, e.g., $f(x) = x^4,\ x \in \mathbb{R}$, which is strictly convex on $I = (-\infty, +\infty)$; however, being $f''(x) = 12x^2$, we have $f''(0) = 0$.

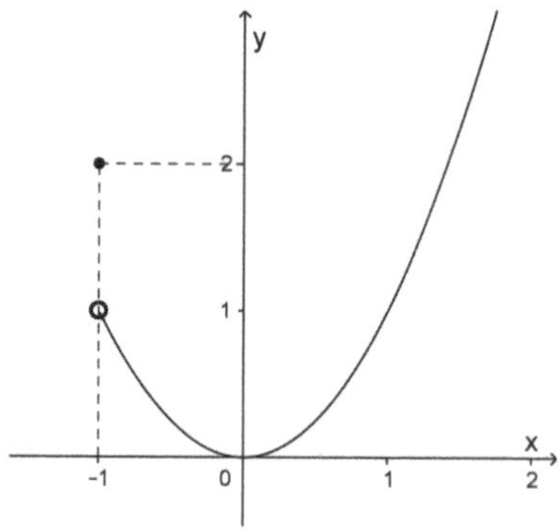

Fig. 3.7 Graph of a convex function $f(x)$, but discontinuous at $x = -1$

A convex (concave) function over a convex set $X \subset \mathbb{R}^n$ is not necessarily continuous on X. For instance, consider the function $f : \mathbb{R} \longrightarrow \mathbb{R}$ defined as follows

$$f(x) = \begin{cases} 2, & \text{for } x = -1, \\ x^2, & \text{for } x > -1 \end{cases}$$

which is obviously convex on the interval $I = \{x \in \mathbb{R} : x \geqq -1\}$, but it is not continuous on I. See Fig. 3.7. Note that the unique point where $f(x)$ is not continuous is the boundary point $x = -1$.

Theorem 3.2.6 *Let $f : X \longrightarrow \mathbb{R}$ be convex (concave) on the convex set $X \subset \mathbb{R}^n$. Let be* $\text{int}(X) \neq \emptyset$. *Then f is continuous on* $\text{int}(X)$; *hence if X is open, then f is continuous on X.*

The proof of the above theorem is a bit laborious and will be omitted. For a proof, see, e.g., [21]. However, the reader is invited to think to the case of functions of one variable: in this case the fact that if the function has a point of non continuity at the interior of its domain, necessarily some line segment intersects the diagram and so the function cannot be convex nor concave. If the points of non continuity are boundary points, this fact does not necessarily happen. See Fig. 3.8.

We have remarked that a convex (concave) function has to be defined on a convex set $X \subset \mathbb{R}^n$. However, there exists a stricter link between convex functions and convex sets. We introduce the following definition.

Definition 3.2.7 Let be given $f : X \longrightarrow \mathbb{R}$, with $X \subset \mathbb{R}^n$. The *epigraph* or *epigraphical set* of f is the set of points of $\mathbb{R}^{n+1}$ which lie "above" the graph of f, that is thc sct

$$\text{epi}\, f = \left\{(x, \alpha) \in X \times \mathbb{R} \subset \mathbb{R}^{n+1} : \alpha \geqq f(x)\right\}.$$

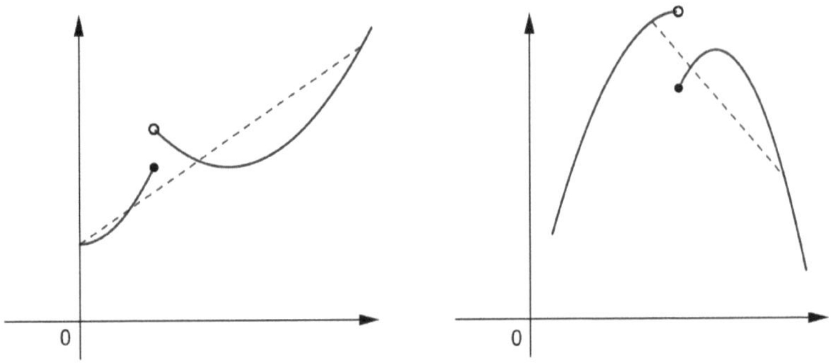

Fig. 3.8 Functions with a point of non continuity at the interior of its domain are neither convex nor concave

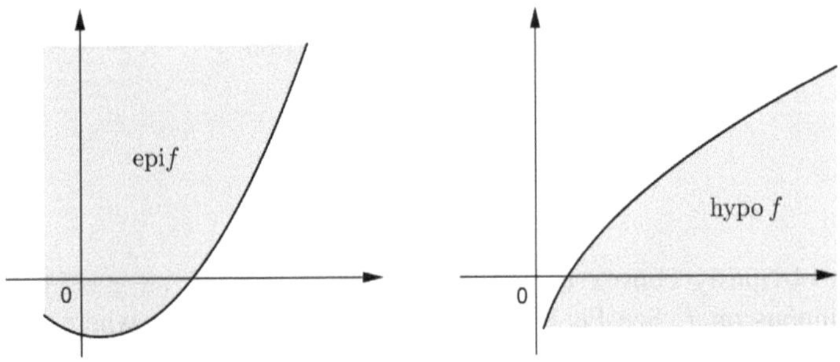

Fig. 3.9 Epigraph and hypograph of a function

The *hypograph* or *hypographical set* of f is the set of points of $\mathbb{R}^{n+1}$ which lie "below" the graph of f, that is the set

$$\text{hypo}\, f = \left\{(x, \alpha) \in X \times \mathbb{R} \subset \mathbb{R}^{n+1} : \alpha \leqq f(x)\right\}.$$

See Fig. 3.9.

We have the following basic result.

Theorem 3.2.8 *Let be given* $f : X \longrightarrow \mathbb{R}$, *with* $X \subset \mathbb{R}^n$ *a convex set. Then* f *is convex over* X *(concave over* X*) if and only if* epi f *(*hypo f*) is a convex set.*

Proof Let us consider any two elements (x^1, y_1) and (x^2, y_2) of the epigraph of f and consider the points $(\lambda x^1 + (1-\lambda)x^2,\ \lambda y_1 + (1-\lambda)y_2)$, $\forall \lambda \in [0, 1]$. Being f convex, it follows

$$f(\lambda x^1 + (1-\lambda)x^2) \leqq \lambda f(x^1) + (1-\lambda) f(x^2) \leqq \lambda y_1 + (1-\lambda)y_2.$$

Hence, the points $(\lambda x^1 + (1-\lambda)x^2,\ \lambda y_1 + (1-\lambda)y_2) \in epi\ f$, $\forall \lambda \in [0, 1]$. Hence, epi f is convex.

Vice-versa. if epi f is convex, from

$$\lambda x^1 + (1-\lambda)x^2, \lambda y_1 + (1-\lambda)y_2 \in \text{epi}\, f$$

it follows

$$f(\lambda x^1 + (1-\lambda)x^2) \leqq \lambda y_1 + (1-\lambda)y_2$$

for every y_1 and y_2 such that $y_1 \geqq f(x^1)$ and $y_2 \geqq f(x^2)$. By putting $y_1 = f(x^1)$ and $y_2 = f(x^2)$,we obtain that $f(x)$ is convex. The proof for the concave case is similar. □

Now we define another set related to real functions of several real variables.

Definition 3.2.9 Let be given $f : X \longrightarrow \mathbb{R}$, with $X \subset \mathbb{R}^n$. The *lower level set* of level α of the function f is the set of points of X such that the value of f is less or equal than α:

$$\text{lev}_{\leqq\alpha} f = \left\{x \in X : f(x) \leqq \alpha\right\}.$$

The *upper level set* of level α of the function f is the set of points of X such that the value of f is greater or equal than α:

$$\text{lev}_{\geqq\alpha} f = \left\{x \in X : f(x) \geqq \alpha\right\}.$$

See Fig. 3.10, where $\text{lev}_{\geqq\alpha} f$ is given by the union of two intervals and it is not therefore a convex set.

Theorem 3.2.10 *Let $f : X \longrightarrow \mathbb{R}$ be convex (concave) on the convex set $X \subset \mathbb{R}^n$. Then the set $\text{lev}_{\leqq\alpha} f$ ($\text{lev}_{\geqq\alpha} f$) is a convex set, for any $\alpha \in \mathbb{R}$.*

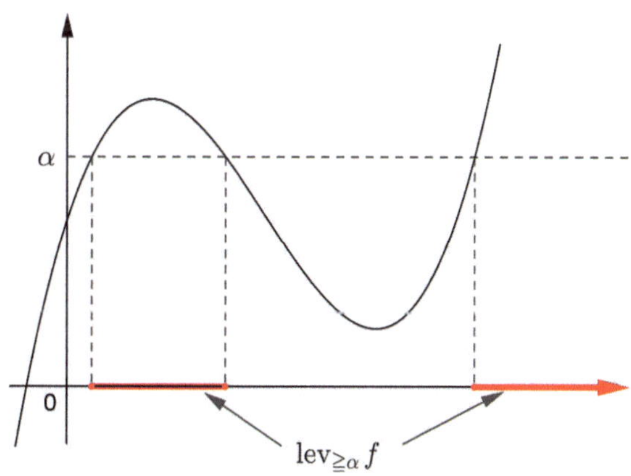

Fig. 3.10 The set $\text{lev}_{\geqq\alpha} f$ is drawn in red

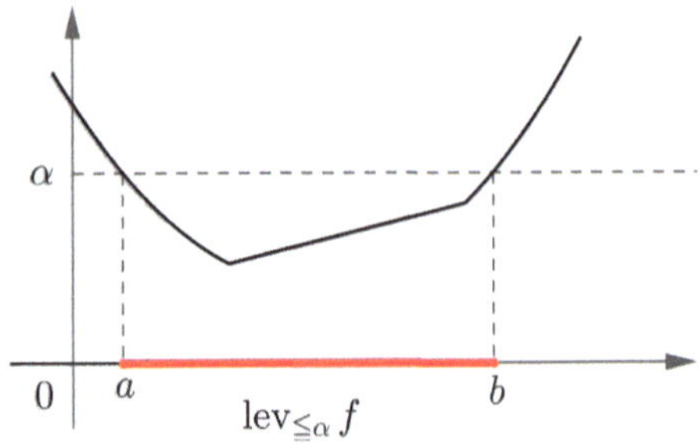

Fig. 3.11 The level set $\mathrm{lev}_{\leqq\alpha} f = [a, b]$ is a convex set

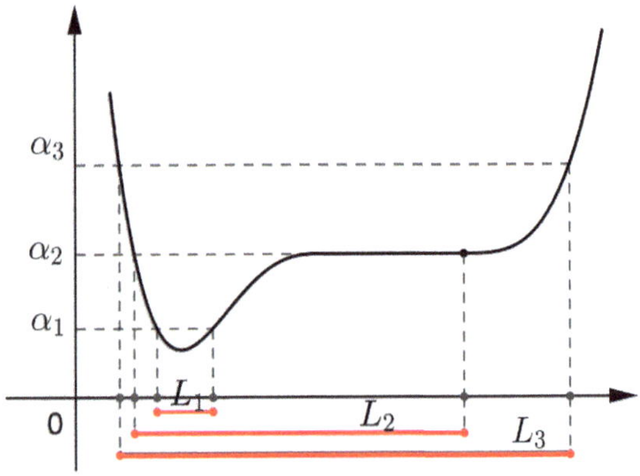

Fig. 3.12 Several level sets for different values of α, all of them are convex. However, f is not convex

Proof The set $\mathrm{lev}_{\leqq\alpha} f$ is surely convex if it consists of a unique point (i.e. if it is a singleton) or if it is the empty set (convex by definition). Let us now suppose that $\mathrm{lev}_{\leqq\alpha} f$ contains at least two points x^1 and x^2, i.e. it holds $f(x^1) \leqq \alpha,\ f(x^2) \leqq \alpha$. We must prove that also $\lambda x^1 + (1-\lambda)x^2 \in \mathrm{lev}_{\leqq\alpha} f,\ \forall\lambda \in [0, 1]$. The thesis follows at once, being

$$f(\lambda x^1 + (1-\lambda)x^2) \leqq \lambda f(x^1) + (1-\lambda) f(x^2) \leqq \lambda\alpha + (1-\lambda)\alpha = \alpha.$$

□

Theorem 3.2.10 is illustrated in Fig. 3.11.

We point out that the previous theorem gives only *necessary conditions* for f to be convex (concave); in other words, the vice-versa does not hold: if, for instance, the sets $\mathrm{lev}_{\leqq\alpha} f(x)$ are convex for all $\alpha \in \mathbb{R}$, not necessarily f is convex. See Fig. 3.12.

For instance, the function $f(x_1, x_2) = e^{-(x_1)^2-(x_2)^2}$ has, as upper level sets, the following ones:

$$\mathrm{lev}_{\geqq\alpha} f = \left\{(x_1, x_2) : (x_1)^2 + (x_2)^2 \leqq -\log\alpha\right\},$$

which are nonempty convex sets for $0 < \alpha \leqq 1$, but $f(x_1, x_2)$ is not concave on $\mathbb{R}^2$.

In spite of various available criteria to decide on the convexity (concavity) of a real function of several variables, often we have to meet long and tedious computations. Sometimes, the difficulties can be overcome thanks to the following results.

Theorem 3.2.11

(1) If $f_1(x), f_2(x), \ldots, f_k(x)$ are convex (concave) functions on the convex set $X \subset \mathbb{R}^n$, then

$$g(x) = \sum_{i=1}^{k} f_i(x)$$

is convex (concave) over X. If there exists an index i such that $f_i(x)$ is strictly convex (strictly concave), then $g(x)$ is strictly convex (strictly concave).

(2) Under the same assumptions of the point 1) also the linear combination

$$g(x) = \sum_{i=1}^{k} \lambda_i f_i(x)$$

with $\lambda_i > 0$, $i = 1, \ldots, k$, is a convex function (or a strictly convex function, if there exists an index i such that $f_i(x)$ is strictly convex). Similarly for the concave case.

(3) If $f(x)$ is convex (strictly convex) on the convex set $X \subset \mathbb{R}^n$ and $g : \mathbb{R} \longrightarrow \mathbb{R}$ is convex and increasing (strictly increasing), then $g[f(x)]$ is convex (strictly convex). Similarly for the concave case, mutatis mutandis.

For instance, the function $f(x_1, x_2, x_3) = e^{(x_1)^2+(x_2)^2+(x_3)^2}$ is strictly convex, as $\varphi(x) = (x_1)^2 + (x_2)^2 + (x_3)^2$ is strictly convex and $z = e^t$ is strictly convex and strictly increasing. For instance, the function $f(x_1, x_2, x_3) = \log(3x_1 + 2x_2 - 6x_3)$ is strictly concave on its domain, being given by the composition of the logarithmic function $g = \log y$ (strictly concave and strictly increasing) and of the linear function $z(x_1, x_2, x_3) = 3x_1 + 2x_2 - 6x_3$. However, if we consider, e.g., the function $f(x) = e^{-x^2}$, $x \in \mathbb{R}$, on the grounds of the previous properties we are not able to conclude about its convexity or concavity, as $y = x^2$ is a strictly convex function, but $z = e^{-t}$ is a strictly decreasing convex function. Indeed, we shall see in the next pages that $f(x) = e^{-x^2}$, $x \in \mathbb{R}$, is a "pseudoconcave function". We point out that the convexity of g is essential in the above property 3). For example, if $f(x) = x$, $x \in \mathbb{R}$, both convex and concave, and $g(x) = x^3$, $x \in \mathbb{R}$, which is strictly increasing but *not* convex, we have $g(f(x)) = x^3$, *not* convex.

Moreover, if $f(x_1, x_2, \ldots, x_n) = g(x_1, x_2 \ldots, x_p)$, with $p < n$, and g is convex (resp. concave) on $\mathbb{R}^p$, also f is convex (resp. concave) on $\mathbb{R}^n$. This result is useful to verify the convexity (concavity) of functions obtained by summing convex or concave functions which depend from a lesser number of variables. In particular, if f_i, $i = 1, \ldots, n$, are n functions of one real variable, which are convex (concave) on an interval $[a_i, b_i]$ and α_i, $i = 1, \ldots, n$, are nonnegative scalars, then the function

$$f(x_1, \ldots, x_n) = \sum_{i=1}^{n} \alpha_i f_i(x_i)$$

is convex (concave) on the convex set $X = [a_1, b_1] \times [a_2, b_2] \times \cdots \times [a_n, b_n]$.

For example, if

$$f(x_1, x_2, x_3) = x_1 \log x_1 + x_2 \log x_2 + x_3 \log x_3,$$

we have $f(x_1, x_2, x_3) = h(x_1)+h(x_2)+h(x_3)$, with $h(x) = x \log x$. Being $h''(x) = \frac{1}{x} > 0$, $\forall x > 0$, we have that $h(x)$ is strictly convex on its domain. Hence also $f(x_1, x_2, x_3)$ is strictly convex on $\text{int}(\mathbb{R}^3_+)$.

We have seen, in Chap. 2, Sect. 2.4, the concept of *directional derivative* in the direction v, of right-hand sided directional derivative, $D^+ f(x, v)$, of left-hand sided directional derivative, $D^- f(x, v)$. It is possible to prove the following results.

Theorem 3.2.12 *Let $f : X \subset \mathbb{R}^n \longrightarrow \mathbb{R}$ be convex (concave) on the open convex set X. Then:*

(i) f is continuous on X (this is a result already given: see Theorem 3.2.6).
(ii) f admits at every point $x \in X$ right-hand sided directional derivative $D^+ f(x, v)$ and left-hand sided directional derivative $D^- f(x, v)$, in every direction v. Moreover, $D^+ f(x, v)$ and $D^- f(x, v)$ are positively homogeneous convex functions of v and it holds

$$D^+ f(x, v) \geqq D^- f(x, v).$$

(iii) At all points $x \in X$ where f admits all n partial derivatives, then f is of class C^1 (and hence differentiable).
(iv) f is differentiable on $X \setminus B$, where B is a subset of X, with $\text{int}(B) = \emptyset$.

With reference to the previous point (iv) of Theorem 3.2.12, let us consider, for example, the function $f : \mathbb{R} \longrightarrow \mathbb{R}$, defined by $f(x) = |x|$. This function is convex on $\mathbb{R}$ (but not strictly convex!), continuous on $\mathbb{R}$ and differentiable on $\mathbb{R} \setminus \{0\}$. More generally, (see [12]): if $f(x)$ is convex on an open convex set $X \subset \mathbb{R}^n$, it is differentiable with continuous partial derivatives everywhere in X except for a set of measure zero in the Lebesgue sense. The notion of Lebesgue measure is beyond the aims of the present book. The reader is referred to any good book on advanced mathematical analysis, such as, for example, [1, 4, 26].

Now we see how differentiable convex (concave) functions can be characterized.

Theorem 3.2.13 *Let be given $f : X \longrightarrow \mathbb{R}$, where $X \subset \mathbb{R}^n$ is open and convex, and let f be differentiable on X. Then f is convex on X if and only if*

$$f(x) - f(y) \geqq \nabla f(y)^\top (x - y), \quad \forall x, y \in X. \tag{3.4}$$

(f is concave on X if and only if relation (3.4) *holds with the inequality $\leqq$).*

Proof

(i) Let f be convex:

$$f(\lambda x + (1-\lambda)y) \leqq \lambda f(x) + (1-\lambda) f(y), \ \forall x, y \in X, \ \forall \lambda \in [0, 1].$$

This relation can be rewritten in the form

$$f(y + \lambda(x-y)) \leqq \lambda f(x) + (1-\lambda) f(y), \ \forall x, y \in X, \ \forall \lambda \in [0, 1],$$

that is

$$f(y + \lambda(x-y)) - f(y) \leqq \lambda [f(x) - f(y)], \ \forall x, y \in X, \ \forall \lambda \in [0, 1].$$

By dividing by λ, $0 < \lambda \leqq 1$, and letting λ go to zero, we obtain (the first member converges to the directional derivative of f in the direction $x - y$)

$$\nabla f(y)^\top (x - y) \leqq f(x) - f(y), \ \forall x, y \in X.$$

(ii) Let us suppose that the inequality $\nabla f(y)^\top (x-y) \leqq f(x) - f(y)$ holds for all $x, y \in X$ and consider the point $\lambda x + (1-\lambda)y$, $0 \leqq \lambda \leqq 1$. Let us rewrite this inequality, with reference to the points x and $\lambda x + (1-\lambda)y$ and with reference to the points y and $\lambda x + (1-\lambda)y$:

(1) $\nabla f(\lambda x + (1-\lambda)y)^\top (1-\lambda)(x-y) \leqq f(x) - f(\lambda x + (1-\lambda)y)$.
(2) $\nabla f(\lambda x + (1-\lambda)y)^\top \lambda(y-x) \leqq f(y) - f(\lambda x + (1-\lambda)y)$.

By multiplying the first inequality by λ, with $\lambda \in (0, 1)$, the second inequality by $(1-\lambda)$, and then by adding, we obtain

$$0 \leqq \lambda f(x) + (1-\lambda) f(y) - f(\lambda x + (1-\lambda)y),$$

that is, being x and y arbitrary, the convexity of f. □

Theorem 3.2.13 is not to be studied by heart (this holds for several concepts of Mathematics!), but it must be understood in its geometric interpretation. We begin to recall the case of a differentiable real function of one real variable (differentiable on an interval $I \subset \mathbb{R}$). Such a function admits a tangent line t at every point of its diagram and everybody is ready to get that f is convex if and only if at every point of the diagram the value of the tangent line never lies above the corresponding

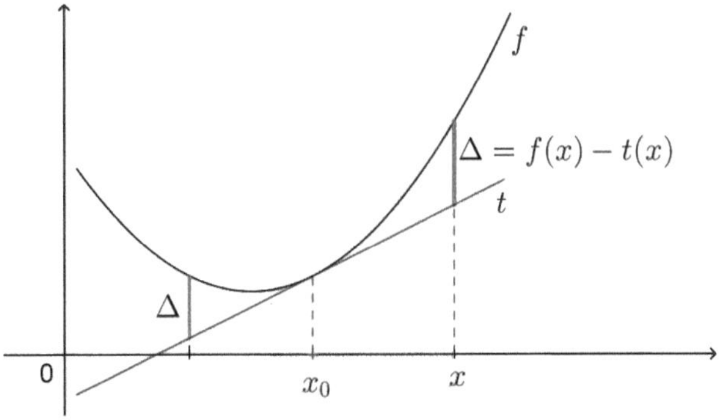

Fig. 3.13 Graph of $y = f(x)$ and its tangent line at x_0. $\Delta = f(x) - t(x) = f(x) - [f(x_0) + f'(x_0)(x - x_0)]$

value of the function (obviously the argument is reversed for differentiable concave functions of one real variable). See Fig. 3.13.

Hence, $f : \mathbb{R} \longrightarrow \mathbb{R}$ is convex if and only if for each x_0, x, it holds $\Delta \geqq 0$, where Δ denotes the difference between $f(x)$ and the corresponding value of the tangent line t:

$$\Delta = f(x) - y \Longrightarrow \Delta = f(x) - \left[f(x_0) + f'(x_0)(x - x_0)\right].$$

Hence f is convex if and only if $\Delta \geqq 0, \ \forall x, \ x_0$, that is

$$f(x) - f(x_0) \geqq f'(x_0)(x - x_0), \quad \forall x, \ x_0.$$

For differentiable real functions of several variables, we make reference to the notion of *tangent hyperplane*. We recall that the equation of the tangent hyperplane to the point $(x^0, f(x^0))$, being $f : \mathbb{R}^n \longrightarrow \mathbb{R}$ differentiable, it holds

$$z - f(x^0) = \nabla f(x^0)^\top (x - x^0).$$

We have that f is convex on the open convex set $X \subset \mathbb{R}^n$ if and only if $\Delta \geqq 0$, for every $x, x^0 \in X$, that is

$$f(x) - f(x^0) \geqq \nabla f(x^0)^\top (x - x^0), \quad \forall x, x^0 \in X,$$

that is relation (3.4). Hence, for a differentiable convex function f on the convex set $X \subset \mathbb{R}^n$, the linearization $f(x^0) + \nabla f(x^0)^\top (x - x^0)$ never overestimates $f(x)$ for every $x, x^0 \in X$. For a differentiable concave function the converse interpretation holds true.

It is then easy to prove that $f : X \longrightarrow \mathbb{R}$, differentiable on the open convex set $X \subset \mathbb{R}^n$, is *strictly convex* on X if and only if

$$f(x) - f(x^0) > \nabla f(x^0)^\top (x - x^0), \quad \forall x, x^0 \in X, \ x \neq x^0.$$

The function f is strictly concave on X if the previous inequality holds in the opposite sense, i.e. with $<$.

Another characterization of convex (or concave) differentiable functions is the following one, which puts into evidence the "monotonicity" of the gradient.

Theorem 3.2.14 *Let be given $f : X \longrightarrow \mathbb{R}$, where $X \subset \mathbb{R}^n$ is open and convex, and let f be differentiable on X. Then f is convex on X if and only if*

$$[\nabla f(x) - \nabla f(y)]^\top (x - y) \geqq 0, \ \forall x, y \in X.$$

The function f is concave on X if and only if

$$[\nabla f(x) - \nabla f(y)]^\top (x - y) \leqq 0, \ \forall x, y \in X.$$

Proof We have already seen that if f is convex, we have

$$\nabla f(y)^\top (x - y) \leqq f(x) - f(y), \quad \forall x, y \in X.$$

By exchanging x with y (and vice-versa), we have

$$\nabla f(x)^\top (y - x) \leqq f(y) - f(x), \quad \forall x, y \in X.$$

Now we add member to member:

$$[\nabla f(x) - \nabla f(y)]^\top (x - y) \geqq 0, \ \forall x, y \in X,$$

i.e. the first part of the theorem is proved.

For what concerns the sufficiency, let be $x, y \in X$. Then, for $0 \leqq \lambda \leqq 1$ we have $\lambda x + (1 - \lambda)y \in X$. Now, by the Mean Value Theorem, we have for some $\bar{\lambda}, 0 < \bar{\lambda} < 1$,

$$f(x) - f(y) = \nabla f\left[y + \bar{\lambda}(x - y)\right]^\top (x - y).$$

But, by assumption

$$\left[\nabla f\left[y + \bar{\lambda}(x - y)\right] - \nabla f(y)\right]^\top \bar{\lambda}(x - y) \geqq 0$$

or

$$\nabla f\left[y + \bar{\lambda}(x - y)\right]^\top (x - y) \geqq \nabla f(y)^\top (x - y).$$

Hence $f(x) - f(y) \geqq \nabla f(y)^\top (x - y)$. □

Now we see the characterization of convex (concave) functions of class $\mathcal{C}^2$. This characterization is rather important, also for "operative" reasons.

Theorem 3.2.15 *Let be given $f : X \longrightarrow \mathbb{R}$, of class $\mathcal{C}^2$ on the open convex set $X \subset \mathbb{R}^n$. Then f is convex on X if and only if*

$$y^\top Hf(x)y \geqq 0, \quad \forall y \in \mathbb{R}^n, \ \forall x \in X,$$

that is if and only if, for every $x \in X$, the Hessian matrix $Hf(x)$ is positive semidefinite on X. The function f is concave on X if and only if $Hf(x)$ is negative semidefinite on X.

Proof

(i) Suppose that f is convex on X and let $x \in X$. Then, by Taylor's formula we have, for any versor $y \in \mathbb{R}^n$ and for $\lambda \in \mathbb{R}$ such that $x + \lambda y \in X$,

$$f(x + \lambda y) = f(x) + \lambda(\nabla f(x)^\top y) + \frac{1}{2}\lambda^2 y^\top Hf(x)y + o(\lambda^2).$$

Being f convex, by Theorem 3.2.13 we have

$$f(x + \lambda y) - f(x) - \lambda(\nabla f(x)^\top y) \geqq 0.$$

This clearly implies

$$\frac{1}{2}\lambda^2 y^\top Hf(x)y + o(\lambda^2) \geqq 0.$$

Dividing by λ^2 and taking $\lambda \longrightarrow 0$ we have that

$$y^\top Hf(x)y \geqq 0.$$

Since y is arbitrary this shows that $Hf(x)$ is positive semidefinite on X.

(ii) Conversely, assume that $Hf(x)$ is positive semidefinite on X. Then, for any $x, y \in X$ we have the Taylor's expansion (with remainder in the form of Lagrange):

$$f(y) = f(x) + \nabla f(x)^\top (y - x) + \frac{1}{2}(y - x)^\top Hf(z)(y - x),$$

where $z \in (x, y)$. Since $Hf(z)$ is positive semidefinite, it is clear from the above equation that

$$f(y) - f(x) - \nabla f(x)^\top (y - x) \geqq 0.$$

This shows that f is convex on X by Theorem 3.2.13.

□

Similarly, it can be proved the following result.

Theorem 3.2.16 *Let be given* $f : X \longrightarrow \mathbb{R}$*, of class* $\mathcal{C}^2$ *on the open convex set* $X \subset \mathbb{R}^n$*. If* $Hf(x)$ *is positive definite (negative definite) on* X*, then* f *is strictly convex (strictly concave) on* X*.*

Remark 3.2.17 Note that, in opposition to Theorems 3.2.15 and Theorem 3.2.16 gives only *sufficient* conditions. Indeed, the vice-versa does not hold: for instance, $f : \mathbb{R} \longrightarrow \mathbb{R}$ given by $f(x) = x^4$ is strictly convex on $\mathbb{R}$ but $f''(0) = 0$. The same holds for $f(x, y) = x^4 + y^4$, strictly convex on $\mathbb{R}^2$, but for which $Hf(0, 0) = [0]$. It can be proved that the subclass of strictly convex functions (of class $\mathcal{C}^2$) *characterized* by a positive definite Hessian matrix is the class of *strongly convex functions,* i.e. those functions $f : X \longrightarrow \mathbb{R}$, differentiable on the open convex set $X \subset \mathbb{R}^n$ and such that there exists a real number $\rho > 0$, for which

$$f(y) - f(x) \geqq \nabla f(x)^\top (y - x) + \rho \|y - x\|^2, \ \forall x, y \in X.$$

Summing up:

$$Hf(x) \text{ positive semidefinite on } X \iff f(x) \text{ convex on } X.$$

$$Hf(x) \text{ positive definite on } X \implies f(x) \text{ strictly convex on } X.$$

An exception is the case of quadratic forms and quadratic functions; see Example 3.2.19.

Now we present some examples, in order to show the practical utility of the last results.

Example 3.2.18

(a) Find the region of the plane xOy where the following function

$$f(x, y) = x^3 + y^3 - 3xy$$

is convex.

We have

$$\frac{\partial f}{\partial x} = 3x^2 - 3y; \ \frac{\partial f}{\partial y} = 3y^2 - 3x; \ \frac{\partial^2 f}{\partial x^2} = 6x; \ \frac{\partial^2 f}{\partial y^2} = 6y; \ \frac{\partial^2 f}{\partial x \partial y} = \frac{\partial^2 f}{\partial y \partial x} = -3.$$

$$Hf(x, y) = \begin{bmatrix} 6x & -3 \\ -3 & 6y \end{bmatrix}.$$

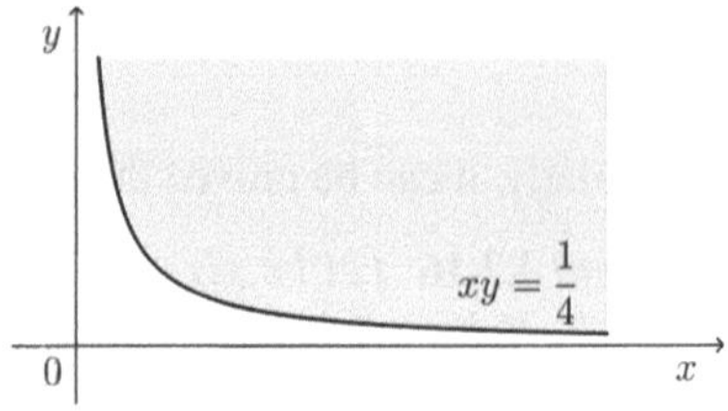

Fig. 3.14 Region $xy \geqq \frac{1}{4}$

Hence f is convex if and only if $x \geqq 0$, $y \geqq 0$, and $6x6y - 9 \geqq 0$. From the last inequality we have $9(4xy - 1) \geqq 0$, that is $xy \geqq \frac{1}{4}$. Therefore the function is convex on the region shaded in Fig. 3.14.

(b) Consider the following function

$$f(x_1, x_2, x_3) = 2(x_1)^2 - 2x_1x_2 + 2(x_2)^2 + 4(x_3)^2.$$

The Hessian matrix of f is

$$Hf(x) = \begin{bmatrix} 4 & -2 & 0 \\ -2 & 4 & 0 \\ 0 & 0 & 8 \end{bmatrix}.$$

We have $\Delta_1 = 4 > 0$; $\Delta_2 = 12 > 0$; $\Delta_3 = 96 > 0$. The associated quadratic form is therefore positive definite on the whole $\mathbb{R}^3$ and hence f is strictly convex on $\mathbb{R}^3$.

(c) Consider the following function

$$f(x_1, x_2, x_3) = (x_1)^2 + \frac{1}{3}(x_2)^3 + \frac{1}{2}(x_3)^2 - x_1x_2 - x_1x_3.$$

The Hessian matrix of f is

$$Hf(x) = \begin{bmatrix} 2 & -1 & -1 \\ -1 & 2x_2 & 0 \\ -1 & 0 & 1 \end{bmatrix}.$$

We have $\Delta_1 = 2 > 0$, $\Delta_2 = 4x_2 - 1$, $\Delta_3 = 2x_2 - 1$. The associated quadratic form is positive definite for $x_2 > \frac{1}{2}$, indefinite for $x_2 < \frac{1}{2}$. For $x_2 = \frac{1}{2}$ it is positive semidefinite, hence f is convex for $x_2 \geqq \frac{1}{2}$ (actually strictly convex for $x_2 > \frac{1}{2}$).

(d) Decide about the convexity or concavity of the following function:

$$f(x, y) = x \log x + y \log y - (x + y) \log(x + y),$$

defined on $\mathrm{int}(\mathbb{R}^2_+)$.

We have

$$\frac{\partial f}{\partial x} = \log x - \log(x+y); \quad \frac{\partial f}{\partial y} = \log y - \log(x+y).$$

$$\frac{\partial^2 f}{\partial x^2} = \frac{y}{x(x+y)}; \; \frac{\partial^2 f}{\partial y^2} = \frac{x}{y(x+y)}; \; \frac{\partial^2 f}{\partial x \partial y} = \frac{\partial^2 f}{\partial y \partial x} = -\frac{1}{x+y}.$$

$$Hf(x,y) = \begin{bmatrix} \frac{y}{x(x+y)} & -\frac{1}{x+y} \\ -\frac{1}{x+y} & \frac{x}{y(x+y)} \end{bmatrix}.$$

We have $\frac{y}{x(x+y)} > 0$ and $\frac{x}{y(x+y)} > 0$ in $\mathrm{dom}(f)$, however

$$|Hf(x,y)| = \frac{xy}{xy(x+y)^2} - \frac{1}{(x+y)^2} = 0, \quad \forall x, y \in \mathrm{dom}(f).$$

Hence f is convex on $\mathrm{int}(\mathbb{R}^2_+)$, but not strictly convex.

(e) Let be given the function

$$f(x_1, x_2) = -(x_1+1)^3 + \alpha x_1 x_2 - \tfrac{1}{4}(x_2)^2.$$

Establish the values of $\alpha \in \mathbb{R}$ for which this function is concave on $\mathbb{R}^2_+$. The Hessian matrix is

$$Hf(x_1, x_2) = \begin{bmatrix} -6(x_1+1) & \alpha \\ \alpha & -\frac{1}{2} \end{bmatrix}.$$

We have $\Delta_1 = -6(x_1+1) < 0$ on $\mathbb{R}^2_+$. If (and only if) $|Hf(x_1, x_2)| \geqq 0$ on $\mathbb{R}^2_+$, the function is concave on the same set. Hence it must hold

$$3(x_1+1) - \alpha^2 \geqq 0,$$

that is $-\sqrt{3} \leqq \alpha \leqq \sqrt{3}$. Note that if $\alpha \in (-\sqrt{3}, \sqrt{3})$, then $f(x_1, x_2)$ is strictly concave on $\mathbb{R}^2_+$.

(f) Consider the function

$$f(x_1, x_2) = \frac{(x_1)^2}{x_2}$$

defined on the convex set $X = \{(x_1, x_2) \in \mathbb{R}^2 : x_2 > 0\}$. We have

$$\frac{\partial f}{\partial x_1} = \frac{2x_1}{x_2}; \; \frac{\partial f}{\partial x_2} = (x_1)^2 \left(-\frac{1}{(x_2)^2}\right) = -\frac{(x_1)^2}{(x_2)^2}.$$

$$\frac{\partial^2 f}{\partial x_1^2} = \frac{2}{x_2}; \quad \frac{\partial^2 f}{\partial x_2^2} = \frac{2(x_1)^2}{(x_2)^3}; \quad \frac{\partial^2 f}{\partial x_1 \partial x_2} = \frac{\partial^2 f}{\partial x_2 \partial x_1} = -\frac{2x_1}{(x_2)^2}.$$

$$Hf(x) = 2\begin{bmatrix} \frac{1}{x_2} & -\frac{x_1}{(x_2)^2} \\ -\frac{x_1}{(x_2)^2} & \frac{(x_1)^2}{(x_2)^3} \end{bmatrix}.$$

We have $\frac{1}{x_2} > 0,\ \forall x_2 \in X;\ \frac{(x_1)^2}{(x_2)^3} \geqq 0,\ \forall x_1, x_2 \in X,\ |Hf(x)| = \frac{1}{x_2}\frac{(x_1)^2}{(x_2)^3} - \frac{(x_1)^2}{(x_2)^4} = 0,\ \forall x_1, x_2 \in X$. Hence $Hf(x)$ is positive semidefinite on X and therefore f is convex on X.

Example 3.2.19 We wish to check, by means of Theorem 3.2.15, that the Cobb-Douglas function of the type

$$f(x, y) = \alpha_0 x^\alpha y^\beta, \quad \alpha_0, \alpha, \beta > 0, \ x, y > 0,$$

is concave if and only if $\alpha + \beta \leqq 1$.

Let us consider, for simplicity, $\alpha_0 = 1$. We have

$$\frac{\partial f}{\partial x} = \alpha x^{\alpha-1} y^\beta; \quad \frac{\partial f}{\partial y} = x^\alpha \beta y^{\beta-1}; \quad \frac{\partial^2 f}{\partial x^2} = \alpha(\alpha-1)x^{\alpha-2}y^\beta;$$

$$\frac{\partial^2 f}{\partial y^2} = x^\alpha \beta(\beta-1)y^{\beta-2}, \quad \frac{\partial^2 f}{\partial x \partial y} = \frac{\partial^2 f}{\partial y \partial x} = \alpha x^{\alpha-1}\beta y^{\beta-1}.$$

$$Hf(x, y) = \begin{bmatrix} \alpha(\alpha-1)x^{\alpha-2}y^\beta & \alpha x^{\alpha-1}\beta y^{\beta-1} \\ \alpha x^{\alpha-1}\beta y^{\beta-1} & x^\alpha\beta(\beta-1)y^{\beta-2} \end{bmatrix}.$$

The Hessian matrix must be negative semidefinite for every $x, y > 0$. We have
$\alpha(\alpha-1)x^{\alpha-2}y^\beta \leqq 0$ for $\alpha \leqq 1$;
$x^\alpha\beta(\beta-1)y^{\beta-2} \leqq 0$ for $\beta \leqq 1$;
$|Hf(x, y)| = x^{2\alpha-2}y^{2\beta-2}\{\alpha\beta(1-\alpha-\beta)\} \geqq 0$ for $(1-\alpha-\beta) \geqq 0$, that is for $\alpha + \beta \leqq 1$.

Example 3.2.20 Establish convexity and concavity of a generic quadratic form $Q(x) = x^\top Ax$ (A symmetric, of order n).

We have $\nabla Q(x) = 2x^\top A = 2Ax;\ HQ(x) = 2A$. Hence $Q(x)$ is convex (resp. concave) if and only if A is positive semidefinite (resp. negative semidefinite). Likewise, we have that $Q(x)$ is strictly convex (resp. strictly concave) if and only if A is positive definite (resp. negative definite).

We can prove the above assertion in a more direct way. More generally, the same holds true for a quadratic function of the form

$$\varphi(x) = x^\top Ax + p^\top x.$$

We have seen that $\varphi(x)$ is convex if and only if the symmetric matrix A is positive semidefinite. Indeed, we have

$$\begin{aligned}\varphi\left[\lambda x+(1-\lambda)y\right] &= \lambda^2 x^\top Ax + 2\lambda(1-\lambda)x^\top Ay + (1-\lambda)^2(y^\top Ay) \\ &\quad +\lambda p^\top x + (1-\lambda)p^\top y \\ &= \lambda\varphi(x) + (1-\lambda)\varphi(y) - \lambda(1-\lambda)(x-y)^\top A(x-y),\end{aligned}$$

and for $\lambda \in (0, 1)$ we have

$$\lambda(1-\lambda)(x-y)^\top A(x-y) \geqq 0$$

if and only if A is positive semidefinite. It is evident that $\varphi(x)$ will be strictly convex if and only if A is positive definite.

3.3 Quasiconvex Functions and Pseudoconvex Functions

We see now two important generalizations of convex (and concave) functions. The first generalization is due to the Italian mathematician Bruno de Finetti [9] and is the class of *quasiconvex functions.* We have previously remarked that the condition

$$\operatorname{lev}_{\leqq\alpha} f \text{ is a convex set for every } \alpha \in \mathbb{R}$$

is only a *necessary conditions* for the convexity of f. Starting from this result de Finetti introduces the following definition.

Definition 3.3.1 Let be given $f : X \longrightarrow \mathbb{R}$, being $X \subset \mathbb{R}^n$ a convex set. The function f is *quasiconvex* on X if and only if $\operatorname{lev}_{\leqq\alpha} f$ is a convex set for every $\alpha \in \mathbb{R}$.

Therefore the class of quasiconvex functions is that class of functions such as the lower-level set are convex for any value of $\alpha \in \mathbb{R}$. For instance, the function $f : \mathbb{R} \longrightarrow \mathbb{R}$ drawn in Fig. 3.12 or in Fig. 3.15 is quasiconvex.

Likewise, $f : X \longrightarrow \mathbb{R}$, being $X \subset \mathbb{R}^n$ a convex set, is quasiconcave on X if and only if all upper-level sets $\operatorname{lev}_{\geqq\alpha} f$ are convex for any value of $\alpha \in \mathbb{R}$.

We have the following basic result.

Theorem 3.3.2 *The function $f : X \longrightarrow \mathbb{R}$ is quasiconvex on the convex set $X \subset \mathbb{R}^n$ if and only if*

$$f(x^1) \leqq f(x^2) \implies f(\lambda x^1 + (1-\lambda)x^2) \leqq f(x^2),\ \forall x^1, x^2 \in X,\ \forall \lambda \in [0, 1],$$

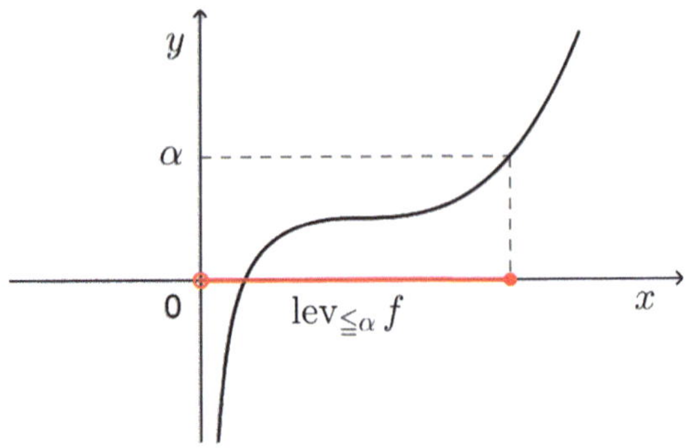

Fig. 3.15 A quasiconvex function

that is, if and only if

$$f(\lambda x^1 + (1-\lambda)x^2) \leqq \max\left\{f(x^1), f(x^2)\right\}, \quad \forall x^1, x^2 \in X, \ \forall \lambda \in [0,1].$$

Proof Let f be quasiconvex on the convex set X and let us consider two points $x, y \in X$ and an arbitrary scalar $\lambda \in [0,1]$. Let be, for example, $f(x) \leqq f(y)$. Hence x and y belong to $\text{lev}_{\leqq f(y)} f$. Being this last set convex, we get that also $\lambda x + (1-\lambda)y \in \text{lev}_{\leqq f(y)} f$:

$$f(\lambda x + (1-\lambda)y) \leqq f(y) = \max\{f(x), f(y)\}.$$

Vice-versa if the relation

$$f(\lambda x + (1-\lambda)y) \leqq f(y) = \max\{f(x), f(y)\}, \ \forall x, y \in X, \forall \lambda \in [0,1]$$

holds true, let us consider $x, y \in \text{lev}_{\leqq \alpha} f$, for a certain scalar α. Then we have

$$f(\lambda x + (1-\lambda)y) \leqq \max\{f(x), f(y)\} \leqq \alpha, \ \forall \lambda \in [0,1].$$

This shows that also $(\lambda x + (1-\lambda)y) \in \text{lev}_{\leqq \alpha} f$, $\forall \lambda \in [0,1]$, i.e. $\text{lev}_{\leqq \alpha} f$ is a convex set. □

In case of quasiconcave functions we have the characterization

$$f(\lambda x^1 + (1-\lambda)x^2) \geqq \min\left\{f(x^1), f(x^2)\right\}, \ \forall x^1, x^2 \in X, \ \forall \lambda \in [0,1].$$

Therefore, f is quasiconcave if and only if $-f$ is quasiconvex.

It is easy to see that, for the case of functions of one real variable, if $f : (a,b) \longrightarrow \mathbb{R}$ is *monotone* on (a,b), then f is both quasiconvex and quasiconcave on (a,b). In this case it is also said that f is *quasilinear.* Therefore, for instance, the function $f(x) = \log x$, which is strictly concave, is quasilinear on its domain.

The following result is useful in utility theory because it shows that quasiconvexity (or quasiconcavity) is preserved by any increasing transformation (which, unlike the convex case, is no longer required to be itself a convex function).

- Let $f(x)$ be defined on a convex set $X \subset \mathbb{R}^n$ and let $g : \mathbb{R} \longrightarrow \mathbb{R}$ such that $f(X) \subset \text{dom}(g)$.

(a) If $f(x)$ is quasiconvex (quasiconcave) and g is increasing, then $g(f(x))$ is quasiconvex (quasiconcave).
(b) If $f(x)$ is quasiconvex (quasiconcave) and g is decreasing, then $g(f(x))$ is quasiconcave (quasiconvex).

We recall again the example of $f(x) = x,\ x \in \mathbb{R}$, and $g(x) = x^3,\ x \in \mathbb{R}$, which is increasing, but not convex nor concave. We have $g(f(x)) = x^3$, which is both quasiconvex and quasiconcave.

The next result, due to the economists [2], is a characterization of differentiable quasiconvex (quasiconcave) functions.

Theorem 3.3.3 *Let be given $f : X \longrightarrow \mathbb{R}$, with $X \subset \mathbb{R}^n$ open and convex; let f be differentiable on X. Then f is quasiconvex (quasiconcave) on X if and only if*

$$f(x^1) \leqq f(x^2) \Longrightarrow \nabla f(x^2)^\top (x^1 - x^2) \leqq 0, \quad \forall x^1, x^2 \in X.$$
$$(f(x^1) \geqq f(x^2) \Longrightarrow \nabla f(x^2)^\top (x^1 - x^2) \geqq 0, \quad \forall x^1, x^2 \in X).$$

Proof

(a) Let us suppose that f is quasiconvex and it holds $f(x) \leqq f(y)$. Then we have, for all $\lambda \in [0, 1]$,

$$f\left[\lambda x + (1 - \lambda)y\right] \leqq \max\{f(x), f(y)\} = f(y),$$

i.e.

$$f\left[\lambda x + (1 - \lambda)y\right] - f(y) \leqq 0.$$

Let us multiply by $\frac{1}{\lambda}$, $\lambda \in (0, 1]$, and then take the limit for $\lambda \longrightarrow 0^+$. We obtain

$$D^+ f(y, x - y) \leqq 0.$$

But, as f is differentiable, it will hold

$$D^+ f(y, x - y) = \nabla f(y)^\top (x - y) \leqq 0.$$

(b) Conversely, let $x, y \in X$ and let $f(y) \leqq f(x)$. It is enough to show that for all $\lambda \in (0, 1)$ it holds $f((1 - \lambda)x + \lambda y) \leqq f(x)$.

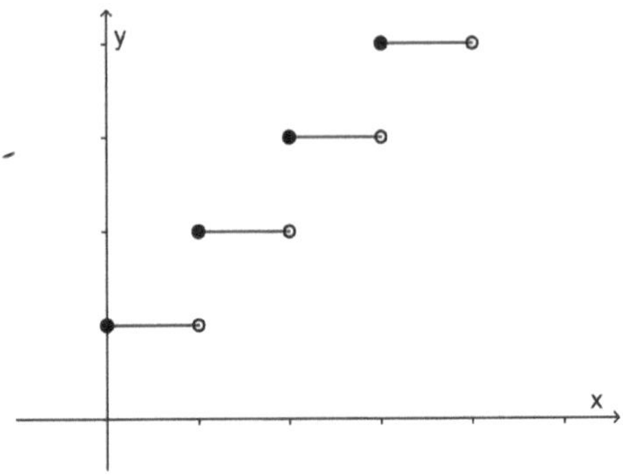

Fig. 3.16 A quasiconvex function which is discontinuous at interior points of its domain

Define $h : [0, 1] \longrightarrow \mathbb{R}$ by $h(\lambda) = f((1-\lambda)x+\lambda y)$. Note that h is differentiable and hence continuous. Clearly $h(1) \leqq h(0)$. We need to show that $h(\lambda) \leqq h(0)$ for all $\lambda \in (0, 1)$. Suppose on the contrary that there exists $\lambda_0 \in (0, 1)$ such that not only $h(\lambda_0) > h(0)$ but also such that $h'(\lambda_0) < 0$. Being h continuous (and differentiable) such a point surely will exist, by the Mean Value Theorem (or Lagrange Theorem). Then we have

$$h'(\lambda_0) = \nabla f(z^0)^\top (y - x) < 0$$

where $z^0 = (1 - \lambda_0)x + \lambda_0 y = x + \lambda_0(y - x)$, and so $x - z^0 = -\lambda_0(y - x)$. Therefore, $\nabla f(z^0)^\top (x - z^0) = -\lambda_0 \nabla f(z^0)^\top (y - x) > 0$. By assumption, this implies that $f(x) > f(z^0)$, i.e. $h(0) > h(\lambda_0)$ which is a contradiction. □

We point out that quasiconvex (quasiconcave) functions may be discontinuous also at *interior* points of their domain. See Fig. 3.16.

From the previous discussion we obtain that a convex function is also quasiconvex, but the converse is not true. A functional class "intermediate" between convex functions and quasiconvex functions, is given by *pseudoconvex functions,* introduced by O. L. Mangasarian in 1965, for differentiable functions. The definition of Mangasarian has then be extended to non differentiable functions and to functions having some type of directional derivatives, but here we shall follow the original definition. For other types of generalized convexity see [10, 16–18, 20, 27].

Definition 3.3.4 Let be given $f : X \subset \mathbb{R}^n \longrightarrow \mathbb{R}$, differentiable on the open convex set X. This function is *pseudoconvex* on X if

$$\nabla f(x^2)^\top (x^1 - x^2) \geqq 0 \Longrightarrow f(x^1) \geqq f(x^2),\ \forall x^1, x^2 \in X.$$

By means of the contrapositive law ($p \Longrightarrow q$ is equivalent to $\bar{q} \Longrightarrow \bar{p}$) the previous relation may be written in the equivalent form

$$f(x^1) < f(x^2) \Longrightarrow \nabla f(x^2)^\top (x^1 - x^2) < 0,\ \ \forall x^1, x^2 \in X,$$

which is useful to call it to mind: it is the characterization of differentiable quasiconvex functions, modified with strict inequalities.

The corresponding definition of *pseudoconcave functions* is immediate, taking into account that, also for this case, f is pseudoconcave if and only if $-f$ is pseudoconvex.

The following theorem clarifies the connections between (differentiable) convex functions, pseudoconvex functions and quasiconvex functions.

Theorem 3.3.5 *Let $f : X \longrightarrow \mathbb{R}$ be differentiable on the open convex set $X \subset \mathbb{R}^n$. Then we have:*

(a) If f is convex on X, then f is pseudoconvex on X;
(b) If f is pseudoconvex on X, then f is quasiconvex on X;
(c) If f is quasiconvex on X and if $\nabla f(x) \neq [0]$, $\forall x \in X$, then f is pseudoconvex on X.

Proof

(a) If f is differentiable and convex on X by Theorem 3.2.13 we have

$$f(x) - f(y) \geqq \nabla f(y)^\top (x-y), \quad \forall x, y \in X.$$

If it results $\nabla f(y)^\top (x-y) \geqq 0$, then we have $f(x) \geqq f(y)$, i.e. Definition 3.3.4 holds true.

(b) Let us suppose absurdly that f is pseudoconvex on X, but not quasiconvex on X. Then, there will exist a pair $x, y \in X$, and a scalar $\lambda^* \in [0,1]$ such that $f(x) \leqq f(y)$ and $f(\lambda^* x + (1-\lambda^*)y) > f(y)$. Let us denote by $\lambda_0 \in [0,1]$ the point such that

$$f(\lambda_0 x + (1-\lambda_0)y) = \max_{\lambda \in [0,1]} f(\lambda x + (1-\lambda)y).$$

This maximum point λ_0 cannot be 0 nor 1: if $\lambda_0 = 0$ we would have $f(y) \geqq f(\lambda x + (1-\lambda)y)$, $\forall \lambda \in [0,1]$, against the assumptions; if $\lambda_0 = 1$, we would have $f(x) \geqq f(\lambda x + (1-\lambda)y)$, $\forall \lambda \in [0,1]$, again against the assumptions. Hence λ_0 will be an interior point of the interval $[0,1]$ and therefore for the function $\varphi(\lambda) = f(\lambda x + (1-\lambda)y)$ it holds

$$\varphi'(\lambda_0) = \nabla f(\lambda_0 x + (1-\lambda_0)y)^\top (x-y) = 0.$$

By noting that $x - (\lambda_0 x + (1-\lambda_0)y) = (1-\lambda_0)(x-y)$, the last equality may be written as

$$\nabla f(\lambda_0 x + (1-\lambda_0)y)^\top [x - (\lambda_0 x + (1-\lambda_0)y)] = 0.$$

But being f pseudoconvex, it follows $f(x) \geqq f(\lambda_0 x + (1-\lambda_0)y)$, which is a contradiction, since

$$f(\lambda_0 x + (1-\lambda_0)y) \geqq f(\lambda^* x + (1-\lambda^*)y) > f(y) \geqq f(x).$$

(c) Suppose that f is quasiconvex on X and that $\nabla f(x) \neq [0]$, $\forall x \in X$. To prove that f is pseudoconvex, we assume that $\nabla f(y)^\top (x-y) \geqq 0$ and prove that $f(x) \geqq f(y)$. If $\nabla f(y)^\top (x-y) > 0$, then we have $f(x) > f(y)$ by Theorem 3.3.3 (the contrapositive law). We need only rule out the case

$$\nabla f(y)^\top (x-y) = 0 \text{ and } f(x) < f(y).$$

Assume that the above case holds true and let v be the nonzero vector such that $v = \nabla f(y)$. For all $t > 0$ we get

$$\begin{aligned}\nabla f(y)^\top (x + tv - y) &= \nabla f(y)^\top (tv + x - y)\\ &= t\nabla f(y)^\top v + \nabla f(y)^\top (x-y) = t\,\|\nabla f(y)\|^2 > 0.\end{aligned}$$

Since f is continuous at x, we can choose a nonzero t small enough so that $f(x+tv) < f(y)$ and $\nabla f(y)^\top (x+tv-y) > 0$. If we put $x' = x + tv$, we have

$$\nabla f(y)^\top (x'-y) > 0 \text{ and } f(x') < f(y),$$

a contradiction to the characterization of quasiconvexity. Hence the case

$$\nabla f(y)^\top (x-y) = 0 \text{ and } f(x) < f(y)$$

cannot hold and therefore f is pseudoconvex.

□

On the grounds of point (c) of the previous theorem, we have, for instance, that $f(x) = \log x$, $x \in \mathbb{R}$, which is both quasiconvex and quasiconcave, having no stationary points, is both pseudoconvex and pseudoconcave.

Summing up, if we suppose that $f : X \longrightarrow \mathbb{R}$ is differentiable on the open convex set $X \subset \mathbb{R}^n$, we have the following implications:

$$f \text{ strictly convex on } X \Longrightarrow f \text{ convex on } X \Longrightarrow f \text{ pseudoconvex on } X$$

$$\Longrightarrow f \text{ quasiconvex on } X.$$

The reverse implications in general do not hold. Obviously, the same implications hold for concave functions and generalized concave functions.

Consider, for instance, the function $f(x) = x^3$, $x \in \mathbb{R}$, which is both quasiconvex and quasiconcave, but not a convex function, nor a concave function.

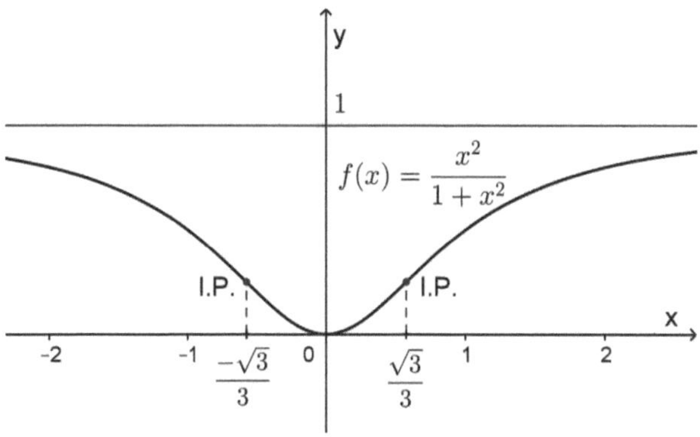

Fig. 3.17 Graph of the function $f(x) = \frac{x^2}{1+x^2}$, which is pseudoconvex. I.P. = inflexion point

Furthermore, this function is not pseudoconvex: for example, we have $f'(0)(-5 - 0) = 0 \nRightarrow f(-5) \geqq f(0)$. For example, the function

$$f(x) = \frac{x^2}{1+x^2}, \quad x \in \mathbb{R},$$

is pseudoconvex on $\mathbb{R}$ (and hence also quasiconvex), but it is not convex (see Fig. 3.17).

We point out that pseudoconvex functions of one real variable cannot have inflection points with an horizontal tangent (i.e. with $f'(x) = 0$ at that point). The reason will be clear when, in the next chapter, we shall treat the optimality properties of pseudoconvex functions. The same holds with reference to pseudoconcave functions. An important example of pseudoconcave function is the probability density of a *normal (or Gaussian) distribution,* here considered to be standardized, i, e, with mean equal to zero and variance equal to one:

$$f(x) = \frac{1}{\sqrt{2\pi}} e^{-\frac{x^2}{2}}.$$

See Fig. 3.18.

It must be pointed out that, in opposition to what holds for convex functions, the sum of quasiconvex functions is not, in general, a quasiconvex function. For instance, $f(x) = x^3$, $x \in \mathbb{R}$, is quasiconvex, and $g(x) = -x$, $x \in \mathbb{R}$, is quasiconvex, but $z(x) = f(x) + g(x) = x^3 - x$ is *not* quasiconvex. Sec Fig. 3.19.

Similar remarks hold for the sum of pseudoconvex functions.

Second-order characterizations of quasiconvex and pseudoconvex functions are available; see, e.g., [3, 6, 8].

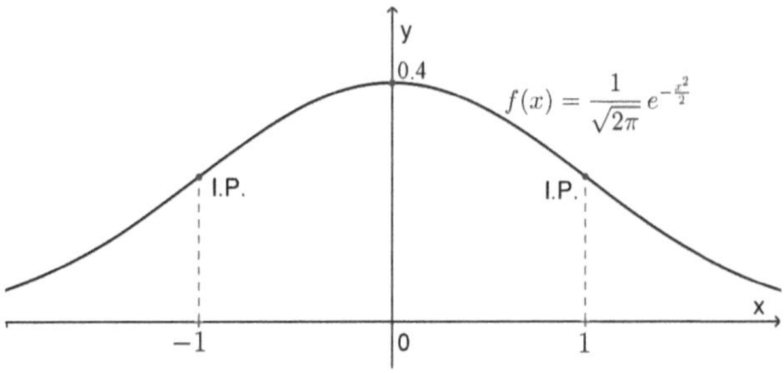

Fig. 3.18 Graph of the normal density function

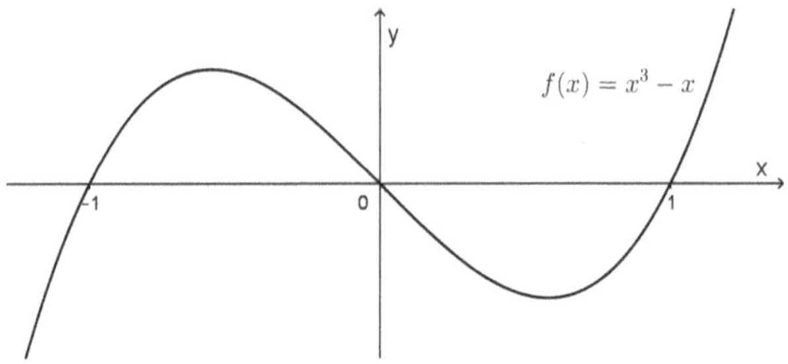

Fig. 3.19 Graph of the function $y = x^3 - x$

Theorem 3.3.6 *Let $f : X \longrightarrow \mathbb{R}$ be of class C^2 on the open convex set $X \subset \mathbb{R}^n$. Then f is pseudoconvex on X if and only if the following conditions (a) and (b) hold true:*

(a) $x \in X,\ v \in \mathbb{R}^n,\ \nabla f(x)^\top v = 0 \Longrightarrow v^\top Hf(x)v \geqq 0;$
(b) $x \in X,\ \nabla f(x) = [0] \Longrightarrow$ *f admits a (global) minimum point at x.*

Theorem 3.3.7 *Let $f : X \longrightarrow \mathbb{R}$ be of class C^2 on the open convex set $X \subset \mathbb{R}^n$. Then f is quasiconvex on X if and only if the following conditions (1) and (2) hold true:*

(1) Condition (a) of Theorem 3.2.6 holds;
(2) If $\nabla f(x) = [0]$, then for all vectors $h \in \mathbb{R}^n$ such that $x + th \in X,\ t \in \mathbb{R}$, the function $\varphi(t) \equiv f(x+th)$ is quasiconvex on the interval $I = \{t : x + th \in X\}$.

Note that Theorems 3.3.6 and 3.3.7 involve constrained quadratic forms, where the constraint is given by $\nabla f(x)^\top v = 0$. Furthermore, from Theorems 3.3.6 and 3.3.7 it appears that a sufficient condition for a twice-continuously differentiable function $f : X \longrightarrow \mathbb{R}$ to be pseudoconvex (and hence quasiconvex) on the open convex set $X \subset \mathbb{R}^n$ is that the bordered matrix (of order $n + 1$)

$$M = \begin{bmatrix} 0 & \nabla f(x)^\top \\ \nabla f(x) & Hf(x) \end{bmatrix}$$

has its leading principal minors of order $3, \ldots, n+1$, all negative, for all $x \in X$. If the said leading principal minors alternate in sign, for all $x \in X$, beginning with the leading principal minor of order 3, of positive sign, then f is pseudoconcave (and hence quasiconcave).

Finally, we give some useful results which relate homogeneous functions and quasiconvex functions.

Theorem 3.3.8 *Let $f : C \longrightarrow \mathbb{R}$ be a homogeneous function of degree $\alpha \geqq 1$, defined on the convex cone $C \subset \mathbb{R}^n$. If f is quasiconvex on C and if $f(x) > 0$ for all $x \in C$, $x \neq [0]$, then $f(x)$ is convex on C. If f is homogeneous of degree $0 < \alpha \leqq 1$, if f is quasiconcave on C and $f(x) > 0$ for all $x \in C$, $x \neq [0]$, then f is concave on C.*

We have seen in Example 3.2.19, that a Cobb-Douglas function of the type

$$f(x, y) = \alpha_0 x^\alpha y^\beta, \quad \alpha_0, \alpha, \beta > 0, \ x, y > 0$$

is concave if and only if $\alpha+\beta \leqq 1$. It can be proved that this function, homogeneous of degree $\alpha + \beta$, is quasiconcave for all $\alpha, \beta > 0$, concave for $\alpha + \beta \leqq 1$ (hence quasiconcave, but not concave for $\alpha + \beta > 1$) and actually strictly concave for $\alpha + \beta < 1$.

Another useful relationship between homogeneous functions and convex functions (or concave functions) is given by the following considerations:

- A function $f : \mathbb{R}^n \longrightarrow \mathbb{R}$ such that

(i) $f(x + y) \leqq f(x) + f(y)$, $\forall x, y \in \mathbb{R}^n$ (*subadditive property*);
(ii) $f(\lambda x) = \lambda f(x)$, $\forall x \in \mathbb{R}^n$, $\forall \lambda > 0$ (*positive homogeneity property*),

is a convex function.

The functions satisfying (i) and (ii) are called *sublinear functions* and are quite important in some questions of nonsmooth analysis (i.e. questions of mathematical analysis in absence of differentiability). Sublinear functions are therefore a subclass of convex functions. It is however possible to show that if $f : \mathbb{R}^n \longrightarrow \mathbb{R}$ is a convex and positively homogeneous function with $f([0]) = 0$, then f is a sublinear function.

The properties of convex (concave) and generalized convex (generalized concave) functions related to optimization problems, will be examined in the next chapter.

3.4 Separation Theorems and Theorems of the Alternative

Several important results in optimization , mathematical economics, games theory, etc. can be obtained by means of the so-called *separation theorems* for convex sets. A hyperplane H in $\mathbb{R}^n$ *separates* two convex sets S_1 and S_2 of $\mathbb{R}^n$ if S_1 is entirely contained in one of the two closed half-spaces associated to H and S_2 is entirely contained in the other one. Then we say that H is a *separating hyperplane* of the two sets S_1 and S_2. We say that a hyperplane H in $\mathbb{R}^n$ is a *supporting hyperplane* for a (non empty) convex set S of $\mathbb{R}^n$ if S is entirely contained in one of the two closed half-spaces associated to H and, furthermore, at least one point of S belongs to H. The closed half-space which contains S is also called *supporting half-space.* More formally, we have the following definition.

Definition 3.4.1 The hyperplane $H = \left\{x \in \mathbb{R}^n : c^\top x = \beta,\ c \neq [0]\right\}$ *separates* the two sets S_1 and S_2 if $c^\top x \leqq \beta,\ \ \forall x \in S_1$ and $c^\top x \geqq \beta,\ \ \forall x \in S_2$, being S_1 and S_2 two non empty sets of $\mathbb{R}^n$. The hyperplane H *strictly separates* S_1 and S_2 if $c^\top x > \beta,\ \forall x \in S_1$ and $c^\top x < \beta,\ \forall x \in S_2$.

We have to note that two separable sets not necessarily are disjoint sets, but two strictly separable sets must be disjoint. However, the vice-versa does not hold: not always two disjoint and separable sets are strictly separable. See the following pictures in Fig. 3.20.

With reference to separation and supporting hyperplanes, there are several results, whose proofs are usually non trivial, even if the same results may appear trivial, at least for what concerns their geometrical meaning in $\mathbb{R}^2$. The following

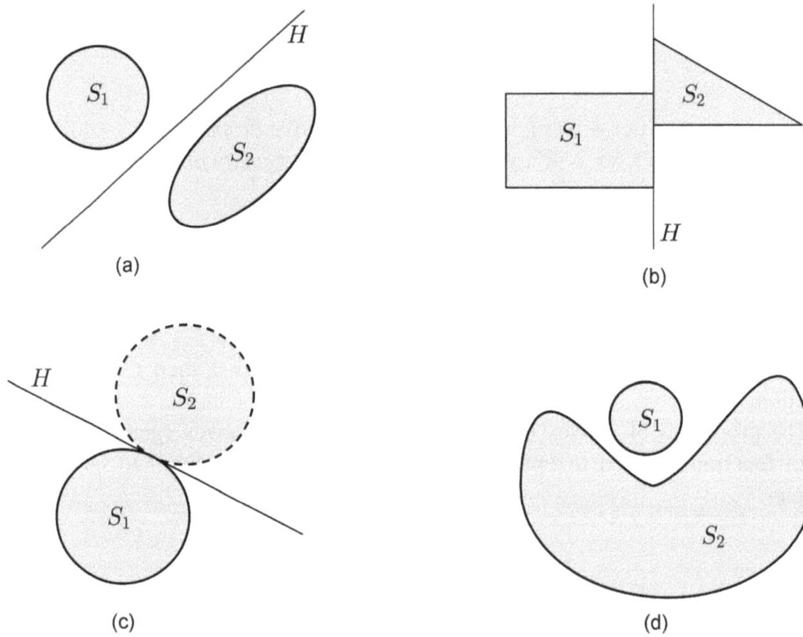

Fig. 3.20 (**a**) Strictly separable sets. (**b**) Non disjoint and separable sets. (**c**) Disjoint and separable sets, but they are not strictly separable. (**d**) Non separable sets

two results make reference to the separation of a convex set and a convex set given by a singleton, precisely the origin of $\mathbb{R}^n$.

Theorem 3.4.2 *Let C be a non empty, closed and convex set of $\mathbb{R}^n$, such that $[0] \notin C$. Then, there exists a hyperplane which strictly separates C and the origin of $\mathbb{R}^n$, i.e. the zero vector $[0]$. In other words, there exist $c \neq [0]$ and $\alpha > 0$ such that $c^\top x > \alpha, \ \forall x \in C$.*

The basic assumption of the previous theorem on strict separation is that C is closed and convex. If we consider convex set, not necessarily closed, it is possible to obtain a weaker separation theorem.

Theorem 3.4.3 *Let C be a non empty convex set of $\mathbb{R}^n$, such that $[0] \notin C$. Then, there exists a hyperplane which separates C and the origin of $\mathbb{R}^n$, that is there exists $c \neq [0]$ such that $c^\top x \geqq 0, \ \forall x \in C$.*

The following two separation results, sometimes accredited to the German mathematician H. Minkowski, are consequences of the previous theorems.

Theorem 3.4.4 (Separation Theorem) *Let be given two non empty, disjoint and convex sets S_1 and S_2 of $\mathbb{R}^n$. Then, there exists a hyperplane H which separates S_1 and S_2. In other words, there exist $c \neq [0]$ and $\alpha \in \mathbb{R}$ such that $c^\top x \leqq \alpha, \ \forall x \in S_1$ and $c^\top x \geqq \alpha, \ \forall x \in S_2$.*

Theorem 3.4.5 (Strict Separation Theorem) *Let be given two non empty, disjoint, closed and convex sets S_1 and S_2 of $\mathbb{R}^n$; one of these sets is also bounded (i.e. it is convex and compact). Then, there exists a hyperplane H which strictly separates S_1 and S_2. In other words, there exist $c \neq [0]$ and $\alpha \in \mathbb{R}$ such that $c^\top x < \alpha, \forall x \in S_1$ and $c^\top x > \alpha, \forall x \in S_2$.*

For the proof of the previous results, see, e.g., [21, 22]. As a consequence of these results we have the following properties:

(A) If x^* is a boundary point for a closed and convex set $S \subset \mathbb{R}^n$, there exists at least one supporting hyperplane for S which contains the point x^*.
(B) A convex and closed set $S \subset \mathbb{R}^n$ is given by the intersection of all its supporting hyperplanes.

The so-called *theorems of the alternative* may be considered the "algebraic forms" of separations theorems for convex sets; this kind of results will be our "battle horse" for the study of several questions of mathematical programming (Chap. 4).

Consider two systems involving mathematical relations containing unknowns, say

$$\mathbf{S} \text{ and } \mathbf{S}^*.$$

We say that systems **S** and **S*** "are in alternative" or that a "theorem of the alternative" holds for the same, if

S admits a solution if and only if **S*** does not admit a solution,

that is, in an equivalent assertion,

S does not admit a solution if and only if **S*** admits a solution.

It results that either both **S** and **S*** admit a solution or that both **S** and **S*** admit no solution. In order to prove a certain theorem of the alternative, usually we have to prove that **S** and **S*** cannot have both a solution. Then, in order to finish the proof, we have to show that if one of the two systems does not admit a solution, the other one has a solution (this part of the proof is usually not trivial). One of the most quoted theorems of the alternative for *linear systems,* is the celebrated *theorem (or lemma) of Farkas or of Farkas-Minkowski.*

Theorem 3.4.6 (Theorem or Lemma of Farkas or Farkas-Minkowski) *Let be given a matrix A of order (m, n) and a vector $c \in \mathbb{R}^n$. Then the system $Ax \leqq [0]\,,\ x \in \mathbb{R}^n$, implies $c^\top x \leqq 0$, if and only if there exists $y \geqq [0]\,,\ y \in \mathbb{R}^m$, such that $y^\top A = c^\top$. Equivalently: the relations*

$$\left\{Ax \leqq [0]\,,\ \ c^\top x > 0\right\}$$

admit a solution if and only if the relations

$$\left\{y^\top A = c^\top,\ \ y \geqq [0]\right\}$$

do not admit a solution $y \in \mathbb{R}^m$.

Proof

(a) If $y^\top A = c^\top$, then $y^\top Ax = c^\top x$. Hence the relations $y \geqq [0]\,,\ Ax \leqq [0]$ imply $c^\top x \leqq 0$.
(b) Conversely, if the relations $y^\top A = c^\top,\ y \geqq [0]$ have no solution, then if we consider the set

$$C_1 = \left\{z \in \mathbb{R}^n : y^\top A = z^\top,\ \ y \geqq [0]\right\}$$

and the set $C_2 = \{c\}$, it results that C_1 and C_2 are disjoint sets. Obviously C_2 is a convex and compact set; we have previously noted that C_1 is a convex polyhedral cone, cone which can be also described in the form $\left\{z \in \mathbb{R}^n : Bz \leqq [0]\right\}$, with B suitable matrix of order (m, n) (Theorem of Minkowski-Weyl). This confirms the fact that C_1 is a *closed* convex set. For a direct proof that C_1 is a closed set, the reader may consult [5, 15, 24]. We can therefore apply

Theorem 3.4.5 on strict separation: there exists a nonzero vector, here denoted by x and a scalar $\beta \in \mathbb{R}$ such that $c^\top x > \beta$ and $y^\top Ax < \beta$ for some $x \in \mathbb{R}^n$ and for any $y \geqq [0]$. If some component of Ax were positive, by choosing the corresponding component of y sufficiently large, we would obtain $y^\top Ax > \beta$. Hence $Ax \leqq [0]$. If we take $y = [0]$, we have $y^\top Ax = 0 < \beta$ and hence $c^\top x > \beta > 0$. Therefore the implication $Ax \leqq [0] \Longrightarrow c^\top x \leqq 0$ does not hold for any $x \in \mathbb{R}^n$, unless $y^\top A = c^\top$ for some $y \geqq [0]$, $y \in \mathbb{R}^m$. The theorem is proved. We now prove that the thesis of the theorem is equivalent to the following systems "in alternative":

$$\mathbf{S} : \begin{cases} Ax \leqq [0] \\ c^\top x > 0, \end{cases} \quad \mathbf{S}^* : \begin{cases} y^\top A = c^\top \\ y \geqq [0] . \end{cases}$$

Indeed, if $\mathbf{S}$ has no solution, then $\mathbf{S}^*$ has a solution: for every x such that $Ax \leqq [0]$ it follows $c^\top x \leqq 0$; therefore there exists (by the first version of the theorem) $y \geqq [0]$ such that $y^\top A = c$ and hence $\mathbf{S}^*$ admits a solution. If, conversely, $\mathbf{S}^*$ admits a solution, then $\mathbf{S}$ is inconsistent: from $y^\top Ax = c^\top$, $y \geqq [0]$, we get $y^\top Ax = c^\top x$, that is

$$y_1 A_1 x + y_2 A_2 x + \cdots + y_m A_m x = c^\top x,$$

and if $c^\top x > 0$, for some x, then we would have $A_i x > 0$ for at least an index i, and therefore $\mathbf{S}$ cannot admit a solution. □

Obviously $\mathbf{S}$ and $\mathbf{S}^*$ can be written in other equivalent forms, for example,

$$\mathbf{S} = \left\{ A^\top u \geqq [0] , \ b^\top u < 0 \right\}; \quad \mathbf{S}^* = \left\{ Ax = b, \ x \geqq [0] \right\} .$$

In the Appendix at the end of this section, it will be given a proof of the Farkas theorem, based only on algebraic methods, and following this formulation.

From the theorem of Farkas we can obtain a lot of theorems of the alternative for linear systems: indeed, this theorem is a spring from which many rivers flow. Now we obtain the following *theorem of Motzkin*.

Theorem 3.4.7 (Theorem of the Alternative of Motzkin) *Let be given the matrices A, B and C, of order, respectively, (m_1, n), (m_2, n), (m_3, n). Then the system*

$$Ax < [0]$$

$$Bx \leqq [0]$$

$$Cx = [0]$$

admits no solution $x \in \mathbb{R}^n$ if and only if the system

$$u^\top A + v^\top B + w^\top C = [0] , \ u \geq [0] , \ v \geqq [0] , \ w \in \mathbb{R}^{m_3},$$

admits a solution. (The matrices B, C may do not appear in the previous relations).

Proof

(a) Let us suppose $u^\top A + v^\top B + w^\top C = [0]$ for some $u \geq [0]$, $v \geqq [0]$, $w \in \mathbb{R}^{m_3}$. Then, we have, for every $x \in \mathbb{R}^n$, $u^\top Ax + v^\top Bx + w^\top Cx = 0$. If $Bx \leqq [0]$ then $v^\top Bx \leqq 0$ and if $Cx = [0]$, then $w^\top Cx = 0$. Then $u^\top Ax \geqq 0$. Being $u \geq [0]$, we have therefore $Ax < [0]$.
(b) If the system $Ax < [0]$, $Bx \leqq [0]$, $Cx = [0]$ does not admit a solution, obviously the system

$$\begin{aligned} Ax + \xi e &\leqq [0] \\ Bx &\leqq [0] \\ Cx &= [0] \\ \xi &> 0, \end{aligned}$$

where $e = [1, 1, \dots, 1]^\top$, does not admit a solution $(x, \xi) \in \mathbb{R}^{n+1}$. Hence for all vectors $[x, \xi]^\top$ we have

$$\begin{bmatrix} A & e \\ B & [0] \\ C & [0] \\ -C & [0] \end{bmatrix} \begin{bmatrix} x \\ \xi \end{bmatrix} \leqq [0] \text{ implies } \xi \leqq 0.$$

By applying the theorem of Farkas, with $c^\top = [0, 0, \dots, 0, 1] \in \mathbb{R}^{n+1}$, we have that there exists $y^\top = \left[y^1, y^2, y^3, y^4\right] \in \mathbb{R}^{m_1+m_2+m_3+m_3}$ such that

$$(y^1)^\top A + (y^2)^\top B + (y^3)^\top C - (y^4)^\top C = [0], \quad (y^1)^\top e = 1.$$

Now we put $y^1 = u$, $y^2 = v$, $y^3 - y^4 = w$ and observe that $(y^1)^\top e = 1$ implies that y^1 cannot be the zero vector: hence $y^1 = u \geq [0]$ which ends the proof. □

By removing in the theorem of Motzkin matrices B and C, that is by considering only matrix A, we obtain the following theorem of the alternative, which is perhaps the first published result of this kind, due to the German mathematician P. Gordan.

Theorem 3.4.8 (Theorem of the Alternative of Gordan) *Let be given the matrix A of order (m, n); the system*

$$Ax < [0]$$

admits no solution $x \in \mathbb{R}^n$ if and only the following system

$$y^\top A = [0], \quad y \geq [0] \quad (y \in \mathbb{R}^n)$$

admits a solution y.

We recall that $y^\top A$ means $\sum_{i=1}^{m} y_i A_i$, being A_i the i-th row of A. This will be useful for some developments in the next chapter.

Now we consider also the following theorem of the alternative for a nonlinear system, a theorem that will be used in the next chapter and that may be considered the nonlinear version of the theorem of Gordan.

Theorem 3.4.9 (Theorem of Fan-Glicksberg-Hoffman [10]) *Let*

$$g_1(x), g_2(x), \ldots, g_m(x)$$

be real valued convex functions defined on the convex set $X \subset \mathbb{R}^n$. The system

$$g(x) < [0]$$

admits no solution in X if and only if there exists a row vector $p^\top \geq [0]$ such that

$$p^\top g(x) \geqq 0, \quad \forall x \in X.$$

Proof Suppose that the system $g(x) < [0]$ admits no solution $x \in X$. Let us define the sets

$$H(x) = \left\{h \in \mathbb{R}^n : x \in X,\ g(x) < h\right\},$$

and denote by H the set of all $H(x)$:

$$H = \underset{x \in X}{\cup} H(x).$$

Now we show that H is a nonempty convex set, not containing the origin.

(i) Consider two generic elements of H, say h^1 and h^2; first, we have to show that $\lambda h^1 + (1-\lambda)h^2 \in H$ for all $\lambda \in [0, 1]$. Since h^1 and h^2 belong to H, each one of them belongs at least to one of the sets $H(x)$: $h^1 \in H(x^1)$, $h^2 \in H(x^2)$. Then, by the convexity of g, we have

$$\begin{aligned} \lambda h^1 + (1-\lambda)h^2 &> \lambda g(x^1) + (1-\lambda)g(x^2) \\ &\geqq g\left[\lambda x^1 + (1-\lambda)x^2\right], \quad \forall \lambda \in [0, 1]. \end{aligned}$$

Hence, being $\lambda x^1 + (1-\lambda)x^2 \in X$,

$$\lambda h^1 + (1-\lambda)h^2 \in H(\lambda x^1 + (1-\lambda)x^2) \subset H.$$

(ii) Now we show that H does not contain the origin. Indeed, if this is not the case, the origin would belong to some $H(x)$ and hence it would hold $g(x) < [0]$ for some $x \in X$, which is excluded by assumption.

Being H convex and being $[0] \notin H$, by Theorem 3.4.3 (separation theorem), there exists a vector $p \neq [0]$ such that

$$p^\top h \geqq p^\top [0] = 0, \quad \forall h \in H.$$

As the elements h_i can be taken arbitrary large (they are not upper bounded), it must hold $p \geq [0]$. Let be $\varepsilon > 0$ (arbitrary). For all $x \in X$, denote by u the vector whose i-th component is $u_i = g_i(x) + \varepsilon, \ i = 1, \ldots, m$. Hence $u \in H(x) \subset H$ and

$$p^\top u = \sum_{i=1}^{m} p_i g_i(x) + \sum_{i=1}^{m} p_i \varepsilon \geqq 0, \quad \forall x \in X.$$

By putting $\sum_{i=1}^{m} p_i \varepsilon = a, \ a > 0$, we get $\sum_{i=1}^{m} p_i g_i(x) + a \geqq 0, \ \forall x \in X$ and as this inequality must hold for all $a > 0$, we must have

$$\sum_{i=1}^{m} p_i g_i(x) \geqq 0, \quad \forall x \in X.$$

Indeed, if $\sum_{i=1}^{m} p_i g_i(x) < 0$ for one $x \in X$, it would be possible to find a neighbourhood of zero where $\sum_{i=1}^{m} p_i g_i(x) + a < 0$, against the result previously obtained. Conversely, if for some $\bar{x} \in X$ we have $g(\bar{x}) < [0]$, then $p^\top g(\bar{x}) < 0$, for every $p \geq [0]$. □

Note that with $X = \mathbb{R}^n$ and $g(x) = Ax$, we have that the system $Ax < [0]$ admits no solution $x \in \mathbb{R}^n$ if and only if there exists $p \geq [0]$ such that $p^\top Ax \geqq 0, \ \forall x \in \mathbb{R}^n$. But then, as it must hold also $p^\top A\bar{x} \geqq 0, \ \forall \bar{x} = -x$, necessarily it will hold $p^\top A = [0]$. This shows that the theorem of Gordan is indeed a particular case of the theorem of Fan-Glicksberg-Hoffman.

3.5 Extremal Points of a Convex Set

A point x^* of a (non empty) convex set $S \subset \mathbb{R}^n$ is said to be an *extreme point of* S if, even when S is deprived of the point x^*, it remains a convex set, i.e. the set $S \setminus \{x^*\}$ is still a convex set, in other words, there do not exist in S two distinct points such that x^* is an interior point of the segment joining the said two points. More formally: x^* is an extreme point of S if it is *not* possible to express x^* in the form

$$x^* = \lambda x^1 + (1 - \lambda)x^2, \text{ with } x^1 \neq x^2, \ x^1, x^2 \in S, \ 0 < \lambda < 1.$$

(Note the strict inequalities for the scalar λ).

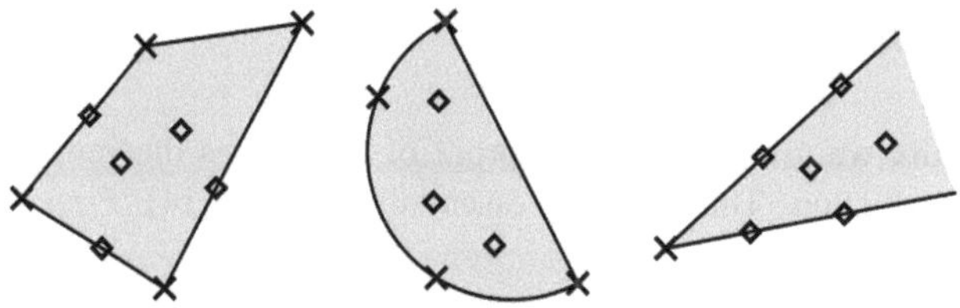

Fig. 3.21 Extreme and non extreme points

If S is a singleton, then its unique element is considered an extreme point of S. Note that if x^* is an extreme point of S, then surely x^* belongs to the boundary of S, but the vice-versa does not hold in general: we may have boundary points of S which are not extreme points of S. See the following Fig. 3.21.

Furthermore, there may exist convex sets (also bounded) which have no extreme point; for example, this is the case of the open bounded convex set formed by all interior points of a circle (open ball). Any half-space has no extreme points.

We have previously seen the definition of a *convex polyhedron;* it is possible to show that a convex polyhedron is a convex set with a *finite number* of extreme points (hence this characterization is equivalent to the one given in Example 3.1.2). Convex polyhedra have a particular importance in several applications, such as, for instance, Linear Programming (see Chap. 5). Extreme points of a convex polyhedron are also called *vertices* of the polyhedron.

There are several results concerning extreme points of convex sets. We recall only the following ones. First we observe that by means of the partial ordering introduced for vectors of $\mathbb{R}^n$ (Sect. 1.2), it is possible to give the notion of lower bounded set and upper bounded set, also for sets of $\mathbb{R}^n$. More precisely:

- The set $A \subset \mathbb{R}^n$ is lower bounded if there exists a vector $x^* \in \mathbb{R}^n$ such that, for all $x \in A$, it holds $x^* \leqq x$. The set A is said to be upper bounded if there exists $x^{**} \in \mathbb{R}^n$ such that, for all $x \in A$, it holds $x^{**} \geqq x$.

Theorem 3.5.1 *If $S \subset \mathbb{R}^n$ is closed, convex, lower and/or upper bounded, then every supporting hyperplane of S contains at least one extreme point of S.*

For the proof, see, e.g., [19].

Theorem 3.5.2 (Theorem of Krein-Milman) *If $S \subset \mathbb{R}^n$ is closed, convex and bounded, then S is given by the convex hull of its extreme points. If S is a convex and bounded polyhedron (that is, a polytope), every point of S is given by the convex combination of finitely many extreme points of S.*

3.6 Appendix

In this appendix we give an algebraic proof of the Farkas theorem, based on the principle of induction. This proof is essentially due to [14]. First we need the following result.

Lemma 3.6.1 *Let be given the linear system (A matrix of order* (m, n)*)*

$$Ax = b. \tag{3.5}$$

This system admits a solution if and only if the system

$$\begin{cases} A^\top u = [0] \\ b^\top u = k, \quad k \neq 0 \end{cases} \tag{3.6}$$

does not admit a solution.

Proof If system (3.5) admits a solution, then we have

$$\text{rank}\begin{pmatrix} A^\top \\ b^\top \end{pmatrix} = \text{rank}(A^\top) < 1 + \text{rank}(A^\top) = \text{rank}\begin{pmatrix} A^\top & [0] \\ b^\top & k \end{pmatrix}$$

and hence system (3.6) does not admit as solution.

Similarly, if system (3.5) does not admit a solution, then

$$\text{rank}\begin{pmatrix} A^\top \\ b^\top \end{pmatrix} = 1 + \text{rank}(A^\top) = \text{rank}\begin{pmatrix} A^\top & [0] \\ b^\top & k \end{pmatrix}$$

and therefore system (3.6) admits a solution. □

Lemma 3.6.1 is also known as the *Fredholm theorem of the alternative.*

Now we recall the Farkas theorem (or Farkas lemma or Farkas-Minkowski lemma):

- Either the system

$$\begin{cases} Ax = b \\ x \geqq [0] \end{cases} \tag{3.7}$$

has a solution, or the system

$$\begin{cases} A^\top u \geqq [0] \\ b^\top u < 0 \end{cases} \tag{3.8}$$

has a solution, but never both.

Proof (Gale) First we prove that systems (3.7) and (3.8) cannot have both a solution. Indeed, if there exist $\bar{x}$ and $\bar{u}$ solutions of, respectively, (3.7) and (3.8), then we obtain

$$u^\top A\bar{x} = \bar{u}^\top b < 0$$
$$u^\top A\bar{x} = (A^\top \bar{u})^\top x \geqq 0,$$

i.e. a contradiction.

In order to complete the proof, we have now to show that if system (3.7) has no solution, then system (3.8) has a solution. We observe that if system (3.7) has no solution, then we have two possibilities:

1. The system $Ax = b$ has no solution. In this case, from Lemma 3.6.1, it results that (3.6) has a solution for any $k \neq 0$. If we choose $k < 0$, it results that (3.8) has a solution.
2. The system $Ax = b$ has a solution, but no nonnegative solutions. By the induction method, applied to the number n of the columns of A, we prove that also in this case, (3.8) admits a solution.

For $n = 1$, if $A^1 x = b$ admits a solution $x < 0$, then (3.8) admits the solution $u = -b$, as

$$u^\top A = -b^\top A^1 = -\frac{1}{x} b^\top b > 0.$$
$$u^\top b = -b^\top b < 0,$$

being $b^\top b > 0$, if $b \neq [0]$. If $b = [0]$ the result is trivial: system (3.7) admits obviously the solution $x = [0]$ and hence system (3.8) cannot have a solution, being $b^\top u = 0$, for any u.

Let us suppose that the property holds for a matrix A with $(n-1)$ columns. The system described by (3.7) can then be written as

$$\sum_{j=1}^{n-1} A^j x_j = b, \tag{3.9}$$

whereas (3.8) can be rewritten as

$$\begin{cases} u^\top A^j \geqq 0, \ \ j = 1, \ldots, n-1 \\ \qquad u^\top b < 0. \end{cases} \tag{3.10}$$

The induction hypothesis is that (3.9) has a solution, but not a nonnegative solution, and (3.10) has a solution. Let u_0 be a solution of (3.10). If $u_0^\top A^n \geqq 0$, the proof is finished. If $u_0^\top A^n < 0$, let be

$$\bar{A}^j = A^j + \lambda_j A^n, \quad j = 1, \ldots, n-1,$$
$$\bar{b} = b + \lambda_0 A^n,$$

where

$$\lambda_j = -\frac{u_0^\top A^j}{u_0^\top A^n}, \quad j = 1, \ldots, n-1,$$

$$\lambda_0 = -\frac{u_0^\top b}{u_0^\top A^n}.$$

It results $\lambda_j \geqq 0, \ j = 1, \ldots, n-1$, and $\lambda_0 < 0$. The following system, in the $(n-1)$ variables y_j

$$\sum_{j=1}^{n-1} \bar{A}^j y_j = \bar{b} \tag{3.11}$$

that is

$$\sum_{j=1}^{n-1} A^j y_j + \left(\sum_{j=1}^{n-1} \lambda_j y_j - \lambda_0 \right) A^n = b \tag{3.12}$$

cannot have a nonnegative solution, as then we have

$$\left(\sum_{j=1}^{n-1} \lambda_j y_j - \lambda_0 \right) \geqq 0$$

and hence, from (3.12) it results that (3.7) has a nonnegative solution, in contradiction with our assumptions. Then, by the induction hypothesis once again (from system (3.11)), it results that there exists $\bar{u} \in \mathbb{R}^m$ such that

$$\begin{cases} \bar{u}^\top \bar{A}^j \geqq 0, \quad j = 1, \ldots, n-1, \\ \qquad \bar{u}^\top \bar{b} < 0. \end{cases}$$

Now we prove that

$$\tilde{u} = \bar{u} - \frac{\bar{u}^\top A^n}{u_0^\top A^n} u_0$$

is a solution of system (3.8). Indeed, we have

$$\tilde{u}^\top A^j = \bar{u}^\top A^j - \frac{\bar{u}^\top A^n}{u_0^\top A^n} u_0^\top A^j = \bar{u}^\top \bar{A}^j \geqq 0, \quad j = 1, \ldots, n-1;$$

$$\tilde{u}^\top A^n = \bar{u}^\top A^n - \frac{\bar{u}^\top A^n}{u_0^\top A^n} u_0^\top A^n = 0;$$

$$\tilde{u}^\top b = \bar{u}^\top \left(b - \frac{u_0^\top b}{u_0^\top A^n} A^n \right) = \bar{u}^\top \bar{b} < 0.$$

The theorem is proved. □

From the lemma of Farkas-Minkowski it is easy to get the following generalization of the same.

- Either the system

$$\begin{cases} A_{11}x^1 + A_{12}x^2 + A_{13}x^3 \geqq b^1 \\ A_{21}x^1 + A_{22}x^2 + A_{23}x^3 = b^2 \\ A_{31}x^1 + A_{32}x^2 + A_{33}x^3 \leqq b^3 \\ x^1 \geqq [0]\,,\ x^2 \text{ unconstrained, } x^3 \leqq [0] \end{cases}$$

admits a solution, or the system

$$\begin{cases} -A_{11}^\top u^1 - A_{21}^\top u^2 - A_{31}^\top u^3 \geqq [0] \\ -A_{12}^\top u^1 - A_{22}^\top u^2 - A_{32}^\top u^3 = [0] \\ -A_{13}^\top u^1 - A_{23}^\top u^2 - A_{33}^\top u^3 \leqq [0] \\ u^1 \geqq [0]\,,\ u^2 \text{ unconstrained, } u^3 \leqq [0] \\ (b^1)^\top u^1 + (b^2)^\top u^2 + (b^3)^\top u^3 > 0 \end{cases}$$

admits a solution, but never both.

Indeed, we can make the substitutions $x^2 = x^4 - x^5$ and $x^3 = -x^6$, with $x^4 \geqq [0]$, $x^5 \geqq [0]$ and $x^6 \geqq [0]$ and, furthermore,

$$x^7 = A_{11}x^1 + A_{12}x^2 + A_{13}x^3 - b^1,$$
$$x^8 = b^3 - A_{31}x^1 - A_{32}x^2 - A_{33}x^3.$$

The first system can then be rewritten as

$$\begin{cases} A_{11}x^1 + A_{12}x^4 - A_{12}x^5 - A_{13}x^6 - x^7 = b^1 \\ A_{21}x^1 + A_{22}x^4 - A_{22}x^5 - A_{23}x^6 = b^2 \\ A_{31}x^1 + A_{32}x^4 - A_{32}x^5 - A_{33}x^6 + x^8 = b^3 \\ x^1 \geqq [0]\,,\ x^4 \geqq [0]\,,\ x^5 \geqq [0]\,,\ x^6 \geqq [0]\,,\ x^7 \geqq [0]\,,\ x^8 \geqq [0]\,. \end{cases}$$

The corresponding system "in alternative" is then (I_1 and I_2 are identity matrices):

$$\begin{cases} \begin{bmatrix} A_{11}^\top & A_{21}^\top & A_{31}^\top \\ A_{12}^\top & A_{22}^\top & A_{32}^\top \\ -A_{12}^\top & -A_{22}^\top & -A_{32}^\top \\ -A_{13}^\top & -A_{23}^\top & -A_{33}^\top \\ -I_1 & [0] & [0] \\ [0] & [0] & I_2 \end{bmatrix} \begin{bmatrix} u^1 \\ u^2 \\ u^3 \end{bmatrix} \leqq [0] \\ (b^1)^\top u^1 + (b^2)^\top u^2 + (b^3)^\top u^3 > 0, \end{cases}$$

which is the second system written above.

This chapter reviews the main tools of Convex Analysis necessary for later chapters. The interested reader can find more information in references [7, 11, 13, 23, 25, 26, 28].

References

1. T.M. Apostol, *Mathematical Analysis* (Addison-Wesley, London, 1974)
2. K.J. Arrow, A.C. Enthoven, Quasiconcave programming. Econometrica **29**, 522–552 (1961)
3. M. Avriel, W.E. Diewert, S. Schaible, I. Zang, *Generalized Concavity* (Plenum, New York, 1988)
4. R.G. Bartle, *The Elements of Real Analysis* (Wiley, New York, 1976)
5. D.P. Bertsekas, *Nonlinear Programming*, 2nd edn. (Athena Scientific, Belmont, 1999)
6. A. Cambini, L. Martein, *Generalized Convexity and Optimization. Theory and Applications* (Springer, New York, 2009)
7. A.C. Chiang, K. Wainwright, *Fundamental Methods of Mathematical Economics* (McGraw-Hill, New York, 2004)
8. J.P. Crouzeix, J.A. Ferland, Criteria for quasiconvexity and pseudoconvexity; relationships and comparisons. Math. Program. **23**, 193–205 (1982)
9. B. de Finetti, Sulle stratificazioni convesse. Ann. Mat. Pura Appl. **30**, 173–183 (1949)
10. W.E. Diewert, M. Avriel, I. Zang, Nine kinds of quasiconcavity and concavity. J. Econ. Theory **25**, 397–420 (1981)
11. K. Fan, I. Glicksberg, A.J. Hoffman, Systems of inequalities involving convex functions. Proc. Am. Math. Soc. **8**, 617–622 (1957)
12. W. Fenchel, *Convex Cones, Sets and Functions.* Mimeographed Lecture Notes (Princeton University, Princeton, 1951)

13. M. Florenzano, C. Le Van, *Finite Dimensional Convexity and Optimization* (Springer, Berlin, 2001)
14. D. Gale, *The Theory of Linear Economic Models* (McGraw-Hill, New York, 1960)
15. G. Giorgi, B. Jimenez, V. Novo, *Basic Mathematical Programming Theory* (Springer, Berlin, 2023)
16. H. Greenberg, W.P. Pierskalla, A review of quasi-convex functions. Oper. Res. **19**, 1553–1570 (1971)
17. A. Guerraggio, E. Molho, The origins of quasi-concavity: a development between mathematics and economics. Historia Math. **31**, 62–75 (2004)
18. N.H. Hadjisavvas, S. Komlósi, S. Schaible (Eds.), *Handbook of Generalized Convexity and Genetalized Monotonicity* (Springer, New York, 2005)
19. G. Hadley, *Linear Algebra* (Addison-Wesley, Reading, 1969)
20. O.L. Mangasarian, Pseudo-convex functions. SIAM J. Control **3**, 281–290 (1965)
21. O.L. Mangasarian, *Nonlinear Programming* (McGraw-Hill, New York, 1969)
22. B. Martos, *Nonlinear Programming* (North Holland, Amsterdam, 1975)
23. P. Michel, *Cours de Mathématiques pour Economistes* (Economica, Paris, 1984)
24. J. Nocedal, S.J. Wright, *Numerical Optimization*, 2nd edn. (Springer, New York, 2006)
25. R.T. Rockafellar, *Convex Analysis* (Princeton University Press, Princeton, 1970)
26. W. Rudin, *Principles of Mathematical Analysis*, 3rd edn. (McGraw-Hill, New York, 1976)
27. W.A. Thompson, D.V. Parke, Some properties of generalized convex functions. Oper. Res. **21**, 305–313 (1973)
28. R.V. Vohra, *Advanced Mathematical Economics* (Routledge, London, 2005)

Chapter 4
Static Optimization

4.1 Unconstrained and Constrained Optimization Problems

Optimization problems represent one of the most prominent aspects in Economic Theory, Statistics, Management Sciences, Operations Research, modern Finance Theory, etc. Examples abound, so that some economists (usually of non Marxian vocation) have actually identified Economic Theory as that science which teaches how to choose "in an optimal way" the use of scanty economic resources. It is the case, for example, of the English economist L. Robbins, who in "An Essay on the Nature and Significance of Economic Sciences" (1932) writes: "Economics is the science which studies human behaviour as a relationship between ends and scarce means which have alternative uses" .

We recall now some basic notions, perhaps already known from previous mathematical courses.

Let be given a real function of several real variables, defined on a set $T \subset \mathbb{R}^n$, that is, $f : T \longrightarrow \mathbb{R}$.

Definition 4.1.1 The point $x^* \in T$ is said to be a *local minimum point* or *local minimizer* for f if there exists a neighbourhood $U(x^*)$ of x^* such that

$$f(x^*) \leqq f(x), \quad \forall x \in U(x^*) \cap T.$$

Definition 4.1.2 The point $x^* \in T$ is said to be a *global minimum point* or *global minimizer* for f if

$$f(x^*) \leqq f(x), \quad \forall x \in T.$$

A global minimum point is, even more, a local minimum point, but obviously the converse is not true. If the previous inequalities hold with the strict sign: $f(x^*) < f(x), \forall x \neq x^*$, we speak of *strict local (or global) minimum point.*

G. Giorgi et al., *Lectures on Mathematics for Economic and Financial Analysis*,
https://doi.org/10.1007/978-3-031-83339-7_4

In Definitions 4.1.1 and 4.1.2 we may then add the specification "weak local (or global) minimum point" whenever there is a possibility to make confusion with the corresponding concepts of strict local (or global) minimum points. If the above inequalities hold with the sign $\geqq$ (respectively, $>$, with $x \neq x^*$), then obviously x^* is a local (or global) *maximum point* (respectively: a *strict* local or global maximum point). When we do not wish to operate a distinction between minimum and maximum points, we speak also of *optimum points, extremal points, extremizer points, optimizer points.*

Obviously, a strict global optimum point is necessarily *unique* (ubi maior…). A very important theorem which assures (under certain conditions) the existence, for a real function, of optimum points is the well known (already previously mentioned) Theorem of K. Weierstrass.

Theorem 4.1.3 (Theorem of Weierstrass) *Let $X \subset \mathbb{R}^n$ be a closed and bounded set (that is, a compact set) and let $f : X \longrightarrow \mathbb{R}$ be continuous over X. Then f admits over X both minimum and maximum values. That is, there exists at least one $x^* \in X$ such that $f(x^*) \leqq f(x)$, $\forall x \in X$ and at least one $x^{**} \in X$ such that $f(x^{**}) \geqq f(x)$, $\forall x \in X$.*

Hence, with the word "maximum value" or simply "maximum" and with the word "minimum value" or simply "minimum" , we mean that *unique* value (if there exists) $f(x^*)$ for which the inequality $f(x^*) \geqq f(x)$, $\forall x \in X$ (in case of maximum) holds true or for which the inequality $f(x^*) \leqq f(x)$, $\forall x \in X$ (in case of minimum) holds true, maybe in correspondence of several points $x^* \in X$ (maybe also infinite points x^*). Do not make confusion between the two concepts!

Furthermore, the Theorem of Weierstrass gives only *sufficient conditions.* Consider, for instance, the function $f(x) = \sin x$, $x \in \mathbb{R}$, which admits mimimum m ($m = -1$) and maximum M ($M = 1$), for example, on the interval $X = (0, +\infty)$, which is not a bounded nor a closed set.

We observe that minimum points, local or global (respectively, maximum points, local or global) of a function $f : \mathbb{R}^n \longrightarrow \mathbb{R}$ coincide with the corresponding maximum points (respectively, with the corresponding minimum points) of $-f$. Hence, it is equivalent to look for the minimum points of f and to look for the maximum points of $-f$ (and obviously it is equivalent to look for the maximum points of f and to look for the minimum points of $-f$). For what concerns the minimum and maximum values, we have the obvious relation

$$m \equiv \text{ minimum value of } f = - \text{ maximum value of } (-f) \equiv M.$$

We illustrate this relation in Fig. 4.1.

Another concept related to the existence of global optimum points is that of *coercive functions.*

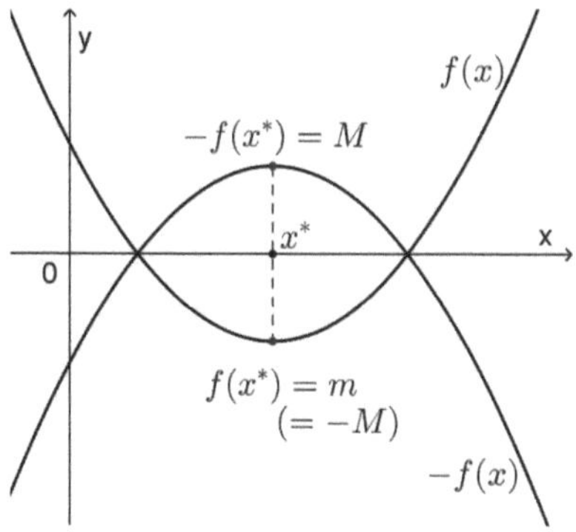

Fig. 4.1 Relation between the minimum of f and the maximum of $-f$

Definition 4.1.4 Let be given $f : A \subset \mathbb{R}^n \longrightarrow \mathbb{R}$, with A not bounded. Then f is *coercive* on A if

$$\lim_{x \in A,\ \|x\| \longrightarrow +\infty} f(x) = +\infty.$$

In other words, for any $M > 0$ there exists δ_M such that $x \in A,\ \|x\| > \delta_M \Longrightarrow f(x) > M$.

We have the following result.

Theorem 4.1.5 *Let be $f : A \subset \mathbb{R}^n \longrightarrow \mathbb{R}$, with A closed and not bounded; let f be continuous and coercive on A. Then f admits minimum on A.*

By recalling what said above, it is easy to obtain conditions for the existence of the maximum value in terms of coercive functions.

Furthermore, we point out the following remarks.

- Isolated points of $\text{dom}(f) = T \subset \mathbb{R}^n$ are both minimum and maximum points for f, at least in the local sense.
- We have previously remarked that global strict minimum (maximum) points are necessarily unique. However, we may have infinite global weak minimum (maximum) points: for instance, the function $f(x) = \sin x,\ x \in \mathbb{R}$, has infinite maximum points at every $x^* = \frac{\pi}{2} \pm 2k\pi,\ k \in \mathbb{N}$. None of these points is a (global) strict maximum point, as in all these points we have $\sin x^* = 1$; However, each of these points is a *local* strict maximum point. Obviously, the same reasoning holds with respect to minimum points of $f(x) = \sin x$.
- We may have infinite local strict minimum points (maximum points) and no global minimum point (no global maximum point): for instance, the function $f(x) = x \sin x,\ x \in \mathbb{R}$, has infinite local strict minimum points and infinite local strict maximum points, but no global minimum point, nor global maximum point.

The reader perhaps has noticed that in the previous definitions we have not been concerned with the case of *vector functions.* Indeed, it is just within economic theory that the concepts of vector optimum points or Pareto optimum points (from the Italian economist and sociologist Vilfredo Pareto, 1848–1923) were born. If we take into consideration, for example, the inequality $f(x^*) \geqq f(x)$, $\forall x \in X \subset \mathbb{R}^n$, which characterized a global maximum point for $f : X \longrightarrow \mathbb{R}$, it is useful to remark that this definition is equivalent to require that *no point $x \in X$ exists such that $f(x) > f(x^*)$*. This is the way that may be followed in order to extend to the vector case the optimality notions given for the scalar case. For example, let be $X \subset \mathbb{R}^n$ and $f : X \longrightarrow \mathbb{R}^p$. Then we say that $x^0 \in X$ is a *vector maximum point* or a *Pareto maximum point* for f on X if there is *no* $x \in X$ such that

$$f(x^0) \leq f(x), \text{ i.e. } f(x^0) \leqq f(x) \text{ and } f(x^0) \neq f(x),$$

that is such that

$$f_i(x^0) \leqq f_i(x), \text{ for all } i = 1, 2, \ldots, p, \text{ and } f_j(x^0) < f_j(x) \text{ for some } j.$$

In other words, a Pareto maximum point concerns a situation in which no one can be made better-off without someone else becoming worse-off. The mathematical literature on vector extremum problems is very huge, however we shall not be concerned with this type of problems in the present book.

We operate the following classification of the static optimization problems we study in the present chapter: *unconstrained optimization problems* and *constrained optimization problems.*

(1) *Unconstrained optimization problems.* This type of problems are concerned with the determination of the maximum and/or minimum points of a real function $f : T \subset \mathbb{R}^n \longrightarrow \mathbb{R}$ (i.e. the extremum points, the optimum points) which are *interior points* of the domain T of f. We remark that in this case (which surely occurs if the domain T is an *open set*), the definition of a *local* minimum or local maximum point of f can be rewritten in the simpler form $f(x^*) \leqq f(x)$ (resp. $f(x^*) \geqq f(x)$),$\forall x \in U(x^*)$, as it is always possible to consider $U(x^*) \subset T$.
(2) *Constrained optimization problems.* This type of problems are concerned with the determination of the extremum points of $f : T \subset \mathbb{R}^n \longrightarrow \mathbb{R}$ on a subset K of T, K not an open set. Usually K is given by the solutions of a system of equalities and/or inequalities. For example:

$$K = \left\{x \in \mathbb{R}^n : h_j(x) = 0,\ j = 1, 2, \ldots, r,\ r < n\right\};$$
$$K = \left\{x \in \mathbb{R}^n : g_i(x) \leqq 0,\ i = 1, 2, \ldots, m\right\};$$
$$K = \left\{x \in \mathbb{R}^n : g_i(x) \leqq 0,\ i = 1, 2, \ldots, m; \right.$$
$$\left. h_j(x) = 0,\ j = 1, 2, \ldots, r,\ r < n\right\}.$$

These types of problems are also called *mathematical programming problems* or also *nonlinear programming problems.* The function f is called the *objective function,* whereas the functions which generate K are the *constraints or constraint functions* of the problem. The set K is the *feasible set,* or *opportunity set,* etc.

4.2 Unconstrained Optimization Problems

For these types of problems the following result is basic.

Theorem 4.2.1 (Fermat Theorem) *Let be given* $f : T \subset \mathbb{R}^n \longrightarrow \mathbb{R}$*, endowed with all n partial derivatives on* T*; let* $x^0 \in \text{int}(T)$ *be a local minimum point or a local maximum point for* f*. Then it holds*

$$\nabla f(x^0) = [0],$$

that is, x^0 *is a stationary point (or critical point) for* f*.*

Proof Let us suppose that x^0 is an interior local maximum point for f; then the function (of one variable!)

$$f(x_1, x_2^0, x_3^0, \ldots, x_n^0)$$

obtained by keeping constant the $(n-1)$ variables $x_2, x_3, \ldots, x_n$, has a maximum point at x_1^0. But then, thanks to the classical Fermat theorem for functions of one real variable, it will hold

$$\frac{\partial f(x^0)}{\partial x_1^0} = 0.$$

By repeating this reasoning for all variables, we obtain the thesis. □

The above result, similarly to the same result for functions of one real variable, gives *necessary conditions* such as an interior point of the domain of f, say x^0, under the assumptions that f has at x^0 all its first-order partial derivatives, is a local optimum point: the point must be a stationary point, i.e. $\nabla f(x^0) = [0]$. Hence, stationary points may be optimum points or not. Without further investigations we have no information on the optimality of stationary points, optimality that may not occur. We have an information on the points that are *not stationary points* (i.e. $\nabla f(x^0) \neq [0]$): they surely are not (unconstrained) optimum points. In order to find the stationary points (if any) of a function $f : \mathbb{R}^n \longrightarrow \mathbb{R}$, we have therefore to solve the system

$$\begin{cases} \frac{\partial f}{\partial x_1} = 0 \\ \frac{\partial f}{\partial x_2} = 0 \\ \quad\vdots \\ \frac{\partial f}{\partial x_n} = 0. \end{cases}$$

These conditions are also called (mainly in the economic literature) *necessary first-order optimality conditions.*

Example 4.2.2 Find the stationary points (if any) of the following functions.

(*a*) $f(x_1, x_2, x_3) = -(x_1)^2 - 2(x_2)^2 - 3(x_3)^2 + \frac{1}{2}x_1x_2 - \frac{1}{2}x_2x_3$.
(*b*) $f(x_1, x_2, x_3) = x_1 \log x_2 \cdot e^{-x_3}$.
(*c*) $f(x, y) = x \log(y^2 + x)$.
(*d*) $f(x, y) = \log(2x - y) + xy$.
(*e*) $f(x, y, z) = 5x - x^3 + 2y^2 - z + e^{x+y+z}$.
(*f*) $f(x, y) = 3y^3 \log(x^2 + 2) + y^5 - 5$.

We have:

(*a*)

$$\frac{\partial f}{\partial x_1} = -2x_1 + \frac{1}{2}x_2.$$
$$\frac{\partial f}{\partial x_2} = -4x_2 + \frac{1}{2}x_1 - \frac{1}{2}x_3.$$
$$\frac{\partial f}{\partial x_3} = -6x_3 - \frac{1}{2}x_2.$$

Hence we must consider the following linear homogeneous system

$$\begin{cases} -2x_1 + \frac{1}{2}x_2 = 0 \\ \frac{1}{2}x_1 - 4x_2 - \frac{1}{2}x_3 = 0 \\ -\frac{1}{2}x_2 - 6x_3 = 0. \end{cases}$$

Being

$$\begin{vmatrix} -2 & \frac{1}{2} & 0 \\ \frac{1}{2} & -4 & -\frac{1}{2} \\ 0 & -\frac{1}{2} & -6 \end{vmatrix} = -46 \neq 0,$$

we deduce that the unique stationary point is $x^* = [0, 0, 0]^\top$.

(*b*) We have to consider the following system, where $x_2 > 0$,

$$\begin{cases} \log x_2 \cdot e^{-x_3} = 0 \\ \frac{x_1}{x_2}e^{-x_3} = 0 \\ -x_1 \log x_2 \cdot e^{-x_3} = 0. \end{cases}$$

From the second equation we deduce $x_1 = 0$, from the first equation we deduce $x_2 = 1$ and hence, from the third equation, we deduce that x_3 can be any real number. We have therefore infinite stationary points:

$$x^* = [0, 1, \alpha]^\top, \quad \alpha \in \mathbb{R}.$$

(c) It must hold $y^2 + x > 0$, i.e. $x > -y^2$. Then we have

$$\frac{\partial f}{\partial x} = \log(y^2 + x) + \frac{x}{y^2 + x}; \quad \frac{\partial f}{\partial y} = \frac{2xy}{y^2 + x}.$$

From the last relation we deduce that it must hold either $x = 0$ and $y \neq 0$ or $y = 0$ and $x > 0$. If $x = 0$, from $\log y^2 = 0$ we deduce $y = \pm 1$. Now let be $x > 0$ and $y = 0$. From $\log x + \frac{x}{x} = \log x + 1 = 0$, we deduce $x = e^{-1}$. Hence we have three stationary points, all belonging to $\mathrm{dom}(f) : A(0, 1),\ B(0, -1)$ and $C = (e^{-1}, 0)$.

(d) It must hold $2x - y > 0$, hence the domain is given by that portion of the plane xOy which is above the straight line $y = 2x$. Then we have

$$\frac{\partial f}{\partial x} = 0 \Longrightarrow \frac{2}{2x - y} + y = 0; \quad \frac{\partial f}{\partial y} = 0 \Longrightarrow \frac{-1}{2x - y} + x = 0 \Longrightarrow \frac{1}{2x - y} = x.$$

By substituting into the first equation we get $2x + y = 0$, from which $y = -2x$. From $\frac{1}{2x-y} = x$ we have therefore $\frac{1}{4x} = x$, that is

$$\frac{1 - 4x^2}{4x} = 0,$$

from which $4x^2 = 1$, i.e. $x = \pm\frac{1}{2}$. From $y = -2x$ we obtain therefore

$$y = \mp 1.$$

We have two points which make the gradient equal to the zero vector:

$$A(\tfrac{1}{2}, -1), \quad B(-\tfrac{1}{2}, 1).$$

However, the point B *does not belong* to the domain of f. Hence we have only one stationary point, i.e. $A(\frac{1}{2}, -1)$.

(e) We have

$$(i) \quad \frac{\partial f}{\partial x} = 5 - 3x^2 + e^{x+y+z} = 0;$$

$$(ii) \quad \frac{\partial f}{\partial y} = 4y + e^{x+y+z} = 0;$$

$$(iii) \quad \frac{\partial f}{\partial z} = -1 + e^{x+y+z} = 0.$$

From (iii) we get $e^{x+y+z} = 1 \Longrightarrow x + y + z = 0$. By substituting into $(i) : 5 - 3x^2 + 1 = 0 \Longrightarrow x^2 = 2 \Longrightarrow x = \pm\sqrt{2}$. From (ii) we get $4y + 1 = 0 \Longrightarrow y = -\frac{1}{4}$. From $x + y + z = 0$ we obtain:

(1) $\sqrt{2} - \frac{1}{4} + z = 0 \Longrightarrow z = \frac{1}{4} - \sqrt{2}$;
(2) $-\sqrt{2} - \frac{1}{4} + z = 0 \Longrightarrow z = \frac{1}{4} + \sqrt{2}$.

Hence we have the following two stationary points:

$$A(\sqrt{2}, -\tfrac{1}{4}, \tfrac{1}{4} - \sqrt{2}); \quad B(-\sqrt{2}, -\tfrac{1}{4}, \tfrac{1}{4} + \sqrt{2}).$$

(f) We have

$$\frac{\partial f}{\partial x} = 3y^3 \frac{1}{x^2+2} 2x = 0;$$
$$\frac{\partial f}{\partial y} = 9y^2 \log(x^2 + 2) + 5y^4 = 0.$$

The first partial derivative is zero either for $x = 0$ and any y or for $y = 0$ and any x. By substituting into the second partial derivative, we obtain in the first case $x = 0$, $y = 0$, and in the second case $y = 0$ and any x.Therefore the horizontal axis $(y = 0)$ gives the infinite stationary points of the function.

Theorem 4.2.3 (Second-Order Optimality Conditions) *Let be given* $f : T \subset \mathbb{R}^n \longrightarrow \mathbb{R}$ *of class* C^2 *on* T *and let* $x^0 \in \text{int}(T)$.

(i) *(Necessary second-order optimality conditions). If* x^0 *is a local maximum point (resp. a local minimum point) for* f, *then*

$$\nabla f(x^0) = [0]\,,$$
$$y^\top Hf(x^0)y \leqq 0 \ (\textit{resp.} \ \geqq 0),\ \forall y \in \mathbb{R}^n.$$

In other words, x^0 *is a stationary point (and this is not a novelty) and the Hessian matrix of* f, *evaluated at* x^0, *is negative semidefinite (resp. positive semidefinite).*

(ii) *(Sufficient second-order optimality conditions). Let* x^0 *be a stationary point for* f, *i.e.* $\nabla f(x^0) = [0]$.

- *If* $Hf(x^0)$ *is negative definite, then* x^0 *is an unconstrained strict local maximum point for* f.
- *If* $Hf(x^0)$ *is positive definite, then* x^0 *is an unconstrained strict local minimum point for* f.

Proof

(*i*) By the Taylor's formula, with the remainder in the Peano's form, we can write

$$\begin{aligned} f(x^0+h) - f(x^0) &= \nabla f(x^0)^\top h + \frac{1}{2}h^\top Hf(x^0)h + o(\|h\|^2) \\ &= \frac{1}{2}h^\top Hf(x^0)h + o(\|h\|^2), \quad \text{for } h \longrightarrow [0], \end{aligned}$$

being x^0 a stationary point. Let be, for instance, x^0 a local maximum point: $f(x^0+h) - f(x^0) \leqq 0$ for $h \in U([0])$, where $U([0]$ is a convenient neighbourhood (n-dimensional) of $h = [0]$. Then we have

$$\frac{1}{2}h^\top Hf(x^0h + o(\|h\|^2) \leqq 0$$

for $h \in U([0])$. If $h^\top Hf(x^0)h > 0$ for some $h \in \mathbb{R}^n$, it will hold also $\frac{1}{2}h^\top Hf(x^0)h + o(\|h\|^2) > 0$ for a convenient h and hence we get $f(x^0+h) - f(x^0) > 0$, against the assumption that x^0 is a local maximum point. Hence it must hold $y^\top Hf(x^0)y \leqq 0$ for all $h \in \mathbb{R}^n$.

(*ii*) If $h^\top Hf(x^0)h < 0$ for all $h \in \mathbb{R}^n$, $h \neq [0]$, we have

$$\frac{1}{2}h^\top Hf(x^0)h + o(\|h\|^2) < 0$$

for $h \in U([0])$ and hence $f(x^0+h) - f(x^0) < 0$ for $h \in U([0])$, that is x^0 is a strict local maximum point for f.

□

We recall that a symmetric matrix is negative (respectively:positive) definite if and only if all its eigenvalues (which are real numbers!) are negative (respectively:positive). Equivalently: it is negative (positive) definite if and only if all its leading principal minors alternate in sign, beginning with the negative sign (respectively: they are all positive).

Other results which stem from the above theorem, are the following ones:

- If at the stationary point x^0 it holds $Hf(x^0)$ *indefinite,* we deduce that x^0 is not a maximum point, nor a minimum point: it is a so-called *saddle point,* that is a point that in several cases turns out to be a minimum point, with respect to some directions, and a maximum point, with respect to other directions. This situation in $\mathbb{R}^2$ is often depicted as a surface which looks like a horse saddle.
- If at the stationary point x^0 it holds $Hf(x^0)$ *semidefinite,* any decision must be taken after further considerations. This is the so-called "doubtful case" and hence any situation may occur (recall that (*i*) of Theorem 4.2.3 concerns only *necessary optimality conditions!*). Generally speaking, one should try to understand how is the behaviour of the function on a neighbourhood of the said point x^0, that is to

try to evaluate the difference Δf between the values that the function assumes on a neighbourhood of x^0 and the value $f(x^0)$.

Example 4.2.4 Let be given $f(x, u) = x^2 - axy + y^2$, $a \in \mathbb{R}$. The system $\nabla f(x^0) = [0]$ is given by

$$\begin{cases} 2x - ay = 0 \\ -ax + 2y = 0. \end{cases}$$

The associated coefficients matrix is

$$A = \begin{bmatrix} 2 & -a \\ -a & 2 \end{bmatrix}$$

with $\det(A) = 4 - a^2$.

For $a \neq \pm 2$, the unique solution of the system is $(0, 0)$, which is therefore, in this case, the unique stationary point. We have

$$Hf(x) = \begin{bmatrix} 2 & -a \\ -a & 2 \end{bmatrix}$$

which is positive definite if $4 - a^2 > 0$, i.e. for $|a| < 2$; it is positive semidefinite for $a = \pm 2$ and indefinite for $|a| > 2$. If $a = 2$, the function becomes $f(x, y) = x^2 - 2xy + y^2 = (x - y)^2$, which assumes global minimum points at all points of the straight line $y = x$ (the minimum being equal to zero). If $a = -2$, the function becomes $f(x, y) = x^2 + 2xy + y^2 = (x + y)^2$, which assumes global minimum points at all points of the straight line $y = -x$ (the minimum being equal to zero).

Hence:

If $-2 < a < 2$, the function assumes minimum at $(0, 0)$.

If $a = 2$, the function assumes minimum at every point (k, k), $k \in \mathbb{R}$.

If $a = -2$, the function assumes minimum at every point $(k, -k)$, $k \in \mathbb{R}$.

If $|a| > 2$, $Hf(0, 0)$ is indefinite, hence $(0, 0)$ is a saddle point.

Example 4.2.5

(a) Discuss the type of stationary points of the function

$$z = x^y$$

defined on $T = \{(x, y) \in \mathbb{R}^2 : 0 < x < +\infty;\ -\infty < y < +\infty\}$.

We have

$$\begin{cases} \frac{\partial z}{\partial x} = yx^{y-1} \\ \frac{\partial z}{\partial y} = x^y \log x. \end{cases}$$

Hence, the unique stationary point is $A(1,0) \in T$. Then we have

$$\frac{\partial^2 z}{\partial x^2} = y(y-1)x^{y-2};$$

$$\frac{\partial^2 z}{\partial x \partial y} = yx^{y-1}\log x + x^y\frac{1}{x} = yx^{y-1}\log x + x^{y-1} = x^{y-1}(y\log x + 1);$$

$$\frac{\partial^2 z}{\partial y \partial x} = x^{y-1} + yx^{y-1}\log x = x^{y-1}(1 + y\log x) = \frac{\partial^2 z}{\partial x \partial y};$$

$$\frac{\partial^2 z}{\partial y^2} = x^y \log x \cdot \log x = x^y \log^2 x.$$

$$Hz(x,y) = \begin{bmatrix} y(y-1)x^{y-2} & x^{y-1}(y\log x + 1) \\ x^{y-1}(y\log x + 1) & x^y \log^2 x \end{bmatrix};$$

$$Hz(1,0) = \begin{bmatrix} 0 & 1 \\ 1 & 0 \end{bmatrix}.$$

Therefore $Hz(1,0)$ is *indefinite:* the point $(1,0)$ is a saddle point for f.

(b) Same question as before for the function

$$f(x,y) = \log(xy) - \frac{1}{2}x^2 + y.$$

It must hold $xy > 0$. Then we have

$$\frac{\partial f}{\partial x} = \frac{y}{xy} - \frac{1}{2}2x = \frac{1}{x} - x = 0 \Longrightarrow 1 - x^2 = 0 \Longrightarrow x = \pm 1.$$

$$\frac{\partial f}{\partial y} = \frac{x}{xy} + 1 = \frac{1+y}{y} = 0 \Longrightarrow 1 + y = 0 \Longrightarrow y = -1.$$

We have only one stationary point: $P(-1,-1)$; the point $(1,-1)$ is not acceptable!

$$\frac{\partial^2 f}{\partial x^2} = -\frac{1}{x^2} - 1; \quad \frac{\partial^2 f}{\partial x \partial y} = \frac{\partial^2 f}{\partial y \partial x} = 0; \quad \frac{\partial^2 f}{\partial y^2} = -\frac{1}{y^2}.$$

$$Hf(x,y) = \begin{bmatrix} -\frac{1}{x^2} - 1 & 0 \\ 0 & -\frac{1}{y^2} \end{bmatrix};$$

$$Hf(-1,-1) = \begin{bmatrix} -2 & 0 \\ 0 & -1 \end{bmatrix}.$$

The point $P(-1,-1)$ is a strict local maximizer.

Example 4.2.6

(1) Discuss the type of stationary points of the function $f(x, y) = x(y - x)^2$.
We have

$$\frac{\partial f}{\partial x} = (y - x)^2 - 2x(y - x) = (y - x)(y - x - 2x) = (y - x)(y - 3x).$$
$$\frac{\partial f}{\partial y} = 2x(y - x).$$

Hence we have the following system

$$\begin{cases} (y - x)(y - 3x) = 0 \\ 2x(y - x) = 0. \end{cases}$$

We deduce that there are infinite stationary points: all points of the straight line $y = x$. Then we have

$$\frac{\partial^2 f}{\partial x^2} = -(y - 3x) - 3(y - x) = -y + 3x - 3y + 3x = 6x - 4y;$$
$$\frac{\partial^2 f}{\partial y^2} = 2x;$$
$$\frac{\partial^2 f}{\partial x \partial y} = \frac{\partial^2 f}{\partial y \partial x} = 2y - 4x.$$

$$Hf(x, y) = \begin{bmatrix} 6x - 4y & 2y - 4x \\ 2y - 4x & 2x \end{bmatrix};$$

$$Hf(x, x) = \begin{bmatrix} 2x & -2x \\ -2x & 2x \end{bmatrix}$$

and therefore $\det(Hf(x, x)) = 0$. Therefore:

- $Hf(x, x)$ is positive semidefinite for $x > 0$;
- $Hf(x, x)$ is negative semidefinite for $x < 0$;
- $Hf(0, 0) = [0]$, which is both positive semidefinite and negative semidefinite.

Which is the behaviour of f outside the straight line of the stationary points? Taking into account that $f(x, x) = 0$, we investigate the sign of the difference

$$\Delta f(x, y) = f(x, y) - f(x, x) = f(x, y) - 0 = f(x, y) = x(y - x)^2.$$

Hence we have to study the sign of the function; consider Fig. 4.2.

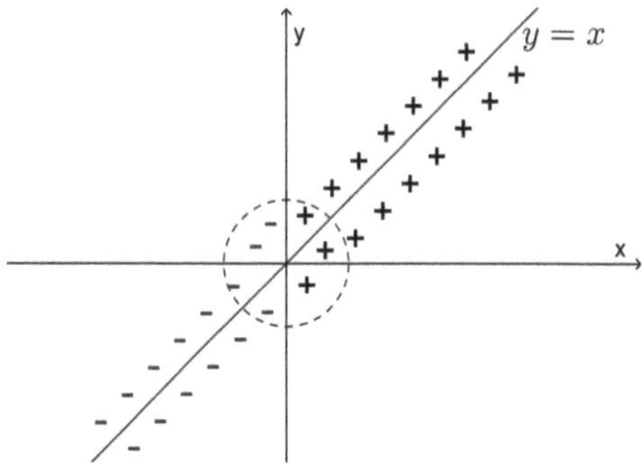

Fig. 4.2 Sign of the function $\Delta f(x, y) = x(y - x)^2$

We deduce that:

(*a*) For $x > 0$ the stationary points are minimum points.
(*b*) For $x < 0$ the stationary points are maximum points.
(*c*) At $(0, 0)$ there is a saddle point.

(2) The same question as before for the function $f(x, y) = x^3 + x^2y + 2y^2$. We have

$$\frac{\partial f}{\partial x} = 0 \Longrightarrow 3x^2 + 2xy = 0; \quad \frac{\partial f}{\partial y} = 0 \Longrightarrow x^2 + 4y = 0 \Longrightarrow y = -\frac{x^2}{4}.$$

Now substitute the last expression of y into the first equation; we obtain $x = 0$ and $x = 6$. The two stationary points are therefore

$$A(0, 0); \quad B(6, -9).$$

Then we have

$$\frac{\partial^2 f}{\partial x^2} = 6x + 2y; \quad \frac{\partial^2 f}{\partial x \partial y} = \frac{\partial^2 f}{\partial y \partial x} = 2x; \quad \frac{\partial^2 f}{\partial y^2} = 4,$$

$$Hf(0, 0) = \begin{bmatrix} 0 & 0 \\ 0 & 4 \end{bmatrix},$$

which is a positive semidefinite matrix ("doubtful case").

$$Hf(6, -9) = \begin{bmatrix} 18 & 12 \\ 12 & 4 \end{bmatrix}.$$

As $\det(Hf(6, -9)) = -72 < 0$, it turns out that the point $B(6, -9)$ is a saddle point.

Now let us try to understand the behaviour of the function on a neighbourhood of $A(0, 0)$. Consider the restriction of f on the horizontal axis, i.e. on

$y = 0$. We have $f(x, 0) = x^3$. This function (a function of one real variable) has an inflection point at the origin, hence it follows that $A(0, 0)$ cannot be a maximum point, nor a minimum point for f : it is a saddle point for f.

(3) The same question as before for the function $f(x, y) = (2y - x^2)(y - x^2)$.
We have

$$\begin{aligned}\frac{\partial f}{\partial x} &= -2x(y - x^2) + (2y - x^2)(-2x)\\ &= -2xy + 2x^3 - 4xy + 2x^3 = 4x^3 - 6xy.\\ \frac{\partial f}{\partial y} &= 2(y - x^2) + (2y - x^2) = 2y - 2x^2 + 2y - x^2 = 4y - 3x^2.\end{aligned}$$

$$\begin{cases} 2x(x^2 - 3y) = 0 \Longrightarrow x = 0 \text{ or } (x^2 - 3y) = 0; \\ -3x^2 + 4y = 0 \Longrightarrow 3x^2 = 4y. \end{cases}$$

$(x^2 - 3y) = 0 \Longrightarrow x^2 = 3y \Longrightarrow 3x^2 = 9y$. Hence from

$$\begin{cases} 3x^2 = 4y \\ 3x^2 = 9y \end{cases}$$

we have $x = 0,\ \ y = 0$. Then we have

$$\frac{\partial^2 f}{\partial x^2} = 12x^2 - 6y; \quad \frac{\partial^2 f}{\partial y^2} = 4; \quad \frac{\partial^2 f}{\partial x \partial y} = \frac{\partial^2 f}{\partial y \partial x} = -6x.$$

$$Hf([0]) = \begin{bmatrix} 0 & 0 \\ 0 & 4 \end{bmatrix}.$$

$Hf([0])$ is positive semidefinite, with $f([0]) = 0$. Then we have

$f(x, y) = 0$, for $y = x^2$ and for $y = \frac{1}{2}x^2$;
$f(x, y) > 0$ for $y > x^2$ and for $y < \frac{1}{2}x^2$;
$f(x, y) < 0$ for $y < x^2$ and for $y > \frac{1}{2}x^2$.

See Fig. 4.3.

We deduce that $(0, 0)$ is a saddle point. This is an example of a function that achieves a minimum at the origin *along every straight line starting from the origin,* yet nevertheless does not achieve a local minimum at the origin. This example is attributed to the Italian mathematician G. Peano and evidences that a minimum (or a maximum) on lines is not necessarily a minimum (resp. a maximum) for the function in question.

(4) We propose to the willing reader the following exercise. Precise the type of stationary points for the function $f(x, y, z) = x^2y + 2z^2 + \frac{4}{3}y^3 - yz^2$. (There

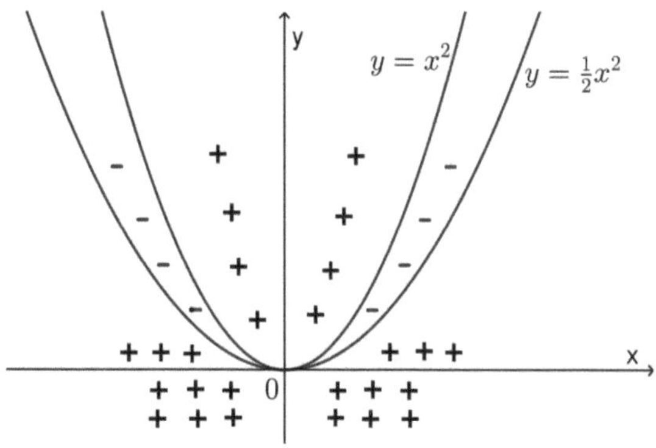

Fig. 4.3 Sign of the function $f(x, y) = (2y - x^2)(y - x^2)$

are three stationary points: $A(0, 0, 0)$; $B(0, 2, 4)$; $C(0, 2, -4)$. All three points are saddle points).

We wish to stress again that (i) of Theorem 4.2.9 is only a necessary optimality condition and that (ii) of the same theorem is only a sufficient optimality conditions. Other examples supporting these assertions are:

- Consider $f(x_1, x_2) = (x_1)^4 - (x_2)^2$ and the point $x^0 = [0, 0]^\top$. Being $f(x^0) = 0$, $f(0, x_2) = -(x_2)^2 < 0$, $\forall x_2 \neq 0$, and $f(x_1, 0) = (x_1)^4 > 0$, $\forall x_1 \neq 0$, we deduce that $x^0 = [0, 0]^\top$ cannot be an extremum point (it is in fact a saddle point); yet we have

$$\nabla f(x_1, x_2) = \begin{bmatrix} 4(x_1)^3 \\ -2x_2 \end{bmatrix},$$

so $\nabla f(x^0) = [0]$, i.e. x^0 is a stationary point, moreover,

$$Hf(x^0) = \begin{bmatrix} 0 & 0 \\ 0 & -2 \end{bmatrix},$$

which is negative semidefinite.
- Consider $f(x_1, x_2) = -(x_1)^2 - \frac{1}{12}(x_2)^4$. The point $x^0 = [0, 0]^\top$ is a stationary point and

$$Hf(x^0) = \begin{bmatrix} -2 & 0 \\ 0 & 0 \end{bmatrix}$$

which is negative semidefinite. However, $f(x) < 0 = f(x^0) = f(0, 0)$, $\forall x \in \mathbb{R}^2$, $x \neq [0]$. Hence, $x^0 = [0, 0]^\top$ is a strict global maximum point for f.

Remark 4.2.7

(1) Let be given the *strictly increasing* function $\varphi : \mathbb{R} \longrightarrow \mathbb{R}$; it is quite easy to prove that to find the maximum (minimum) of the composite function $\varphi(f(x))$, with $f : \mathbb{R}^n \longrightarrow \mathbb{R}$, is equivalent to find the maximum (resp. the minimum) of $f(x)$ on the domain of $\varphi(f(x))$. Therefore, we have, for instance,

$$\max \{\log f(x)\} \iff \max \{f(x)\}, \quad \text{with } f(x) > 0.$$

$$\min \left\{\sqrt{f(x)}\right\} \iff \min \{f(x)\}, \quad \text{with } f(x) \geqq 0.$$

$$\max \left\{e^{f(x)}\right\} \iff \max \{f(x)\}.$$

(2) Let be given the *strictly decreasing* function $\varphi : \mathbb{R} \longrightarrow \mathbb{R}$; then, to find the minimum of the composite function $\varphi(f(x))$, with $f : \mathbb{R}^n \longrightarrow \mathbb{R}$, is equivalent to find the maximum of $f(x)$ on the domain of $\varphi(f(x))$. To find the maximum of the composite function $\varphi(f(x))$ is equivalent to find the minimum of $f(x)$ on the domain of $\varphi(f(x))$. Therefore, we have, for instance,

$$\min \left\{e^{-f(x)}\right\} \iff \max \{f(x)\}.$$

$$\max \left\{\frac{1}{\sqrt{f(x)}}\right\} \iff \min \{f(x)\}, \quad \text{with } f(x) > 0.$$

In other words, this simple remark often allows a simplification of the research of the unconstrained optimum points of a function, by deleting the strictly monotonic functions φ which form the composite function $\varphi(f(x))$; obviously, we have to perform the research on the domain of the composite function, which may be narrower than the domain of $f(x)$.

Example 4.2.8 Find the extremum points of

$$f(x, y) = e^{x^2+2y^2-2y+1}.$$

In order to reduce the computations, it is convenient to consider the function $g(x, y) = x^2 + 2y^2 - 2y + 1$, defined on $\mathbb{R}^2$. We have

$$\frac{\partial g}{\partial x} = 2x; \quad \frac{\partial g}{\partial y} = 4y - 2.$$

$$\begin{cases} 2x = 0 \\ 4y - 2 = 0 \end{cases} \Longrightarrow x = 0, \; y = \frac{1}{2}.$$

$$\frac{\partial^2 g}{\partial x^2} = 2; \quad \frac{\partial^2 g}{\partial x \partial y} = \frac{\partial^2 g}{\partial y \partial x} = 0; \quad \frac{\partial^2 g}{\partial y^2} = 4.$$

$$Hg(x, y) = \begin{bmatrix} 2 & 0 \\ 0 & 4 \end{bmatrix}.$$

The matrix $Hg(x, y)$ is therefore positive definite on $\mathbb{R}^2$; we deduce that g is strictly convex on $\mathbb{R}^2$ and that $(0, \frac{1}{2})$ is the unique strict global minimum point for g and hence also for f, being e^z a strictly increasing function.

The last sentence of the previous example let suspect that the local optimum results, if referred to convex or concave functions, can be "ameliorated" . Indeed, we have the following basic result.

Theorem 4.2.9 *Let $f : X \longrightarrow \mathbb{R}$ be a convex (resp. concave) function on the convex set $X \subset \mathbb{R}^n$.*

(i) Every local minimum point (resp. local maximum point) of f is also a global minimum point (resp. a global maximum point).
(ii) The set of all minimum points (resp. of all maximum points) is a convex set.
(iii) If, moreover, f is differentiable on the open convex set X, then every stationary point of f is a global minimum point (resp. a global maximum point) for f over X.

Proof

(i) Let x^* be a local minimum point and let y be *any* other point of the convex set X. For $\lambda > 0$ and sufficiently near to 1, we have $\lambda x^* + (1 - \lambda)y \in X$ (we recall that X is a convex set) and furthermore $\lambda x^* + (1 - \lambda)y \in U(x^*) \cap X$. As x^* is a local minimum point, we have

$$f(x^*) \leqq f(\lambda x^* + (1 - \lambda)y).$$

Being f a convex function, we have

$$f(x^*) \leqq f(\lambda x^* + (1 - \lambda)y) \leqq \lambda f(x^*) + (1 - \lambda)f(y),$$

from which

$$f(x^*) \leqq \lambda f(x^*) + (1 - \lambda)f(y)$$

that is

$$f(x^*) - \lambda f(x^*) \leqq (1 - \lambda)f(y)$$

that is

$$(1-\lambda)f(x^*) \leqq (1-\lambda)f(y)$$

and hence $f(x^*) \leqq f(y)$, but being y an arbitrary point of X, the thesis follows. The proof for the maximum points is similar.

(ii) Let x^* and x^{**} be two minimum points (necessarily global minimum points): $f(x^*) = f(x^{**})$. We have to show that $\lambda x^* + (1-\lambda)x^{**}$ is a global minimum point, for all $\lambda \in (0, 1)$. We have

$$f(\lambda x^* + (1-\lambda)x^{**}) \leqq \lambda f(x^*) + (1-\lambda) f(x^{**}) = f(x^*), \quad \forall \lambda \in (0, 1).$$

It cannot be $f(\lambda x^* + (1-\lambda)x^{**}) < f(x^*)$, as x^* is by assumption a global minimum point for f; it will hold therefore $f(\lambda x^* + (1-\lambda)x^{**}) = f(x^*)$, $\forall \lambda \in (0, 1)$, that is $\lambda x^* + (1-\lambda)x^{**}$ is a global minimum point of f, for every $\lambda \in (0, 1)$.

(iii) From Fermat's theorem we know that

$$x^* \text{ interior local minimum point of } f \Longrightarrow \nabla f(x^*) = [0]\,.$$

We have to prove the converse implication. Being f differentiable and convex on the open convex set $X \subset \mathbb{R}^n$, we have

$$f(x) - f(x^*) \geqq \nabla f(x^*)^\top (x - x^*), \ \forall x, x^* \in X.$$

But being x^* a stationary point, it will hold $f(x) - f(x^*) \geqq 0, \ \forall x \in X$, that is $f(x^*) \leqq f(x), \ \forall x \in X$. □

The thesis (iii) of the previous theorem asserts that the condition $\nabla f(x^*) = [0]$, which is in general only a necessary condition for x^* to be an interior local optimum point (Fermat's theorem), for convex or concave functions it becomes also a sufficient condition for x^* to be a global optimum point. The thesis (ii) is also known as "Theorem of local/global minimum (maximum) points" . If f is *strictly convex (strictly concave),* then we have

$$x^* \in X \text{ local minimum (local maximum) point of } f$$

$$\Longrightarrow x^* \textit{unique global minimum point(maximum point)} \text{ for } f$$

$$\text{(and hence, necessarily "strict").}$$

For completeness we prove the above last property.

Theorem 4.2.10 *Let $f : X \longrightarrow \mathbb{R}$ be a strictly convex (resp. strictly concave) function on the convex set $X \subset \mathbb{R}^n$. If f admits a minimum (resp. maximum) point over the set X, this point is unique.*

Proof Assume, absurdly, that there exist two distinct minimum points, say x^0 and x^{00}. Being f strictly convex, it would hold

$$f\left(\frac{1}{2}x^0 + \frac{1}{2}x^{00}\right) < \frac{1}{2}f(x^0) + \frac{1}{2}f(x^{00}) = f(x^0),$$

which is absurd, being $\frac{1}{2}x^0 + \frac{1}{2}x^{00} \in X$, with x^0 global minimum point of f. □

Remark 4.2.11 When the Hessian matrix $Hf(x)$ *does not depend on* x and is positive definite or positive semidefinite (resp. negative definite or negative semidefinite), then these properties hold on the whole domain of f and hence we are in the presence of the case of a strictly convex function or of a convex function (respectively, a strictly concave function or a concave function). With reference to its related stationary points, the previous results therefore apply.

Example 4.2.12

(a) Discuss the type of stationary points of the function

$$f(x, y) = (x - y)^2 + 3(x - y).$$

We have

$$\frac{\partial f}{\partial x} = 2(x - y) + 3; \quad \frac{\partial f}{\partial y} = -2(x - y) - 3.$$

$$\begin{cases} 2(x - y) + 3 = 0 \\ -2(x - y) - 3 = 0 \end{cases} \Longrightarrow x - y = -\frac{3}{2}.$$

All points of the straight line $y = x + \frac{3}{2}$ are stationary points.

$$\frac{\partial^2 f}{\partial x^2} = 2; \quad \frac{\partial^2 f}{\partial y^2} = 2; \quad \frac{\partial^2 f}{\partial x \partial y} = \frac{\partial^2 f}{\partial y \partial x} = -2.$$

$$Hf(x, y) = \begin{bmatrix} 2 & -2 \\ -2 & 2 \end{bmatrix}.$$

Therefore $\det(Hf(x, y)) = 0$. The Hessian matrix $Hf(x, y)$ does not depend on x, y and it is positive semidefinite for every $(x, y) \in \mathbb{R}^2$. Hence f is convex on $\mathbb{R}^2$: every point of the straight line $y = x + \frac{3}{2}$ is a global minimum point for f.

(b) Discuss, the type of stationary points of the following function, where α is a real parameter,

$$f(x_1, x_2, x_3) = -\frac{1}{2}(x_1)^2 - \frac{5}{2}(x_2)^2 - (x_3)^2 + x_1x_2 + x_1x_3 + \alpha x_2x_3.$$

We have

$$\frac{\partial f}{\partial x_1} = -2\frac{1}{2}x_1 + x_2 + x_3;$$
$$\frac{\partial f}{\partial x_2} = x_1 - 2\frac{5}{2}x_2 + \alpha x_3;$$
$$\frac{\partial f}{\partial x_3} = x_1 + \alpha x_2 - 2x_3.$$

Therefore we write the system

$$\begin{cases} -x_1 + x_2 + x_3 = 0 \\ x_1 - 5x_2 + \alpha x_3 = 0 \\ x_1 + \alpha x_2 - 2x_3 = 0. \end{cases}$$

The determinant of the matrix of this system is $\alpha^2+2\alpha-3$. We have $\alpha^2+2\alpha-3 = 0$ for $\alpha = 1$ and for $\alpha = -3$. Hence, for $\alpha = 1$ and for $\alpha = -3$ we have infinite stationary points. For $\alpha \neq 1$ and for $\alpha \neq -3$ there is a unique stationary point, i.e. the zero vector $x^* = [0] \in \mathbb{R}^3$. The Hessian matrix is

$$Hf(x) = \begin{bmatrix} -1 & 1 & 1 \\ 1 & -5 & \alpha \\ 1 & \alpha & -2 \end{bmatrix}.$$

Its first North-West principal minor is $-1 < 0$, its North-West principal minor of order two is

$$\begin{vmatrix} -1 & 1 \\ 1 & -5 \end{vmatrix} = 4 > 0.$$

The determinant of $Hf(x)$ is $\alpha^2 + 2\alpha - 3$.

Case (i) If $\alpha \in (-3, 1)$ the Hessian matrix $Hf(x)$ is negative definite for all nonzero $x \in \mathbb{R}^3$, hence $x^* = [0]$ is the unique global maximum point for f, being f in this case a strictly concave function.

Case (ii) If $\alpha \in (-\infty, 3) \cup (1, +\infty)$ the Hessian matrix $Hf(x)$ is indefinite and hence $x^* = [0]$ is a saddle point.

Case (iii) If $\alpha = 1$ we have the system

$$\begin{cases} -x_1 + x_2 + x_3 = 0 \\ x_1 - 5x_2 + x_3 = 0 \\ x_1 + x_2 - 2x_3 = 0, \end{cases}$$

which admits the infinite solutions

$$x^* = \left[\frac{3}{2}\bar{x}_3, \frac{1}{2}\bar{x}_3, \bar{x}_3\right]^\top, \quad \bar{x}_3 \in \mathbb{R}.$$

The related Hessian matrix is

$$Hf(x) = \begin{bmatrix} -1 & 1 & 1 \\ 1 & -5 & 1 \\ 1 & 1 & -2 \end{bmatrix}.$$

Its North-West principal minors of order one are: $-1; -5; -2$. Its North-West principal minors of order two are:

$$\begin{vmatrix} -1 & 1 \\ 1 & -5 \end{vmatrix} = 4 > 0; \quad \begin{vmatrix} -1 & 1 \\ 1 & -2 \end{vmatrix} = 1 > 0; \quad \begin{vmatrix} -5 & 1 \\ 1 & -2 \end{vmatrix} = 9 > 0.$$

Being $\det(Hf(x)) = 0$, it turns out that $Hf(x)$ is negative semidefinite on $\mathbb{R}^3$. Therefore f is concave and the infinite stationary points

$$x^* = \left[\frac{3}{2}\bar{x}_3, \frac{1}{2}\bar{x}_3, \bar{x}_3\right]^\top, \quad \bar{x}_3 \in \mathbb{R}$$

are global maximum points for f.

Case (iv) If $\alpha = -3$ we have the system

$$\begin{cases} -x_1 + x_2 + x_3 = 0 \\ x_1 - 5x_2 - 3x_3 = 0 \\ x_1 - 3x_2 - 2x_3 = 0, \end{cases}$$

which admits the infinite solutions

$$x^{**} = \left[\frac{1}{2}\bar{x}_3, -\frac{1}{2}\bar{x}_3, \bar{x}_3\right]^\top, \quad \bar{x}_3 \in \mathbb{R}.$$

The related Hessian matrix is

$$Hf(x) = \begin{bmatrix} -1 & 1 & 1 \\ 1 & -5 & -3 \\ 1 & -3 & -2 \end{bmatrix}.$$

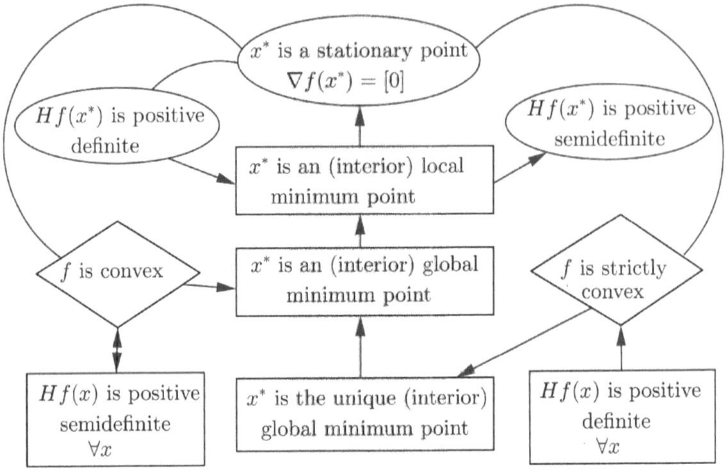

Fig. 4.4 Relations among various optimality conditions. Unconstrained case

Its North-West principal minors of order one are: $-1; -5; -2$. Its North-West principal minors of order two are:

$$\begin{vmatrix} -1 & 1 \\ 1 & -5 \end{vmatrix} = 4 > 0; \quad \begin{vmatrix} -1 & 1 \\ 1 & -2 \end{vmatrix} = 1 > 0; \quad \begin{vmatrix} -5 & -3 \\ -3 & -2 \end{vmatrix} = 1 > 0.$$

Being $\det(Hf(x)) = 0$, it turns out that $Hf(x)$ is negative semidefinite on $\mathbb{R}^3$. Therefore f is concave and the infinite stationary points

$$x^{**} = \left[\frac{1}{2}\bar{x}_3, -\frac{1}{2}\bar{x}_3, \bar{x}_3\right]^\top, \quad \bar{x}_3 \in \mathbb{R}$$

are global maximum points for f.

The diagram in Fig. 4.4 points into evidence the links between the various optimality conditions for unconstrained optimization problems.

We have seen that a convex (concave) function which admits local minimum (maximum) points, really admits global minimum (maximum) points. If, furthermore, the function is also differentiable on the open and convex set $X \subset \mathbb{R}^n$, then all its stationary points are global minimum (resp. global maximum) points for the function. These important properties hold also for *pseudoconvex functions (pseudoconcave functions)*, however not for quasiconvex (quasiconcave) functions.

Theorem 4.2.13 *Let $f : X \longrightarrow \mathbb{R}$ be pseudoconvex on the open convex set $X \subset \mathbb{R}^n$. Then:*

(*i*) *Every its stationary point is a global minimum point for f.*
(*ii*) *Every local minimum point for f is also a global minimum point.*

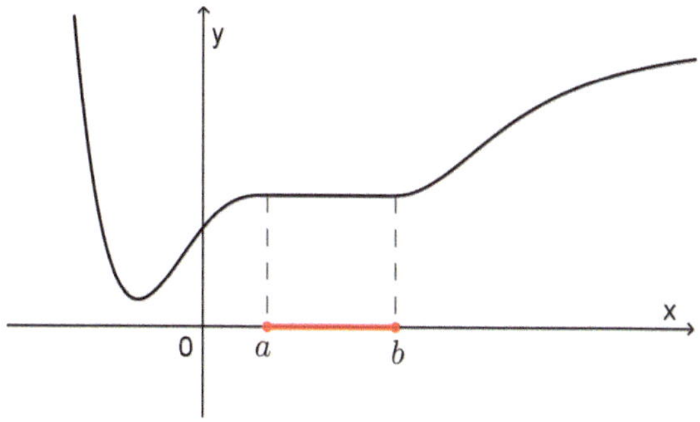

Fig. 4.5 A quasiconvex function. Each point of the interval (a, b) is a local minimum but it is not a global minimum

Proof

(i) Let x^0 be a stationary point for f (i.e. $\nabla f(x^0) = [0]$). From the definition of pseudoconvex functions we deduce the following relation

$$\nabla f(x^0)^\top (x - x^0) = [0]^\top (x - x^0) = 0 \Longrightarrow f(x) \geqq f(x^0), \ \forall x \in X.$$

(ii) If x^0 is a point of local minimum for f, from Fermat's theorem we have that x^0 is a stationary point for f and from (i) of the present theorem we have that the same point is a global minimum point for f.

□

Again we point out that the two previous properties do not hold for quasiconvex (quasiconcave) functions. Consider, for example, $f(x) = x^3, \ x \in \mathbb{R}$, which has at zero an inflexion point with a horizontal tangent line. Consider, for example, the quasiconvex function drawn in Fig. 4.5 which has local minimum points, which are not global ones.

We have previously remarked that pseudoconvex (pseudoconcave) functions $f : \mathbb{R} \longrightarrow \mathbb{R}$ cannot have (in opposition to which may happen for quasiconvex or quasiconcave functions) inflection points with a horizontal tangent line. On the other hand, quasiconvex functions (quasiconcave functions), similarly to convex (concave) functions have the property that the sets of all their global minimum points (global maximum points) are *convex sets.*

Theorem 4.2.14 *Let $f : X \longrightarrow \mathbb{R}$ be a quasiconvex function on the convex set $X \subset \mathbb{R}^n$. The set of all global minimum points of f is a convex set.*

Proof Let x^0 and x^{00} be two global minimum points for f, that is $f(x^0) = f(x^{00})$. It holds

$$f(\lambda x^0 + (1 - \lambda)x^{00}) \leqq \max\left\{f(x^0), f(x^{00})\right\}, \quad \forall \lambda \in [0, 1]$$

and hence $f(\lambda x^0 + (1-\lambda)x^{00}) = f(x^0)$ (we cannot have $f(\lambda x^0 + (1-\lambda)x^{00}) < f(x^0)$, as x^0 is a global minimum point for f). Hence all points $\lambda x^0 + (1-\lambda)x^{00}$, $\lambda \in [0,1]$, are global minimum points for f and the thesis is proved. □

Furthermore, we have the following result.

Theorem 4.2.15 *Let $f : X \longrightarrow \mathbb{R}$ be a quasiconvex function on the convex set $X \subset \mathbb{R}^n$. If x^0 is a strict local minimum point for f, then x^0 is the unique strict global minimum point for f over X.*

For what concerns *maximum points* of convex functions, we have the following result.

Theorem 4.2.16 *Let $f : X \subset \mathbb{R}^n \longrightarrow \mathbb{R}$ be a convex function, non constant, on the convex set X. If $x^* \in X$ is a maximum point for f over X, then x^* is a boundary point of X, i.e. $x^* \in \partial X$.*

Proof If x^* would be an interior point of X, being f non constant on X, there would exist a point $y \in X$ such that $f(y) < f(x^*)$. Moreover, $f(x) \leqq f(x^*)$, $\forall x \in X$. Being $x^* \in \text{int}(X)$, it is always possible to find a vector $z \in X$ such that

$$x^* = \lambda y + (1-\lambda)z.$$

Hence

$$f(x^*) = f(\lambda y + (1-\lambda)z) \leqq \lambda f(y) + (1-\lambda) f(z) < f(x^*),$$

which is absurd. □

We conclude the present section with some hints on the concept of *subgradient* for a convex function, which is essential in modern Convex Analysis. The basic reference book for these questions is [32].

We have seen that a convex function $f : X \subset \mathbb{R}^n \longrightarrow \mathbb{R}$ is continuous at every point $x^0 \in \text{int}(X)$, however a convex function need not be differentiable, even at points $x^0 \in \text{int}(X)$. It is true that at these points convex functions admit (finite) directional derivatives with respect to every direction $v \in \mathbb{R}^n$, but as previously observed in Chap. 2, directional differentiability does not imply differentiability. Think, for example, to the simple function $f(x) = |x|$, $x \in \mathbb{R}$, convex for all $x \in \mathbb{R}$, but not differentiable at $x_0 = 0$. It is evident that even if $f(x) = |x|$ does not admit a tangent line at $(0,0)$, there are infinite straight lines which pass through $(0,0)$ and whose values are less or equal than the value of $f(x) = |x|$. In other words, there exist infinite straight lines $y = f(x_0) + m(x - x_0)$, with $x_0 = 0$, such that $f(x) \geqq f(x_0) + m(x - x_0)$. We can give the following definition.

Definition 4.2.17 Let $f : X \subset \mathbb{R}^n \longrightarrow \mathbb{R}$ be a convex function on the convex set X. A vector $p \in \mathbb{R}^n$ such that

$$f(x) \geqq f(x^0) + p^\top (x - x^0), \ \forall x \in X,$$

is called a *subgradient* of $f(x)$ at $x^0 \in X$ ("supergradient" if f is concave and the above inequality holds with $\leqq$). The set of all subgradients of $f(x)$ at a point x^0 is called the *subdifferential* of $f(x)$ at x^0 and usually denoted by $\partial f(x^0)$.

Hence, with reference to $f(x) = |x|$, we have that $\partial f(0) = [-1, 1]$. If $\partial f(x^0) \neq \emptyset$, we say that $f(x)$ is *subdifferentiable at* x^0. It can be proved that if $x^0 \in \text{int}(X)$, then $\partial f(x^0) \neq \emptyset$. Moreover, $\partial f(x^0)$ is a closed and convex set (in $\mathbb{R}^n$).

Theorem 4.2.18 *Let $f : X \longrightarrow \mathbb{R}$ be defined on the open convex set $X \subset \mathbb{R}^n$; then f is convex on X if and only if f has a subgradient at each point $x^0 \in X$, that is for every point $x^0 \in X$ there must exist $p \in \mathbb{R}^n$ such that*

$$f(x) \geqq f(x^0) + p^\top (x - x^0), \ \ \forall x \in X.$$

If $f(x)$ is differentiable at $x^0 \in \text{int}(X)$, and if $p \in \mathbb{R}^n$ is a vector satisfying the inequality of Definition 4.2.17, then p is unique and it holds $p = \nabla f(x^0)$, i.e. $\partial f(x^0) = \left\{\nabla f(x^0)\right\}$. Hence, if a convex function is differentiable at $x^0 \in \text{int}(X)$, its gradient at x^0 is the unique subgradient. But even in the case of $f(x)$ not differentiable at $x^0 \in \text{int}(X)$, we have the following interesting result, which characterizes subgradients by means of directional derivatives.

Theorem 4.2.19 *Let $f : X \subset \mathbb{R}^n \longrightarrow \mathbb{R}$ be a convex function on the convex set X and let $x^0 \in \text{int}(X)$; then $p \in \partial f(x^0)$ if and only if*

$$D^+ f(x^0, v) \geqq p^\top v, \ \forall v \in \mathbb{R}^n.$$

A more complete treatment of these questions, within "nonsmooth" Convex Analysis, is beyond the aims of the present book. We give only the following simple, but basic result.

Theorem 4.2.20 *Let $f : X \subset \mathbb{R}^n \longrightarrow \mathbb{R}$ be a convex function on the convex set X and let $x^0 \in \text{int}(X)$; then x^0 is a global minimum point of $f(x)$ over X if and only if $[0] \in \partial f(x^0)$.*

Proof If $f(x^0) \leqq f(x)$, $\forall x \in X$, then the inequality of Definition 4.2.17 is satisfied by $p = [0]$, i.e. $[0] \in \partial f(x^0)$. The vice-versa is also immediate. □

4.3 Optimization Problems with an Abstract Constraint

In the present section we take into consideration an optimization problem with a "set constraint" or "abstract constraint", that is the feasible set $K \subset \mathbb{R}^n$ is not necessarily generated by a system of equalities and/or inequalities and it is not (obviously) an open set. Hence the optimum point x^0 may belong to the boundary of K (if $x^0 \in \text{int}(K)$ we come back obviously to the case of unconstrained optimization problems). We need the notion of *feasible directions.* Let be $v \in \mathbb{R}^n$ (if $v \neq [0]$

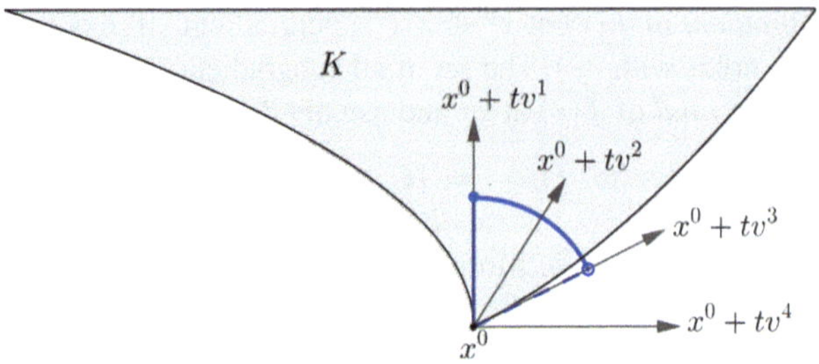

Fig. 4.6 The cone $F(K, x^0)$ is marked in blue. Note that $v^1, v^2 \in F(K, x^0)$, but $v^3, v^4 \notin F(K, x^0)$. One has $F(K, x^0) = \{\lambda_1 v^1 + \lambda_2 v^3 : \lambda_1 > 0, \lambda_2 \geqq 0\} \cup \{[0]\}$

we may suppose that $\|v\| = 1$, i.e. v is a "versor") and let $x^0 \in K \subset \mathbb{R}^n$ (K not necessarily open nor closed). The vector v is a *feasible direction at* x^0 with respect to K if there exists $\bar{t} > 0$ such that $x^0 + tv \in K,\ \forall t \in \left(0, \bar{t}\right]$. From a geometric point of view, a feasible direction v is such that, if $x^0 \in K$, every point $x^0 + tv$ of the segment joining x^0 and $x^0 + \bar{t}v$ belongs to K. We note that if $x^0 \in \operatorname{int}(K)$, every direction is feasible for K. If x^0 is an isolated point and if we consider $v \neq [0]$, no direction is feasible. The set of all feasible directions at $x^0 \in K$, with respect to the set K, is a cone, said "cone of feasible directions" (at x^0 with respect to K) and denoted by

$$F(K, x^0) = \left\{ v \in \mathbb{R}^n : \exists \bar{t} > 0 \text{ s.t. } x^0 + tv \in K,\ \forall t \in \left(0, \bar{t}\right] \right\}.$$

This cone is illustrated in Fig. 4.6.

Theorem 4.3.1 *Let $x^0 \in K$ be a local minimum (resp. maximum) point for f over K, with $f : \mathbb{R}^n \longrightarrow \mathbb{R}$, f differentiable at x^0 and $K \subset \mathbb{R}^n$. Then it holds*

$$\nabla f(x^0)^\top v \geqq 0, \quad \forall v \in F(K, x^0)$$

$$(\nabla f(x^0)^\top v \leqq 0, \quad \forall v \in F(K, x^0)).$$

Proof Being f differentiable at x^0, we have

$$f(x^0 + tv) - f(x^0) = t\nabla f(x^0)^\top v + o(t),$$

with $x^0 + tv \in K$ for all feasible vectors v at x^0 with respect to K. Let x^0 be, e.g., a local minimum point of f over K. If $\nabla f(x^0)^\top v < 0$, for $t > 0$ and convenient, it would hold also

$$f(x^0 + tv) - f(x^0) < 0,$$

against the assumption that x^0 is a local minimum point of f over K. □

Remark 4.3.2

(*i*) Recall that, being f differentiable at x^0, the quantity $\nabla f(x^0)^\top v$ is the *directional derivative of* f *at* x^0, in the (feasible) direction v. Theorem 4.3.1 just says that, under the assumptions of the same, it must hold $Df(x^0, v) \geqq 0$ ($Df(x^0, v) \leqq 0$) for all $v \in F(K, x^0)$. The meaning should be clear: if x^0 is, for instance, a local maximum point, any feasible direction, does not increase the value of f : the related directional derivative, with respect to the said directions, is less or equal to zero.

(*ii*) Theorem 4.3.1 is valid also under the assumptions that f admits at x^0 directional derivatives with respect to all feasible directions for the set K. The thesis is the same, with $Df(x^0, v)$ instead of $\nabla f(x^0)^\top v$.

(*iii*) Theorem 4.3.1 is a true generalization of Fermat's Theorem. Indeed, if $x^0 \in \operatorname{int}(K)$ and f is differentiable at x^0, as any direction from x^0 is a feasible direction for K, we have, for example in the case of a minimum point x^0,

$$\begin{cases} \nabla f(x^0)^\top v \geqq 0 \\ \nabla f(x^0)^\top(-v) \geqq 0 \iff \nabla f(x^0)v \leqq 0 \end{cases}$$

for all $v \in \mathbb{R}^n$. That is $\nabla f(x^0)^\top v = 0$, for all $v \in \mathbb{R}^n$, that is $\nabla f(x^0) = [0]$.

Example 4.3.3

(1) Let be $f(x) = x$, $x \in \mathbb{R}$, and $K = [0, 1]$. Obviously, f assumes its global maximum point at $x^0 = 1$; the unique feasible directional derivative at $x^0 = 1$ is the left-hand side derivative; indeed we have $f'_-(x^0)(-1) < 0$. At $x^{00} = 0$ the function assumes its global minimum and we have $f'_+(x^{00})(1) > 0$.

(2) Consider the function $f(x) = (x_1)^2 - x_1 + x_2 + x_1x_2$, and the constraint set $K = \mathbb{R}^2_+$.

We have

$$\frac{\partial f}{\partial x_1} = 2x_1 - 1 + x_2; \quad \frac{\partial f}{\partial x_2} = 1 + x_1.$$

Hence we have the stationary point $(-1, 3) \notin K$. The minimum point of f over K is $x^0 = \left[\frac{1}{2}, 0\right]^\top$. The feasible directions at this point, with respect to K, are given by $[v_1, v_2]^\top$, with $v_1 \in \mathbb{R}$ and $v_2 \geqq 0$. Then we have

$$\nabla f(x^0) = \begin{bmatrix} 0 \\ \frac{3}{2} \end{bmatrix},$$

and $\nabla f(x^0)^\top v = 0 \cdot v_1 + \frac{3}{2}v_2 = \frac{3}{2}v_2 \geqq 0$.

Remark 4.3.4 By recalling the definition of *polar cone* C^* of a cone $C \subset \mathbb{R}^n$ (Definition 3.1.6), it is possible to rewrite the thesis of Theorem 4.3.1 in the form

$$-\nabla f(x^0) \in (F(K, x^0))^*,$$

when x^0 is a local minimum point of f over K, or in the form

$$\nabla f(x^0) \in (F(K, x^0))^*,$$

when x^0 is a local maximum point of f over K. We have to remark that in some (perhaps "pathological") cases the above necessary optimality conditions become useless. For example, if

$$K = \left\{(x_1, x_2) \in \mathbb{R}^2 : (x_1)^2 + (x_2)^2 = 1\right\},$$

then the only feasible direction to K at any $x^0 \in K$ is the zero vector. If we pretend $v \neq [0]$, we have $F(K, x^0) = \emptyset$. In any case, the above optimality conditions become of no utility. In general, if we have K given by *linear constraints,* it is quite easy to compute the feasible directions at a point $x^0 \in K$ (with respect to K). But if the constraints are *not linear,* and we have $h_j(x^0) = 0,\ j = 1, \ldots, r < n$, in general there exists no vector $v \neq [0]$ such that $h_j(x^0+tv) = 0,\ \forall t \in \left(0, \bar{t}\right],\ \forall j = 1, \ldots, r$. This is the reason of introducing other "local cone approximations" which describe better than $F(K, x^0)$ the set K around a point $x^0 \in K$. One of the most used of these local cone approximations is the *tangent cone* or *Bouligand tangent cone,* called also *contingent cone,* to K at $x^0 \in K$, defined as

$$T(K, x^0) = \left\{ \begin{array}{c} v \in \mathbb{R}^n : \exists \{\lambda_n\} \subset \mathbb{R}_+,\ \exists \{x^n\} \subset K,\ \text{with } x^n \longrightarrow x^0, \\ \text{such that } \lambda_n(x^n - x^0) \longrightarrow v. \end{array} \right\}$$

$T(K, x^0)$ is a nonempty closed cone and if $x^0 \in \operatorname{int}(K)$, then $T(K, x^0) = \mathbb{R}^n$. Being always $F(K, x^0) \subset T(K, x^0)$ and even $\operatorname{cl}(F(K, x^0)) \subset T(K, x^0)$, we have $(T(K, x^0))^* \subset (F(K, x^0))^*$. Furthermore, it can be shown that, under the assumptions of Theorem 4.3.1, we have

$$-\nabla f(x^0) \in (T(K, x^0))^*$$

$$(\text{resp.} : \nabla f(x^0) \in (T(K, x^0))^*),$$

which are therefore necessary optimality conditions sharper than the ones of Theorem 4.3.1. However, we shall not pursue, in the present section, this approach to optimality conditions by means of the Bouligand tangent cone. For further insights, see, e.g., [7], [14], [15], and [17].

If K is a *convex set,* the thesis of Theorem 4.3.1 is simplified into the form

$$\nabla f(x^0)^\top(x - x^0) \geqq 0, \ \forall x \in K,$$

when x^0 is a local minimum point of f over K, and into the form

$$\nabla f(x^0)^\top(x - x^0) \leqq 0, \ \forall x \in K,$$

when x^0 is a local maximum point of f over K.

If f is *convex* or even *pseudoconvex* (resp. *concave* or even *pseudoconcave*) on the convex set K, the previous conditions become also *sufficient optimality conditions* (so that x^0 is a global minimum point or respectively a global maximum point of f over K). Indeed, this result is immediate, by recalling that a differentiable convex function is also pseudoconvex and that we have therefore

$$x, x^0 \in K, \ \nabla f(x^0)^\top(x - x^0) \geqq 0 \Longrightarrow f(x) \geqq f(x^0).$$

(Similarly for the case of f pseudoconcave).

For the optimization problem treated in the present section, we have also *second order necessary optimality conditions:*

Theorem 4.3.5 *Let $f : A \subset \mathbb{R}^n \longrightarrow \mathbb{R}$ be of class C^2 on the open set A and let $K \subset A$. If $x^0 \in K$ is a local minimum point (resp. a local maximum point) of f over K, then we have, for every feasible direction v at x^0, with respect to K,*

(1) $\nabla f(x^0)^\top v \geqq 0$ (resp. $\leqq 0$);
(2) if $\nabla f(x^0)^\top v = 0$, then $v^\top Hf(x^0)v \geqq 0$ (resp. $\leqq 0$).

4.4 Constrained Optimization Problems with Equality Constraints

In the present section we suppose that the feasible set K of an optimization problem is determined by the solutions of a system of equations, i.e. by the solution of a vector equation of the type $h(x) = [0]$, with $h : \mathbb{R}^n \longrightarrow \mathbb{R}^m$ and with $m < n$ (this last request is motivated by the fact that usually every equation of the system $h(x) = [0]$ reduces of one unity the "degrees of freedom" , and therefore, without the said request, it is very likely that the feasible set can shrink to a singleton or to the empty set, surely not interesting cases). Hence we take into consideration the following formulations:

$$\begin{cases} \min f(x) \\ \text{subject to: } h(x) = [0] \end{cases}; \quad \begin{cases} \max f(x) \\ \text{subject to: } h(x) = [0] \end{cases},$$

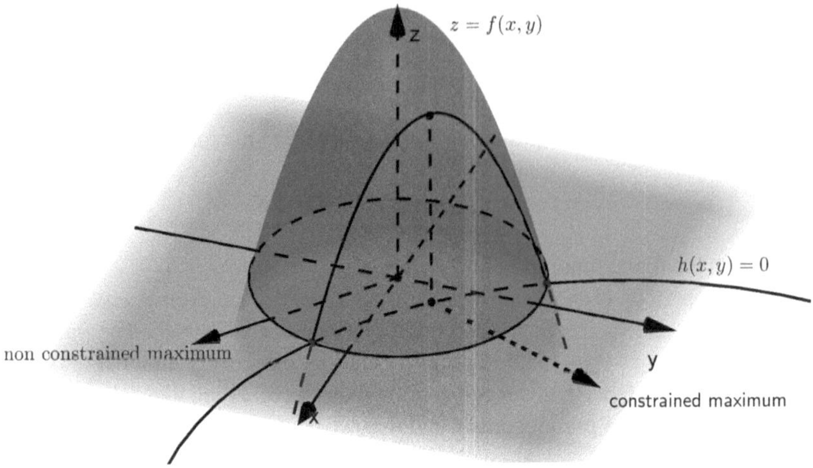

Fig. 4.7 Maximum of f non-constrained and maximum constrained to $h = 0$

where $f : \mathbb{R}^n \longrightarrow \mathbb{R}$ and $h : \mathbb{R}^n \longrightarrow \mathbb{R}^m$. These problems are also called "classical constrained optimization problems" , as they were introduced by J. L. Lagrange in the second half of the eighteenth century. Our formulation is quite general: if some constraint is of the type $h_j(x) = b_j$, with $b_j \neq 0$, it will suffice to put $g_j(x) = h_j(x) - b_j$ in order to obtain the above formulation. We recall that the set

$$K = \{x \in \mathbb{R}^n : h(x) = [0]\}$$

is the *feasible set*, the function $f(x)$ is the *objective function* and the functions

$$h_1(x), h_2(x), \ldots, h_m(x),$$

are the *constraints* of the problem.

We begin to consider the simplest case, i.e. the one where $n = 2$ and $m = 1$, i.e. we have two variables in the objective function and in the constraint, and we have one constraint (see Fig. 4.7):

$$\begin{cases} \quad \min(\max) f(x, y) \\ \text{subject to: } h(x, y) = 0. \end{cases}$$

In other words, we have to look for the extremum points (if any) of a function of two real variables on the zero level set ($\text{lev}_{=\alpha} h$, with $\alpha = 0$) described by the constraint of the problem. For the moment, let us suppose that f and h are at least differentiable functions on their domains. We take into consideration three methods for solving the above simple problems.

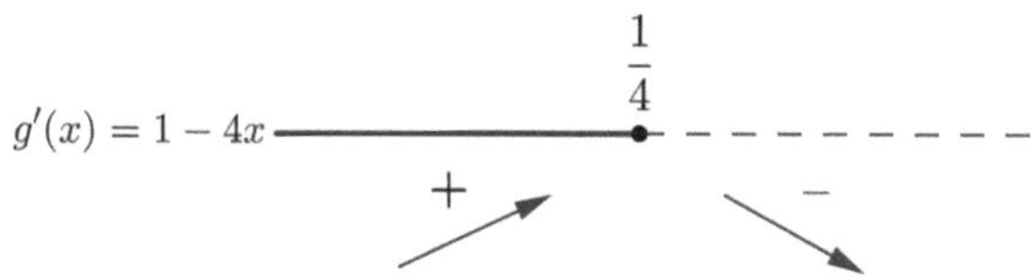

Fig. 4.8 Sign of $g'(x)$. The arrows indicate if g is increasing or decreasing

(A) *Elicitation or substitution method.*

We try to express a variable as a function of the remaining variable, in the constraint $h(x, y) = 0$. Then, by substituting the explicit form of the variable so obtained, into the objective function, we transform the constrained problem into an unconstrained problem. Some examples will clarify the procedure.

Example 4.4.1

(a) Find the extremum points, if any, of $f(x, y) = xy$ over the feasible set

$$K = \left\{(x, y) \in \mathbb{R}^2 : 2x + y - 1 = 0\right\}.$$

Form the constraint we get $y = 1 - 2x$. By substituting into the objective function, we have

$$g(x) = x(1 - 2x) = x - 2x^2;$$
$$g'(x) = 1 - 4x.$$

The study of the sign of $g'(x)$ is made in Fig. 4.8.

The point $\bar{x} = \frac{1}{4}$ is the maximum point for g. From $y = 1 - 2x$ we get $\bar{y} = 1 - 2\frac{1}{4} = \frac{1}{2}$. Hence the point $\left(\frac{1}{4}, \frac{1}{2}\right)$ is the unique constrained maximum pointy of f on K.

(b) Find, if any, the extremum points of $z = f(x, y) = \log x + \log y - xy$ and find the extremum points of the same function, subject to the constraint: $y - x^2 = 0$.

We have

$$\frac{\partial f}{\partial x} = \frac{1}{x} - y; \quad \frac{\partial f}{\partial y} = \frac{1}{y} - x.$$

Therefore we have $xy = 1$, with x and y lying on the first Cartesian quadrant. Then we have

$$\frac{\partial^2 f}{\partial x^2} = -\frac{1}{x^2}; \quad \frac{\partial^2 f}{\partial x \partial y} = \frac{\partial^2 f}{\partial y \partial x} = -1; \quad \frac{\partial^2 f}{\partial y^2} = -\frac{1}{y2}$$

and hence

$$Hf(x, \frac{1}{x}) = \begin{bmatrix} -\frac{1}{x^2} & -1 \\ -1 & -x^2 \end{bmatrix}.$$

It holds $\Delta_1 < 0,\ \Delta_2 = \det(Hf(x, \frac{1}{x})) = -\frac{1}{x^2}(-x^2) - 1 = 0$. The Hessian matrix is therefore negative semidefinite on all stationary points. However, for $x > 0$ and $y > 0$ we have $z = \log(xy) - xy$ which becomes, at all stationary points, $z = \log 1 - 1 = -1$. Then we have $\Delta f = \log(xy) - xy - (-1) = \log(xy) - xy + 1$. Putting $xy = t$ we have that it holds $\log t \leqq t - 1$, hence all points $xy = 1$ $(x > 0,\ y > 0)$ are maximum points for f.

Now we consider the constraint $y = x^2$. The objective function becomes

$$g(x) = \log x + \log x^2 - x^3,$$

and, as it holds,

$$g'(x) = \frac{3 - 3x^3}{x},$$

we conclude that $\bar{x} = 1$ is a maximum point for g and hence $(1, 1)$ is a constrained maximum point for z.

(*c*) Find, if any, the extremum points of $f(x, y) = 2x - y + 1$ under the constraint $y - 2x^3 = 0$.

We have that the constraint can be expressed as $y = 2x^3$; therefore the objective function becomes

$$\varphi(x) = 2x - 2x^3 + 1.$$

We have $\varphi'(x) = 2(1 - 3x^2)$, hence $x = -\frac{1}{\sqrt{3}},\ y = -\frac{2}{3\sqrt{3}}$ is a local constrained minimum point and $x = \frac{1}{\sqrt{3}},\ y = \frac{2}{3\sqrt{3}}$ is a local constrained maximum point (the function $\varphi(x)$ is unbounded and also $f(x, y)$ is unbounded on the feasible set).

This method must be applied with a bit of caution: the explicitation makes sense only for those values of the independent variable for which the dependent variable is defined and the constraint is verified. If we have, for instance, the problem

$$\begin{cases} \min(\max) f(x, y) = x^2 - xy \\ \quad \text{subject to: } xy = 1 \end{cases}$$

and we substitute the constraint into the objective function, we get $g(x) = x^2 - 1$, which assumes its minimum point at $\bar{x} = 0$, where, however, the constraint is not verified (really, the problem admits no maximum points, nor minimum points over the above constraint).

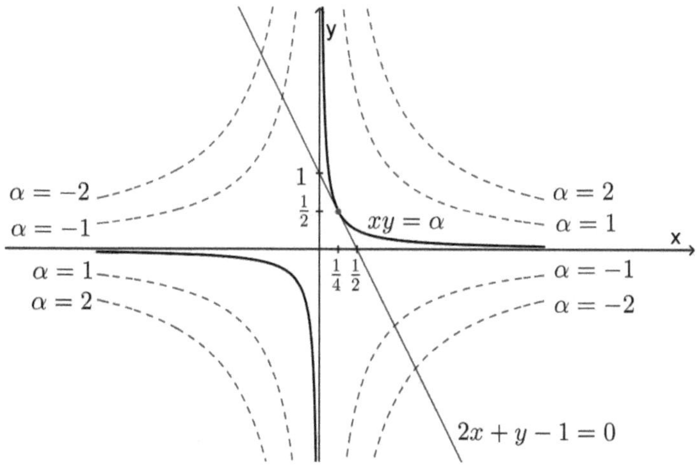

Fig. 4.9 Geometric method applied to Example 4.4.1. Some level lines of f and level line $xy = \alpha$ (in black) tangent to the straight line $2x + y - 1 = 0$

(B) *Geometric method by means of level sets or level lines.*

Also this method may be useful in the case of two variables and one constraint. We recall that, given $z = f(x, y)$, the level line equal to α is given by the orthogonal projection on the plane xOy of all sets of points of the surface (generated by f) and having value $z = \alpha$, i.e., with the previous notation, $\text{lev}_{=\alpha} f = \{(x, y) \in \mathbb{R}^2 : f(x, y) = \alpha\}$.

Now we wish to solve the problem of Example 4.4.1, point (a), by the "geometric method" . The level lines of the objective function are given by $xy = \alpha$. With $\alpha > 0$ they are equilateral hyperbolas lying on the first and third Cartesian quadrant; with $\alpha < 0$ the said hyperbolas belong to the second and fourth Cartesian quadrant. Finally, if $\alpha = 0$, we obtain the two Cartesian axes. See Fig. 4.9, where some level lines of f are shown, together with the constraint, given by the straight line $2x + y - 1 = 0$.

Hence we have to find the value of $\alpha > 0$ such that the line $xy = \alpha$ is tangent to the straight line generated by the constraint (obviously if $\alpha \leqq 0$ the problem admits no solution). Thus we have

$$\begin{cases} xy = \alpha \\ 2x + y - 1 = 0 \end{cases} \Longrightarrow \begin{cases} x(1 - 2x) = \alpha \\ y = 1 - 2x \end{cases} \Longrightarrow -2x^2 + x - \alpha = 0.$$

Being $\Delta = 1 - 8\alpha$, from $1 - 8\alpha = 0$ we obtain $\alpha = \frac{1}{8}$. From $x(1 - 2x) = \frac{1}{8}$ we obtain $-(4x - 1)^2 = 0$, that is $x = \frac{1}{4}$. Therefore we obtain that the point $A(\frac{1}{4}, \frac{1}{2})$ is a constrained maximum point for f over K.

(C) *Method of Lagrange multipliers.*

This method is a general method, that is it works also for classical constrained optimization problems, with $n > 2$ and $m > 1$. We give first the necessary optimality conditions for the case under examination, i.e. for $n = 2$, $m = 1$. Subsequently, we shall give the general theorem.

Theorem 4.4.2 *Let $f(x, y)$ be differentiable and $h(x, y)$ be of class C^1 on the open set $A \subset \mathbb{R}^2$. Let (x_0, y_0) be a local maximum or a local minimum point of f over*

$$K = \left\{(x, y) \in \mathbb{R}^2 : h(x, y) = 0\right\}$$

and let be $\nabla h(x_0, y_0) \neq [0]$. Then, there exists a (unique) scalar $\lambda \in \mathbb{R}$, said Lagrange multiplier, such that

$$\nabla f(x_0, y_0) - \lambda \nabla h(x_0, y_0) = [0]. \tag{4.1}$$

Proof Let be, for instance, $(\partial h(x_0, y_0)/\partial y) \neq 0$. By the Implicit Function Theorem (Theorem 2.6.2), we have respected the related assumptions), there exists a neighbourhood $I(x_0)$ where $h(x, y)$ implicitly defines a function $y = \varphi(x)$ such that

$$y_0 = \varphi(x_0); \quad \varphi'(x) = -\frac{\partial h/\partial x}{\partial h/\partial y}.$$

In the said neighbourhood we have also

$$z = f(x, y) = f(x, \varphi(x)) = F(x).$$

Therefore it must hold $F'(x_0) = f'(x_0, \varphi(x_0)) = 0$. Being

$$F'(x) = \frac{\partial f}{\partial x} + \frac{\partial f}{\partial y}\varphi'(x),$$

we obtain

$$F'(x) = \frac{\partial f}{\partial x} + \frac{\partial f}{\partial y} \cdot \frac{\frac{-\partial h}{\partial x}}{\frac{\partial h}{\partial y}}.$$

Hence, at (x_0, y_0) it will hold

$$\frac{\partial f}{\partial x} - \left(\frac{\partial f}{\partial y} / \frac{\partial h}{\partial y}\right) \frac{\partial h}{\partial x} = 0.$$

On the other hand, we have the following identity:

$$\frac{\partial f}{\partial y} - \left(\frac{\partial f}{\partial y} / \frac{\partial h}{\partial y}\right) \frac{\partial h}{\partial y} = 0.$$

Now we put

$$\left(\frac{\partial f}{\partial y} / \frac{\partial h}{\partial y}\right) = \lambda.$$

Therefore we have, at (x_0, y_0),

$$\begin{cases} \frac{\partial f}{\partial x} - \lambda \frac{\partial h}{\partial x} = 0 \\ \frac{\partial f}{\partial y} - \lambda \frac{\partial h}{\partial y} = 0 \end{cases},$$

that is

$$\nabla f(x_0, y_0) - \lambda \nabla h(x_0, y_0) = [0].$$

□

Remark 4.4.3

(1) As, in the above theorem, there are no restrictions on the sign of the multiplier λ, obviously the thesis of the theorem may be rewritten in the equivalent form

$$\nabla f(x_0, y_0) + \lambda \nabla h(x_0, y_0) = [0].$$

(2) The function $\mathcal{L} : \mathbb{R}^3 \longrightarrow \mathbb{R}$ defined as follows

$$\mathcal{L}(x, y, \lambda) = f(x, y) \pm \lambda h(x, y)$$

is said *Lagrangian function*. It is immediate to note that the sentence "the point $(x_0, y_0) \in K$ satisfies relation (4.1)" is equivalent to the sentence:

"The triplet (x_0, y_0, λ) is a stationary point, with respect to x, y, λ, of the Lagrangian function $\mathcal{L}$, that is it must hold

$$\begin{cases} \nabla_{x,y}\mathcal{L}(x, y, \lambda) = [0] \\ \nabla_{\lambda}\mathcal{L}(x, y, \lambda) = 0." \end{cases}$$

In a more concise way:

$$\nabla \mathcal{L}(x, y, \lambda) = [0]. \tag{4.2}$$

(3) We stress that the condition of Theorem 4.4.2 is only a necessary optimality condition; moreover, it is evident that it holds both for minimizers and for maximizers.

It may be interesting to give a geometric interpretation of the result of Theorem 4.4.2. Let us suppose that both $f(x, y)$ and $h(x, y)$ are of $\mathcal{C}^1$class on the open set $A \subset \mathbb{R}^2$. Let be, for instance, $(\partial h(x_0, y_0)/\partial y) \neq 0$ and suppose that in (4.1) we have $\lambda \neq 0$. From

$$\frac{\partial f(x_0, y_0)}{\partial x} = \lambda \frac{\partial h(x_0, y_0)}{\partial x}$$
$$\frac{\partial f(x_0, y_0)}{\partial y} = \lambda \frac{\partial h(x_0, y_0)}{\partial y}$$

it follows, under the above assumptions,

$$\frac{\frac{\partial f(x_0, y_0)}{\partial x}}{\frac{\partial f(x_0, y_0)}{\partial y}} = \frac{\frac{\partial h(x_0, y_0)}{\partial x}}{\frac{\partial h(x_0, y_0)}{\partial y}}. \tag{4.3}$$

As $(\partial h(x_0, y_0)/\partial y) \neq 0$, the equation of the level line $f(x, y) = f(x_0, y_0)$ defines implicitly, thanks to the Implicit Function Theorem of Dini (Theorem 2.6.2), a function $y = \psi(x)$, whose derivative, evaluated at x_0, is given by

$$\psi'(x_0) = -\frac{\frac{\partial f(x_0, y_0)}{\partial x}}{\frac{\partial f(x_0, y_0)}{\partial y}},$$

that is, the first member of (4.3), changed of sign. Moreover, we know that under the assumptions made, the constraint $h(x, y) = 0$ defines implicitly on a neighbourhood of (x_0, y_0), a function $y = \varphi(x)$, whose derivative, evaluated at x_0, is given by

$$\varphi'(x_0) = -\frac{\frac{\partial h(x_0, y_0)}{\partial x}}{\frac{\partial h(x_0, y_0)}{\partial y}},$$

that is, the second member of (4.3), changed of sign. Then we have, thanks to (4.3), $\psi'(x_0) = \varphi'(x_0)$. As $y_0 = \varphi(x_0) = \psi(x_0)$, from a geometric point of view, this means that at the optimum point (x_0, y_0) the constraint line and the level line of the objective function have a common tangent line. See Fig. 4.10.

Think over the above figure. At all points where the constraint crosses a level line of the objective function, there cannot exist maximum nor minimum points. At a (constrained) maximum or minimum point the constraint "touches" a level line of the objective function, but without crossing the same. If $\lambda = 0$ this geometric interpretation may not hold.

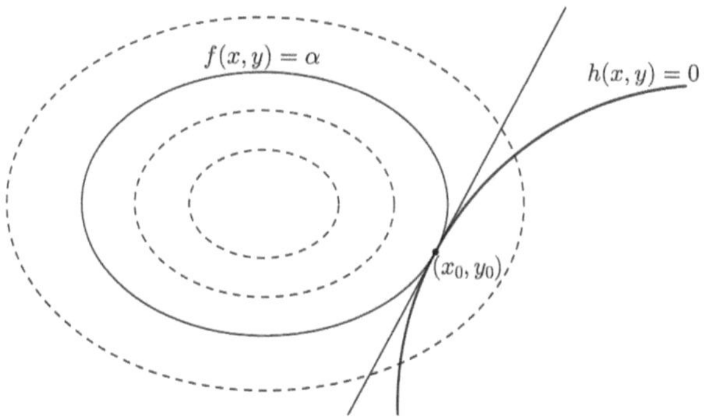

Fig. 4.10 Level lines of f and the line $h(x, y) = 0$. At the optimum, the constraint and the level line of f have a common tangent line

Now we give the second-order sufficient optimality conditions, always for the case under consideration, i.e. $n = 2,\ m = 1$.

Theorem 4.4.4 *Let $f : \mathbb{R}^2 \longrightarrow \mathbb{R}$ and $h : \mathbb{R}^2 \longrightarrow \mathbb{R}$ be of class C^2 on the open set $A \subset \mathbb{R}^2$. Let (x_0, y_0, λ_0) a triplet which satisfies relation* (4.2). *If the quadratic form*

$$z^\top H_{x,y}\mathcal{L}(x_0, y_0, \lambda_0)z \tag{4.4}$$

is positive definite (resp. negative definite) for all vectors $z \in \mathbb{R}^2$, $z \neq [0]$, such that $\nabla h(x_0, y_0)^\top z = 0$, then (x_0, y_0) is a strict local minimum point (resp. a strict local maximum point) of f over $K = \left\{(x, y) \in \mathbb{R}^2 : h(x, y) = 0\right\}$.

This theorem will be proved for the general case (Theorem 4.4.7). The theorem imposes the examination of the sign of the quadratic form (4.4), constrained to the vectors z which are orthogonal to the gradient $\nabla h(x_0, y_0)$. By recalling the criterion on quadratic forms, constrained by a system of homogeneous linear relations, we obtain the following result:

If the triplet (x_0, y_0, λ_0) satisfies relation (4.2), and if we consider the bordered matrix

$$\Delta = \begin{bmatrix} 0 & \nabla h(x_0, y_0) \\ (\nabla h(x_0, y_0))^\top & H_{x,y}\mathcal{L}(x_0, y_0, \lambda_0) \end{bmatrix},$$

then if it holds $\det(\Delta) > 0$, then (x_0, y_0) is a constrained strict local minimum point. If $\det(\Delta) < 0$, then (x_0, y_0) is a constrained strict local maximum point.

Example 4.4.5 Find, by using the Lagrange multipliers method, the extremum points, if any, of $f(x, y) = x + y$ subject to $h(x, y) = x^2 + 2y^2 - 6 = 0$.

We have $\mathcal{L}(x, y, \lambda) = x + y - \lambda(x^2 + 2y^2 - 6)$,

$$\begin{cases} \frac{\partial \mathcal{L}}{\partial x} = 1 - \lambda 2x = 0 \\ \frac{\partial \mathcal{L}}{\partial y} = 1 - \lambda 4y = 0 \\ \frac{\partial \mathcal{L}}{\partial \lambda} = -(x^2 + 2y^2 - 6) = 0. \end{cases}$$

As both x and y cannot be zero (otherwise they would betray the first two equations of the above system), we have

$$\lambda = \frac{1}{2x} = \frac{1}{4y}$$

from which we obtain $x = 2y$. Then, from the third equation we obtain $4y^2 + 2y^2 = 6$, that is $6y^2 = 6$, that is $y^2 = 1$, that is $y = \pm 1$; $x = \pm 2$.

Hence we have the two triplets:

$$A = (2, 1, \frac{1}{4}); \quad B = (-2, -1, -\frac{1}{4}).$$

We have $\frac{\partial \mathcal{L}}{\partial x} = 1 - \lambda 2x$ and $\frac{\partial \mathcal{L}}{\partial y} = 1 - \lambda 4y$, hence

$$H_{x,y}\mathcal{L}(x, y, \lambda) = \begin{bmatrix} -2\lambda & 0 \\ 0 & -4\lambda \end{bmatrix}.$$

Now we consider the following bordered matrix

$$\Delta = \begin{bmatrix} 0 & \frac{\partial h}{\partial x}; \frac{\partial h}{\partial y} \\ \frac{\partial h}{\partial x} & \\ \frac{\partial h}{\partial y} & H_{x,y}\mathcal{L}(x, y, \lambda) \end{bmatrix}$$

that is

$$\Delta = \begin{bmatrix} 0 & 2x & 4y \\ 2x & -2\lambda & 0 \\ 4y & 0 & -4\lambda \end{bmatrix}.$$

At $A = (2, 1, \frac{1}{4})$ we have

$$\Delta = \begin{bmatrix} 0 & 4 & 4 \\ 4 & -\frac{1}{2} & 0 \\ 4 & 0 & -1 \end{bmatrix}; \quad \det(\Delta) = 24 > 0.$$

Therefore the point (2, 1) is a constrained strict local maximum point. At $B = (-2, -1, -\frac{1}{4})$ we have

$$\Delta = \begin{bmatrix} 0 & -4 & -4 \\ -4 & \frac{1}{2} & 0 \\ -4 & 0 & 1 \end{bmatrix}; \quad \det(\Delta) = -24 < 0.$$

Therefore the point $(-2, -1)$ is a constrained strict local minimum point.

Now we consider the general case, i.e. where $f : \mathbb{R}^n \longrightarrow \mathbb{R}$ and $h : \mathbb{R}^n \longrightarrow \mathbb{R}^m$. We remark at once that the two first methods previously considered, for the case $n = 2,\ m = 1$, are in general no longer useful, whereas the method of Lagrange multipliers has a validity also for the general case.

Theorem 4.4.6 (Necessary First-Order Optimality Conditions) *Let be given the problem*

$$\begin{aligned} &\min(\max)\, f(x) \\ &\textit{subject to: } x \in K = \left\{x \in \mathbb{R}^n : h(x) = [0]\right\}, \end{aligned}$$

with $f : A \longrightarrow \mathbb{R}$, $h : A \longrightarrow \mathbb{R}^m$, $A \subset \mathbb{R}^n$, A open, $m < n$, f differentiable on A, h of class C^1 on A. Let x^0 be a local minimum or a local maximum point of f over K and let the Jacobian matrix $\nabla h(x^0)$ be of full rank (i.e. $\operatorname{rank}(\nabla h(x^0)) = m$*, i.e. the gradients $\nabla h_1(x^0), \ldots, \nabla h_m(x^0)$ are linearly independent; in this case the point x^0 is called a regular point). Then, there exists a unique vector $\lambda^0 \in \mathbb{R}^m$ ("vector of Lagrange multipliers") such that (x^0, λ^0) is a stationary point for the Lagrangian function, i.e. it holds*

$$\nabla f(x^0) - (\lambda^0)^\top \nabla h(x^0) = [0]. \tag{4.5}$$

Proof We give a proof similar to the one given for the simple case of $n = 2,\ m = 1$ (Theorem 4.4.2). If necessary, without loss of generality, renumber the variables so that the last m columns of the Jacobian matrix $\nabla h(x^0)$ are linearly independent. The vector x will be therefore decomposed into two sub-vectors, say x^1 and x^2, with x^1 having $(n-m)$ elements and x^2 having m elements: $x = \left[x^1, x^2\right]^\top$. By the Implicit Function Theorem of Dini (Theorem 2.6.2), there exists a suitable neighbourhood of x^0 where it holds

$$x^2 = g(x^1)$$

where g is a vector function with m components. The initial problem may therefore be written, at least locally, as an unconstrained optimization problem:

$$\min(\max)\, F(x^1) = f(x^1, g(x^1)).$$

Now we denote by $\frac{\partial f}{\partial x^1}$ the gradient of f with respect to x^1 (vector of $\mathbb{R}^{n-m}$), by $\frac{\partial f}{\partial x^2}$ the gradient of f with respect to x^2 (vector of $\mathbb{R}^m$), and by $\frac{\partial g}{\partial x^1}$ the Jacobian matrix of g with respect to x^1 (matrix of order $(m, n-m)$). Therefore we have the following necessary condition for the existence of a local optimum point for F :

$$\frac{\partial F}{\partial x^1} = \frac{\partial f}{\partial x^1} + \left(\frac{\partial f}{\partial x^2}\right)^\top \frac{\partial g}{\partial x^1} = [0] .$$

As the constraints can be written, at least locally, in the form

$$h(x^1, g(x^1)) = [0] ,$$

we have ("chain rule")

$$\frac{\partial h}{\partial x^1} + \frac{\partial h}{\partial x^2}\frac{\partial g}{\partial x^1} = [0] ,$$

where the Jacobian matrix $\frac{\partial h}{\partial x^1}$ is of order $(m, n-m)$ and the Jacobian matrix $\frac{\partial h}{\partial x^2}$, of order m, is nonsingular by assumption. Hence

$$\frac{\partial g}{\partial x^1} = -\left(\frac{\partial h}{\partial x^2}\right)^{-1}\left(\frac{\partial h}{\partial x^1}\right).$$

The condition $\frac{\partial F}{\partial x^1} = [0]$ can be therefore rewritten in the form

$$\frac{\partial f}{\partial x^1} - \left(\frac{\partial f}{\partial x^2}\right)^\top \left(\frac{\partial h}{\partial x^2}\right)^{-1}\left(\frac{\partial h}{\partial x^1}\right) = [0] .$$

On the other hand, the following identity holds true

$$\frac{\partial f}{\partial x^2} - \left(\frac{\partial f}{\partial x^2}\right)^\top \left(\frac{\partial h}{\partial x^2}\right)^{-1}\left(\frac{\partial h}{\partial x^2}\right) = [0] .$$

Hence, if we put

$$\lambda = \frac{\partial f}{\partial x^2}\left(\frac{\partial h}{\partial x^2}\right)^{-1} , \quad \lambda \in \mathbb{R}^m,$$

we get the thesis of the theorem. □

Theorem 4.4.7 (Sufficient Second-Order Optimality Conditions) *Let f and h be of class C^2 on the open set $A \subset \mathbb{R}^n$; If the pair (x^0, λ^0) satisfies relation* (4.5), *with $x^0 \in K$ and if the following quadratic form*

$$z^\top H_x \mathcal{L}(x^0, \lambda^0) z$$

is positive definite (resp. negative definite) for every vector $z \in \mathbb{R}^n,\ z \neq [0]$, such that

$$\nabla h_i(x^0)^\top z = 0, \quad i = 1, \ldots, m,$$

the point x^0 is a strict local minimum (resp. a strict local maximum) of f over K.

Proof Let us suppose that x^0 is not a constrained strict local minimum point; then there exists a sequence $\left\{x^k\right\} \subset K$ such that

$$x^k \longrightarrow x^0, \ \ x^k \neq x^0$$

and

$$f(x^k) \leqq f(x^0), \ \ \forall k.$$

As $z^k = (x^k - x^0)/\left\|x^k - x^0\right\|$ is a "normalized" sequence, i.e. $\left\|z^k\right\| = 1$, it is possible to obtain a subsequence, for simplicity indexed as the previous one, which converges:

$$z^k = \frac{x^k - x^0}{\left\|x^k - x^0\right\|} \longrightarrow z \neq [0], \text{ for } k \longrightarrow +\infty.$$

Put $t_k = \left\|x^k - x^0\right\|$, then $x^k = x^0 + t_k z^k$. On the one hand, as $x^k \in K$, one has

$$h_i(x^k) = 0, \ \ \forall k, \ i = 1, \ldots, m.$$

From here and since $h_i(x^0) = 0, i = 1, \ldots, m$, we derive that

$$0 = h_i(x^0 + t_k z^k) - h_i(x^0) = t_k \nabla h_i(x^0)^\top z^k + t_k \|z^k\| \eta_k, \ \ \forall k, \ i = 1, \ldots, m,$$

with $\eta_k \longrightarrow 0$ for $k \longrightarrow +\infty$. Dividing by t_k and taking the limit, we deduce that

$$\nabla h_i(x^0)^\top z = 0, \quad i = 1, \ldots, m.$$

On the other hand, we have

$$\mathcal{L}(x^0 + t_k z^k, \lambda^0) = \mathcal{L}(x^0, \lambda^0) + t_k \nabla_x \mathcal{L}(x^0, \lambda^0) z^k + \frac{t_k^2}{2} (z^k)^\top H\mathcal{L}(x^0, \lambda^0) z^k$$
$$+ t_k^2 \left\| z^k \right\|^2 \varepsilon_k,$$

with $\varepsilon_k \longrightarrow 0$ for $k \longrightarrow +\infty$.

But

$$\mathcal{L}(x^0 + t_k z^k, \lambda^0) = f(x^k);$$
$$\mathcal{L}(x^0, \lambda^0) = f(x^0)$$

and

$$\nabla_x \mathcal{L}(x^0, \lambda^0) z^k = 0.$$

Hence

$$0 \geqq \frac{f(x^k) - f(x^0)}{t_k^2/2} = (z^k)^\top H\mathcal{L}(x^0, \lambda^0) z^k + 2 \left\| z^k \right\|^2 \varepsilon_k.$$

When $k \longrightarrow +\infty$, the limit in the right-hand side is of course

$$z^\top H\mathcal{L}(x^0, \lambda^0) z > 0$$

by assumption since $\nabla h_i(x^0)^\top z = 0$, $i = 1, \dots, m$. Hence the contradiction, being $z \neq [0]$. □

In view of applications, it is convenient to recall the criterion for establishing the sign of a constrained quadratic form, with constraints given by an homogeneous linear system of the form $Bx = [0]$ (Theorem 1.7.13).

• We build the bordered matrix, of order $(m + n)$,

$$M = \begin{bmatrix} [0] & \nabla h(x^0) \\ \nabla h(x^0)^\top & H_x \mathcal{L}(x^0, \lambda^0) \end{bmatrix}$$

(here $\nabla h(x^0)$ is the Jacobian matrix of $h(x)$ evaluated at x^0) and we consider the North-West principal minors of M, of order $2m + 1, \dots, m + n :\ \Delta_h,\ h = 2m + 1, \dots, m + n$.

(*a*) If the said North-West principal minors have the sign of $(-1)^m$, that is $(-1)^m \Delta_h > 0,\ h = 2m + 1, \dots, m + n$, then the constrained quadratic form of Theorem 4.4.7 is positive definite, and hence x^0 is a constrained strict local minimum point.

(*b*) If the said North-West principal minors alternate in sign, by beginning with the sign of $(-1)^{m+1}$, then the constrained quadratic form of Theorem 4.4.7 is negative definite, and hence x^0 is a constrained strict local maximum point.

As previously observed, if $m = 1$ and $n = 2$, we have to check only the sign of $\det(M)$:

- If $\det(M) < 0$, then x^0 is a constrained strict local minimum point;
- If $\det(M) > 0$, then x^0 is a constrained strict local maximum point.

Example 4.4.8 Consider the problem

$$\min(\max)\ 5x_1 + 2x_2 - x_3$$

$$\text{subject to: } x_1x_2 = 3$$
$$x_1x_3 = 1.$$

We write the Lagrangian function in the form

$$\mathcal{L}(x, \lambda) = 5x_1 + 2x_2 - x_3 + \lambda_1(3 - x_1x_2) + \lambda_2(1 - x_1x_3)$$

and the system $\nabla\mathcal{L}(x, \lambda) = [0]$:

$$\begin{cases} 3 - x_1x_2 = 0 \\ 1 - x_1x_3 = 0 \\ 5 - \lambda_1x_2 - \lambda_2x_3 = 0 \\ 2 - \lambda_1x_1 = 0 \\ -1 - \lambda_2x_1 = 0. \end{cases}$$

This system has two solutions (stationary points of the Lagrangian function):

$$(x^0, \lambda^0) = (1, 3, 1, 2, -1); \quad (x^{00}, \lambda^{00}) = (-1, -3, -1. - 2, 1).$$

It results

$$M = \Delta = \begin{bmatrix} 0 & 0 & -x_2 & -x_1 & 0 \\ 0 & 0 & -x_3 & 0 & -x_1 \\ -x_2 & -x_3 & 0 & -\lambda_1 & -\lambda_2 \\ -x_1 & 0 & -\lambda_1 & 0 & 0 \\ 0 & -x_1 & -\lambda_2 & 0 & 0 \end{bmatrix}.$$

Being $n = 3$, $m = 2$, we have to check the sign of the North-West principal minors of Δ, by starting from the one of order $2m + 1 = 5$, hence we have to check only the sign of $\det(\Delta)$. For (x^0, λ^0) we have $\det(\Delta) = 10$, hence $\det(\Delta)$ has the sign of $(-1)^m = (-1)^2$ and therefore $x^0 = (1, 3, 1)$ is a constrained strict local minimum point. For (x^{00}, λ^{00}) we find $\det(\Delta) = -10$, hence $\det(\Delta)$ has the sign of $(-1)^{m+1} = (-1)^3$ and therefore $x^{00} = (-1, -3, -1)$ is a constrained strict local maximum point.

Remark 4.4.9

(*a*) The previous sufficient second-order optimality conditions yield strict local optimum points. This may be of scarce interest in many applications, especially of economic and financial type. It is not difficult to prove that if $\mathcal{L}(x, \lambda)$ is convex with respect to x (or even pseudoconvex, with respect to x) and $x^0 \in K$ satisfies relation (4.5), then x^0 is a *global* minimum point of f over K (global maximum point if $\mathcal{L}(x, \lambda)$ is concave or even pseudoconcave, with respect to x). For instance, if we write the Lagrangian function in the form $\mathcal{L}(x, \lambda) = f(x) - \lambda h(x)$, if f is convex, $\lambda > [0]$ and every h_i, $i = 1, \ldots, m$, is concave, then $\mathcal{L}$ is convex. If, for instance, f is convex and h is linear (and hence both convex and concave), then $\mathcal{L}$ is convex, regardless the sign of the multipliers vector λ (see Theorem 3.2.11). Indeed, consider $x^0 \in K$ and (x^0, λ^0) such that $\nabla_x \mathcal{L}(x^0, \lambda^0) = [0]$. Let be, for instance, $\mathcal{L}(x, \lambda)$ pseudoconvex with respect to x, i.e.

$$\nabla_x \mathcal{L}(x^0, \lambda^0)^\top (x - x^0) \geqq 0 \Longrightarrow \mathcal{L}(x, \lambda^0) \geqq \mathcal{L}(x^0, \lambda^0), \ \forall x, x^0.$$

If the first-order optimality conditions hold at (x^0, λ^0), the first member of the above implication is verified and then it will hold, for every $x \in K$, $f(x) \geqq f(x^0)$, being $\lambda^0 h(x) = 0$, for every $x \in K$.

For example, if we have the problem of finding the extremum points of

$$f(x, y, z) = x^2 + y^2 + z^2$$

subject to:

$$\begin{cases} h_1(x, y, z) = x + y - 1 = 0 \\ h_2(x, y, z) = x - z = 0, \end{cases}$$

we write the Lagrangian function in the form

$$\mathcal{L}(x, y, z, \lambda_1, \lambda_2) = x^2 + y^2 + z^2 - \lambda_1(x + y - 1) - \lambda_2(x - z).$$

The first-order necessary conditions are

$$\begin{cases} 2x - \lambda_1 - \lambda_2 = 0 \\ 2y - \lambda_1 = 0 \\ 2z + \lambda_2 = 0 \\ x + y - 1 = 0 \\ x - z = 0. \end{cases}$$

The unique solution of this system is $(x, y, z) = (\frac{1}{3}, \frac{2}{3}, \frac{1}{3})$, with multipliers $\lambda_1 = \frac{4}{3}$, $\lambda_2 = -\frac{2}{3}$.

Being f convex and the constraints linear functions, the Lagrangian function is convex (with respect to the variables) and we can conclude that the stationary point we have found is a constrained global minimum point for f.

(*b*) The first-order condition

$$\nabla f(x^0) - \lambda^\top \nabla h(x^0) = [0]$$

is obviously equivalent to

$$\nabla f(x^0) = \sum_{i=1}^{m} \lambda_i \nabla h_i(x^0),$$

that is, $\nabla f(x^0)$ is given by the linear combination of the gradients $\nabla h_i(x^0)$, $i = 1, \ldots, m$, with coefficients $\lambda_1, \lambda_2, \ldots, \lambda_m$.

(*c*) The theorem on Lagrange multipliers may not "work" if its assumptions are not respected (similarly to what happens for all theorems). Consider the following example.

Example 4.4.10 Solve the problem

$$\begin{cases} \min(x^2 + y^2) \\ \text{subject to: } (x-1)^3 - y^2 = 0 \end{cases}$$

and show that the method of Lagrange multipliers does not work in this case.

We solve the problem by means of level lines. The level lines of the objective function $x^2 + y^2 = \alpha$, with $\alpha > 0$ represent concentric circumferences with center $(0, 0)$ and radius $\sqrt{\alpha}$. We plot the diagram of the constraint

$$y = \pm\sqrt{(x-1)^3}.$$

First we consider the function $y_1 = \sqrt{(x-1)^3}$; its domain is $T = [1, +\infty)$; we have $y_1(1) = 0$, $\lim_{x \longrightarrow +\infty} y_1 = +\infty$; $y_1' = \frac{3}{2}(x-1)^{\frac{1}{2}}$.

Hence it holds $y_1' \geqq 0$ for $x \geqq 1$. Then we have $y_1'' = \frac{3}{4}\frac{1}{(x-1)^{1/2}}$. Hence it holds $y_1'' > 0$ for $x > 1$.

The function $y_2 = -\sqrt{(x-1)^3}$ is symmetric of the previous one, with respect to the horizontal axis. See Fig. 4.11.

The solution is therefore at $(1, 0)$, with $f(1, 0) = 1$.

Now we try to use the multipliers Lagrange method.

$$\mathcal{L}(x, y, \lambda) = x^2 + y^2 + \lambda\left[(x-1)^3 - y^2\right].$$

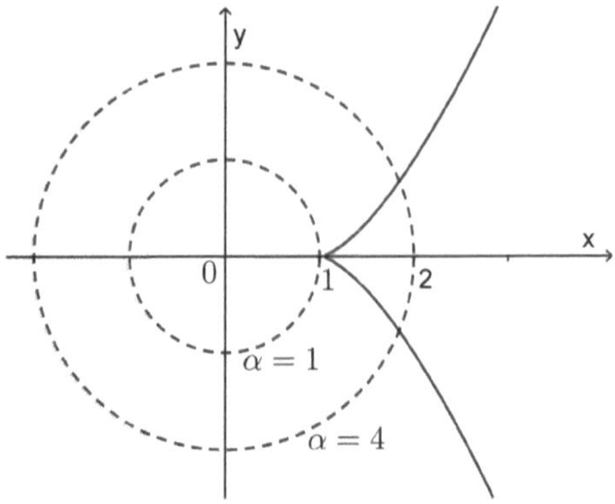

Fig. 4.11 Example 4.4.10: Graph of the curve $(x-1)^3 - y^2 = 0$ and level lines of the objective function

$$\begin{cases} \frac{\partial \mathcal{L}}{\partial x} = 2x + 3\lambda(x-1)^2 = 0 \\ \frac{\partial \mathcal{L}}{\partial y} = 2y - 2\lambda y = 0 \\ \frac{\partial \mathcal{L}}{\partial \lambda} = (x-1)^3 - y^2 = 0. \end{cases}$$

From the second equation we have $2y(1-\lambda) = 0$, from which it follows either $y = 0$ for every λ or $\lambda = 1$ for every y. With $y = 0$, from the third equation we obtain $x = 1$. By substituting into the first equation we obtain $2 \cdot 1 + 3\lambda \cdot 0 = 0$, which is absurd. With $\lambda = 1$ we have, in the first equation, $2x + 3(x^2 - 2x + 1) = 0$, that is $3x^2 - 4x + 3 = 0$, which has no solution in $\mathbb{R}$. Therefore the above system has no real solutions. The catch is due to the fact that the gradient of the constraint, i.e. the vector $\left[3(x-1)^2; -2y\right]^\top$, is equal to $[0]$ at the optimal solution $(1, 0)$ and hence the assumptions of the theorem on Lagrange multipliers are not all respected.

Remark 4.4.11 Not always we have to use the sufficient optimality conditions for the problems in question. When the feasible set K is *closed and bounded (i.e. compact)* and f is continuous on K (if f is differentiable on an open set containing K, it is obviously continuous on K), then the Weierstrass theorem holds and it is sufficient to evaluate the objective function at the stationary points found, in order to have all the information on the existence of maximum and minimum points.

Example 4.4.12 (*a*) Find the extremum points of $f(x_1, x_2) = (x_1)^2 - x_1x_2 + (x_2)^2$ under the constraint $h(x_1, x_2) = (x_1)^2 + (x_2)^2 - 1 = 0$.

We have

$$\mathcal{L}(x_1, x_2, \lambda) = (x_1)^2 - x_1x_2 + (x_2)^2 - \lambda\left[(x_1)^2 + (x_2)^2 - 1\right].$$

$$\begin{cases} \frac{\partial \mathcal{L}}{\partial x_1} = 2x_1 - x_2 - 2\lambda x_1 = 0 \\ \frac{\partial \mathcal{L}}{\partial x_2} = -x_1 + 2x_2 - 2\lambda x_2 = 0 \\ \frac{\partial \mathcal{L}}{\partial \lambda} = -\left[(x_1)^2 + (x_2)^2 - 1\right] = 0. \end{cases}$$

We consider the first and the second equation:

$$\begin{cases} 2x_1 - x_2 - 2\lambda x_1 = 0 \\ -x_1 + 2x_2 - 2\lambda x_2 = 0 \end{cases} \Leftrightarrow \begin{matrix} 2(1-\lambda)x_1 - x_2 = 0 \\ -x_1 + 2(1-\lambda)x_2 = 0. \end{matrix}$$

We do not take into consideration the zero solution, as it does not verify the constraint. Then we impose the condition

$$\begin{vmatrix} 2(1-\lambda) & -1 \\ -1 & 2(1-\lambda) \end{vmatrix} = 0,$$

that is $4\lambda^2 - 8\lambda + 3 = 0$, from which $\lambda_1 = \frac{1}{2}$, $\lambda_2 = \frac{3}{2}$. With $\lambda_1 = \frac{1}{2}$ we have

$$\begin{cases} 2x_1 - x_2 - 2 \cdot \frac{1}{2}x_1 = 0 \\ -x_1 + 2x_2 - 2 \cdot \frac{1}{2}x_2 = 0 \end{cases} \Longrightarrow \begin{cases} x_1 - x_2 = 0 \\ -x_1 + x_2 = 0, \end{cases}$$

that $x_1 = x_2$. By substituting into the constraint we have

$$(x_2)^2 + (x_2)^2 - 1 = 0 \Longrightarrow 2(x_2)^2 = 1 \Longrightarrow x_2 = \pm\sqrt{\frac{1}{2}}.$$

Therefore

$$x_1 = \pm\sqrt{\frac{1}{2}}.$$

With $\lambda_2 = \frac{3}{2}$ we have $x_1 = -x_2$ and hence

$$x_2 = \pm\sqrt{\frac{1}{2}}; \quad x_1 = \mp\sqrt{\frac{1}{2}}.$$

Therefore we have 4 stationary points:

$$A\left(\sqrt{\frac{1}{2}}, \sqrt{\frac{1}{2}}\right); \; B\left(-\sqrt{\frac{1}{2}}, -\sqrt{\frac{1}{2}}\right); \; C\left(\sqrt{\frac{1}{2}}, -\sqrt{\frac{1}{2}}\right); \; D\left(-\sqrt{\frac{1}{2}}, \sqrt{\frac{1}{2}}\right).$$

The feasible set K is a circumference in $\mathbb{R}^2$ and hence it is a closed and bounded set. Owing to the Theorem of Weierstrass, it is sufficient to evaluate the objective function at the above 4 stationary points and then to operate the comparison among the results obtained. We have that $f(A) = f(B) = \frac{1}{2}$ and $f(C) = f(D) = \frac{3}{2}$. Hence the points A and B are global minimizers and the points C and D are global maximizers.

(*b*) Consider the problem

$$\max(\min) f(x, y) = xy,$$

$$(x, y) \in K = \left\{(x, y) \in \mathbb{R}^2 : (x-2)^2 + y^2 = 4\right\}.$$

The Lagrangian function is

$$\mathcal{L} = xy - \lambda\left[(x-2)^2 + y^2 - 4\right].$$

Hence we have thc following system of first-order conditions

$$\begin{cases} \frac{\partial \mathcal{L}}{\partial x} = y - 2\lambda(x-2) = 0 \\ \frac{\partial \mathcal{L}}{\partial y} = x - 2\lambda y = 0 \\ \frac{\partial \mathcal{L}}{\partial \lambda} = -\left[(x-2)^2 + y^2 - 4\right] = 0. \end{cases}$$

From the second equation we obtain $x = 2\lambda y$, and by substituting into the first equation we get

$$y - 2\lambda(2\lambda y - 2) = 0, \text{ i.e. } y(1 - 4\lambda^2) + 4\lambda = 0.$$

It cannot be $1 - 4\lambda^2 = 0$, i.e. $\lambda = \pm\frac{1}{2}$, as in this case we would get $4\left(\pm\frac{1}{2}\right) = 0$, which is absurd. Hence it must hold $1 - 4\lambda^2 \neq 0$, which allows to write $y = \frac{4\lambda}{4\lambda^2 - 1}$. Hence

$$x = 2\lambda y = \frac{8\lambda^2}{4\lambda^2 - 1}.$$

By substituting these values into the third equation, we obtain

$$16\lambda^2(3 - 4\lambda^2) = 0.$$

We consider two cases:

(1) $\lambda = 0$; we have in this case $x = 0,\ y = 0$.
(2) $3 = 4\lambda^2$, i.e. $\lambda = \pm\frac{\sqrt{3}}{2}$. With $\lambda = \frac{\sqrt{3}}{2}$, we have $x = 3,\ y = \sqrt{3}$. With $\lambda = -\frac{\sqrt{3}}{2}$, we have $x = 3,\ y = -\sqrt{3}$.

We have therefore three stationary points:

$$A(3, \sqrt{3}),\ B(3, -\sqrt{3}),\ C(0, 0).$$

As $f(A) = 3\sqrt{3},\ f(B) = -3\sqrt{3},\ f(C) = 0$, and the feasible set K is closed and bounded (it is a circumference of center $(2, 0)$ and radius 2), we deduce that A is the global maximizer and B is the global minimizer. The point C is a saddle point, as the objective function assumes positive values on the first quadrant and negative values on the fourth quadrant.

(c) We propose to the willing reader the following problem.

$$\min(\max) f(x, y) = x^2 + y^2$$

$$(x, y) \in K = \left\{(x, y) \in \mathbb{R}^2 : y^2 - \log(5 - x^2) = 0\right\}.$$

(The feasible set is closed and bounded and we have four stationary points: $A(0, \sqrt{\log 5}),\ B(0, -\sqrt{\log 5}),\ C(-2, 0),\ D(2, 0)$).

We have previously pointed out the role of the regularity assumption on the constraints in order to obtain the first-order necessary optimality conditions in terms of Lagrange multipliers: the Jacobian matrix $\nabla h(x^0)$ must be of full rank. However, it is possible to obtain a weaker result, without assuming the said regularity condition and by introducing the so-called *augmented Lagrangian function:*

$$\mathcal{L}(x, \mu, \lambda) = \mu f(x) - \lambda h(x),$$

where $\mu \in \mathbb{R}$ and $\lambda \in \mathbb{R}^m$. Note that with $\mu = 1$ obviously we obtain the usual Lagrangian function and that with $\mu = 0$ the objective function in fact disappears from the Lagrangian function, which is not too much desirable. By means of the augmented Lagrangian function we can reformulate Theorem 4.4.6 without imposing any regularity conditions on the constraints.

Theorem 4.4.13 *Let $x^0 \in K$ be a local optimum point of f on the feasible set*

$$K = \left\{x \in \mathbb{R}^n : h(x) = [0]\right\}.$$

Let f be differentiable and h of class C^1 on the open set $A \subset \mathbb{R}^n$. Then, there exists a nonzero vector $(\mu_0, \lambda^0) \in \mathbb{R}^{m+1}$, such that

$$\mu_0 \nabla f(x^0) - (\lambda^0)^\top \nabla h(x^0) = [0].$$

Proof If $\nabla h(x^0)$ has full rank, it is sufficient to consider $\mu_0 = 1$ and the thesis is obtained, by Theorem 4.4.6. If $\nabla h(x^0)$ has not full rank, this means that the gradients $\nabla h_1(x^0), \ldots, \nabla h_m(x^0)$ are linearly dependent. By choosing $\mu_0 = 0$, this means that we can write

$$\sum_{i=1}^{m} \lambda_i^0 \nabla h_i(x^0) = [0]$$

with coefficients λ_i^0, $i = 1, \ldots, m$, *not all zero.* Also in this case the thesis is obtained. □

Example 4.4.14 Consider the problem

$$\max f(x, y) = x$$

subject to the constraints (note that we have $m = n = 2$)

$$h_1(x, y) = y - x^2 = 0$$
$$h_2(x, y) = y + x^2 = 0.$$

It is easy to see that the feasible set is given by the origin of $\mathbb{R}^2 : (0, 0)$. We have

$$\nabla h_1(0, 0) = \nabla h_2(0, 0) = (0, 1)^\top .$$

Hence the two gradients of the constraints are linearly dependent at (0, 0); in fact we have no solution of the equation

$$\nabla f(0, 0) - \lambda_1 \nabla h_1(0, 0) - \lambda_2 \nabla h_2(0, 0) = [0] .$$

But we have a nonzero solution of

$$\mu_0 \nabla f(0, 0) - \lambda_1 \nabla h_1(0, 0) - \lambda_2 \nabla h_2(0, 0) = [0] ,$$

namely $\mu_0 = 0, \ \lambda_1 = -1, \ \lambda_2 = 1.$

Theorem 4.4.13, due to the German-Greek mathematician C. Caratheodory, has a more general version, for optimization problems with inequality constraints, version known as the *Fritz John Theorem.* We shall examine this result in the next pages, when we shall treat optimization problems with inequality constraints; in this case those conditions, which assure that the multiplier associated to the gradient of the objective function (in the Lagrangian function of the problem) is different from zero, are called *constraint qualifications.*

Besides first-order necessary optimality conditions, we have, for classical optimization problems, also second-order necessary optimality conditions, mainly of analytic interest. For the reader's convenience we give these conditions.

Theorem 4.4.15 *Let $x^0 \in K$ be a local minimum point (resp. a local maximum point) of f over K, let f and every h_i, $i = 1, \ldots, m$, be of class C^2 on the open set $A \subset \mathbb{R}^n$. Let the gradients $\nabla h_1(x^0), \ldots, \nabla h_m(x^0)$ be linearly independent. Then it holds*

$$\nabla f(x^0) - \sum_{i=1}^{m} \lambda_i^0 \nabla h_i(x^0) = [0]$$

and, moreover, it holds

$$h^\top H\mathcal{L}(x^0, \lambda^0)h \geqq 0 \ (resp. \ \leqq 0)$$

for all vectors $h \in \mathbb{R}^n$ *such that*

$$\nabla h_i(x^0)^\top h = 0, \ i = 1, \ldots, m.$$

4.5 Sensitivity Analysis and Economic Interpretation of the Lagrange Multipliers

We consider again the "classical" optimization problem, i.e., for instance, the following problem

$$(P): \qquad \begin{cases} \min f(x) \\ \text{subject to: } h(x) = [0], \end{cases}$$

where $f : \mathbb{R}^n \longrightarrow \mathbb{R}$ and $h : \mathbb{R}^n \longrightarrow \mathbb{R}^m$, $m < n$, f and h of class C^2 on an open set $A \subset \mathbb{R}^n$. We suppose that the point x^0 (regular) is a local solution of this problem and let λ^0 be the associated Lagrange multipliers vector. We consider the Lagrangian function in the form

$$\mathcal{L}(x, \lambda) = f(x) - \lambda^\top h(x).$$

Now we consider a problem similar to (P), where the vector function which describes the constraints has been "perturbed" , in the sense that the zero vector which appears at the right-hand side, has been substituted by a nonzero vector $b \in \mathbb{R}^m$:

$$P(b): \qquad \begin{cases} \min f(x) \\ \text{subject to: } h(x) = b. \end{cases}$$

It is rather intuitive to think that, for values of b "near" to the zero vector, the perturbed problem $P(b)$ will admit a solution $x^0(b)$ "near" to $x^0([0]) = x^0$. Indeed, under suitable assumptions, also the optimum value $f(x^0(b))$ is a continuous function of b and, when it is a differentiable function, the components of its gradient, with respect to b, assume the meaning of "speed of change" of the same optimum value, with respect to every unit of variation of the right-hand side of the constraints of the perturbed problem $P(b)$. More precisely, we have the following result.

Theorem 4.5.1 (Sensitivity Theorem) *Consider the perturbed problem* $P(b)$, *where the functions* f *and* h *are of class* C^2 *on an open set* $A \subset \mathbb{R}^n$. *Suppose* *that for* $b = [0]$ *the problem* $P([0]) = (P)$ *admits* x^0 *as a regular solution point (i.e. the Jacobian matrix* $\nabla h(x^0)$ *is of full rank). Let* λ^0 *be the associated Lagrange multipliers vector and let be satisfied the second-order sufficient optimality conditions of Theorem 4.4.7. Then, there exists a neighbourhood* $U([0])$ *such that for all* $b \in U([0])$ *there exist the vector functions* $x^0(b)$ *and* $\lambda^0(b)$, *which are of class* C^1, *with respect to* b. *Moreover, it results* $x^0([0]) = x^0$ *and* $x^0(b)$ *is a strict local minimum point for* $P(b)$. *Furthermore the optimum value function* $f(x^0(b))$ *is differentiable with respect to* b *and it holds*

$$\nabla_b f(x^0(b)) = \lambda^0(b), \quad \forall b \in U([0]). \text{ In particular, } \nabla_b f(x^0(b))\Big|_{b=[0]} = \lambda^0.$$

Proof Consider the system of equations

$$\begin{aligned} \nabla f(x) - \lambda^\top \nabla h(x) &= [0] \\ h(x) - b &= [0]\,. \end{aligned}$$

By hypothesis there is a solution (x^0, λ^0) to this system when $b = [0]$. The Jacobian matrix of this system at this solution is

$$M = \begin{bmatrix} H_x\mathcal{L}(x^0, \lambda^0) & \nabla h(x^0)^\top \\ \nabla h(x^0) & [0] \end{bmatrix}.$$

Because by assumption x^0 is a regular point and $H_x\mathcal{L}(x^0, \lambda^0)$ is positive definite on $\{z \in \mathbb{R}^n : \nabla h(x^0)z = [0]\}$, it follows that this matrix is nonsingular. Indeed, suppose

$$M \begin{bmatrix} s \\ u \end{bmatrix} = [0]\,.$$

Then $\nabla h(x^0)s = [0]$. Moreover,

$$\begin{bmatrix} s^\top, u^\top \end{bmatrix} M \begin{bmatrix} s \\ u \end{bmatrix} = s^\top H_x\mathcal{L}(x^0, \lambda^0)s + 2s^\top (\nabla h(x^0))^\top u = s^\top H_x\mathcal{L}(x^0, \lambda^0)s.$$

If $s \neq [0]$, the second-order conditions imply the positivity of the last expression, a contradiction. Thus $s = [0]$, but then

$$M \begin{bmatrix} s \\ u \end{bmatrix} = (\nabla h(x^0))^\top u = [0]$$

and the linear independence of the lines of $\nabla h(x^0)$ implies that $u = [0]$.

The Implicit Function Theorem (Theorem 2.6.2) applies and for sufficiently small $\|b\|$ there exist solutions $x^0(b)$, $\lambda^0(b)$ to the system, which are in fact continuously differentiable, with $x^0 = x^0([0])$ and $\lambda^0 = \lambda^0([0])$. We can therefore use the chain rule to find the derivative of $f(x(b))$, with respect to b. We find

$$\nabla_b f(x^0(b))\Big|_{b=[0]} = \nabla_x f(x^0)^\top \nabla_b x^0([0])$$

and

$$\nabla_b h(x^0(b))\Big|_{b=[0]} = \nabla h(x^0)\nabla_b x^0([0]).$$

But, being $h(x) - b = [0]$, the second of these is equal to the identity matrix I of $\mathbb{R}^m$, while this, in view of the first-order condition on the Lagrangian, implies that the first can be written as

$$\nabla_b f(x^0(b))\Big|_{b=[0]} = \lambda^0.$$

□

Theorem 4.5.1 is a particular version of a more general result due to [12]. Problem $P(b)$ is an example of "parametric optimization problems". The "sensitivity analysis" of these kinds of problems studies the effects caused by the variations of the parameters on the optimum value function and on other quantities of the parametric problem. We observe that in several economic treatments of these questions, the derivative of the optimum value function (called also, sometimes, the "marginal function") is performed, without regards to conditions that assure the differentiability of this function.

Theorem 4.5.1 has also a nice economic interpretation, as the Lagrange multipliers, in addition to be a method for solving a classical constrained optimization problem, also provide a sensitivity analysis for the same, as they show how sensitive the optimum value of the objective function is with respect to changes in the constraint constants. For example, if any Lagrange multiplier is equal to zero at the solution, then small changes in the corresponding constraint constant would not affect the optimum value of the objective function.

For example, if the objective function has the meaning of a value (profits, revenue, costs) and the constraints specify a given value for a certain quantity (e.g. input), then the Lagrange multiplier measures the sensitivity of a value to changes in a quantity and hence represents a price, often called in economic analysis, a *shadow price* (of the input).

Example 4.5.2 Consider the problem

$$\min f(x, y) = x^2 + 2y^2$$
$$\text{subject to} : x - y = b.$$

By using the method of Lagrange multipliers, we find that the point $(\frac{2}{3}b, -\frac{1}{3}b)$ is the solution of the problem. The associated Lagrange multiplier is $\lambda(b) = \frac{4}{3}b$. Obviously this function is continuous with respect to b. Moreover, the optimum value function

$$V(b) = f(\frac{2}{3}b, -\frac{1}{3}b) = \left(\frac{2}{3}b\right)^2 + 2\left(-\frac{1}{3}b\right)^2 = \frac{2}{3}b^2$$

and obviously it holds $V'(b) = \frac{4}{3}b = \lambda(b)$.

4.6 Constrained Optimization Problems with Inequality Constraints: Sensitivity and Envelope Theorem

In the present section we consider the case of a feasible set given by a system of (weak) inequalities; this is the case of the modern *mathematical programming problems,* problems often encountered in Economic Analysis, Operations Research, Engineering, etc. These problems are also called *nonlinear programming problems,* as the functions involved may be (not: "must be") also nonlinear. We choose, at least for the moment, to treat the following problem

$$\min_{x \in K} f(x)$$

with $K = \{x \in \mathbb{R}^n : g_i(x) \leqq 0, \ i = 1, \dots, m\}$, being $f, \ g_i : \mathbb{R}^n \longrightarrow \mathbb{R}, i = 1, \dots, m$.

Note that in the above formulation no restriction on the number of the constraints is imposed. Note, furthermore, that if all inequalities were given as strict inequalities and the constraints are continuous, usually the feasible set K would be an open set, and hence we would have an unconstrained optimization problem. The above formulation is, in spite of its appearances, rather general and the results we shall obtain on the same, can be easily fitted to other formulations. Indeed, we remark that:

(*a*) If we have a maximization problem, we recall that to find the maximum of $f(x)$ is equivalent to find the minimum of $-f(x)$, and it results, as previously remarked in Sect. 4.1,

$$f(x^0) \equiv \text{maximum value of } f(x) = -\text{minimum value of } \{-f(x)\}$$

For example, the function $f(x) = x^2 - 2x + 2, \ x \in \mathbb{R}$, has a (global) minimizer at $x_0 = 1$, with $f(x_0) = 1$. If we look for the maximizer of $-f(x) = -x^2 + 2x - 2 \equiv f_1(x)$, we find $x_0 = 1$ (as before), with $f_1(x_0) = -1$, hence $f_1(1) = -f(1)$.

(*b*) If we have inequalities of the type $g_k(x) \geqq 0$, obviously we can write the same as $-g_k(x) \leqq 0$. If we have some constraints of the type $g_k(x) \leqq b_k$, we can write the same as $g_k(x) - b_k \leqq 0$, and, by denoting with $\varphi_k(x)$ the function $g_k(x) - b_k$, we come back the original formulation of the problem.

(*c*) It is true that an equality constraint can be written as two "opposed" inequalities: one with the sign $\leqq$ and the other one with the sign $\geqq$. However, this device can generate some drawbacks (for example, we increase the number of the constraints), hence the equality constraints (if present) must be treated in their original form. It is then obvious that if a mathematical programming problem presents both inequality and equality constraints, it results of a formulation more general than the one considered above and we shall give only some basic results on this kind of problems, where essentially the results of the previous Sect. 4.4 and the ones of the present section, are used "jointly".

(*d*) The introduction of *nonnegative variables,* that is the requirement $x \geqq [0]$, is rather usual in economic and financial applications. Even if we can rewrite this requirement as

$$\begin{aligned} -x_1 &\leqq 0 \\ -x_2 &\leqq 0 \\ &\vdots \\ -x_n &\leqq 0 \end{aligned}$$

it is better do not operate this device and to treat this case in a specific way. See further.

The so-called *linear programming problems* (L. P.) are very important, both from theoretical points of view and on the grounds of their countless applications. In these problems the objective function f and the constraint function g are both linear functions (or affine functions). They can be formulated, for example, in the version

$$\begin{cases} \min c^\top x \\ Ax \leqq [0] \end{cases} \quad \text{or also} \quad \begin{cases} \min c^\top x \\ Ax \leqq b \end{cases},$$

where $x, c \in \mathbb{R}^n$, $c \neq [0]$, A is a matrix of order (m, n) and $b \in \mathbb{R}^m$. Usually, in these problems we have also the sign constraint on the vector $x : x \geqq [0]$. See the next chapter.

We consider again the original problem of the present section, i.e.

$$(P): \qquad \min_{x \in K} f(x),$$

where $K = \left\{x \in \mathbb{R}^n : g(x) \leqq [0]\right\}$, $f : A \longrightarrow \mathbb{R}$, $g : A \longrightarrow \mathbb{R}^m$ and $A \subset \mathbb{R}^n$ is an open set. For the moment, we distinguish two classes of problems:

(*i*) *Differentiable mathematical programming problems,* when f and every g_i, $i = 1, \ldots, m$, are differentiable on A.
(*ii*) *Convex mathematical programming problems,* when f and every g_i, $i = 1, \ldots, m$, are convex functions on the convex set A.

Now we shall treat the first class of the above problems, that is mathematical programming problems (with inequality constraints) under differentiability assumptions on the functions involved in the problem.

Necessary Optimality Conditions Under Differentiability Assumptions

We consider the mathematical programming problem

$$(P): \qquad \begin{cases} \min f(x) \\ g_i(x) \leqq 0, \\ i = 1, \ldots, m \end{cases}$$

with $f : \mathbb{R}^n \longrightarrow \mathbb{R}$, $g_i : \mathbb{R}^n \longrightarrow \mathbb{R}$, $i = 1, \ldots, m$; f and every g_i, $i = 1, \ldots, m$, differentiable functions on the open set $A \subset \mathbb{R}^n$. Let us consider $x^0 \in K$ (feasible set of (P)); the constraints g_i such that $g_i(x^0) = 0$ are called *active or effective or binding constraints at* x^0. Those constraints for which $g_i(x^0) < 0$ are called *non active, non effective, non binding constraints at* x^0. The set

$$I(x^0) = \left\{i : g_i(x^0) = 0\right\}$$

is the *set of the indices of active constraints at* x^0. Obviously $I(x^0) \subset \{1, \ldots, m\}$. For instance, if the constraints are the following ones

$$g_1(x) = x_2 - (x_1)^2 \leqq 0$$
$$g_2(x) = x_1 - 1 \leqq 0$$
$$g_3(x) = -x_1 \leqq 0$$
$$g_4(x) = -x_2 \leqq 0,$$

at $x^0 = (0, 0)$ we have $I(x^0) = \{1, 3, 4\}$;
at $x^0 = (1, 0)$ we have $I(x^0) = \{2, 4\}$;
at $x^0 = (\frac{7}{8}, \frac{1}{2})$ we have $I(x^0) = \emptyset$.
The feasible set K is drawn in Fig. 4.12.

We can remark that active constraints at a feasible point x^0 have usually a more important role than non active constraints. Indeed, if a constraint g_i is non active at

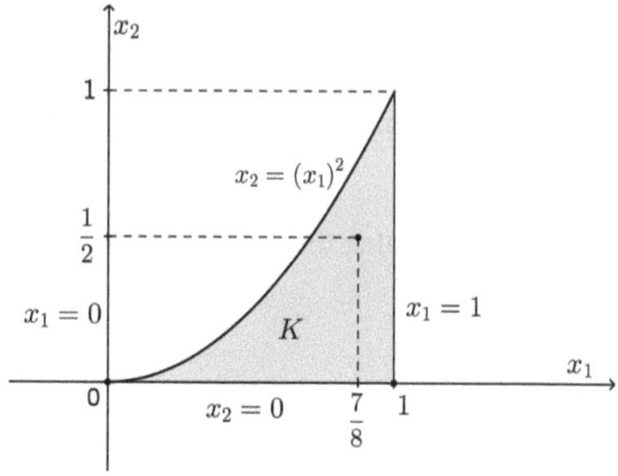

Fig. 4.12 Feasible set. Active indices

a point $x^0 \in K$ and is continuous, it will be possible to move the point on a suitable neighbourhood of the same, without "betraying" the constraint.

Now we present three necessary optimality conditions for problem (P). We obtain these conditions subsequently, starting form the next theorem, even if the historical course of the said optimality conditions has been different. We begin with a "negative" optimality condition, therefore not too useful from an operational point of view, but very useful in obtaining the next necessary optimality conditions.

Theorem 4.6.1 (Abadie Theorem or Abadie Linearization Lemma) *See [1]. Let $x^0 \in K = \{x : g_i(x) \leqq 0,\ i = 1, \ldots, m\}$ be a local solution of problem (P). Then the linear system*

$$\begin{cases} \nabla f(x^0)^\top z < 0 \\ \nabla g_i(x^0)^\top z < 0,\ i \in I(x^0) \end{cases} \tag{4.6}$$

has no solutions $z \in \mathbb{R}^n$. In other words, the system $Az < [0]$, with

$$A = \begin{bmatrix} \nabla f(x^0)^\top \\ \nabla g_I(x^0) \end{bmatrix},$$

where $\nabla g_I(x^0)$ is the Jacobian matrix of the active constraints at x^0, has no solutions $z \in \mathbb{R}^n$.

Proof We suppose absurdly that system (4.6) admits a solution $\bar{z}$. Then:

(*a*) For all $i \notin I(x^0)$, that is for all i such that $g_i(x^0) < 0$, it results, for t convenient,

$$g_i(x^0 + t\bar{z}) < 0,$$

as all constraints are continuous (preserving-sign theorem).

(b) For all $i \in I(x^0)$, that is for all i such that $g_i(x^0) = 0$, it results

$$g_i(x^0 + t\bar{z}) = \underbrace{g_i(x^0)}_{=0} + t \underbrace{\nabla g_i(x^0)^\top \bar{z}}_{<0 \text{ by assumption}} + o(t\,\|\bar{z}\|).$$

It follows that for $t > 0$ and convenient, it holds

$$g_i(x^0 + t\bar{z}) < 0.$$

(c) It results

$$f(x^0 + t\bar{z}) - f(x^0) = t \underbrace{\nabla f(x^0)^\top \bar{z}}_{<0 \text{ by assumption}} + o(t\,\|\bar{z}\|).$$

It follows that for $t > 0$ and convenient, it holds

$$f(x^0 + t\bar{z}) - f(x^0) < 0.$$

Hence there exists $t > 0$ and convenient, such that $x^0 + t\bar{z} \in K$ (that is, it is a feasible point) and such that $f(x^0 + t\bar{z}) < f(x^0)$, against the assumptions that x^0 is a local minimum point for f over K. □

Now we consider the following second result on necessary optimality conditions for (P) : it is a theorem due to [20], and it may be considered the first printed result specifically concerning optimality conditions for a nonlinear programming problem, with inequality constraints.

Theorem 4.6.2 (Theorem of Fritz John) *Let $x^0 \in K$ be an optimal solution for (P). Then, there exist multipliers ("John multipliers") $y_0, y_1, \dots, y_m$, such that*

(1) $y_0 \nabla f(x^0) + \sum_{i=1}^m y_i \nabla g_i(x^0) = [0]$;
(2) $\sum_{i=1}^m y_i g_i(x^0) = 0 \iff y_i g_i(x^0) = 0,\ i = 1, \dots, m$;
(3) $y_0 \geqq 0,\ y_1 \geqq 0, \dots, y_m \geqq 0,\ y_0, y_1, \dots, y_m$ *not all zero, That is we have*

$$\begin{bmatrix} y_0 \\ y \end{bmatrix} \geq [0].$$

Proof By the previous result of Abadie (Theorem 4.5.1; note that Theorem 4.5.1 and the present theorem have the same assumptions), the system

$$\begin{bmatrix} \nabla f(x^0)^\top \\ \nabla g_I(x^0) \end{bmatrix} z < [0]$$

has no solutions $z \in \mathbb{R}^n$. But then, by the theorem of the alternative of Gordan (Theorem 3.4.8), the "dual" system

$$[y_0, y_I]^\top \begin{bmatrix} \nabla f(x^0)^\top \\ \nabla g_I(x^0) \end{bmatrix} = [0]$$

has a solution $[y_0, y_I]^\top \geq [0]^\top$. This means that there exist multipliers $y_0,\ y_i,\ i \in I(x^0)$, *nonnegative and not all zero,* such that

$$y_0 \nabla f(x^0) + \sum_{i \in I(x^0)} y_i \nabla g_i(x^0) = [0]\,.$$

Now if we choose $y_i = 0,\ i \notin I(x^0)$, we can write

$$y_0 \nabla f(x^0) + \sum_{i=1}^{m} y_i \nabla g_i(x^0) = [0]\,.$$

Being $g_i(x^0) = 0,\ \forall i \in I(x^0)$ and being $y_i = 0, \forall i \notin I(x^0)$, relation (2) of the theorem is immediate. □

The equalities $y_i g_i(x^0) = 0,\ i = 1, \ldots, m$, appearing in (2) of Theorem 4.6.2 (equivalently $\sum_{i=1}^m y_i g_i(x^0) = 0$) are called *complementary slackness conditions:* if $g_i(x^0) = 0$, the associated multiplier y_i may be either zero or positive; if a multiplier y_i is *positive,* surely the associated constraint is zero at x^0 ($y_i > 0 \Longrightarrow g_i(x^0) = 0$). If $g_i(x^0) < 0$, then the associated multiplier y_i must be zero ($g_i(x^0) < 0 \Longrightarrow y_i = 0$). If $y_i > 0$ for *all active constraints at* x^0, we say that the *strict complementary slackness conditions* hold.

Note that the theorem of Fritz John involves an "augmented" Lagrangian function, of the type

$$\mathcal{L}(x, y_0, y) = y_0 f(x) + y^\top g(x)$$

and hence $\mathcal{L} : \mathbb{R}^{n+m+1} \longrightarrow \mathbb{R}$. Finally, we note that if in relation (1) of Theorem 4.6.2 it holds $y_0 \neq 0$ (i.e. $y_0 > 0$, by relation (3) of the same theorem), then (1) can be rewritten in the form

$$\nabla f(x^0) + \sum_{i=1}^{m} \lambda_i \nabla g_i(x^0) = [0]\,,$$

where $\lambda_i = y_i / y_0,\ i = 1, \ldots, m$. Unfortunately, it may occur $y_0 = 0$; in this case we would have a necessary optimality condition where the objective function plays no role and which is therefore independent from the objective function. This troublesome situation puts into evidence, in a sense, the presence of an "anomaly" in the related problem.

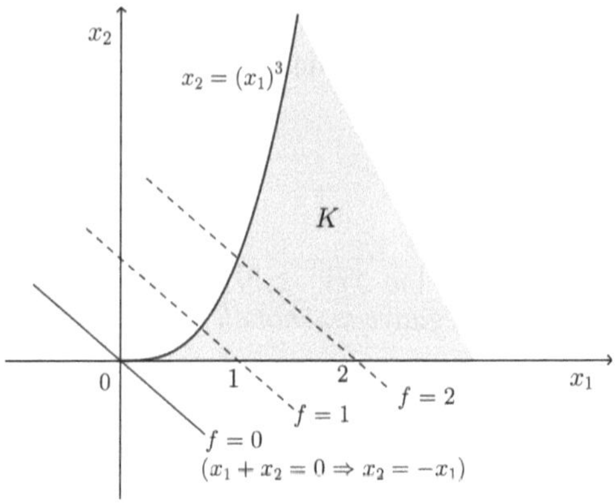

Fig. 4.13 Example 4.6.3: feasible set and level lines of f

Example 4.6.3 Consider the following problem

$$\min f(x_1, x_2) = x_1 + x_2$$
$$\text{subject to}: \quad -(x_1)^3 + x_2 \leqq 0,$$
$$-x_2 \leqq 0.$$

See Fig. 4.13.

It appears that the minimum point of f over K is at $x^0 = (0, 0)$. The Fritz John conditions are

$$y_0 \begin{bmatrix} 1 \\ 1 \end{bmatrix} + y_1 \begin{bmatrix} -3(x_1)^2 \\ 1 \end{bmatrix} + y_2 \begin{bmatrix} 0 \\ -1 \end{bmatrix} = \begin{bmatrix} 0 \\ 0 \end{bmatrix}$$

$$y_1((-x_1)^3 + x_2) = 0$$
$$y_2(-x_2) = 0$$

$$\begin{bmatrix} y_0 \\ y_1 \\ y_2 \end{bmatrix} \geq [0] .$$

At $x^0 = (0, 0)$ we have

$$y_0 \begin{bmatrix} 1 \\ 1 \end{bmatrix} + y_1 \begin{bmatrix} 0 \\ 1 \end{bmatrix} + y_2 \begin{bmatrix} 0 \\ -1 \end{bmatrix} = \begin{bmatrix} 0 \\ 0 \end{bmatrix},$$

that is

$$\begin{cases} y_0 + y_1 \cdot 0 + y_2 \cdot 0 = 0 \\ y_0 + y_1 - y_2 = 0 \end{cases}$$

from which $y_0 = 0,\ y_1 = y_2$. Hence we have $y_0 = 0$ and $y_1 = y_2 = \alpha > 0$. The conditions of Fritz John are in fact satisfied with these multipliers by *any objective function* differentiable at $(0, 0)$, independently from the fact that this point is or not a local minimum point on K for the said functions. When this fact occurs, the same makes the Fritz John conditions not too "selective" in order to make a first "screening" among the candidates to the solution of the problem. That's why it is important that the multiplier y_0 is different from zero: in all applications of mathematical programming problems, we expect that the related optimality conditions depend from the structure of the objective function, and obviously also from the set of the constraints.

It is therefore important to have conditions assuring that in the Fritz John theorem it holds $y_0 > 0$. The "price to pay" is of imposing on the constraints some "regularity condition" (the reader will remember the regularity conditions imposed on the Jacobian matrix of the constraints in a classical optimization problem). With reference to mathematical programming problems (i.e. with inequality constraints), the said conditions are usually called *constraint qualifications.* There are many constraint qualifications in the literature of mathematical programming problems, some of them quite general and difficult to be checked. Evidently, in Example 4.6.3, no constraint qualification is satisfied at the optimum point $x^0 = (0, 0)$. We shall consider only few constraint qualifications, rather easy to check.

(A) *Arrow-Hurwicz-Uzawa first constraint qualifications* (A.H.U. c.q. (1). See [5]. It is expressed as follows.

Let $x^0 \in K$; the system

$$\nabla g_i(x^0)^\top z < 0, \quad i \in I(x^0),$$

admits a solution $z \in \mathbb{R}^n$. In other words, the system

$$\nabla g_I(x^0) z < [0]$$

admits a solution, and hence, by the theorem of the alternative of Gordan, it is inconsistent the system

$$y^\top \nabla g_I(x^0) = [0]\,, \quad y \geq [0]\,.$$

That is, there do not exists scalars y_i, $i \in I(x^0)$, *all nonnegative and not all zero,* such that

$$\sum_{i \in I(x^0)} y_i \nabla g_i(x^0) = [0],$$

that is, the gradients $\nabla g_i(x^0), i \in I(x^0)$, are *positively linearly independent.*

We recall the notion of independent vectors (of $\mathbb{R}^n$): a set of k vectors of $\mathbb{R}^n$, x^1, $x^2, \ldots, x^k$, is linearly independent if there do not exist scalars λ_1, $\lambda_2, \ldots, \lambda_k$, *not all zero,* such that

$$\sum_{i=1}^{k} \lambda_i x^i = [0].$$

This means that the "control" on the multipliers has to be made non only for nonnegative multipliers (and not all zero), but also for nonpositive multipliers and for multipliers having elements with mixed sign (and with not all zero elements). Hence, if the gradients $\nabla g_i(x^0), i \in I(x^0)$ are linearly independent, they are also positively linearly independent, but the vice-versa does not hold. We can therefore state the following constraint qualification, less general that the one of Arrow-Hurwicz-Uzawa described above, in the sense that the following one implies the Arrow-Hurwicz-Uzawa c. q. 1, but the vice-versa does not hold.

(B) *Linear independence constraint qualification.*

The vectors $\nabla g_i(x^0), i \in I(x^0)$ are linearly independent. Note the similarity with the regularity assumption made when we introduced the Lagrange multiplier method for optimization problems with equality constraints.

(C) *Slater constraint qualification* (Slater c. q.). It is expressed as follows:

The constraints g_i, $i = 1, \ldots, m$, or even g_i, $i \in I(x^0)$, are convex on the convex set $A \subset \mathbb{R}^n$, with $K \subset A$, and there exists $\bar{x} \in K$ such that $g(\bar{x}) < [0]$.

Now we show that the Slater c. q. implies the Arrow-Hurwicz-Uzawa c. q. 1. Indeed, if every constraint g_i, $i \in I(x^0)$ is convex, it will hold, $\forall i \in I(x^0)$,

$$\nabla g_i(x^0)^\top (\bar{x} - x^0) \leqq \underbrace{g_i(\bar{x})}_{<0} - \underbrace{g_i(x^0)}_{=0}$$

and hence

$$\nabla g_i(x^0)^\top (\bar{x} - x^0) < 0, \ \forall i \in I(x^0).$$

By putting $(\bar{x} - x^0) = z$ we obtain the implication in which we are interested: the system

$$\nabla g_i(x^0)^\top z < 0, \ \forall i \in I(x^0),$$

admits a solution $z \in \mathbb{R}^n$.

It results therefore that the Arrow-Hurwicz-Uzawa c. q. 1 is the most general of the above three constraint qualifications:

$$\boxed{\nabla g_i(x^0), i \in I(x^0), \text{l. i.}} \Longrightarrow \boxed{\text{A.H.U. c.q. 1}} \Longleftarrow \boxed{\text{Slater c. q.}}$$

Now we can state and prove the most celebrated result on optimality conditions for problem (P), that is the Kuhn-Tucker theorem [22], theorem that however was almost entirely anticipated by W. Karush in his Master Thesis, discussed in 1939 at the Department of Mathematics of Chicago University. See [13]. It is therefore more correct to speak of the *Theorem of Karush-Kuhn-Tucker,* what indeed is done in modern treatments of mathematical programming problems.

Theorem 4.6.4 (Theorem of Karush-Kuhn-Tucker) *Let $x^0 \in K$ be a local solution of (P) and assume that at x^0 a constraint qualification is satisfied. More specifically, we assume that the A.H.U. c.q. 1 holds true. Then, there exist multipliers λ_i, $i = 1, \ldots, m$, such that:*

(a) $\nabla f(x^0) + \sum_{i=1}^m \lambda_i \nabla g_i(x^0) = [0]$;
(b) $\sum_{i=1}^m \lambda_i g_i(x^0) = 0 \iff \lambda_i g_i(x^0) = 0, \ i = 1, \ldots, m$;
(c) $\lambda_i \geqq 0, \ i = 1, \ldots, m$.

Proof Let us suppose that the A. H. U. c. q. is verified and that in relation (1) of Theorem 4.6.2 (Theorem of Fritz John) we have $y_0 = 0$. Hence the said relation becomes

$$\sum_{i=1}^m y_i \nabla g_i(x^0) = [0]$$

and, by choosing $y_i = 0$, for all $i \notin I(x^0)$, we have

$$\sum_{i \in I(x^0)} y_i \nabla g_i(x^0) = [0]$$

where $y_i \geqq 0$, $\forall i \in I(x^0)$, y_i *not all zero.* But then, by the theorem of the alternative of Gordan, the system

$$\nabla g_i(x^0)^\top z < 0, \ i \in I(x^0),$$

has no solution, in contrast with the assumption on the validity of the A.H.U. c. q. 1. □

The theorems of Abadie, Fritz John and the A. H. U. c. q. 1 can be given in a more general form which is suitable for further developments and considerations. We begin by operating a distinction in the set of the indices of active constraints $I(x^0)$:

$$I_C(x^0) = \left\{ i : i \in I(x^0),\ g_i \text{ is a } concave \text{ function} \right\};$$

$$\begin{aligned} I_{NC}(x^0) &= \left\{ i : i \in I(x^0),\ g_i \text{ is not a } concave \text{ function} \right\} \\ &= \left\{ i \in I(x^0) \setminus I_C(x^0) \right\}. \end{aligned}$$

Theorem 4.6.1 of Abadie can then be generalized in the following way (this is indeed the original version proposed by Abadie).

Theorem 4.6.5 (Theorem of Abadie 2) *Let* $x^0 \in K$ *be a local solution of* (P). *Then the system*

$$\begin{cases} \nabla f(x^0)^\top z < 0 \\ \nabla g_{NC}(x^0) z < [0] \\ \nabla g_C(x^0) z \leqq [0] \end{cases}$$

has no solution $z \in \mathbb{R}^n$.

Proof It is the same of Theorem 4.6.1, with the addition that for $i \in I_C(x^0)$ we can reason as follows: as for every $i \in I_C(x^0)$ the related constraint is a concave function (differentiable, by assumption), we have, with $t > 0$ and convenient,

$$g_i(x^0 + t\bar{z}) - g_i(x^0) \leqq t \nabla g_i(x^0)^\top \bar{z} \leqq 0.$$

Therefore we are able to find a feasible point, different from x^0 and in a neighborhood of x^0, such that the objective function assumes there a value lower than $f(x^0)$, against the assumption that x^0 is a local minimum point of f over K. □

Theorem 4.6.5 can be further generalized, by making the partition of the set $I(x^0)$ into indices related to active constraints which are *pseudoconcave* and indices related to active constraints *not pseudoconcave*; see, e.g. [24].

From Theorem 4.6.5 we can better specify the sign of the Fritz John multipliers. We denote by y^I_{NC} the vector of those Fritz John multipliers related to the constraints whose indices are in I_{NC}. We have the following result.

Theorem 4.6.6 (Theorem of Fritz John 2) *Let $x^0 \in K$ be a local solution of (P). Then, there exist nonnegative multipliers $y_0, y_1, \ldots, y_m$ such that:*

(1) $y_0 \nabla f(x^0) + \sum_{i=1}^m y_i \nabla g_i(x^0) = [0]$;
(2) $\sum_{i=1}^m y_i g_i(x^0) = 0 \iff y_i g_i(x^0) = 0, \; i = 1, \ldots, m$;
(3) $\begin{bmatrix} y_0 \\ y_{NC}^I \end{bmatrix} \geq [0]$.

Proof It is the same of Theorem 4.6.2. Here we consider the fact that the system

$$\begin{cases} \nabla f(x^0)^\top z < 0 \\ \nabla g_{NC}(x^0) z < [0] \\ \nabla g_C(x^0) z \leqq [0] \end{cases}$$

admits no solution; then we apply to this system the theorem of the alternative of Motzkin, without matrix C, appearing in the same theorem (Theorem 3.4.7). □

Also the A. H. U. c. q. 1 can be given in a more general form, which is just the original version considered by Arrow et al. [5]:

- A. H. U. c. q. 2: the system

$$\begin{cases} \nabla g_i(x^0)^\top z < 0, \; i \in I_{NC}(x^0) \\ \nabla g_i(x^0)^\top z \leqq 0, \; i \in I_C(x^0) \end{cases}$$

  admits a solution $z \in \mathbb{R}^n$.

Note that if $g_i, \; i \in I(x^0)$, is a *linear function* (or also a *linear affine function*), the A. H. U. c. q. 2 is surely verified, as a linear (affine) function is both convex and concave and in this case the above system becomes

$$\nabla g_i(x^0)^\top z \leqq 0, \; i \in I(x^0),$$

which obviously admits always at least the solution $z = [0]$. The same considerations hold if the constraints $g_i, \; i \in I(x^0)$, are all *concave* (this case is called by Arrow, Hurwicz and Uzawa "reverse constraint qualification"). It results that in the *linear programming problems* we can skip the problem of the constraint qualifications, as the constraints are in this case qualified by default!

Finally, we remark that also the A. H. U. c. q. 2 can be further generalized, by making the distinction between *pseudoconcave* (active) constraints and *non pseudoconcave* constraints. See [24].

Now we prove the Theorem of Karush-Kuhn-Tucker, under the A. H. U. c. q. 2.

Theorem 4.6.7 (Theorem of Karush-Kuhn-Tucker 2) *Let $x^0 \in K$ be a local solution of problem (P) and assume that at x^0 the A. H. U. c. q. 2 is verified. Then the relations a), b) and c) of Theorem 4.6.4 hold true.*

Proof We know from Theorem 4.6.6 (Theorem of Fritz John 2) that there exist multipliers

$$\begin{bmatrix} y_0 \\ y_{NC}^I \end{bmatrix} \geq [0] .$$

Hence, if $I_{NC}(x^0) = \emptyset$, we have $y_0 > 0$ and the theorem is proved. Suppose therefore $I_{NC} \neq \emptyset$ and suppose absurdly that $y_0 = 0$. Then there will exist multipliers $y_i \geqq 0$, $i = 1, \ldots, m$, *not all zero,* such that

$$\sum_{i \in I_{NC}(x^0)} y_i \nabla g_i(x^0) + \sum_{i \in I_C(x^0)} y_i \nabla g_i(x^0) = [0] , \tag{4.7}$$

being furthermore

$$y_{NC}^I \geq [0] .$$

Being the A. H. U. c. q. 2 verified, it will exist a vector $\bar{z} \in \mathbb{R}^n$ such that

$$\nabla g_{I_{NC}}(x^0)\bar{z} < [0] , \quad \nabla g_{I_C}(x^0)\bar{z} \leqq [0] .$$

Now we multiply these systems by y_{NC}^I and y_C^I , respectively, and we sum the related results, obtaining

$$(y_{NC}^I)^\top \nabla g_{I_{NC}}(x^0)\bar{z} + (y_C^I)^\top \nabla g_{I_C}(x^0)\bar{z} < 0,$$

in contradiction with (4.7). Hence it holds $y_0 > 0$. □

Remark 4.6.8 The constraint qualifications we have examined are not the most general among the ones proposed in the literature. They are however rather easy to be described and checked. [5] take into consideration other constraint qualifications, rather general, among which the constraint qualifications used by Kuhn and Tucker in their pioneering paper of 1951. For a survey of various constraint qualifications proposed for a mathematical programming problem, the reader may consult [7, 14, 15, 24, 31].

Sufficient Optimality Conditions Under Differentiability Assumptions

Theorem 4.6.9 *Let $x^0 \in K$; let f be pseudoconvex and let every g_i, $i \in I(x^0)$, be quasiconvex (on an open convex set X containing K). Let the conditions a), b), and c) of the Theorem of Karush-Kuhn-Tucker (Theorem 4.6.4) be satisfied. Then x^0 is a solution of (P).*

Proof As every constraint $g_i,\ i \in I(x^0)$, is quasiconvex (and differentiable), by Theorem 3.3.3 we have, $\forall i \in I(x^0),\ \forall x \in K$:

$$\underbrace{g_i(x)}_{\leqq 0} - \underbrace{g_i(x^0)}_{=0} \leqq 0 \Longrightarrow \nabla g_i(x^0)^\top (x - x^0) \leqq 0.$$

Therefore, being $\lambda_i \geqq 0,\ \forall i \in I(x^0)$, we have, for every $x \in K$:

$$\sum_{i \in I(x^0)} \lambda_i \nabla g_i(x^0)^\top (x - x^0) \leqq 0.$$

Being $\lambda_i = 0,\ \forall i \notin I(x^0)$, we can write

$$\sum_{i \notin I(x^0)} \lambda_i \nabla g_i(x^0)^\top (x - x^0) = 0$$

and hence

$$\sum_{i=1}^{m} \lambda_i \nabla g_i(x^0)^\top (x - x^0) \leqq 0, \quad \forall x \in K.$$

Consider now relation (a) of Theorem 4.6.4 of Karush-Kuhn-Tucker:

$$\nabla f(x^0) + \sum_{i=1}^{m} \lambda_i \nabla g_i(x^0) = [0]\,.$$

Now we multiply both sides of this last relation by $(x - x^0)$:

$$\left[\nabla f(x^0) + \sum_{i=1}^{m} \lambda_i \nabla g_i(x^0) \right]^\top (x - x^0) = 0,$$

that is

$$\nabla f(x^0)^\top (x - x^0) + \underbrace{\sum_{i=1}^{m} \lambda_i \nabla g_i(x^0)^\top (x - x^0)}_{\leqq 0,\ \forall x \in K} = 0.$$

It follows that it holds, for every $x \in K$,

$$\nabla f(x^0)^\top (x - x^0) \geqq 0.$$

But, being f pseudoconvex, it results

$$f(x) - f(x^0) \geqq 0, \ \forall x \in K,$$

that is $f(x^0) \leqq f(x), \ \forall x \in K.$ □

Arrow and Enthoven [3] remark, by a counterexample, that the quasiconvexity of the objective function (together with the quasiconvexity of the constraints) does not imply that the Karush-Kuhn-Tucker conditions are sufficient for x^0 to be a solution of (P). However, if we add the assumption that $\nabla f(x) \neq [0], \ \forall x \in X$, then the quasiconvexity of the objective function qualifies the same as a pseudoconvex function (see what said after Definition 3.3.4). Hence, in this case Theorem 4.6.9 applies. It is possible to obtain sufficient optimality conditions also by assuming that f is quasiconvex and that $\nabla f(x^0) \neq [0]$ (i.e. the previous condition holds at x^0 only), and that the remaining assumptions of Theorem 4.6.9 hold. Indeed, we prove that if $f(x)$ is quasiconvex and $\nabla f(x^0) \neq [0]$, then for all x, x^0 we have

$$f(x) - f(x^0) < 0 \Longrightarrow \nabla f(x^0)^\top (x - x^0) < 0.$$

Suppose $f(x) < f(x^0)$ and choose $\alpha > 0$ so small that $f(x + \alpha \nabla f(x^0)) \leqq f(x^0)$. Then we have, thanks to the quasiconvexity of $f(x)$,

$$\nabla f(x^0)^\top (x + \alpha \nabla f(x^0) - x^0) \leqq 0,$$

so

$$\nabla f(x^0)^\top (x - x^0) \leqq -\alpha \nabla f(x^0)^\top \nabla f(x^0) < 0,$$

being $\nabla f(x^0) \neq [0], \ \alpha > 0$. We have proved the second member of the above implication: $\nabla f(x^0)^\top (x - x^0) < 0$. The remaining part of the proof is the same as before; taking the contrapositive law of the said implication, we have

$$\nabla f(x^0)^\top (x - x^0) \geqq 0 \Longrightarrow f(x) - f(x^0) \geqq 0.$$

For all $x \in K$, being every $g_i, \ i \in I(x^0)$, quasiconvex, and being the Karush-Kuhn-Tucker conditions verified at $x^0 \in K$, the first member of the last implication holds true, and hence it hold true also the second member.

The previous results provide sufficient global optimality conditions for (P) on the grounds of the Karush-Kuhn-Tucker conditions and by means of suitable generalized convexity assumptions on the functions involved in (P). There are also sufficient second-order optimality conditions for (P), which do not require (generalized) convexity assumptions, but which yield only local results. The more known optimality conditions of this type are due to [26] (see also [13]). We denote by $I^+(x^0)$ the set of the indices of active constraints at x^0 for which $\lambda_i > 0$ in the related optimality conditions of Karush-Kuhn-Tucker:

$$I^+(x^0) = \left\{ i : i \in I(x^0),\ \lambda_i > 0 \right\}.$$

Theorem 4.6.10 (Sufficient Second-Order Optimality Conditions for (P)) *Let the functions f and g_i, $i = 1, \ldots, m$, be of class C^2 on an open set X containing K. Let $x^0 \in K$ a point which satisfies the conditions (a), (b) and (c) of Theorem 4.6.4 of Karush-Kuhn-Tucker, with λ^0 as related multipliers vector. Furthermore, it holds*

$$z^\top H_x \mathcal{L}(x^0, \lambda^0) z > 0$$

for all $z \neq [0]$ such that

$$z \in Z^+(x^0) = \left\{ z \in \mathbb{R}^n : \nabla g_i(x^0)^\top z = 0,\ \forall i \in I^+(x^0) \right\},$$

where $\mathcal{L}(x, \lambda) = f(x) + \lambda g(x)$, $\lambda \geqq [0]$. Then x^0 is a point of strict local minimum of $f(x)$ over K.

For the proof see, e.g., [6] and [26]. We remark that in the above theorem we have to study the sign of a constrained quadratic form (see Sect. 1.7, Theorems 1.7.5 and 1.7.8). It can be proved a weaker sufficient second-order optimality condition (which is the condition considered by McCormick): the Hessian matrix $H_x \mathcal{L}(x^0, \lambda^0)$ must be positive definite on the subcone of $Z^+(x^0)$ defined as

$$Z(x^0) = \left\{ \begin{array}{l} z \in \mathbb{R}^n : \nabla g_i(x^0)^\top z = 0,\ \forall i \in I^+(x^0), \\ \nabla g_i(x^0)^\top z \leqq 0,\ \forall i \in I(x^0) \setminus I^+(x^0) \end{array} \right\}.$$

Obviously, if $I^+(x^0) = I(x^0)$, i.e. the *strict complementary slackness conditions* hold, then $Z^+(x^0) = Z(x^0)$.

There are also necessary second-order optimality conditions for (P) : see e.g., [6] and [26].

Sensitivity and the Envelope Theorem

We give now some basic notions on sensitivity for a parametric nonlinear programming problem with inequality constraints and on the so-called "envelope theorem". Indeed, sensitivity results and their economic interpretations ("shadow prices") are available also for these types of problems, and appear in several textbooks of Mathematical Economics, not always expressed in a correct way. We just state the following theorem, referring the reader to the basic book of [12] for further results and generalizations. We consider the problem

$$(P(b)): \quad \begin{cases} \min f(x) \\ g_i(x) \leqq b_i, \ i = 1, \ldots, m, \end{cases}$$

where $f : \mathbb{R}^n \longrightarrow \mathbb{R}$ and $g_i : \mathbb{R}^n \longrightarrow \mathbb{R}$, $i = 1, \ldots, m$, are of class C^2 on their domain.

Theorem 4.6.11 *Let $x^0 \in K$ be a local minimum point for $P([0])$, that is for the non parametric problem (P), let the gradients $\nabla g_i(x^0)$, $i \in I(x^0)$, be linearly independent and let $\lambda_i^0 > 0$, $\forall i \in I(x^0)$, where λ^0 is the related multipliers vector in the Karush-Kuhn-Tucker conditions for (P) (in other words, we require that the strict complementary slackness conditions hold: $I(x^0) = I^+(x^0)$). Furthermore, assume that at x^0 the sufficient second-order optimality conditions of Theorem 4.6.9 are verified. Then, there exists a neighbourhood $U([0]) \subset \mathbb{R}^m$ where the functions $\hat{x}(b)$ and $\hat{\lambda}(b)$ are defined, such that $\hat{x}(b)$ is a local minimum point for $(P(b))$, being $\hat{x}([0]) = x^0$, and $\hat{\lambda}(b)$ is the related vector of multipliers, with $\hat{\lambda}([0]) = \lambda^0$. Moreover, the functions $\hat{x}(b)$ and $\hat{\lambda}(b)$ are of class C^1 in $U([0])$ and it holds*

$$\nabla_b f(\hat{x}(b)) = -\hat{\lambda}(b), \ \forall b \in U([0]).$$

In the previous parametric problem, and also in the problem considered in Sect. 4.5, the parameters appear as right-hand side members of the constraints. More generally, it is possible to "incorporate" the parameter or the vector of parameters into the constraints and also into the objective function. For example, we can consider the problem

$$\min f(x, \varepsilon)$$

$$\text{subject to: } g_i(x, \varepsilon) \leqq 0, \ i = 1, \ldots, m;$$

$$h_j(x, \varepsilon) = 0, \ j = 1, \ldots, r,$$

where $f, g_i, h_j : \mathbb{R}^n \times \mathbb{R}^k \longrightarrow \mathbb{R}$, being $\varepsilon \in \mathbb{R}^k$ a vector of parameters. This is just the type of parametric mathematical programming problems examined by Fiacco [12] in his classical book on sensitivity analysis, to which the interested reader is referred.

We shall give some insights on the *unconstrained* parametric mathematical programming problem of the type described above, as this kind of problems has been extensively considered in the economic literature, in connection with the so-called "envelope theorem" . We consider therefore the problem

$$\min \ (\text{or} \max) f(x, \varepsilon),$$

with $x \in \mathbb{R}^n$ and $\varepsilon \in \mathbb{R}^k$, $f : \mathbb{R}^n \times \mathbb{R}^k \longrightarrow \mathbb{R}$. For each ε suppose we have found the minimum (or the maximum) of $f(x, \varepsilon)$. The minimum value (the maximum value) usually depends on ε. We denote this value by $f^*(\varepsilon)$ and call f^* the *optimum value function* (sometimes f^* is called the "marginal function"). The vector x which minimizes (maximizes) $f(x, \varepsilon)$ usually depends on ε and will be denoted by $x^*(\varepsilon)$. Then $f^*(\varepsilon) = f(x^*(\varepsilon), \varepsilon)$.

How $f^*(\varepsilon)$ vary as the i-th parameter ε_i changes? *Provided that f^* is differentiable* (this is not generally granted, even if f is differentiable!), we have the following so-called *envelope theorem.* It holds

$$\frac{\partial f^*(\varepsilon)}{\partial \varepsilon_i} = \left[\frac{\partial f(x, \varepsilon)}{\partial \varepsilon_i}\right]_{x=x^*(\varepsilon)}, \quad i = 1, \ldots, k.$$

This result can be proved as follows: because $x = x^*(\varepsilon)$ minimizes (or maximizes) $f(x, \varepsilon)$ with respect to x, all partial derivatives $\partial f(x^*(\varepsilon), \varepsilon)/\partial x_i$ must be zero. Hence

$$\begin{aligned}\frac{\partial f^*(\varepsilon)}{\partial \varepsilon_i} &= \frac{\partial}{\partial \varepsilon_i}(f(x^*(\varepsilon), \varepsilon)) \\ &= \sum_{i=1}^{n} \frac{\partial f(x^*(\varepsilon), \varepsilon)}{\partial x_i} \frac{\partial x_i^*(\varepsilon)}{\partial \varepsilon_i} + \left[\frac{\partial f(x, \varepsilon)}{\partial \varepsilon_i}\right]_{x=x^*(\varepsilon)} = \left[\frac{\partial f(x, \varepsilon)}{\partial \varepsilon_i}\right]_{x=x^*(\varepsilon)}.\end{aligned}$$

Hence, when the parameter ε_i changes, $f^*(\varepsilon)$ changes for two reasons. First, a change in ε_i changes $x^*(\varepsilon)$. Second, $f(x^*(\varepsilon), \varepsilon)$ changes directly because the variable ε_i changes. The above envelope theorem claims that the first effect is zero. This curious name comes from the classical mathematical analysis. We recall that if we are given a one-dimensional parameterized family of curves, each depending from a real parameter α which runs over some interval, curves described by $f_\alpha : \mathbb{R} \longrightarrow \mathbb{R}$, the curve $h : \mathbb{R} \longrightarrow \mathbb{R}$ is the *envelope* of the family if each point on the curve h is tangent to one of the curves f_α and each curve f_α is tangent to h. See, e.g., [2] or [17]. That is, for each α, there is some t and also for each t, there is some α, satisfying $f_\alpha(t) = h(t)$ and $f_\alpha'(t) = h'(t)$. For example, the family of parabolas of equation

$$y - (x - \alpha)^2 = 0$$

admits as its envelope curve the straight line $y = 0$. In Economic Analysis, the long-run cost curve is the envelope of the short-run cost curves, a result sometimes referred as the "Wong-Viner Theorem" (see, e.g., [33]). The envelope theorem states, in geometric terms, that for $z = f(x, \varepsilon)$, $x \in \mathbb{R}$, and under appropriate conditions, the graph of the optimum value function f^* is the envelope of the family of graphs of $f(x, \varepsilon)$.

But which are the conditions assuring the differentiability of f^*? In the economic literature this question is often skipped. We report a result of [12], who gives also the formula for the second-order derivative of the optimum value function.

Theorem 4.6.12 (Fiacco) *Let us consider the problem*

$$\min f(x, \varepsilon)$$

with $f : \mathbb{R}^n \times \mathbb{R}^k \longrightarrow \mathbb{R}$, $f(x, \varepsilon)$ *twice continuously differentiable in* (x, ε) *near* $(\bar{x}, \bar{\varepsilon})$. *If the second-order sufficient optimality conditions for an unconstrained local minimum of* f *hold at* $(\bar{x}, \bar{\varepsilon})$, *then for* ε *near* $\bar{\varepsilon}$, *there exists a unique once continuously differentiable vector function* $x(\varepsilon)$ *satisfying the second-order sufficient conditions for an unconstrained local minimum of* $f(x, \varepsilon)$, *such that* $x(\bar{\varepsilon}) = \bar{x}$ *and hence* $x(\varepsilon)$ *is a locally unique unconstrained local minimum of* $f(x, \varepsilon)$. *Moreover, the optimum value function* f^* *is twice-continuously differentiable in* (x, ε) *near* $(\bar{x}, \bar{\varepsilon})$ *and in a neighbourhood of* $\varepsilon = \bar{\varepsilon}$ *we have*

(a) $\nabla_\varepsilon f^*(\varepsilon) = \nabla_\varepsilon f(x(\varepsilon), \varepsilon)$;
(b) $\nabla^2_\varepsilon f^*(\varepsilon) = \nabla^2_{x\varepsilon} f \nabla_\varepsilon x + \nabla^2_\varepsilon f \mid_{x=x(\varepsilon)}$.

We remark that it is possible to obtain the directional differentiability of the above optimum value function, under mild conditions, by assuming that in $f(x, \varepsilon)$ the vector x belongs to a compact set $X \subset \mathbb{R}^n$ and that f and the partial derivatives $\partial f / \partial \varepsilon_i$ are continuous. Moreover, it is also possible to obtain a formula to compute the said directional derivative. See the so-called "Danskin's Theorem" in the book of [12]. Without making reference to second-order conditions, but assuming suitable convexity conditions, it is possible to obtain the differentiability of the above optimum value function $f^*(\varepsilon)$. The following result is taken from [34].

Theorem 4.6.13 *Consider the problem* $\min f(x, \varepsilon)$. *Suppose that* $f^*(\varepsilon)$ *is finite and convex in* $\varepsilon \in A \subset \mathbb{R}^k$, *where* A *is an open convex set. Assume that the point* $(\bar{x}, \bar{\varepsilon}) \in \mathbb{R}^n \times A$ *satisfies* $f(\bar{x}, \bar{\varepsilon}) = f^*(\bar{\varepsilon})$ *and that the gradient vector* $\nabla_\varepsilon f$ *exists at* $(\bar{x}, \bar{\varepsilon})$. *Then* $f^*(\varepsilon)$ *is differentiable at* $\bar{\varepsilon}$ *and*

$$\nabla f^*(\bar{\varepsilon}) = \nabla_\varepsilon f(\bar{x}, \bar{\varepsilon}).$$

On the grounds of a result of Fiacco, the optimum value function $f^*(\varepsilon)$ is convex on the convex set $A \subset \mathbb{R}^k$ if $f(x, \varepsilon)$ is *jointly convex*, i.e. it is convex with respect to x and ε.

Example 4.6.14 Let be given a profit function π which depends on the input vector $v \in \mathbb{R}^n$ and the parametric prices $p \in \mathbb{R}$ and $q \in \mathbb{R}^n$ ($p > 0$ and $q > [0]$), with π of the form

$$\pi = pF(v) - q_1 v_1 - \cdots - q_n v_n,$$

where F is the production function.

Let $\pi^*(p, q)$ denote the value function in the problem of maximizing π with respect to v, and let $v^* = v^*(p, q)$ be the associated vector. Then, according to the envelope theorem, assuming the differentiability of the value function, we have

$$\frac{\partial \pi^*(p,q)}{\partial p} = \frac{\partial \pi(v^*, p, q)}{\partial p} = F(v^*);$$
$$\frac{\partial \pi^*(p,q)}{\partial q_i} = \frac{\partial \pi(v^*, p, q)}{\partial q_i} = -v_i^*.$$

These results are known as *Hotelling's Lemma,* after the American statistician Harold Hotelling (1895–1973). In particular, we can say that if the price of the input vector increases, then the firm's maximal profit decreases and we can compute, by a mathematical formula, these movements, if we know the firm's profit function, even if we do not know the firm's production function.

4.7 Other Formulations of Constrained Optimization Problems

On the grounds of what said in the previous pages, it is not difficult to obtain the Karush-Kuhn-Tucker conditions or the Fritz John conditions for other formulations of mathematical programming problems. For instance, consider the problem

$$\begin{cases} \min f(x) \\ g(x) \leqq [0] \\ x \geqq [0], \end{cases}$$

where there is a nonnegativity condition on the vector of variables. Obviously this condition is almost always present in problems of economic or financial nature. We rewrite this problem in the form

$$\begin{cases} \min f(x) \\ r(x) \leqq [0] \end{cases}$$

being

$$r(x) = \begin{bmatrix} g(x) \\ -x \end{bmatrix}, \quad r : \mathbb{R}^n \longrightarrow \mathbb{R}^{m+n}.$$

Let us suppose that the constraints g_i, $i = 1, \ldots, m$, verify some constraint qualification at the local solution point $x^0 \in K$ (note that the constraints $-x_i$, $i = 1, \ldots, n$, are linear and so they are qualified by default; moreover the Jacobian matrix of $-x$ is the matrix $-I$, nonsingular). The Karush-Kuhn-Tucker conditions for the last problem are therefore

(i) $\nabla f(x^0) + \lambda^\top \nabla r(x^0) = [0]$;
(ii) $\lambda^\top r(x^0) = 0$;
(iii) $\lambda \geqq [0]$.

That is,

$$\nabla f(x^0) + (\lambda^1)^\top \nabla g(x^0) + (\lambda^2)^\top (-I) = [0];$$

$$\lambda_i r_i(x^0) = 0,\ i = 1, \ldots, m+n \iff \left\{(\lambda^1)^\top g(x^0) = 0,\ (\lambda^2)^\top (-x^0) = 0\right\}; \tag{4.8}$$

$$\lambda^1 \geqq [0],\ \lambda^2 \geqq [0]. \tag{4.9}$$

That is

$\nabla f(x^0) + (\lambda^1)^\top \nabla g(x^0) = \lambda^2 \geqq [0]$, by the second relation of (4.9);
$(\lambda^1)^\top g(x^0) = 0$;
$(\nabla f(x^0) + (\lambda^1)^\top \nabla g(x^0))^\top x^0 = 0$, by the second relation of (4.8);
$\lambda^1 \geqq [0]$.

Hence, the Karush-Kuhn-Tucker conditions for the problem under examination are:

$\nabla f(x^0) + \lambda^\top \nabla g(x^0) \geqq [0]$;
$(\nabla f(x^0) + \lambda^\top \nabla g(x^0))^\top x^0 = 0$;
$\lambda^\top g(x^0) = 0$;
$\lambda \geqq [0]$.

For example, if the problem

$$\begin{cases} \min f(x) \\ x \geqq [0] \end{cases}$$

has at $x^0 \geqq [0]$ a local solution, the associated Karush-Kuhn-Tucker conditions are:

$\nabla f(x^0) \geqq [0]$;
$\nabla f(x^0)^\top x^0 = 0$.

Now we consider a maximization problem, of the type

$$\begin{cases} \max f(x) \\ g(x) \leqq [0] \\ x \geqq [0]. \end{cases}$$

Taking into account that $\max f(x) \iff \min\{-f(x)\}$, we have:

$$\begin{aligned}
&-\nabla f(x^0)+\lambda^\top \nabla g(x^0) \geqq [0]\,;\\
&(-\nabla f(x^0)+\lambda^\top \nabla g(x^0))^\top x^0 = 0;\\
&\lambda^\top g(x^0) = 0;\\
&\lambda \geqq [0]\,.
\end{aligned}$$

That is

$$\begin{aligned}
&\nabla f(x^0)-\lambda^\top \nabla g(x^0) \leqq [0]\,;\\
&(\nabla f(x^0)-\lambda^\top \nabla g(x^0))^\top x^0 = 0;\\
&\lambda^\top g(x^0) = 0;\\
&\lambda \geqq [0]\,.
\end{aligned}$$

For example, if the problem

$$\begin{cases} \max f(x) \\ x \geqq [0] \end{cases}$$

has at $x^0 \geqq [0]$ a local solution, the associated Karush-Kuhn-Tucker conditions are:

$$\begin{aligned}
&\nabla f(x^0) \leqq [0]\,;\\
&\nabla f(x^0)^\top x^0 = 0.
\end{aligned}$$

We now write, for the reader's convenience, a "summarizing list" of the Karush-Kuhn-Tucker conditions associated to various formulations of a mathematical programming problem.

(a) $\min f(x)$.
(a_1) Constraints of the type $g_i(x) \leqq 0,\ i=1,\dots,m$:
$$\nabla f(x^0)+\sum_{i=1}^m \lambda_i \nabla g_i(x^0) = [0]\,.$$
(a_2) Constraints of the type $g_i(x) \geqq 0,\ i=1,\dots,m$:
$$\nabla f(x^0)-\sum_{i=1}^m \lambda_i \nabla g_i(x^0) = [0]\,.$$

Moreover, the conditions: $\lambda_i g_i(x^0)=0,\ \lambda_i \geqq 0,\ i=1,\dots,m.$

(b) $\max f(x)$.
(b_1) Constraints of the type $g_i(x) \leqq 0,\ i=1,\dots,m$:
$$\nabla f(x^0)-\sum_{i=1}^m \lambda_i \nabla g_i(x^0) = [0]\,.$$
(b_2) Constraints of the type $g_i(x) \geqq 0,\ i=1,\dots,m$:
$$\nabla f(x^0)+\sum_{i=1}^m \lambda_i \nabla g_i(x^0) = [0]\,.$$

Moreover, the conditions: $\lambda_i g_i(x^0)=0,\ \lambda_i \geqq 0,\ i=1,\dots,m.$
Now we consider also the nonnegativity conditions on vector x, i.e. $x \geqq [0]$.

(a) $\min f(x)$.
(a_1) Constraints of the type $g_i(x) \leqq 0,\ i=1,\dots,m,\ x \geqq [0]$:
$$\begin{aligned}
&\nabla f(x^0)+\sum_{i=1}^m \lambda_i \nabla g_i(x^0) \geqq [0]\,;\\
&(\nabla f(x^0)+\sum_{i=1}^m \lambda_i \nabla g_i(x^0))^\top x^0 = 0.
\end{aligned}$$

(a_2) Constraints of the type $g_i(x) \geqq 0,\ i = 1, \ldots, m,\ x \geqq [0]$:
$\nabla f(x^0) - \sum_{i=1}^{m} \lambda_i \nabla g_i(x^0) \leqq [0]$;
$(\nabla f(x^0) - \sum_{i=1}^{m} \lambda_i \nabla g_i(x^0))^\top x^0 = 0$.

Moreover, the conditions : $\lambda_i g_i(x^0) = 0,\ \lambda_i \geqq 0,\ i = 1, \ldots, m$.

(b) max $f(x)$.
(b_1) Constraints of the type $g_i(x) \leqq 0,\ i = 1, \ldots, m,\ x \geqq [0]$:
$\nabla f(x^0) - \sum_{i=1}^{m} \lambda_i \nabla g_i(x^0) \leqq [0]$;
$(\nabla f(x^0) - \sum_{i=1}^{m} \lambda_i \nabla g_i(x^0))^\top x^0 = 0$.
(b_2) Constraints of the type $g_i(x) \geqq 0,\ i = 1, \ldots, m,\ x \geqq [0]$:
$\nabla f(x^0) + \sum_{i=1}^{m} \lambda_i \nabla g_i(x^0) \geqq [0]$;
$(\nabla f(x^0) + \sum_{i=1}^{m} \lambda_i \nabla g_i(x^0))^\top x^0 = 0$.

Moreover, the conditions : $\lambda_i g_i(x^0) = 0,\ \lambda_i \geqq 0,\ i = 1, \ldots, m$.

Example 4.7.1 Consider a firm which produces only one type of goods by means of two production factors and with the following production function

$$f(x_1, x_2) = \log\left(1 + \frac{x_1}{1+x_1} + \frac{x_2}{1+x_2}\right).$$

We want to find the optimal production plan, i.e. the one which maximizes the profit over $\mathbb{R}^2_+$, under the assumption that the unitary costs of the factors are $p_1 = p_2 = \frac{p}{2}$, where $p > 0$ is the unitary price of the good produced. The profit is given by $\pi(p, x_1, x_2) = pf(x_1, x_2) - p_1x_1 - p_2x_2$.

The Karush-Kuhn-Tucker conditions are

$$\frac{1}{1 + \frac{x_1}{1+x_1} + \frac{x_2}{1+x_2}} \cdot \frac{1}{(1+x_1)^2} \leqq \frac{1}{2}$$

$$\frac{1}{1 + \frac{x_1}{1+x_1} + \frac{x_2}{1+x_2}} \cdot \frac{1}{(1+x_2)^2} \leqq \frac{1}{2}$$

$$\left(\frac{1}{1 + \frac{x_1}{1+x_1} + \frac{x_2}{1+x_2}} \cdot \frac{1}{(1+x_1)^2} - \frac{1}{2}\right) x_1 = 0$$

$$\left(\frac{1}{1 + \frac{x_1}{1+x_1} + \frac{x_2}{1+x_2}} \cdot \frac{1}{(1+x_1)^2} - \frac{1}{2}\right) x_2 = 0$$

$$x_1 \geqq 0,\ \ x_2 \geqq 0.$$

Owing to the symmetry of the problem, the solutions of the system follow with $x_1 = x_2$. As the relation $x_1 = x_2 = 0$ does not satisfy the system, it must hold $x_1 = x_2 > 0$. From the equation

$$\frac{1}{1 + \frac{2x_1}{1+x_1}} \cdot \frac{1}{(1+x_1)^2} = \frac{1}{2},$$

we deduce $3x_1^2 + 4x_1 - 1 = 0$, from which

$$x_1 = \frac{\sqrt{7}-2}{3} \text{ and hence } x_2 = \frac{\sqrt{7}-2}{3}$$

(obviously the root $\frac{-\sqrt{7}-2}{3}$ is not acceptable). Being the objective function f a *concave* function on $\mathbb{R}_+^2$, the above points are the solutions of the problem, i.e. they maximize the profit of the firm.

4.8 Constrained Optimization Problems with Both Inequality and Equality Constraints: Some Examples

In the present section we give some insights on constrained optimization problems with both inequality and equality constraints, i.e. with *mixed constraints.* We shall be concerned with the problem

$$(P): \quad \begin{cases} \min f(x) \\ \text{subject to: } g_i(x) \leqq 0, \ i = 1, \dots, m; \\ h_j(x) = 0, \ j = 1, \dots, r < n, \end{cases}$$

where f, every g_i, $i = 1, \dots, m$, and every h_j, $j = 1, \dots, r$, are real-valued functions defined on the open set $X \subset \mathbb{R}^n$. We make the assumptions that f and every g_i are at least differentiable on X and that every h_j is at least continuously differentiable on X.

Let S denote the *feasible set* of (P), i.e.

$$S = \left\{x \in X : g(x) \leqq [0], \ h(x) = [0]\right\}.$$

The set of indices of the *active constraints* at $x^0 \in S$ is, as usual,

$$I(x^0) = \left\{i : g_i(x^0) = 0\right\}.$$

We have for (P) the following Fritz John necessary conditions.

Theorem 4.8.1 *Let $x^0 \in S$ be a local minimum point for (P). Then, there exist multipliers $(u_0, u, v) \in \mathbb{R} \times \mathbb{R}^m \times \mathbb{R}^r$, such that*

$$u_0 \nabla f(x^0) + \sum_{i=1}^{m} u_i \nabla g_i(x^0) + \sum_{j=1}^{r} v_j \nabla h_j(x^0) = [0],$$

$$[u_0, u_1, \dots, u_m]^\top \geqq [0], \ [u_0, u_1, \dots, u_m, v_1, v_2, \dots, v_r]^\top \neq [0]$$

$$u_i g_i(x^0) = 0, \ i = 1, \dots, m.$$

Note that the vector $[u_0, u_1, \ldots, u_m, v_1, v_2, \ldots, v_r]^\top$ is a *nonzero vector,* however it can happen that $u_0 = 0$. Also for (P) we need some *constraint qualification* in order to avoid the case $u_0 = 0$. A constraint qualification for (P) quite easy to check is the *linear independence constraint qualification* (LICQ):

- The gradients

$$\left\{\nabla g_i(x^0),\ i \in I(x^0),\ \nabla h_j(x^0),\ j = 1, \ldots, r\right\}$$

are linearly independent.

This constraint qualification, when satisfied, assures not only $u_0 > 0$ in Theorem 4.8.1, but also that the related multipliers are uniquely determined. Another constraint qualification for (P) which has several important properties (besides the fact that it guarantees $u_0 > 0$ in Theorem 4.8.1), is the *Mangasarian-Fromovitz constraint qualification* (MFCQ) (see [25]) :

- (a) The gradients $\nabla h_j(x^0),\ j = 1, \ldots, r$, are linearly independent;
  (b) The system

$$\begin{cases} \nabla g_i(x^0)^\top z < 0,\ \forall i \in I(x^0) \\ \nabla h_j(x^0)^\top z = 0,\ \forall j = 1, \ldots, r, \end{cases}$$

has a solution $z \in \mathbb{R}^n$.

We can note that this constraint qualification may be considered as an adaptation to problem (P) of the Arrow-Hurwicz-Uzawa constraint qualification previously described. It can be shown that (LICQ) $\Longrightarrow$(MFCQ).

Other constraint qualifications for (P) are:

- All constraints $g_i,\ i = 1, \ldots, m$, and $h_j,\ j = 1, \ldots, r$, are linear affine.
- All constraints $h_j(x),\ j = 1, \ldots, r$, are linear affine and all constraints $g_i,\ i = 1, \ldots, m$, are concave (or even pseudoconcave). This is the *reverse constraint qualification.*
- *Slater constraint qualification:*

Every h_j is linear affine, every $g_i,\ i \in I(x^0)$, is convex on the open convex set $X \subset \mathbb{R}^n$ (or even pseudoconvex) and there exists $\bar{x} \in X$ such that $g_i(\bar{x}) < 0, \forall i \in I(x^0),\ h(\bar{x}) = [0]$.

Theorem 4.8.2 (Karush-Kuhn-Tucker) *Let $x^0 \in S$ be a local minimum point for (P) and let a constraint qualification be satisfied. Then there exist multipliers $(u, v) \in \mathbb{R}^m \times \mathbb{R}^r$, such that*

$$\nabla f(x^0) + \sum_{i=1}^{m} u_i \nabla g_i(x^0) + \sum_{j=1}^{r} v_j \nabla h_j(x^0) = [0]\,,$$

$$u \geqq [0]\,,\ \ u_i g_i(x^0) = 0,\ \ i = 1, \ldots, m.$$

Proof We give the proof assuming the validity of (LICQ). The proof is quite immediate: x^0 must satisfy the Fritz John conditions of Theorem 4.8.1. If $u_0 > 0$ there is nothing to prove: it is sufficient to divide all multipliers by u_0 (and to denote again by u and v the vectors of the new multipliers). If $u_0 = 0$, then we have

$$\sum_{i \in I(x^0)} u_i \nabla g_i(x^0) + \sum_{j=1}^{r} v_j \nabla h_j(x^0) = [0],$$

but then the above gradients are linearly dependent, in contradiction with the assumption of the proof. □

As for what concerns first-order sufficient optimality conditions for (P), we have the following result (see, e.g. [24]).

Theorem 4.8.3 *Let $x^0 \in S$ and suppose that the Karush-Kuhn-Tucker conditions are verified at x^0 :*

$$\nabla f(x^0) + \sum_{i=1}^{m} u_i \nabla g_i(x^0) + \sum_{j=1}^{r} v_j \nabla h_j(x^0) = [0],$$

$$u \geqq [0], \quad u_i g_i(x^0) = 0, \quad i = 1, \ldots, m.$$

Suppose further that $f(x)$ is a pseudoconvex function, every $g_i(x)$, $i = 1, \ldots, m$, is a quasiconvex function and every $h_j(x)$, $j = 1, \ldots, r$, is a linear affine function. Then x^0 is a solution for (P).

Also for (P) we have second-order optimality conditions. To describe the second-order conditions for (P) we define the following *Lagrangian function* (or simply *the Lagrangian*):

$$\mathcal{L}(x, u, v) = f(x) + \sum_{i=1}^{m} u_i g_i(x) + \sum_{j=1}^{r} v_j h_j(x) = f(x) + u^\top g(x) + v^\top h(x).$$

Using the Lagrangian we can rewrite the gradient conditions appearing in the Karush-Kuhn-Tucker theorem as follows:

$$\nabla_x \mathcal{L}(x^0, u.v) = [0]$$

$$u \geqq [0], \quad u^\top g(x^0) = 0.$$

Similarly, $H_x \mathcal{L}(x, u, v)$ or $\nabla_x^2 \mathcal{L}(x, u, v)$ denotes the Hessian matrix of $\mathcal{L}$ with respect to the x vector. Let $x^0 \in S$ and define the following sets:

$$I^+(x^0) = \left\{ i \in I(x^0) : u_i > 0 \right\},$$

where u_i are the multipliers that appear in Theorem 4.8.2, and

$$Z(x^0) = \left\{ \begin{array}{c} z \in \mathbb{R}^n : \nabla g_i(x^0)^\top z = 0, \ i \in I^+(x^0); \\ \nabla g_i(x^0)^\top z \leqq 0, \ i \in I(x^0) \setminus I^+(x^0); \nabla h_j(x^0)^\top z = 0, \ j = 1, \ldots, r \end{array} \right\}$$

Obviously $Z(x^0)$ is a polyhedral cone. We have the following second-order sufficient optimality conditions for (P), essentially due to [26].

Theorem 4.8.4 *If at the feasible point x^0 for (P), there exist $u \geqq [0]$, $v \in \mathbb{R}^r$, such that*

$$\nabla_x \mathcal{L}(x^0, u.v) = [0] ;$$

$$u_i g_i(x^0) = 0, \ i = 1, \ldots, m;$$

$$z^\top \nabla_x^2 \mathcal{L}(x^0, u.v) z > 0, \ \ \forall z \in Z(x^0), \ z \neq [0] ,$$

then x^0 is a strict local minimizer for problem (P).

Note that if the *strict complementary slackness conditions* hold, i.e. $I^+(x^0) = I(x^0)$, then in Theorem 4.8.4 we can use the criteria which establish the positive definiteness of a quadratic form subject to a system of homogeneous linear equations. For second-order necessary optimality conditions for (P), see, e.g., [26] or [15].

Remark 4.8.5 We have stated in the previous pages that the constraint qualifications we have introduced and used are not the most general among the ones appearing in the literature. For the reader's convenience we give here two general constraint qualifications for (P), which frequently appear in the study of optimality conditions for the same problem. Let $x^0 \in S$ (feasible set of (P)); we recall the notion of *Bouligand tangent cone* of S at x^0, previously given in Sect. 4.3:

$$T(S, x^0) = \left\{ \begin{array}{c} v \in \mathbb{R}^n : \exists \{\lambda_n\} \subset \mathbb{R}_+, \ \exists \{x^n\} \subset S, \text{ with } x^n \longrightarrow x^0, \\ \text{such that } \lambda_n(x^n - x^0) \longrightarrow v. \end{array} \right\}$$

The *linearizing cone* of S at x^0 is defined as

$$L(S, x^0) = \Big\{ v \in \mathbb{R}^n : \nabla g_i(x^0)^\top v \leqq 0, \ i \in I(x^0); \ \nabla h_j(x^0)^\top v = 0,$$
$$j = 1, \ldots, r \} .$$

We remark that both the tangent cone and the linearizing cone are in fact closed cones and that the latter is a polyhedral convex cone, which is not necessarily true for the tangent cone. In addition, it can be proved that the inclusion

$$T(S, x^0) \subset L(S, x^0)$$

holds true for all $x^0 \in S$.

• Let $x^0 \in S$. We say that the *Abadie constraint qualification* holds at x^0 if

$$T(S, x^0) = L(S, x^0).$$

• Let $x^0 \in S$. We say that the *Guignard-Gould-Tolle constraint qualification* holds at x^0 if

$$(T(S, x^0))^* = (L(S, x^0))^*,$$

i.e. if the *polar* of the tangent cone equals the polar of the contingent cone.

The Guignard-Gould-Tolle constraint qualification is, in a sense, the weakest constraint qualification for (P). See, e.g., [15] and [16]. Obviously, both the Abadie c. q. and the Guignard-Gould-Tolle c. q. may be difficult to check. Let $x^0 \in S$, then the following chain of implications holds:

$$\text{(LICQ)} \Longrightarrow \text{(MFCQ)} \Longrightarrow \text{(Abadie CQ)} \Longrightarrow \text{(Guignard-Gould-Tolle CQ)}.$$

Karush-Kuhn-Tucker conditions are a basic tool in the construction of several numerical algorithms to solve nonlinear programming problems. However, the direct application, with "paper and pen" , of the said conditions, may result quite cumbersome, also for problems of lower dimensions. The following exercises put into evidence what just said.

Example 4.8.6

(*a*) Solve, by means of the Karush-Kuhn-Tucker conditions, the following problem:

$$\begin{cases} \min\,(x_1-2)^2+(x_2-3)^2 \\ x_1+x_2-2 \leqq 0 \\ (x_1)^2-4 \leqq 0. \end{cases}$$

The constraints are qualified, as they are convex functions and the point $(0, 0) \in K$ satisfies the Slater condition. We write the Lagrangian function of this problem:

$$\mathcal{L}(x, \lambda) = (x_1-2)^2+(x_2-3)^2+\lambda_1(x_1+x_2-2)+\lambda_2((x_1)^2-4).$$

The Karush-Kuhn-Tucker conditions are:

$$\frac{\partial \mathcal{L}}{\partial x_1} = 2(x_1-2)+\lambda_1+2\lambda_2 x_1 = 0;$$

$$\frac{\partial \mathcal{L}}{\partial x_2} = 2(x_2-3)+\lambda_1 = 0;$$

$$\lambda_1(x_1+x_2-2) = 0;$$

$$\lambda_2((x_1)^2 - 4) = 0;$$
$$\lambda_1 \geqq 0,\ \lambda_2 \geqq 0.$$

(1) Suppose $\lambda_1 = \lambda_2 = 0$. From the first equation we obtain $2(x_1 - 2) = 0$ and from the second equation we obtain $2(x_2 - 3) = 0$. Hence $x_1 = 2$ and $x_2 = 3$. However $(2, 3) \notin K$: from $x_1 + x_2 - 2$ we have $2+3-2 = 3 > 0$.
(2) Suppose $\lambda_1 = 0,\ \lambda_2 > 0$. From the complementary slackness conditions we have $(x_1)^2 - 4 = 0$, that is $x_1 = \pm 2$; from the second equation we have $x_2 = 3$. The point $(2, 3) \notin K$, but the point $(-2, 3) \in K$. However, by substituting this last point into the first equation, we obtain $\lambda_2 = -2$, not acceptable.
(3) Suppose $\lambda_1 > 0,\ \lambda_2 = 0$. From the first complementary slackness condition we have $x_1 + x_2 - 2 = 0$, that is $x_2 = 2 - x_1$. From the first equation we have $2(x_1 - 2) + \lambda_1 = 0$; from the second equation we have $2(x_2 - 3) + \lambda_1 = 0$. Hence:

$$\begin{cases} \lambda_1 = -2(x_1 - 2) \\ \lambda_1 = -2(x_2 - 3). \end{cases}$$

But being $x_2 = 2 - x_1$, we can rewrite the second equation of the last system as $\lambda_1 = -2(2 - x_1 - 3)$. We obtain therefore the equation

$$-2(x_1 - 2) = -2(-x_1 + 1),$$

from which we get $x_1 = \frac{1}{2}$ and hence $x_2 = \frac{3}{2}$.

The point $\left(\frac{1}{2}, \frac{3}{2}\right) \in K$ and we have $\lambda_1 = -2(\frac{1}{2} - 2) = 3,\ \lambda_2 = 0$.

(4) Suppose $\lambda_1 > 0,\ \lambda_2 > 0$. From $(x_1)^2 - 4 = 0$ we have $x_1 = \pm 2$ and by substituting these values into $x_1 + x_2 - 2 = 0$, we obtain:
   - By substituting $x_1 = 2$, we obtain $x_2 = 0$, however the first equation is not satisfied, being $\lambda_1 > 0,\ \lambda_2 > 0$.
   - By substituting $x_1 = -2$, we obtain $x_2 = 4$, which falsifies the second equation: $2 + \lambda_1 = 0$ gives $\lambda_1 = -2$, not acceptable.

In conclusion, as the objective function is a strictly convex function, being

$$Hf(x_1, x_2) = \begin{bmatrix} 2 & 0 \\ 0 & 2 \end{bmatrix},$$

it follows that the point $P(\frac{1}{2}, \frac{3}{2})$ is the solution of our problem.

(*b*) Solve the problem

$$\begin{cases} \max f(x_1, x_2) = x_1 + \log(x_2 + 1) \\ x_1 + x_2 - 1 \leqq 0 \\ x_1 \geqq 0, \ x_2 \geqq 0 \end{cases}$$

by means of the Karush-Kuhn-Tucker conditions.

The Karush-Kuhn-Tucker conditions for this problem are:

$$\nabla f(x^0) - \lambda^\top \nabla g(x^0) \leqq [0],$$
$$(\nabla f(x^0) - \lambda^\top \nabla g(x^0))^\top x^0 = 0,$$

$$\lambda_i \geqq 0, \ \forall i; \ \ \lambda_i g_i(x^0) = 0, \ \forall i.$$

The constraints are linear, hence they are automatically qualified. We have

$$\begin{bmatrix} 1 \\ \frac{1}{x_2+1} \end{bmatrix} - \lambda \begin{bmatrix} 1 \\ 1 \end{bmatrix} \leqq \begin{bmatrix} 0 \\ 0 \end{bmatrix},$$

from which

$$\begin{cases} 1 - \lambda \leqq 0 \\ \frac{1}{x_2+1} - \lambda \leqq 0. \end{cases}$$

The complementary slackness conditions with respect to x_1 and x_2 are

$$\begin{cases} (1 - \lambda)x_1 = 0 \\ \left(\frac{1}{x_2+1} - \lambda\right) x_2 = 0, \end{cases}$$

and finally we impose

$$\lambda \geqq 0, \ \lambda(x_1 + x_2 - 1) = 0.$$

From $(1 - \lambda)x_1 = 0$ it follows that $x_1 = 0$ or $\lambda = 1$.

(*i*) If $x_1 = 0$ we have $\lambda(x_2 - 1) = 0$, i.e. $x_2 = 1$ and hence $1\left(\frac{1}{2} - \lambda\right) = 0$, from which $\lambda = \frac{1}{2}$, not acceptable (it must hold $\lambda \geqq 1$).

(*ii*) If $\lambda = 1$, from $\left(\frac{1}{x_2+1} - 1\right) x_2 = 0$ we obtain $\left(\frac{1-x_2-1}{x_2+1}\right) x_2 = 0$, i.e. $x_2 = 0$. So, from $1(x_1 + 0 - 1) = 0$ we have $x_1 = 1$.

In conclusion, with $\lambda = 1$ we have located the feasible point $P(1, 0)$. Note that from $\left(\frac{1}{x_2+1} - \lambda\right) x_2 = 0$ it follows $x_2 = 0$ or $\frac{1}{1+x_2} = \lambda$. If $x_2 = 0$ we derive that $\lambda(x_1 - 1) = 0$, which produces $\lambda = 0$, not acceptable, and again $x_1 = 1$. From $\frac{1}{1+x_2} = \lambda$, being $\lambda \geqq 1$, we find $x_2 = 0,\ \lambda = 1$ and hence we find again the point $P(1, 0)$. Being the objective function a concave function (it is the sum of two concave functions), the point $P(1, 0)$ is the solution of the problem.

(*c*) Consider the problem

$$\min(x_1 - 12)^2 + (x_2 + 6)^2$$

subject to:

$$\begin{aligned} g_1(x) &= (x_1)^2 + 3x_1 + (x_2)^2 - 4.5x_2 - 6.5 \leqq 0 \\ g_2(x) &= (x_1 - 9)^2 + (x_2)^2 - 64 \leqq 0 \\ h_1(x) &= 8x_1 + 4x_2 - 20 = 0. \end{aligned}$$

Determinate whether or not the point $\bar{x} = (\bar{x}_1, \bar{x}_2) = (2, 1)$ is a solution to the problem.

First we check for feasibility:

$$\begin{aligned} g_1(\bar{x}) &= 0; \\ g_2(\bar{x}) &= -14 < 0; \\ h_1(\bar{x}) &= 0. \end{aligned}$$

Then we compute all gradients at $\bar{x}$:

$$\nabla f(\bar{x}) = \begin{pmatrix} -20 \\ 14 \end{pmatrix};\ \nabla g_1(\bar{x}) = \begin{pmatrix} 7 \\ -2.5 \end{pmatrix};\ \nabla g_2(\bar{x}) = \begin{pmatrix} -14 \\ 2 \end{pmatrix};$$
$$\nabla h_1(\bar{x}) = \begin{pmatrix} 8 \\ 4 \end{pmatrix}.$$

Then we have to solve the system

$$\begin{pmatrix} -20 \\ 14 \end{pmatrix} + \begin{pmatrix} 7 \\ -2.5 \end{pmatrix} u_1 + \begin{pmatrix} -14 \\ 2 \end{pmatrix} u_2 + \begin{pmatrix} 8 \\ 4 \end{pmatrix} v_1 = \begin{pmatrix} 0 \\ 0 \end{pmatrix},$$

with $u_1 \geqq 0,\ u_2 \geqq 0$.

The triplet $(4, 0, -1)$ solves the system; $u_1 > 0,\ u_2 = 0,\ u_1 g_1(\bar{x}) = 0$ and $u_2 g_2(\bar{x}) = 0$. Note, moreover, that $f(x),\ g_1(x)$ and $g_2(x)$ are convex functions and that $h_1(x)$ is a linear (affine) function. Therefore $\bar{x} = (2, 1)$ is the solution of the problem.

(*d*) Consider the problem

$$\min f(x)$$

$$\text{subject to}: \ h_1(x) = \frac{(x_1)^2}{2} + \frac{(x_2)^2}{2} + 3\frac{(x_3)^2}{4} + 3\frac{(x_4)^2}{4} - 7 = 0$$

$$g_1(x) = (x_1)^2 + (x_2)^2 + (x_3)^2 + (x_4)^2 - 10 \leqq 0$$

$$g_2(x) = x_1 + x_2 + x_3 + x_4 - 6 \leqq 0.$$

It is seen that the optimal solution is given by

$$\bar{x} = [1, 1, 2, 2]^\top,$$

with $f(\bar{x}) = 0$. At $\bar{x}$ we have $g_1(\bar{x}) = 0, \ g_2(\bar{x}) = 0$. Hence $I(\bar{x}) = \{1, 2\}$. Now

$$\nabla h_1(\bar{x}) = [1, 1, 3, 3]^\top$$

$$\nabla g_1(\bar{x}) = [2, 2, 4, 4]^\top$$

$$\nabla g_2(\bar{x}) = [1, 1, 1, 1]^\top.$$

Clearly the above three vectors are linearly dependent. Hence the linear independence constraint qualification is *not* satisfied. However, by taking

$$z = [-3, -3, 1, 1]^\top$$

we have that $\nabla h_1(\bar{x})^\top z = 0, \ \nabla g_1(\bar{x})^\top z = -4 < 0, \ \nabla g_2(\bar{x})^\top z = -4 < 0$. This means that the Mangasarian-Fromovitz constraint qualification is satisfied.

(*e*) We propose to the willing reader the following exercise. Consider the problem

$$\min f(x, y) = 2x^2 + 2xy + y^2 - 10x - 10y$$

subject to:

$$\begin{cases} x^2 + y^2 \leqq 5 \\ 3x + y \leqq 6. \end{cases}$$

(*i*) Transform the problem into an equivalent maximization problem.
(*ii*) Write the Lagrangian function for the transformed problem.
(*iii*) Write the Karush-Kuhn-Tucker optimality conditions.
(*iv*) Check that $(x, y) = (1, 2)$ satisfies the Karush-Kuhn-Tucker conditions and find the associated multipliers λ_1 and λ_2.
(*v*) Check that $(x, y) = (1, 2)$ is a global solution of the transformed problem of question (a), and hence also of the original problem.

4.9 Some Considerations on Optimization Problems in $\mathbb{R}^2$

We have previously observed that Karush-Kuhn-Ticker conditions, even if basic for theoretic developments and algorithmic developments, may be of not easy application, also for problems of lower dimensions, e.g. in $\mathbb{R}^2$, when only paper and pencil are available. The technique of the level curves may be useful when $n = 2$ and when it is possible to plot, without too many difficulties, the family of the level curves of the objective function.

Example 4.9.1 Solve the problem

$$\begin{cases} \max z = x_1 + x_2 \\ x_1 + 2x_2 \geqq 2 \\ (x_1)^2 + (x_2)^2 \leqq 4 \\ x_1 \geqq 0,\ x_2 \geqq 0. \end{cases}$$

The feasible set and some level curves of the objective function are shown in Fig. 4.14.

In order to find the level h of the tangent line at the point P (solution point of the problem), we formulate the system

$$\begin{cases} x_1 + x_2 = h \\ (x_1)^2 + (x_2)^2 = 4 \end{cases} \implies \begin{cases} x_1 = h - x_2 \\ (h - x_2)^2 + (x_2)^2 = 4. \end{cases}$$

From the second equation we obtain $h^2 - 2hx_2 + (x_2)^2 + (x_2)^2 - 4 = 0$, that is $2(x_2)^2 - 2hx_2 + h^2 - 4 = 0$.

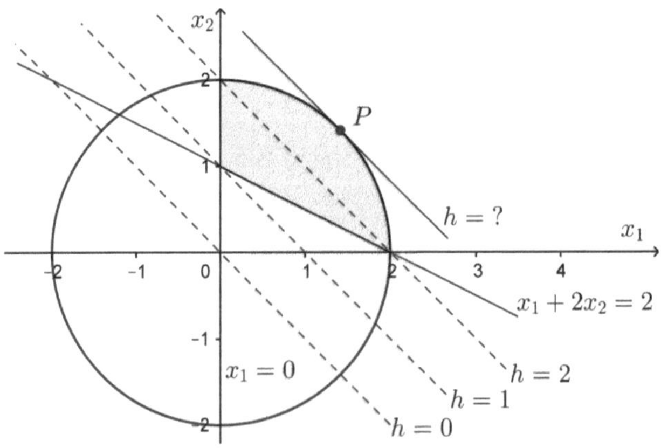

Fig. 4.14 Example 4.9.1: feasible set and level curves of the objective function

Denoting by Δ the "discriminant" of this equation, we have to impose $\Delta = 0$, that is

$$h^2 - 2h^2 + 8 = 0, \text{ that is } -h^2 + 8 = 0.$$

We get $h = \pm 2\sqrt{2}$. Hence the feasible solution is $h = 2\sqrt{2}$. Therefore we have

$$\begin{cases} x_1 = 2\sqrt{2} - x_2 \\ (2\sqrt{2} - x_2)^2 + (x_2)^2 - 4 = 0. \end{cases}$$

By developing the second equation we obtain $4 - 4\sqrt{2}x_2 + 2(x_2) = 0$, that is

$$2(x_2 - \sqrt{2})^2 = 0$$

from which $x_2 = \sqrt{2}$ and hence $x_1 = 2\sqrt{2} - \sqrt{2} = \sqrt{2}$.
The coordinates of P are therefore $x_1 = \sqrt{2},\ x_2 = \sqrt{2}$ and $z(P) = 2\sqrt{2}$.

When the objective function is *continuous* on the feasible set K and K is *closed and bounded,* it is possible, always when $n = 2$, to make use also of the so-called "restrictions method" , mainly in the case the objective function has unwieldy level curves. We recall that under the said assumptions, the Weierstrass theorem assures the existence of minimum and maximum. The restrictions method can be described by the following steps.

(*a*) Determine the type of stationary points (if any) which are interior to K.
(*b*) Consider the restrictions of the objective function f on the boundary of K and determine the minimum and/or maximum points (if any) of f on the various "pieces" of the boundary of K.
(*c*) By comparing the various results obtained with the above procedures sub a) and b), we obtain the global optimum points of the objective function over K.

Example 4.9.2 Find the global optimum points of

$$f(x, y) = 2y - x(y + 1)$$

on the feasible set

$$K = \left\{(x, y) \in \mathbb{R}^2 : |x| \leqq 3,\ |y| \leqq 2\right\}.$$

First we represent the feasible set K in Fig. 4.15.
Now we look for the interior stationary points (if any). As

$$\frac{\partial f}{\partial x} = -y - 1; \quad \frac{\partial f}{\partial y} = 2 - x,$$

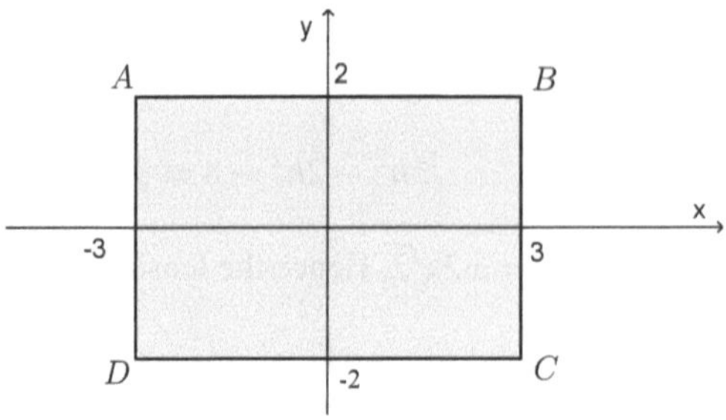

Fig. 4.15 Feasible set of Example 4.9.2

we have

$$\begin{cases} -y-1=0 \\ 2-x=0 \end{cases} \Longrightarrow x=2,\ y=-1.$$

The point $P(2,-1)$ is interior to K. Then we have

$$\frac{\partial^2 f}{\partial x^2}=0,\ \frac{\partial^2 f}{\partial y^2}=0,\ \frac{\partial^2 f}{\partial x \partial y}=\frac{\partial^2 f}{\partial y \partial x}=-1.$$

We deduce that the stationary point $P(2,-1)$ is a saddle point for f.

Now we consider the restrictions of f on the sides of the rectangle K.

(1) Side $\overline{AB}$, that is $y=2,\ -3 \leqq x \leqq 3;\ f(x,2)=4-3x=-3x+4$. We have the equation of a straight line which therefore assumes maximum and minimum at the extremum points of the segment $\overline{AB}$:
$f(-3,2)=13;\ f(3,2)=5$.
(2) Side $\overline{DC}$, that is $y=-2,\ -3 \leqq x \leqq 3;\ f(x,-2)=x-4;\ f(-3,-2)=-7;\ f(3,-2)=-1$.
(3) Side $\overline{DA}$, that is $x=-3,\ -2 \leqq y \leqq 2;\ f(-3,y)=2y+3(y+1)=5y+3$. We have $f(-3,-2)=-7;\ f(-3,2)=13$.
(4) Side $\overline{CB}$, that is $x=3,\ -2 \leqq y \leqq 2;\ f(3,y)=2y-3(y+1)=-y-3$. We have $f(3,-2)=-1;\ f(3,2)=-5$.

By comparison we have that the point $A(-3,2)$ is the point of global maximum, with $f(A)=13$, and that the point $D(-3,-2)$ is the point of global minimum, with $f(D)=-7$. There are no interior points of local minimum nor of local maximum.

Example 4.9.3 (Example Taken from [9, Vol. 2]) Solve the problem

$$\begin{cases} \max(x_1x_2-x_1-x_2+2) \\ x_1x_2 \leqq 6 \\ x_1+x_2 \leqq 5 \\ x_1 \geqq 0,\ x_2 \geqq 0. \end{cases}$$

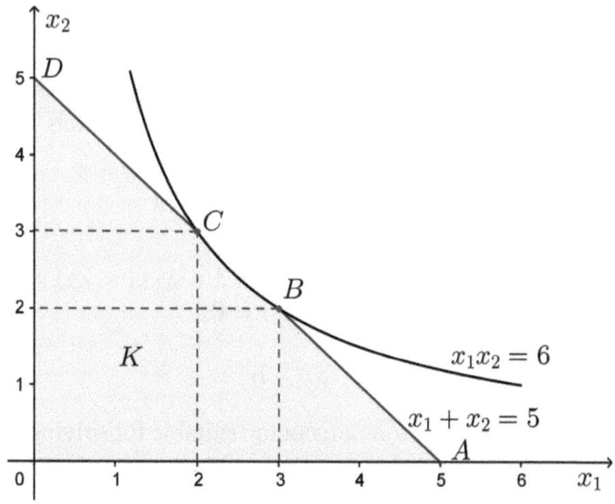

Fig. 4.16 Feasible set of Example 4.9.3

It is convenient to plot the feasible set K (see Fig. 4.16).

First we begin to study the (interior) stationary points, if any. We have $\nabla f(x) = [x_2 - 1;\ x_1 - 1]^\top$ and it holds $\nabla f(x) = [0]$ for $x = [1, 1]^\top$. This point is interior to K. Then we have

$$Hf(1, 1) = \begin{bmatrix} 0 & 1 \\ 1 & 0 \end{bmatrix}$$

and hence the point $P(1, 1)$ is a saddle point.

Now we consider one by one the single lines of the boundary of K.

(*a*) Side $\overline{OA}$. Being $x_2 = 0,\ 0 \leqq x_1 \leqq 5$, the objective function becomes $-x_1+2$, which reaches its maximum at $x_1 = 0$.

(*b*) Side $\overline{OD}$. Being now $x_1 = 0,\ 0 \leqq x_1 \leqq 5$, the objective function becomes $-x_2 + 2$, which reaches its maximum at $x_2 = 0$.

(*c*) Boundary $\overline{AB} \cup \overline{CD}$. Here we have $x_1 + x_2 = 5,\ x_1 \in [0, 2] \cup [3, 5]$. The objective function becomes $x_1(5 - x_1) - 5 + 2 = -(x_1)^2 + 5x_1 - 3,\ x_1 \in [0, 2] \cup [3, 5]$. This function has two maximum points, at $x_1 = 2$ and at $x_1 = 3$.

(*d*) Boundary BC. Here we have $x_1x_2 = 6,\ 2 \leqq x_1 \leqq 3$. The objective function becomes $6 - x_1 - \frac{6}{x_1} + 2$, that is $8 - x_1 - \frac{6}{x_1}$, with $2 \leqq x_1 \leqq 3$, which reaches its maximum at $x_1 = \sqrt{6}$.

Summing up: we have verified that there is no optimum point on $\text{int}(K)$, whereas on the boundary of K there are four maximum points:

$$(0, 0),\ (2, 3), (3, 2),\ (\sqrt{6}, \sqrt{6}).$$

By comparison of these points we obtain information on the global maximum point of f. We have $f(0,0) = 2$; $f(2,3) = 3$; $f(3,2) = 3$; $f(\sqrt{6}, \sqrt{6}) = 8 - 2\sqrt{6} \cong 3.1$. Hence the solution of our problem is the point $(\sqrt{6}, \sqrt{6})$.

If we write the Karush-Kuhn-Tucker conditions for this problem we have the following relations:

$$
\begin{aligned}
&x_2 - 1 - \lambda_1 x_2 - \lambda_2 \leqq 0;\\
&x_1 - 1 - \lambda_1 x_1 - \lambda_2 \leqq 0;\\
&x_1(x_2 - 1 - \lambda_1 x_2 - \lambda_2) + x_2(x_1 - 1 - \lambda_1 x_1 - \lambda_2) = 0,\\
&\lambda_1(6 - x_1 x_2) + \lambda_2(5 - x_1 - x_2) = 0;\\
&x_1 x_2 \leqq 6;\ x_1 + x_2 \leqq 5;\\
&x_1 \geqq 0,\ x_2 \geqq 0,\ \lambda_1 \geqq 0,\ \lambda_2 \geqq 0.
\end{aligned}
$$

These ten conditions are not too suitable for solving the problem in question, with only pen and paper. Indeed, we would take into exam a system of two equations and eight inequations. We take into consideration only some points.

(*i*) $x = [3, 1]^\top$. In order that the two above equations are satisfied (third and fourth Karush-Kuhn-Tucker condition) it must hold $\lambda = [1/6, \ -1/2]$; being $\lambda_2 < 0$ the Karush-Kuhn-Tucker conditions are not verified, hence this point cannot be a maximum nor a minimum point of f over K.

(*ii*) $x = [1, 1]^\top$. In order that the two above equations are satisfied it must be $\lambda = [0]$. With this vector of multipliers, all Karush-Kuhn-Tucker conditions are verified, however the point $x = [1, 1]^\top$ is not a maximum point (local nor global). Remember that the Karush-Kuhn-Tucker conditions, without any suitable assumptions on the objective function and on the constraints, are only *necessary optimality conditions.* Indeed, at this point we have $f(1, 1) = 1$, but if we consider the point $[1 + \alpha, \ 1 + \beta]^\top$, which belongs to K for α and β sufficiently small, if α and β have the same sign, we have $f(1, 1) = 1 + \alpha\beta > 1$.

(*iii*) $x = \left[\sqrt{6}, \ \sqrt{6}\right]^\top$. In order that the two above equations are satisfied it must hold $\lambda = \left[1 - 1/\sqrt{6}, \ 0\right]$. With this vector of multipliers all other Karush-Kuhn-Tucker conditions are satisfied. We already know that this point is the solution of the given problem.

When the feasible set K is not closed and bounded (i.e. compact) we have to be cautious. Consider, with regard to this remark, the following example.

Example 4.9.4 Find the maximum and the minimum of $f(x, y) = x^2 + y^2 + xy + 1$ over the set

$$K = \left\{(x, y) \in \mathbb{R}^2 : xy \geqq -1\right\}.$$

We have

$$\begin{cases} \frac{\partial f}{\partial x} = 2x + y \\ \frac{\partial f}{\partial y} = 2y + x \end{cases}$$

and hence the point $(0, 0)$ is the unique interior critical point. Then we have

$$\frac{\partial^2 f}{\partial x^2} = 2; \ \frac{\partial^2 f}{\partial x \partial y} = \frac{\partial^2 f}{\partial y \partial x} = 1; \ \frac{\partial^2 f}{\partial y^2} = 2.$$

$$Hf(x, y) = \begin{bmatrix} 2 & 1 \\ 1 & 2 \end{bmatrix}.$$

The Hessian matrix is positive definite on $\mathbb{R}^2$ and hence $(0, 0)$ is a point of unconstrained global minimum of the objective function, with $f(0, 0) = 1$. If we consider the boundary of K and apply the Lagrange multipliers rule to the equality constraint $g(x, y) = xy + 1 = 0$, we find two points $(1, -1)$ and $(-1, 1)$. At these points $f(x, y) = 2$. However, these points are not maximum points. The feasible set is closed, but not bounded, so we cannot apply the Weierstrass theorem and assert that, besides $f(0, 0) = 1$, which is the minimum, the objective function has a (constrained) maximum, equal to 2. Indeed, the two points $(1, -1)$ and $(-1, 1)$ are not local extrema of f on K. This is most easily seen by drawing the two branches of the hyperbola $xy = -1$ and noting that f has an extremum on the hyperbola precisely when $h(x, y) = x^2 + y^2$ has an extremum on this set (they coincide). This is the square of the distance from $(0, 0)$ to the curve and can be increased by moving along the curve and decreased by moving off the curve into the feasible set $xy > -1$. These points are local minima of f on the curve $xy = -1$, but *not* local minima over the feasible set K.

4.10 Saddle Point Problems

The problem

$$(P_{\min}) : \quad \begin{cases} \min f(x) \\ g(x) \leqq [0] \\ x \in X, \end{cases}$$

with $f : \mathbb{R}^n \longrightarrow \mathbb{R}$, $g : \mathbb{R}^n \longrightarrow \mathbb{R}^m$, $X \subset \mathbb{R}^n$ a convex set, f a convex function on X and every g_i, $i = 1, \ldots, m$, a convex function on X, is said *convex programming problem.*

The problem

$$(P_{\max}):\qquad \begin{cases} \max f(x) \\ g(x) \geqq [0] \\ x \in X, \end{cases}$$

with $f : \mathbb{R}^n \longrightarrow \mathbb{R}$ a concave function on the convex set $X \subset \mathbb{R}^n$, and every g_i, $i = 1, \ldots, m$, a concave function on X, is said *concave programming problem.*

Under the said assumptions (note that it no differentiability on the functions involved in the said problems is required), it is possible to give necessary and sufficient optimality conditions (global optimality conditions), by exploiting some nonlinear theorems of the alternative. In the next section we shall examine the relationships between the optimality conditions of the present section and the ones of the previous sections, whenever it is assumed that in $(P_{\min})$ or in $(P_{\max})$ the involved functions are differentiable (besides to be convex or concave).

We consider now $(P_{\min})$ and its related *Lagrangian function,* defined in the usual way:

$$\mathcal{L}(x, \lambda) = f(x) + \lambda^\top g(x), \quad \lambda \geqq [0]\,.$$

The point (x^0, λ^0), $x^0 \in X$, $\lambda^0 \geqq [0]$, is said to be a *saddle point* for the Lagrangian function $\mathcal{L}$ and for problem $(P_{\min})$, if it holds

$$\mathcal{L}(x^0, \lambda) \leqq \mathcal{L}(x^0, \lambda^0) \leqq \mathcal{L}(x, \lambda^0), \quad \forall x \in X, \ \forall \lambda \geqq [0]\,.$$

In other words, the value $\mathcal{L}(x^0, \lambda^0)$ represents at the same time the minimum of the function $\mathcal{L}(x, \lambda^0)$, by varying x on the set X, and the maximum of $\mathcal{L}(x^0, \lambda)$, by varying λ on $\mathbb{R}^m_+$. Do not make confusion with the definition of saddle point in unconstrained optimization problems!

If we consider $(P_{\max})$ and its related Lagrangian function, defined as

$$\mathcal{L}(x, \lambda) = f(x) + \lambda^\top g(x), \quad \lambda \geqq [0]\,.$$

the point (x^0, λ^0), $x^0 \in X$, $\lambda^0 \geqq [0]$, is said to be a *saddle point* for $\mathcal{L}$ if:

$$\mathcal{L}(x, \lambda^0) \leqq \mathcal{L}(x^0, \lambda^0) \leqq \mathcal{L}(x^0, \lambda), \quad \forall x \in X, \ \forall \lambda \geqq [0]\,,$$

that is if the value $\mathcal{L}(x^0, \lambda^0)$ represents at the same time the maximum of the function $\mathcal{L}(x, \lambda^0)$, by varying x on the set X, and the minimum of $\mathcal{L}(x^0, \lambda)$, by varying λ on $\mathbb{R}^m_+$.

Clearly a saddle point may never exist and even if it exists, it is not necessarily unique. However, if we have more than one saddle point, for them it holds the property of interchangeability: let (x^1, λ^1) and (x^2, λ^2) be saddle points of the Lagrangian function $\mathcal{L}(x, \lambda)$. Then

$$\mathcal{L}(x^1, \lambda^1) = \mathcal{L}(x^2, \lambda^1) = \mathcal{L}(x^1, \lambda^2) = \mathcal{L}(x^2, \lambda^2).$$

Consequently (x^1, λ^2) and (x^2, λ^1) are also saddle points.

Now we consider $(P_{\min})$, as the results regarding $(P_{\max})$ can be easily obtained from the ones of $(P_{\min})$. The following important theorem gives a necessary optimality condition for $(P_{\min})$.

Theorem 4.10.1 (Theorem of Kuhn-Tucker-Uzawa) *Let x^0 be a point of global minimum for $(P_{\min})$. Then, there exist $m+1$ multipliers, all nonnegative but not all zero, say μ, λ_1^0, $\lambda_2^0, \ldots, \lambda_m^0$, such that*

$$\mu f(x) + (\lambda^0)^\top g(x) \geqq \mu f(x^0), \ \forall x \in X, \tag{4.10}$$

where $\lambda^0 = \left[\lambda_1^0, \lambda_2^0, \ldots, \lambda_m^0\right]^\top$. Moreover, it holds ("complementary slackness conditions")

$$(\lambda^0)^\top g(x^0) = 0. \tag{4.11}$$

Proof As x^0 is a global minimum point for $(P_{\min})$, the system

$$\begin{cases} f(x) - f(x^0) < 0 \\ g(x) \leqq [0] \end{cases}$$

has no solution on X. A fortiori it has no solution on X the system

$$\begin{cases} f(x) - f(x^0) < 0 \\ g(x) < [0] \end{cases}.$$

The function

$$\varphi(x) = \begin{bmatrix} f(x) - f(x^0) \\ g(x) \end{bmatrix}, \quad \varphi : \mathbb{R}^n \longrightarrow \mathbb{R}^{m+1},$$

is convex on the convex set X, being f and g convex on this set. By the Fan-Glicksberg-Hoffman theorem (Theorem 3.4.9), there exist $m+1$ multipliers, *all nonnegative but not all zero,* μ, λ_1^0, $\lambda_2^0, \ldots, \lambda_m^0$, such that

$$\mu(f(x) - f(x^0)) + (\lambda^0)\top g(x) \geqq 0, \ \ \forall x \in X,$$

from which relation (4.10) follows immediately. In order to prove (4.11), let us note that it holds $(\lambda^0)^\top g(x^0) \leqq 0$, being $\lambda^0 \geqq [0]$ and $g(x^0) \leqq [0]$. On the other hand, by substituting x^0 into relation (4.10), we get $(\lambda^0)^\top g(x^0) \geqq 0$, and hence (4.11) must hold. □

If we write the convex programming problem in the form

$$\begin{cases} \min f(x) \\ g(x) \leqq [0] \\ x \geqq [0], \end{cases}$$

Theorem 4.10.1 still holds, by substituting $x \in \mathbb{R}^n_+$ to $x \in X$.

Note that in Theorem 4.10.1 we can have $\mu = 0$, and then, just as in the Fritz John theorem, for the differentiable case, it is important to have conditions which assure that $\mu \neq 0$. In this case it is obviously possible to take $\mu = 1$ and hence to consider the usual Lagrangian function. Also for the problem into consideration, the "price" to pay is the verification of a suitable constraint qualification. The constraint qualification more used for the case in exam (we recall that no differentiability assumption on the functions involved in the problem has been made) is the *Slater condition,* already introduced.

- (S) : The constraint functions g_i, $i = 1, \ldots, m$, are convex on the convex set $X \subset \mathbb{R}^n$, and there exists a vector $\bar{x} \in X$ such that $g(\bar{x}) < [0]$.

Keep in mind that in any case in $(P_{\min})$ every constraint g_i is convex by assumption. We have the following basic result.

Theorem 4.10.2 *Let x^0 be a solution of $(P_{\min})$ and let the above constraint qualification (S) be verified. Then in* (4.10) *it holds $\mu > 0$ and hence it is always possible to choose $\mu = 1$. Moreover, the pair (x^0, λ^0) is a saddle point of the Lagrangian function $\mathcal{L}(x, \lambda)$ and the complementary slackness condition* (4.11) *holds.*

Proof First we show that it holds $\mu > 0$. Indeed, if $\mu = 0$, then in relation (4.10) we have $\lambda^0 \geq [0]$ (i.e. λ^0 semipositive). Then (4.10) becomes

$$(\lambda^0)^\top g(x) \geqq 0, \quad \forall x \in X$$

and in particular we have

$$(\lambda^0)^\top g(\bar{x}) \geqq 0$$

which is absurd, being $\lambda^0 \geq [0]$ and $g(\bar{x}) < [0]$. For what concerns the second part of the theorem, we have

$\mathcal{L}(x, \lambda^0) = f(x) + (\lambda^0)^\top g(x)$;
$\mathcal{L}(x^0, \lambda^0) = f(x^0) + (\lambda^0)^\top g(x^0) = f(x^0)$, by relation (4.11) which continues to be verified;

$\mathcal{L}(x^0, \lambda) = f(x^0) + \lambda^\top g(x^0)$,
and, being $\mu = 1$, we have (relation (4.10)):

$$\mathcal{L}(x^0, \lambda^0) \leqq \mathcal{L}(x, \lambda^0), \quad \forall x \in X.$$

Moreover, it holds

$$\mathcal{L}(x^0, \lambda) \leqq \mathcal{L}(x^0, \lambda^0), \quad \forall \lambda \geqq [0],$$

being $g(x^0) \leqq [0]$. Therefore we have

$$\mathcal{L}(x^0, \lambda) \leqq \mathcal{L}(x^0, \lambda^0) \leqq \mathcal{L}(x, \lambda^0), \quad \forall x \in X, \ \forall \lambda \geqq [0].$$

As previously remarked, the complementary slackness condition (4.11) continues to be verified: on the other hand, we have $(\lambda^0)^\top g(x^0) \leqq 0$, and by putting $\lambda = [0]$ in the first inequality of the saddle point conditions, we get $f(x^0) \leqq f(x^0) + (\lambda^0)^\top g(x^0)$, that is $(\lambda^0)^\top g(x^0) \geqq 0$, and hence it will hold $(\lambda^0)^\top g(x^0) = 0$. □

Another constraint qualification for the saddle point problem is the following one, due to [21]:

- For any $y \geq [0]$ ($y \in \mathbb{R}^m$), there exists $\bar{x} \in X$ such that $y^\top g(\bar{x}) < 0$ (every $g_i(x)$ convex on the convex set X).

It is not difficult to prove that the above Karlin constraint qualification is *equivalent* to the Slater constraint qualification (use the theorem of the alternative of Fan-Glicksberg-Hoffman, i.e. Theorem 3.4.9).

Hence: a necessary condition such that $x^0 \in X$ is a global minimum point for $(P_{\min})$, if condition (S) holds, is that it must exist a vector $\lambda^0 \geqq [0]$ such that the pair (x^0, λ^0) is a saddle point of the Lagrangian function $\mathcal{L}(x, \lambda) = f(x) + \lambda^\top g(x)$, $x \in X$, $\lambda \geqq [0]$. In a quite similar way, if we consider the concave programming problem $(P_{\max})$, whose Lagrangian function is $\mathcal{L}(x, \lambda) = f(x) + \lambda^\top g(x)$, $x \in X$, $\lambda \geqq [0]$, we have that a necessary condition such that $x^0 \in X$ is a global maximum point for $(P_{\max})$, if the Slater c. q. holds (here the constraints are concave and there exists $\bar{x} \in X$ such that $g(\bar{x}) > [0]$), is that it must exist a vector $\lambda^0 \geqq [0]$ such that the pair (x^0, λ^0) is a saddle point for the above Lagrangian function, that is

$$\mathcal{L}(x, \lambda^0) \leqq \mathcal{L}(x^0, \lambda^0) \leqq \mathcal{L}(x^0, \lambda), \quad \forall x \in X, \ \forall \lambda \geqq [0].$$

The vice-versa of the above results (i.e. the sufficient optimality conditions in terms of saddle points of the Lagrangian function) is always true, also for problems non necessarily convex (concave). More precisely, we have the following result.

Theorem 4.10.3 *Let be given the problem*

$$(P): \quad \begin{cases} \min f(x) \\ g(x) \leqq [0] \\ x \in X \subset \mathbb{R}^n. \end{cases}$$

(Note that no particular assumption is made on f, g and X). If there exists a pair (x^0, λ^0), with $x^0 \in X$ and $\lambda^0 \geqq [0]$, which is a saddle point for the Lagrangian function $\mathcal{L}(x, \lambda) = f(x) + \lambda^\top g(x)$ (and with reference to the constrained minimum problem (P)), then x^0 is a solution of (P). Moreover, the complementary slackness condition

$$(\lambda^0)^\top g(x^0) = 0$$

holds true.

Proof Let the pair (x^0, λ^0) be a saddle point for (P); then we have

$$f(x^0) + \lambda^\top g(x^0) \leqq f(x^0) + (\lambda^0)^\top g(x^0) \leqq f(x) + (\lambda^0)^\top g(x), \quad \forall x \in X, \ \forall \lambda \geqq [0].$$

From the first inequality we have

$$(\lambda - \lambda^0)^\top g(x^0) \leqq 0, \quad \forall \lambda \geqq [0].$$

Now we choose $\lambda = \lambda^0 + e^i$ and hence we obtain $(\lambda^0 + e^i - \lambda^0)^\top g(x^0) \leqq 0$, that is $(e^i)^\top g(x^0) \leqq 0$, that is $g_i(x^0) \leqq 0$. By reiterating this operation for every $i = 1, \ldots, m$, we obtain $g(x^0) \leqq [0]$, and hence that x^0 is a feasible point for (P). We already know that the complementary slackness condition holds; in any case we prove again its validity. As $\lambda^0 \geqq [0]$ and $g(x^0) \leqq [0]$, we have $(\lambda^0)^\top g(x^0) \leqq 0$, but from the first above inequality (of the saddle point conditions), by putting $\lambda = [0]$, we obtain $(\lambda^0)^\top g(x^0) \geqq 0$. Hence we have $(\lambda^0)^\top g(x^0) = 0$.

Now let x be any feasible point of (P); from the second inequality of the saddle point conditions, taking the complementary slackness condition into account, we get

$$f(x^0) \leqq f(x) + (\lambda^0)^\top g(x) \leqq f(x),$$

being $(\lambda^0)^\top g(x) \leqq 0$ for any feasible x. Therefore $f(x^0) \leqq f(x)$, for any vector x, feasible for (P). In other words, x^0 is a solution of (P). □

Now we come back to the convex optimization problem $(P_{\min})$. We can summarize the previous results in the following theorem.

Theorem 4.10.4 *In order that $x^0 \in X$ is a solution of $(P_{\min})$ it is sufficient that there exists a vector $\lambda^0 \geqq [0]$ such that the pair (x^0, λ^0) is a saddle point of the Lagrangian function $\mathcal{L}(x, \lambda) = f(x) + \lambda^\top g(x)$; if the constraint qualification (S)*

holds, the previous conditions is also a necessary condition in order that $x^0 \in X$ is a solution of $(P_{\min})$.

We can therefore set up the following scheme.

$$\boxed{(x^0, \lambda^0) \text{ saddle point for } \mathcal{L}(x, \lambda)} \Longrightarrow \boxed{x^0 \text{ solution of } (P_{\min})}$$

$$\boxed{(x^0, \lambda^0) \text{ saddle point for } \mathcal{L}(x, \lambda)} \Longrightarrow \boxed{x^0 \text{ solution of } (P)}$$

$$\boxed{(x^0, \lambda^0) \text{ saddle point for } \mathcal{L}(x, \lambda)} \underset{+Slater}{\Longleftarrow} \boxed{x^0 \text{ solution of } (P_{\min})}$$

Obviously similar propositions hold with reference to $(P_{\max})$.

The Slater constraint qualification (or another suitable constraint qualification) cannot be skipped in the "necessary" part of the previous theorem, unless $(P_{\min})$ or $(P_{\max})$ are linear problems (i.e. the objective function f and the constraint functions g_i, $i = 1, \ldots, m$, are linear functions or linear affine functions). See the next Sect. 4.11 and the next chapter.

Example 4.10.5 Consider the family of concave optimization problems $\max x$, under the constraint $x^2 \leqq b$, b nonnegative real parameter, $x \in \mathbb{R}$.

(*a*) Let $b = 0$. The feasible set is then $K = \{0\}$, the Slater constraint qualification is clearly not satisfied and $x_0 = 0$ is the solution of the problem. It is easy to see that the pair $(0, \lambda_0)$, $\lambda_0 \geqq 0$, cannot be a saddle point for the Lagrangian function of the problem considered.

(*b*) Let $b > 0$. The feasible set is then given by the interval $\left[-\sqrt{b}, \sqrt{b}\right]$ and the point $x^0 = \sqrt{b}$ is the solution of the problem. The Slater constraint qualification is satisfied and it is quite easy to see that the pair $(x_0, \lambda_0) = (\sqrt{b}, \frac{1}{2\sqrt{b}})$ is a saddle point of the Lagrangian function $\mathcal{L}(x, \lambda) = f(x) - \lambda g(x) = x - \lambda\,(x^2 - b)$.

4.11 Relationships Between Saddle Points and Karush-Kuhn-Tucker Conditions

The previous results on saddle points of the Lagrangian function of the optimization problems considered, have been performed without any assumption on differentiability of the functions involved in the said problems. In the present section we examine the relationships between these results (on saddle points) and the Karush-Kuhn-Tucker conditions, by assuming that the objective function $f(x)$ and every constraint $g_i(x)$, $i = 1, \ldots, m$, are differentiable. Indeed, this was the approach considered by H. W. Kuhn and A. W. Tucker in their basic paper of 1951 on nonlinear programming theory.

Consider the problem

$$(P): \qquad \begin{cases} \min f(x) \\ g(x) \leqq [0] \\ x \in X, \end{cases}$$

with $f : \mathbb{R}^n \longrightarrow \mathbb{R}$ and $g : \mathbb{R}^n \longrightarrow \mathbb{R}^m$ differentiable on the open set $X \subset \mathbb{R}^n$.

Theorem 4.11.1 *If the Lagrangian function $\mathcal{L}(x, \lambda)$ of (P) admits a saddle point (x^0, λ^0), $x^0 \in X$, $\lambda^0 \geqq [0]$, then (x^0, λ^0) satisfies the following Karush-Kuhn-Tucker conditions:*

(1) $\nabla_x \mathcal{L}(x^0, \lambda^0) = [0]$, *that is*

$$\nabla f(x^0) + (\lambda^0)^\top \nabla g(x^0) = [0];$$

(2) $\lambda^0 \geqq [0]$; $(\lambda^0)^\top g(x^0) = 0$,
(3) $g(x^0) \leqq [0]$ *(feasibility conditions).*

Proof By the definition of saddle point for problem (P) we have

$$\mathcal{L}(x^0, \lambda) \leqq \mathcal{L}(x^0, \lambda^0) \leqq \mathcal{L}(x, \lambda^0), \quad \forall x \in X, \ \forall \lambda \geqq [0].$$

From the second inequality we have that $\mathcal{L}(x, \lambda^0)$ reaches its minimum at $x = x^0$. By the classical Fermat theorem we have therefore

$$\nabla_x \mathcal{L}(x^0, \lambda^0) = \nabla f(x^0) + (\lambda^0)^\top \nabla g(x^0) = [0].$$

The first inequality in the saddle point conditions is $\mathcal{L}(x^0, \lambda) \leqq \mathcal{L}(x^0, \lambda^0)$, $\forall \lambda \geqq [0]$, that is

$$f(x^0) + \lambda^\top g(x^0) \leqq f(x^0) + (\lambda^0)^\top g(x^0), \ \forall \lambda \geqq [0],$$

that is $(\lambda - \lambda^0)^\top g(x^0) \leqq 0$, $\forall \lambda \geqq [0]$. From the proof of Theorem 4.10.3, it results $g(x^0) \leqq [0]$, $(\lambda^0)^\top g(x^0) = 0$. □

If in (P) we add the assumption that f and g are convex functions on the open convex set $X \subset \mathbb{R}^n$ (that is, we consider problem $(P_{\min})$, where, moreover, the involved functions are differentiable), we obtain the converse result of the previous theorem.

Theorem 4.11.2 *Let in (P) the functions f and every g_i, $i = 1, \ldots, m$, be differentiable and convex on the open convex set $X \subset \mathbb{R}^n$; at the point $x^0 \in K = \{x : x \in X, \ g(x) \leqq [0]\}$, the Karush-Kuhn-Tucker conditions are verified, with a vector of multipliers $\lambda^0 \geqq [0]$. Then the pair (x^0, λ^0) is a saddle point for the Lagrangian function of (P).*

Proof Being f and every g_i, $i = 1, \ldots, m$, convex functions, the Lagrangian function $\mathcal{L}(x, \lambda) = f(x) + \lambda^\top g(x)$, $\lambda \geqq [0]$, is convex. The first Karush-Kuhn-Tucker condition is:

$$\nabla_x \mathcal{L}(x^0, \lambda^0) = \nabla f(x^0) + (\lambda^0)^\top \nabla g(x^0) = [0].$$

Being $\mathcal{L}(x, \lambda)$ convex for every $\lambda \geqq [0]$, by Theorem 4.2.9 the previous relation says that the point (x^0, λ^0) is a global minimum point of $\mathcal{L}(x, \lambda^0)$ and hence

$$\mathcal{L}(x^0, \lambda^0) \leqq \mathcal{L}(x, \lambda^0), \ \forall x \in X.$$

From the second Karush-Kuhn-Tucker condition (complementary slackness condition) and from the fact that $x^0 \in K$ we obtain immediately

$$f(x^0) + \lambda^\top g(x^0) \leqq f(x^0) + (\lambda^0)^\top g(x^0), \ \ \forall \lambda \geqq [0],$$

that is

$$\mathcal{L}(x^0, \lambda) \leqq \mathcal{L}(x^0, \lambda^0), \ \ \forall \lambda \geqq [0],$$

and hence, taking also the previous inequality into account, the thesis. □

The two previous results allow us to make the following considerations. If (P) is a *linear programming problem* (L.P.), then we know that the functions involved in this problem, not only are differentiable but also they are both convex and concave. We know also that the constraints are qualified, as the reverse constraint qualification holds by default. It follows that if x^0 is a solution of (L.P.), the related Karush-Kuhn-Tucker conditions are verified at that point, and from these conditions the saddle point conditions for the Lagrangian function follow, being f and the constraints g_i, $i = 1, \ldots, m$, linear or linear affine. Hence, for a linear programming problem, the necessary optimality conditions, expressed in terms of saddle points of the Lagrangian function, are verified, without imposing a constraint qualification (such as the Slater condition), just as it happens for the necessary optimality conditions expressed in terms of the Karush-Kuhn-Tucker conditions (for a linear programming problem).

For the reader's convenience we sum up with a diagram the main implications previously obtained, both with reference to differentiability assumptions, and for the non differentiable case (see Fig. 4.17). In this diagram (P) is, as usual, the problem

$$\begin{cases} \min f(x) \\ g(x) \leqq [0] \\ x \in X, \end{cases}$$

with $X \subset \mathbb{R}^n$ (X open and/or convex, according to the various cases). The diagram for a maximum problem is easily obtainable.

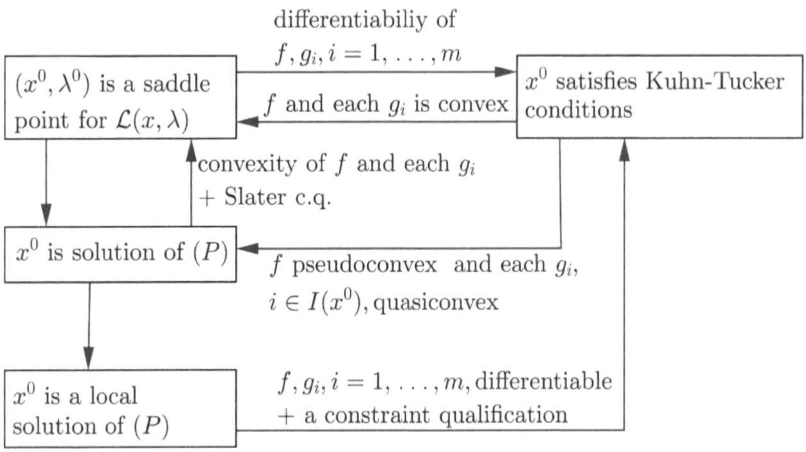

Fig. 4.17 Main implications: differentiable and non differentiable cases

4.12 The Portfolio Selection

The present section is mainly taken from the book of De Giuli et al. [11] and from the book of Bertocchi et al. [8]. The so-called "portfolio selection problem" is essentially due to the economist Harry Markowitz (Nobel prize in Economics in 1980). Consider n alternative investments into as many financial activities (financial securities), having a random return. The investor has at his disposal a capital to invest, by choosing a "portfolio" or "basket" of the said activities. We can always suppose that the available capital is unitary: this means that we are interested only in the shares of the capital to be invested in the various securities, shares we denote by $x_1, x_2, \ldots, x_n$. The share x_i is invested in the i-th activity.

On the grounds of the above assumptions we have $\sum_{i=1}^{n} x_i = 1$, that is, $e^\top x = 1$, where $e^\top = [1, 1, \ldots, 1]$ (summation vector of $\mathbb{R}^n$). We know the *mean values (or expected values)* of the random rates of return of the n securities, values denoted by $r_1, r_2, \ldots, r_n$, and the square matrix V, of order n, which contains the *variances* σ_i^2, $i = 1, \ldots, n$, and the related *covariances* of the said rates:

$$V = \left[v_{ij}\right] = \left[\sigma_i \sigma_j \rho_{ij}\right], \quad i, j = 1, \ldots, n,$$

where ρ_{ij} is the *correlation coefficient* between the rate of return of the i-th activity and the rate of return of the j-th activity. The matrix V, said *matrix of variances-covariances,* is therefore described as follows:

$$V = \begin{bmatrix} \sigma_1^2 & \rho_{12}\sigma_1\sigma_2 & \cdots & \rho_{1n}\sigma_1\sigma_n \\ \rho_{12}\sigma_1\sigma_2 & \sigma_2^2 & \cdots & \rho_{2n}\sigma_2\sigma_n \\ \vdots & \vdots & \cdots & \vdots \\ \rho_{1n}\sigma_1\sigma_n & \rho_{2n}\sigma_2\sigma_n & \cdots & \sigma_n^2 \end{bmatrix}.$$

This matrix is symmetric and positive semidefinite: more precisely, it is possible to prove that, except for the case where between the variances there is a linear relation (in this case V is singular), the matrix V is positive definite. The total variance of the portfolio σ^2 is given by

$$\sigma^2 = x^\top Vx,$$

whereas the average rate of return (expected) of the portfolio is given by

$$\pi = r^\top x,$$

where $r^\top = [r_1, r_2, \dots, r_n]$.

The investor who wishes to minimize his portfolio risk, against an average prefixed rate of return π, has to solve the following constrained optimization problem (it is a *quadratic programming problem*):

$$\min_{x\in\mathbb{R}^n} \sigma^2 = x^\top Vx$$

under the constraints

$$\begin{aligned} r^\top x &= \pi \\ e^\top x &= 1. \end{aligned}$$

This problem admits a unique solution if the matrix V is nonsingular and the vectors r and e are linearly independent. Hence we accept that these conditions hold. By varying π, we obtain a set of solutions, usually called *set of feasible portfolios,* to which it corresponds a set of pairs (σ, π) in the plane which shows in its abscissa the standard deviations and in its ordinate the expected value of the returns.

The necessary optimality conditions for this problem, which are also sufficient, being the objective function $x^\top Vx$ a strictly convex function (V is by assumption positive definite), and the constraints linear functions, are the following one:

$$\nabla\mathcal{L}(x, \lambda) = [0],$$

where

$$\mathcal{L}(x, \lambda) = x^\top Vx - \lambda_1(r^\top x - \pi) - \lambda_2(e^\top x - 1).$$

We recall that the gradient of the objective function of our problem is $2x^\top V$; then the first-order (necessary and sufficient) optimality condition for the same problem is

$$\begin{cases} \frac{\partial \mathcal{L}}{\partial x} = 2x^\top V - \lambda_1 r^\top - \lambda_2 e^\top = [0] \\ \frac{\partial \mathcal{L}}{\partial \lambda_1} = r^\top x - \pi = 0 \\ \frac{\partial \mathcal{L}}{\partial \lambda_2} = e^\top x - 1 = 0. \end{cases}$$

Being V invertible by assumption, we can get x from the first equation

$$x^\top = \frac{1}{2}\lambda_1 r^\top V^{-1} + \frac{1}{2}\lambda_2 e^\top V^{-1}$$

and then substitute this expression into the remaining two equations

$$\begin{cases} r^\top x = \frac{1}{2}\lambda_1 r^\top V^{-1} r + \frac{1}{2}\lambda_2 e^\top V^{-1} r = \pi \\ e^\top x = \frac{1}{2}\lambda_1 r^\top V^{-1} e + \frac{1}{2}\lambda_2 e^\top V^{-1} e = 1. \end{cases}$$

We put:

$\alpha = r^\top V^{-1} r$;
$\beta = r^\top V^{-1} e = e^\top V^{-1} r$ (the equality is due to the symmetry of V^{-1}, in turn due to the symmetry of V);
$\gamma = e^\top V^{-1} e$.

The two last equations of the above system can be therefore written as

$$\begin{cases} \frac{1}{2}\lambda_1 \alpha + \frac{1}{2}\lambda_2 \beta = \pi \\ \frac{1}{2}\lambda_1 \beta + \frac{1}{2}\lambda_2 \gamma = 1. \end{cases}$$

This linear system admits a unique solution: indeed, except for the case of average returns all equal, case we exclude a priori, we have always $\alpha\gamma - \beta^2 > 0$. We prove this assertion. Consider the matrix $C = [r; e]$; this matrix has a rank equal to 2 (we have excluded that all average returns are equal, in which case the rank would be 1). It is easily seen that

$$C^\top V^{-1} C = B = \begin{bmatrix} \alpha & \beta \\ \beta & \gamma \end{bmatrix}$$

and that, putting $y = Cx$ ($x \in \mathbb{R}^2,\ y \in \mathbb{R}^n$), it results $x \neq [0]$ if and only if $y \neq [0]$. Then we have

$$x^\top B x = x^\top C^\top V^{-1} C x = y^\top V^{-1} y > 0,$$

for every $y \neq [0]$, as V^{-1} is positive definite, being V positive definite by assumption. It follows that also B is positive definite, hence $\det(B) = \alpha\gamma - \beta^2 > 0$. By the classical theorem of Cramer on linear systems, our system admits the unique solution

$$\frac{1}{2}\lambda_1 = \frac{\pi\gamma - \beta}{\alpha\gamma - \beta^2}; \quad \frac{1}{2}\lambda_2 = \frac{\alpha - \pi\beta}{\alpha\gamma - \beta^2}.$$

Now we come back to the first equation of the original system and substitute into the same the values of λ_1 and λ_2 just found. We obtain the vector

$$\begin{aligned} x^\top &= \frac{\pi\gamma - \beta}{\alpha\gamma - \beta^2} r^\top V^{-1} + \frac{\alpha - \pi\beta}{\alpha\gamma - \beta^2} e^\top V^{-1} \\ &= \frac{1}{\alpha\gamma - \beta^2}\left[(\pi\gamma - \beta) r^\top V^{-1} + (\alpha - \pi\beta) e^\top V^{-1}\right]. \end{aligned}$$

If we multiply $x^\top$ by V, we obtain

$$x^\top V = \frac{1}{\alpha\gamma - \beta^2}\left[(\pi\gamma - \beta) r^\top + (\alpha - \pi\beta) e^\top\right].$$

Finally, by recalling that the variance of the portfolio is given by $\sigma^2 = x^\top V x$, we have

$$\sigma^2 = x^\top V x = \frac{1}{\alpha\gamma - \beta^2}\left[(\pi\gamma - \beta) r^\top x + (\alpha - \pi\beta) e^\top x\right],$$

that is, by recalling that $r^\top x = \pi$ and $e^\top x = 1$,

$$\begin{aligned} \sigma^2 = x^\top V x &= \frac{1}{\alpha\gamma - \beta^2}\left[(\pi\gamma - \beta)\pi + (\alpha - \pi\beta)\right] \\ &= \frac{\gamma\pi^2 - 2\beta\pi + \alpha}{\alpha\gamma - \beta^2}. \end{aligned}$$

This last equation is the equation of a hyperbola in the plane (σ, π) and with vertex at the point P of coordinates

$$P\left(\frac{1}{\sqrt{\gamma}}, \frac{\beta}{\gamma}\right).$$

The region of the "efficient" portfolios among the feasible ones is given by that branch of the hyperbola with $\pi > \frac{\beta}{\gamma}$ (see Fig. 4.18). Indeed, it is evident that the rationality and caution of the investor make sure that, under the same risk conditions, it is chosen the portfolio with greater π (see, e.g., [8]).

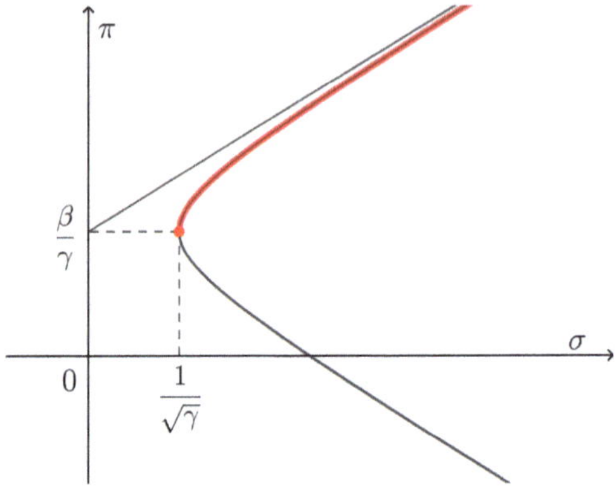

Fig. 4.18 Region of the "efficient" portfolios is given by the branch of the hyperbola with $\pi > \frac{\beta}{\gamma}$

Besides the works quoted along the chapter, we consider the following works useful to the reader to deepen his knowledge on the various subjects: [4, 10, 18, 23, 27–30, 34, 35].

References

1. J. Abadie, On the Kuhn-Tucker theorem, in *Nonlinear Programming*, ed. by J. Abadie (North Holland, Amsterdam, 1967), pp. 19–36
2. T.M. Apostol, *Calculus*, vol. 1 (Blaisdell, Waltham, 1967)
3. K.J. Arrow, A.C. Enthoven, Quasiconcave programming. Econometrica **29**, 522–552 (1961)
4. K.J. Arrow, M.D. Intriligator (Eds.), *Handobook of Mathematical Economics* (North Holland, Amsterdam, 1981)
5. K.J. Arrow, L. Hurwicz, H. Uzawa, Constraint qualifications in maximization problems. Naval. Res. Logist. Quart. **8**, 175–186 (1961)
6. M. Avriel, *Nonlinear Programming* (Prentice-Hall, Englewood Cliffs, 1976)
7. M.S. Bazaraa, C.M. Shetty, *Foundations of Optimization* (Springer, Berlin, 1976)
8. M. Bertocchi, S. Stefani, G. Zambruno, *Matematica per l'Economia e la Finanza* (McGraw-Hill, Milano, 1995)
9. E. Castagnoli, L. Peccati, *Matematica per l'Analisi Economica*, vols. 1–2 (Etas Libri, Milano, 1990)
10. A.C. Chiang, K. Wainwright, *Fundamental Methods of Mathematical Economics* (McGraw-Hill, New York, 2004)
11. M.E. de Giuli, G. Giorgi, M.A. Maggi, U. Magnani, *Matematica per l'Economia e la Finanza* (Zanichelli, Bologna, 2008)
12. A.V. Fiacco, *Introduction to Sensitivity and Stability Analysis in Nonlinear Programming* (Academic Press, New York, 1983)
13. G. Giorgi, T.H. Kjeldsen, *Traces and Emergence of Nonlinear Programming* (Birkhäuser, Basel, 2014)
14. G. Giorgi, A. Guerraggio, J. Thierfelder, *Mathematics of Optimization: Smooth and Nonsmooth Case* (Elsevier, Amsterdam, 2004)
15. G. Giorgi, B. Jiménez, V. Novo, *Basic Mathematical Programming Theory* (Springer, Berlin, 2023)
16. F.J. Gould, J.W. Tolle, A necessary and sufficient qualification for constrained optimization. SIAM J. Appl. Math. **20**, 164–172 (1971)

17. A. Guerraggio, S. Salsa, *Metodi Matematici per l'Economia e le Scienze Sociali* (Giappichelli, Torino, 1997)
18. M.R. Hestenes, *Optimization Theory. The Finite Dimensional Case* (Wiley, New York, 1975)
19. M.D. Intriligator, *Mathematical Optimization and Economic Theory* (Prentice-Hall, Englewood Cliffs, 1971)
20. F. John, Extremum problems with inequalities as subsidiary conditions, in *Studies and Essays, Courant Anniversary Volume*, ed. by K.O. Friedrichs et al. (Interscience Publishers, New York, 1948), pp. 187–204
21. S. Karlin, *Mathematical Methods in Games, Programming and Economics*, vols. 1–2 (Addison-Wesley, Reading, 1959)
22. H.W. Kuhn, A.W. Tucker, Nonlinear programming, in *Proceedings of the Second Berkeley Symposium on Mathematical Statistics and Probability*, ed. by J. Neyman (University of California Press, Berkeley, 1951), pp. 481–492
23. K. Lancaster, *Mathematical Economics* (Macmillan, London, 1968)
24. O.L. Mangasarian, *Nonlinear Programming* (McGraw-Hill, New York, 1969)
25. O.L. Mangasarian, S. Fromovitz, The Fritz John necessary optimality conditions in the presence of equality and inequality constraints. J. Math. Anal. Appl. **15**, 641–652 (1967)
26. G.P. McCormick, Second order conditions for constrained minima. SIAM J. Appl. Math. **15**, 641–652 (1967)
27. G.P. McCormick, *Nonlinear Programming. Theory, Algorithms and Applications* (Wiley, New York, 1983)
28. P. Michel, *Cours de Mathématiques pour Economistes* (Economica, Paris, 1984)
29. W. Novshek, *Mathematics for Economists* (Academic Press, New York, 1993)
30. A.L. Peressini, F.E. Sullivan, J.J. Uhl, *The Mathematics of Nonlinear Programming* (Springer, New York, 1988)
31. D.W. Peterson, A review of constraint qualifications in finite-dimensional spaces. SIAM Rev. **15**, 639–654 (1973)
32. R.T. Rockafellar, *Convex Analysis* (Princeton University Press, Princeton, 1970)
33. P.A. Samuelson, *Foundations of Economic Analysis* (Harvard University Press, Cambridge, 1947)
34. K. Sydsaeter, P. Hammond, A. Seierstad, A. Strom, *Further Mathematics for Economic Analysis*, 2nd edn. (Prentice Hall, London, 2008)
35. A. Takayama, *Analytical Methods in Economics* (University of Michigan Press, Anna Arbor, 1993)
36. W.I. Zangwill, *Nonlinear Programming: A Unified Approach* (Prentice-Hall, Englewood Cliffs, 1969)

Chapter 5
Linear Programming

5.1 Formulation of Linear Programming Problems

As previously mentioned, a *linear programming problem* (L. P. for friends) is characterized by a linear (or linear affine) objective function and by linear (or linear affine) constraints. Being a linear programming problem a particular case of the class of nonlinear programming problems (the involved functions are both convex and concave and also differentiable), all results of the previous chapter hold true for this class of problems, and often with some useful simplifications. In a L. P. problem, the special structure (linearity) of the objective function and of the constraints, allows the possibility to obtain particular solution methods, usually more speed and efficient than the ones available for nonlinear programming problems.

One of the most known and used methods for a L. P. problem is the celebrated "simplex algorithm" developed for the American mathematician G. B. Dantzig in 1947, in which however we shall not be concerned, as the present lectures are not concerned with Operations Research or numerical methods for mathematical programming problems. The reader can see, for these important developments, the books [2, 3]. Other useful books treating L. P. are [1, 5–9, 13, 15–18]. For economic applications the basic textbook is [10].

In a linear programming problem the objective function is usually expressed in the following form

$$f(x) = \sum_{i=1}^{n} c_i x_i = c^\top x, \;\; c \neq [0] \,,$$

and the constraints as inequalities of the type

$$\sum_{j=1}^{n} a_{ij} x_j \leqq b_i, \; i = 1, \ldots, m,$$

G. Giorgi et al., *Lectures on Mathematics for Economic and Financial Analysis*,
https://doi.org/10.1007/978-3-031-83339-7_5

or also

$$\sum_{j=1}^{n} a_{ij}x_j \geqq b_i,\ i = 1, \ldots, m.$$

There is often a sign constraint on the variables, i.e. the components of the vector x are required to be nonnegative: $x \geqq [0]$.

Usually, L. P. problems, in case of maximization problems, are presented in the following formulation, said also "canonical form":

$$\begin{cases} \max c^\top x \\ Ax \leqq b \\ x \geqq [0], \end{cases}$$

where $c,\ x \in \mathbb{R}^n$, $c \neq [0]$, A matrix of order (m, n) and $b \in \mathbb{R}^m$. In case of minimization problems, usually we have the following canonical form

$$\begin{cases} \min c^\top x \\ Ax \geqq b \\ x \geqq [0]. \end{cases}$$

We again remark that the above problems are at the same time convex and concave differentiable optimization problems. These problems can be transformed into problems with equality constraints only (also nonlinear programming problems can be transformed into problems with equality constraints only, but in the case of L. P. the question is more interesting, both from a theoretic and from and algorithmic point of view), by the introduction of suitable variables (obviously the total number of variables increases). More precisely, if the constraints are of the type $Ax \leqq b$, it is possible to introduce a vector $s \geqq [0]$, $s \in \mathbb{R}^m$, of "slack variables" which represent the lacking quantities at the first member. The constraints are therefore transformed as follows:

$$Ax + s = b,\ \ x \geqq [0],\ s \geqq [0].$$

If the constraints are of the type $Ax \geqq b$, by means of nonnegative surplus variables, we have the constraints expressed as

$$Ax - s = b,\ \ x \geqq [0],\ s \geqq [0].$$

The *standard form* of a linear programming problem is the form

$$\begin{cases} \max(\min) c^\top x \\ Ax = b \\ x \geqq [0]. \end{cases}$$

Pay attention! The above terminologies are not... standard! In particular, some classical textbooks call "standard form" what we have called "canonical form" and vice-versa. The two forms (canonical and standard) are in any case *equivalent,* in the sense that one form can be transformed into the other one, by keeping all information contained in the related problem. Usually, the standard form is important for the discussion of the numerical algorithms for the solution of a L. P. problem (e.g. the simplex algorithm).

Example 5.1.1 Transform the following L. P. problem

$$\begin{cases} \min 2x_1 + 3x_2 + 4x_3 \\ 3x_1 + x_2 + 5x_3 \geqq 5 \\ x_1 + x_2 - x_3 = 8 \\ x_1 \geqq 0, x_2 \geqq 0, x_3 \geqq 0 \end{cases}$$

into its standard form.

By means of the slack variable s_1 we obtain the following equivalent problem in standard form:

$$\begin{cases} \min 2x_1 + 3x_2 + 4x_3 + 0s_1 \\ 3x_1 + x_2 + 5x_3 - s_1 = 5 \\ x_1 + x_2 - x_3 = 8 \\ x_1 \geqq 0, x_2 \geqq 0, x_3 \geqq 0, s_1 \geqq 0. \end{cases}$$

The applications of L. P. are countless, in the most different areas, so much that it would be almost impossible to make a list, even if brief. We give only some hints concerning few examples in economic and financial fields.

(*a*) Problems of optimal distribution (or allocation) of limited available resources (production means, raw materials, labour, etc.) in order to optimize an economic outcome (for example, to maximize a profit). Important examples of these types of problems are the ones concerning the so-called *Activity Analysis* considered by the economist and Nobel prize recipient T. C. Koopmans, problems which are in turn related to various linear economic models, such as the Leontief models: there are n production processes ("activities") which produce an output vector from m primary resources. Let $a_{ij} \geqq 0, \forall i, j$ be the quantity of the i-th resource used to put in production the j-th activity, under unit intensities of production. Let $b_i \geqq 0, \forall i$, the available quantity of the i-th resource and let $c_j \geqq 0, \forall j$, the value of the output produced by the j-th activity, under unit intensities of production. Let $x_j \geqq 0, j = 1, \ldots, n$, the level of production intensity of the j-th activity. We wish to choose the vector $x \geqq [0]$ which maximizes the total value of the output produced:

$$\max \sum_{j=1}^{n} c_j x_j$$

by keeping into consideration the restrictions

$$\sum_{j=1}^{n} a_{ij}x_j \leqq b_i,\ i = 1, \ldots, m,\ \ x_j \geqq 0,\ j = 1, \ldots, n.$$

(*b*) Transportation problems. We have to transfer several goods from a set of starting points (for example production factories) towards a set of destinations (for example warehouses, sales stores, etc.), being the transport characterized by different costs for each pair "starting point-destination". We have to find the less expensive solution among all the possible ones (taking into account the effective and limited reception capabilities of the various destination points).

(*c*) Blending problems. As an example of this type of problems we have the so-called "diet problems". When a product is composed by several elements, with several costs, we have to find the less expensive composition, which however assures that the product satisfies certain qualities or technical characteristics (for example, problems of refining of oil products, diet problems, etc.). In the present class of problems we can put also those problems of choosing the best purchase strategy of raw materials.

(*d*) Problems of staff allocation to different activities or types of work, in connection with the various dispositions or skills, in order to optimize the total production or the performance of a department or an office, etc.

Let us consider, for example, a maximization problem in its canonical form. The set $K = \left\{x : x \in \mathbb{R}^n, Ax \leqq b, x \geqq [0]\right\}$ is the *feasible set* (sometimes it is called also the *opportunity set*). We inform the reader that in L. P. it is accepted a terminology alas in use in several writings on this subject, that is to call *feasible solution* every feasible vector x, i.e. every $x \in K$. An *optimal feasible solution* is every vector x which maximizes (or minimizes) the objective function $f(x) = c^\top x$ over the feasible set K.

We recall that the hyperplane H defined by

$$H = \left\{x : x \in \mathbb{R}^n, a^\top x = \alpha,\ a \neq [0]\right\}$$

identifies in $\mathbb{R}^n$ two convex regions, said *affine half-spaces,* of equations $a^\top x \geqq \alpha$ and $a^\top x \leqq \alpha$. These sets are closed sets: the hyperplane H is said the *generating hyperplane* of the two half-spaces. The intersection of a finite number of half-spaces is, as previously said, a *convex polyhedron* (a nonempty, bounded convex polyhedron is said a *polytope*). Hence the feasible set of a L. P. problem, if nonempty, of the type $Ax \leqq b$, $x \geqq [0]$, or $Ax \geqq b$, $x \geqq [0]$, or $Ax = b$, $x \geqq [0]$, is a convex polyhedron, and hence in any case a *closed set.* This polyhedron is therefore determined by the hyperplanes of equation $A_1x = b_1$, $A_2x = b_2, \ldots A_mx = b_m$ (where A_i is the i-th row of A) and by the hyperplanes of equation $x_1 = 0$, $x_2 = 0, \ldots, x_n = 0$. The feasible set K of a L. P. problem may assume the following features.

(i) It may be empty (the constraints are inconsistent).
(ii) It may be unbounded (some variables may assume arbitrarily large values).
(iii) It may be nonempty and bounded (that is a polytope). This is surely the most interesting case and the nearly unique case in practical problems.

If nonempty, the feasible set K of a L. P. problem is, as previously said, a convex polyhedron, here denoted by D (instead of K). We have already seen in Chap. 3, the notion of *supporting hyperplane* for a nonempty convex set S of $\mathbb{R}^n$ and we have already seen the notion of *extreme point* for a nonempty convex set S of $\mathbb{R}^n$. Now let D be a *convex polyhedron* of $\mathbb{R}^n$ and let $H = \{x : x \in \mathbb{R}^n, a^\top x = \alpha, a \neq [0]\}$ be a supporting hyperplane for D. The set

$$H \cap D$$

is said a *face of the polyhedron.* The intersection of several faces of D is itself a face. If we are in $\mathbb{R}^n$, then:

- A face of zero dimension (that is a point) is said a *vertex or corner point* of the polyhedron (it is a point where two or more one-dimensional faces intersect).
- A one-dimensional face is said to be an *edge* of the polyhedron.
- A face of dimension $(n - 1)$ is said to be a *facet or maximal face* of the polyhedron.

For the reader's convenience, we recall that a point x of a convex set $S \subset \mathbb{R}^n$ is an *extreme point* if x is not obtainable as a strict convex combination of distinct pairs of elements of S, that is

$$x \neq \lambda x^1 + (1 - \lambda)x^2, \ \forall x^1, x^2 \in S, \ x^1 \neq x^2, \ \forall \lambda \in (0, 1).$$

The extreme points of a convex polyhedron coincide with its vertices. All faces are in their turn convex polyhedra. Finally, two extreme points of a convex polyhedron are said to be *adjacent* if the line segment joining them is an edge of the polyhedron.

In Fig. 5.1 we have represented three possible feasible sets of a L. P. problem. In (a) and (b) the sets are bi-dimensional, in (a) the feasible set is bounded (and hence it is a polytope), whereas in (b) the polyhedron is unbounded. In (c) the set is three-dimensional. In the first two feasible sets the boundary is given by one-dimensional faces (edges) and by vertices. In the third set the boundary is given by bi-dimensional faces, one-dimensional faces and vertices. The vertices are evidenced by points in bold.

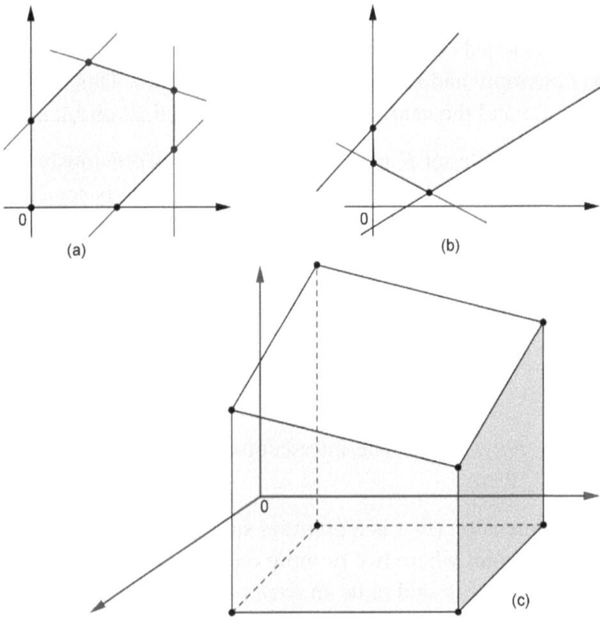

Fig. 5.1 Examples of feasible sets: (**a**) A bounded set in $\mathbb{R}^2$, (**b**) An unbounded set in $\mathbb{R}^2$, (**c**) A bounded set in $\mathbb{R}^3$

5.2 Fundamental Theorems of L. P.

Theorem 5.2.1 (First Fundamental Theorem of L. P.) *A linear programming problem which admits minimum and/or maximum points of its objective function, then it admits only global minimum and/or maximum points. Moreover, these points are not interior points of the feasible set K (hence they are on the boundary of K). The optimum point, if it exists and is unique, is a vertex of K; if there are more than one optimum points, there are in effect infinite optimum points, corresponding to two or more vertices and to all points of the face of K which contains the said vertices.*

Proof We observe first that if the feasible set K is not bounded, the solution of the problem may not exist. If the feasible set K is bounded, being also closed and being the objective function continuous, the classical theorem of Weierstrass assures the existence of a solution. As a linear function or also an affine linear function is both convex and concave, if it admits optimum points, these points cannot be only local optimum points: they must be global optimum points. In any case these points cannot be interior points of K, otherwise it would hold the theorem of Fermat which states that the related gradient evaluated at these points, is the zero vector: however, the gradient of the objective function is given by the constant vector c, with $c \neq [0]$ by assumption. Hence the optimum points, if any, are on the boundary of K. Let be, for instance, $c^\top x \leqq c^\top x^0$, $\forall x \in K$. It is clear that the relation $c^\top x \leqq c^\top x^0$ generates one of the two half-spaces associated to the hyperplane $c^\top x = c^\top x^0$, moreover that this half-space contains the whole K and that the same hyperplane is a supporting hyperplane for K, which is a convex, closed and bounded set, at least from below (if $x \in K$ we have $x \geqq [0]$). Hence Theorem 3.5.1 applies, i.e. we have that every supporting hyperplane for K (and hence also the hyperplane of equation $c^\top x = c^\top x^0$) contains at least one extreme point of K. This allows to say that at least an extreme point of K is an optimum point.

Now we show that if the objective function assumes, for instance, its maximum value in correspondence of more than one extreme point, then the same function assumes the said value at every point belonging to the convex combination of the said maximum points. Let us suppose that $f(x)$ assumes its maximum value on K at the extreme points $x^1, x^2, \ldots, x^q$. Therefore, by denoting the said maximum value by M, we have

$$f(x^1) = f(x^2) = \cdots = f(x^q) = M.$$

If $\bar{x}$ is any point of the convex combination of the above q extreme points, that is if

$$\bar{x} = \lambda_1 x^1 + \lambda_2 x^2 + \cdots + \lambda_q x^q,$$

being $\lambda_k \geqq 0$, $k = 1, 2, \ldots, q$, and $\sum_{k=1}^{q} \lambda_k = 1$, we have, thanks to the linearity of $f(x)$,

$$\begin{aligned} f(\bar{x}) &= f(\lambda_1 x^1 + \lambda_2 x^2 + \cdots + \lambda_q x^q) \\ &= \lambda_1 f(x^1) + \lambda_2 f(x^2) + \cdots + \lambda_q f(x^q) \\ &= M \sum_{k=1}^{q} \lambda_k = M \end{aligned}$$

and hence, also $\bar{x}$ is a maximum point of the objective function over K. Therefore we conclude that if the maximum point is not unique, there are infinite maximum points, given by two or more vertices and by all points of the face of K which contains the said vertices. For minimum problems, obviously the reasoning is quite similar, as well as for problems in standard form. □

Remark 5.2.2 The previous theorem allows to treat a L. P. problem as a problem consisting of a finite sequence of steps: we have to examine the value that the objective function assumes at all vertices of K. Unfortunately, in several practical problems the number of the vertices is so high that this may generate serious difficulties also for a computer. As already said, the vertices are the intersection of the n hyperplanes of equations $x_i = 0$, $i = 1, \dots, n$, as well as of the hyperplanes of equation $A_r x = b_r$, $r = 1, \dots, m$. The number of these intersections is given by

$$\binom{m+n}{n}.$$

Not all the said intersections are vertices, as some intersection point may be outside the feasible set. The above quantity is therefore an upper limitation of the number of vertices of K, however, even for values of m and n not too large, this quantity may be a very large number. For instance, for $m = 50$ and $n = 100$, the said number is about 10^{40}. The computational efficiency of the "simplex method" consists basically to reduce in a drastic way the number of vertices to be checked, in order to find, in a reasonable number of iterations, the solution of a L. P. problem. The simplex method is therefore an iteration procedure, suitable for automatic computations by means of computers, as the procedure begins by considering a starting vertex, pointing out if it is or not an optimum point. If not, the algorithm considers an adjacent vertex, by choosing the one which is the "best", with reference to the optimum point, among all other possible vertices. By going on, the algorithm examines the fastest path among the vertices of the polyhedron, in such a way to reach in the most quickly way the solution, by neglecting to take into consideration those vertices where the value of the objective function departs from the value of the solution. As previously said, we shall not enter into the details of this method (nor of other algorithms).

Remark 5.2.3 Obviously there are L. P. problems which do not admit solution, for instance in the following cases:

- When the constraints are inconsistent and hence $K = \emptyset$. For instance the following inequalities

$$\begin{aligned} x_1 + x_2 &\leqq 2, \\ -x_1 - x_2 &\leqq -6 \end{aligned}$$

  are obviously inconsistent.
- When $K \neq \emptyset$ but K is not bounded and the related objective function is not bounded on K. Consider, e.g., the following example

$$\begin{cases} \max\, x_1 + x_2 \\ -x_1 - x_2 \leqq -1 \\ x_1 \geqq 0,\ x_2 \geqq 0. \end{cases}$$

We have said in several occasions that in the present book we are not interested in algorithmic questions, such as the simplex method, however we give some hints on "algebraic" theoretical results, which are basic in the said method and which allow to reformulate other important results on L. P. problems. We take into consideration the following L. P. problem in its standard form:

$$\begin{cases} \min(\max)\, c^\top x \\ \quad Ax = b \\ \quad x \geqq [0]\,, \end{cases}$$

where A is of order (m, n), $x, c \in \mathbb{R}^n$ ($c \neq [0]$) and $b \in \mathbb{R}^m$. Without loss of generality, we assume the following hypothesis:

- It results $m < n$ and the matrix A has full rank, i.e. $\operatorname{rank}(A) = m$.

At a first glance, this last condition may seem restrictive, as it implies that the number of constraints is less than the number of variables and hence it excludes the case $m \geqq n$. If $m = n$ the system $Ax = b$ has a unique solution (which is feasible if it is nonnegative). If $m > n$ the system $Ax = b$ admits solutions if and only if $(m-n)$ constraints are linear combination of the other equations (i.e. they are "redundant equations"). Then, by removing these redundant equations, we can bring back the system to the case $m < n$. As the rank of A is m, it is always possible to choose, among the n columns of A, m linearly independent columns and, by means of suitable permutations, to obtain that these independent columns are the first m columns (obviously we have to make the same permutations on the elements of the vector x). We obtain the partitioned matrix

$$A = [B;\, N]$$

where B is square, nonsingular and of order m and N is of order $(m, n - m)$. The matrix B is called the *basic matrix* (of A), as its lines are a basis for $\mathbb{R}^m$. Accordingly, the vector x is decomposed as follows

$$x = \begin{bmatrix} x_B \\ x_N \end{bmatrix}$$

where $x_B \in \mathbb{R}^m$ and $x_N \in \mathbb{R}^{n-m}$. The variables of x_B are called *basic variables* and the variables of x_N are said *non basic variables.* We have

$$Ax = b \iff [B;\, N] \begin{bmatrix} x_B \\ x_N \end{bmatrix} b \iff Bx_B + Nx_N = b,$$

from which, being B nonsingular,

$$x_B = B^{-1}b - B^{-1}Nx_N$$

and hence

$$x = \begin{bmatrix} B^{-1}b \\ [0] \end{bmatrix} + \begin{bmatrix} -B^{-1}N \\ I \end{bmatrix} x_N. \tag{5.1}$$

Definition 5.2.4 The solution (5.1) of system $Ax = b$ where $x_N = [0]$ is said *basic solution,* that is the solution

$$x = \begin{bmatrix} B^{-1}b \\ [0] \end{bmatrix}.$$

If, moreover, $x \geqq [0]$, the said solution is said *basic feasible solution* (that is $x \in K$).

Definition 5.2.5 A basic solution with more than $(n - m)$ zero components, is said *degenerate* basic solution. If this solution is also nonnegative, it is called *degenerate basic feasible solution*.

Example 5.2.6 (See [4]) Let us suppose that the system $Ax = b$ is given by

$$\begin{bmatrix} -2 & 0 & 0 & 1 & 0 \\ 0 & 1 & 0 & 0 & 2 \\ 0 & 0 & 1 & 3 & 2 \end{bmatrix} \begin{bmatrix} x_1 \\ x_2 \\ x_3 \\ x_4 \\ x_5 \end{bmatrix} = \begin{bmatrix} 4 \\ 12 \\ 8 \end{bmatrix}.$$

The first three columns of A are linearly independent, hence

$$x_B = \begin{bmatrix} -2 & 0 & 0 \\ 0 & 1 & 0 \\ 0 & 0 & 1 \end{bmatrix}^{-1} \begin{bmatrix} 4 \\ 12 \\ 8 \end{bmatrix} = \begin{bmatrix} -2 \\ 12 \\ 8 \end{bmatrix}.$$

The vector

$$x = \begin{bmatrix} -2 \\ 12 \\ 8 \\ 0 \\ 0 \end{bmatrix}$$

is therefore a basic solution, however not a feasible basic solution, being $x_1 < 0$.

Also the submatrix obtained by taking the fourth, second and third column

$$\begin{bmatrix} 1 & 0 & 0 \\ 0 & 1 & 0 \\ 3 & 0 & 1 \end{bmatrix}$$

is made by linearly independent columns; hence another basic solution is given by

$$x_B = \begin{bmatrix} 1 & 0 & 0 \\ 0 & 1 & 0 \\ 3 & 0 & 1 \end{bmatrix}^{-1} \begin{bmatrix} 4 \\ 12 \\ 8 \end{bmatrix} = \begin{bmatrix} 4 \\ 12 \\ 6 \end{bmatrix}.$$

Therefore $x_4 = 4$, $x_2 = 12$, $x_3 = 6$. Hence the solution

$$x = \begin{bmatrix} 0 \\ 12 \\ 6 \\ 4 \\ 0 \end{bmatrix}$$

is a feasible basic solution.

The notion of basic solution is very important, as the search of optimal solutions of a L. P. problem, expressed in the standard form, can be made among the set of basic feasible solutions.

Theorem 5.2.7 *Let be given a L. P. problem in its standard form:* $K = \{x \in \mathbb{R}^n : Ax = b, \ x \geqq [0]\}$, *with* A *of full rank (i.e. the rank of* A *is equal to the number of rows of* A*) and consider the polyhedron* P *which represents the feasible set. The following conditions are equivalent:*

1. x *is a vertex of* P.
2. x *is an extreme point of* P.
3. x *is a basic feasible solution.*

From this theorem (which will not be proved here; see the textbooks quoted at the beginning of the present chapter) and from Theorem 5.2.1 (first fundamental theorem of L. P.) it is possible to obtain the following result.

Theorem 5.2.8 (Second Fundamental Theorem of L. P.) *Let be given a L. P. problem in its standard form, with* $m < n$ *and* $\operatorname{rank}(A) = m$. *If there exists a feasible solution, then there exists a basic feasible solution. If there exists an optimal solution, then there exists an optimal basic feasible solution.*

5.3 Graphic Solution of L. P. with Two Variables

In particular simple cases, that is in problems with two variables, L. P. problems can be dealt in a geometrical way and solved by means of elementary considerations. It is sufficient to consider the fact that a linear inequality constraint (or a linear affine inequality constraint) with two variables is represented by all points of a half-space in $\mathbb{R}^2$. The feasible set is therefore given by a convex polygon (i.e. a polyhedron in $\mathbb{R}^2$), maybe unbounded; if we fix a value α of the objective function, we obtain the equation of the level line $\mathrm{lev}_{=\alpha} f$. In all points of this line (and only in the said points) the objective function is equal to α. By varying α, we obtain a sheaf of parallel lines. As we know, the interior points of this polygon are not taken into consideration. In case the objective function is not unbounded on the polygon, when we try to improve the level in order to find the solution of the problem, the related level line reaches the boundary of the polygon. Then two cases can occur:

(*a*) The side of the polygon is parallel to the level lines and hence the solution is given by all (infinite) points of the side. We have therefore in this case infinite solution points of the problem.
(*b*) The side of the polygon is not parallel to the level lines and hence in this case the solution is a vertex of the polygon. We have in this case a unique solution of the problem.

Some examples hopefully will clarify the procedure.

Example 5.3.1 Solve the following L. P. problem by means of level lines.

$$\begin{cases} \max(5x_1 + 3x_2) \\ 3x_1 + 5x_2 \leqq 15 \\ 5x_1 + 2x_2 \leqq 10 \\ x_1 \geqq 0, \ x_2 \geqq 0. \end{cases}$$

Let us consider the first constraint

$$3x_1 + 5x_2 \leqq 15,$$

which represents in $\mathbb{R}^2$ the closed half-space containing the origin and bounded by the straight line $3x_1 + 5x_2 = 15$ (this line is denoted by the letter "a" in Fig. 5.2). Similarly, for the second constraint $5x_1 + 2x_2 \leqq 10$, bounded by the straight line $5x_1 + 2x_2 = 10$ (line denoted by the letter "b" in Fig. 5.2). Now consider the family of parallel lines generated by the objective function and of equation

$$5x_1 + 3x_2 = k, \ k \in \mathbb{R}.$$

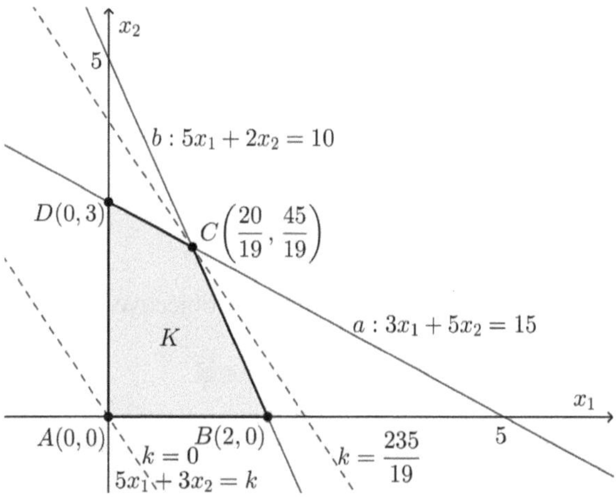

Fig. 5.2 Example 5.3.1: feasible set, level lines $\text{lev}_{=k} f$ and maximum

For $k = 0$ this line passes through the origin and has slope $m = -\frac{5}{3}$. If $k < 0$ the lines in question move into the third quadrant and hence they are not considered, as $x_1 < 0$, $x_2 < 0$. It is therefore obvious that the maximum "feasible" level is the one where the related line has a "contact" with the vertex C of the polygon. The coordinates of C are obtained by resolving the system

$$\begin{cases} 3x_1 + 5x_2 = 15 \\ 5x_1 + 2x_2 = 10. \end{cases}$$

The solution is $x_1 = \frac{20}{19}$, $x_2 = \frac{45}{19}$.
The solution of the problem is therefore

$$x^0 = \left[\frac{20}{19}, \frac{45}{19}\right]^\top.$$

with $c^\top x^0 = \frac{235}{19} \simeq 12,37$.

In a more empirical way, it is sufficient to take into consideration the four vertices of the polygon:

$$(0,0);\ (0,3);\ (2,0);\ (\frac{20}{19}, \frac{45}{19})$$

and to compute the value of the objective function in each one of them. We find, respectively, $0, 9, 10, \frac{235}{19}$. The solution is then immediate (see Fig. 5.2).

Example 5.3.2 Solve the following L. P. problem

$$\begin{cases} \max(2x_1 + x_2) \\ 2x_1 - x_2 \geqq 0 \\ x_1 - x_2 \geqq -1 \\ x_1 - 5x_2 \geqq -20 \\ x_1 \geqq 0,\ x_2 \geqq 0. \end{cases}$$

The sheaf of lines, of level k, generated by the objective function is

$$x_2 = k - 2x_1,\ k \in \mathbb{R}.$$

It is quite immediate to see that the problem admits no solution: the feasible set is unbounded from above and the level of the objective function increases without limits, yet remaining in the feasible set. If we would have considered a *minimum problem* (of the same objective function), instead of a maximum problem, it is immediate to deduce that the solution is then $x^0 = (0, 0)$ (see Fig. 5.3).

Example 5.3.3 Solve the problem

$$\begin{cases} \max(-x_1 + x_2) \\ 2x_1 - x_2 \geqq 0 \\ x_1 - x_2 + 1 \geqq 0 \\ x_1 - 5x_2 + 20 \geqq 0 \\ x_1 \geqq 0,\ x_2 \geqq 0. \end{cases}$$

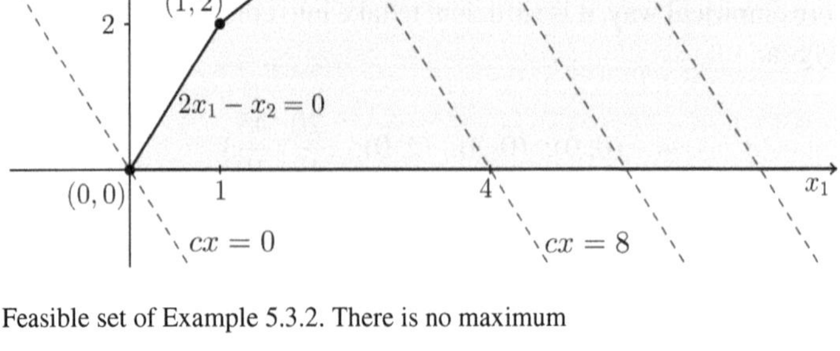

Fig. 5.3 Feasible set of Example 5.3.2. There is no maximum

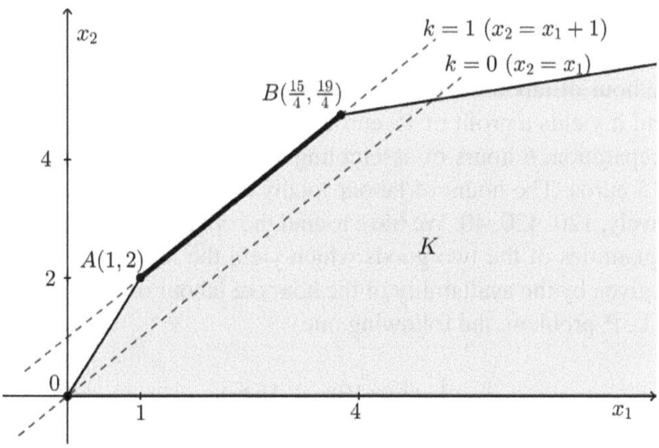

Fig. 5.4 Example 5.3.3: feasible set and maximum

Note that the feasible set K coincides with the one of Example 5.3.2. The objective function is however different and generates the level lines

$$-x_1 + x_2 = k,$$

that is

$$x_2 = x_1 + k.$$

If $k = 0$ we have $x_2 = x_1$; if $k = 1$ we have $x_2 = x_1 + 1$, etc. (see Fig. 5.4).

We deduce that all infinite points of the segment $\overline{AB}$ are solutions of the problem, that is the points

$$x(\lambda) = \lambda \begin{bmatrix} 1 \\ 2 \end{bmatrix} + (1 - \lambda) \begin{bmatrix} \frac{15}{4} \\ \frac{19}{4} \end{bmatrix}, \quad \forall \lambda \in [0, 1].$$

On all these points the objective function assumes a constant value: $c^\top x(\lambda) = 1$. Indeed,

$$\begin{aligned} -x_1 + x_2 &= [-1, 1] \begin{bmatrix} \lambda + (1 - \lambda)\frac{15}{4} \\ 2\lambda + (1 - \lambda)\frac{19}{4} \end{bmatrix} \\ &= -\lambda - \frac{15}{4} + \lambda\frac{15}{4} + 2\lambda + \frac{19}{4} - \lambda\frac{19}{4} \\ &= \frac{-4\lambda - 15 + 15\lambda + 8\lambda + 19 - 19\lambda}{4} = \frac{4}{4} = 1. \end{aligned}$$

Example 5.3.4 A firm is subdivided into three departments (preparation, assembling and testing) and produces two goods, labeled 1 and 2. Each unit of good 1 requires one hour of labour of preparation, five hours of assembling and 0.8 hours of testing and it yields a profit of 10 euros. Each unit of good 2 requires 3 hours of labour of preparation, 6 hours of assembling and 0.4 hours of testing and it yields a profit of 15 euros. The hours of labour totally available in the three departments are, respectively, 120, 420, 40. We have to find the optimal productive combination, that is the quantities of the two goods which yield the maximum profit, under the constraints, given by the availability of the hours of labour of the three departments.

This is a L. P. problem, the following one.

$$\begin{cases} \max(10x_1 + 15x_2) \\ x_1 + 3x_2 \leqq 120 \\ 5x_1 + 6x_2 \leqq 420 \\ 0.8x_1 + 0.4x_2 \leqq 40 \\ x_1 \geqq 0, \ x_2 \geqq 0. \end{cases}$$

The feasible set is represented in Fig. 5.5.

The four vertices of the feasible set are, respectively, $(0, 0)$; $(50, 0)$; $(36, 28)$; $(0, 40)$. The values of the objective function, evaluated at these vertices, are, respectively, $0, 500, 780, 600$. The solution of the problem is therefore $x^0 = [36, 28]^\top$: the optimal production is 36 units of good 1 and 28 units of good 2. It is easy to verify that the hours of the department of preparation and of testing are fully used, whereas the department of assembling uses only 328 hours of the total of 420 hours available (no point of the feasible set makes effective the second constraint).

Now let us suppose that the unit profits are 10 and 30, so that the objective function becomes $10x_1 + 30x_2$, whereas the constraints do not change. Then in the four vertices of the feasible polygon the values of the objective function are, respectively, 0; 500; 1200; 1200. Hence there are two vertices, i.e. $(36, 28)$ and

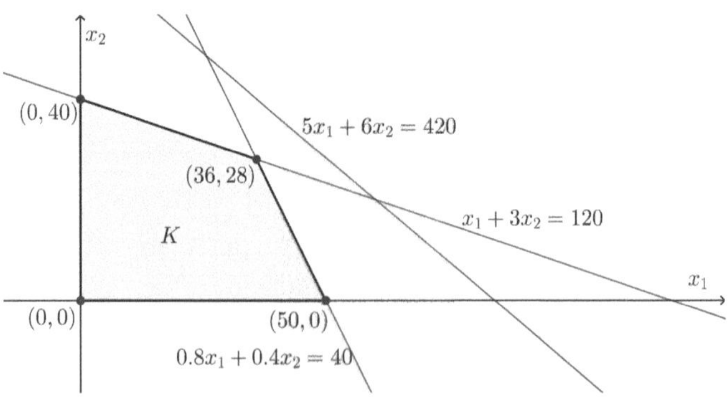

Fig. 5.5 Feasible set of Example 5.3.4. The constraint $5x_1 + 6x_2 = 420$ is redundant

$(0, 40)$, where the objective function assumes the same maximum value. In this case there are therefore infinitely many maximum points, that is all points of the segment joining the said two vertices, that is the infinite points

$$[36\lambda;\ 40 - 12\lambda]^\top,\ 0 \leqq \lambda \leqq 1.$$

5.4 Dual Problems

Several interesting features of a L. P. problem can be pointed out in an indirect way by means of an associated problem, said the "dual problem", which is in turn a L. P, problem, usually denoted by (D), and which is associated with the original L. P. problem (P), by means of precise rules. Dual problems are relevant, not only for their analytical properties but also for computational matters and also for some interpretation matters, arising mainly from economic and financial applications. Finally, it must be observed that duality theory, in the present book examined only in relation to L. P. problems, can be extended to the case of nonlinear programming problems. However, we shall not be concerned in this last case; for these problems see, e.g., [11].

Let us consider the problem

$$(P): \quad \begin{cases} \max c^\top x \\ Ax \leqq b \\ x \geqq [0]. \end{cases}$$

Its dual problem is

$$(D): \quad \begin{cases} \min y^\top b \\ y^\top A \geqq c^\top \\ y \geqq [0] \end{cases}$$

where obviously $y^\top$ is a row vector of $\mathbb{R}^m$.

The original problem (P) is said also "primal problem" (the name was suggested by the father of G. B. Dantzig, the mathematician Tobias Dantzig). If we start from the primal problem

$$(P): \quad \begin{cases} \max c^\top x \\ Ax = b \\ x \geqq [0], \end{cases}$$

that is from a primal problem in a "standard formulation", its dual problem is

$$(D): \quad \begin{cases} \min y^\top b \\ y^\top A \geqq c^\top . \end{cases}$$

Note that in this dual problem the nonnegativity conditions on vector y are absent.

On the grounds of the two above examples, we remark what follows:

(*a*) The dual of (P) is unique.
(*b*) The dual of the dual is the primal problem: in other words, (P) and (D) are, respectively, the dual problem of the other problem ("involutional property").
(*c*) if (P) is a maximization problem, then (D) is a minimization problem and vice-versa; if (P) is a minimization problem, then (D) is a maximization problem and vice-versa.
(*d*) The coefficients of the objective function of (D) are the right-hand side terms of the constraints of (P) and the right-hand side terms of (D) are the coefficients of the objective function of (P).
(*e*) If (P) is in "canonical form", also (D) is in "canonical form", but the constraints of the dual are "reversed", with respect to the ones of the primal.
(*f*) If (P) is in "standard form", then (D) is in "canonical form", *however without the nonnegativity conditions on the variables.*
(*g*) If (P) is in "canonical form", *but lacking of the nonnegativity conditions on the variables,* then (D) is in "standard form".

The problems in cases (*f*) and (*g*) are also called *asymmetric (primal and dual) problems.*

Now we verify property (*b*). We consider the primal (P) in the canonical form. Its dual is therefore

$$\begin{cases} \min y^\top b \\ y^\top A \geqq c^\top \\ y \geqq [0] . \end{cases}$$

This problem can be rewritten in the form

$$\begin{cases} \max(-y^\top b) \\ -y^\top A \leqq -c^\top \\ y \geqq [0] , \end{cases}$$

that is

$$\begin{cases} \max y^\top(-b) \\ y^\top(-A) \leqq -c^\top \\ y \geqq [0] . \end{cases}$$

Now we write the dual of the last formulation obtained:

$$\begin{cases} \min(-c)^\top x \\ (-A)x \geqq -b \\ x \geqq [0] \end{cases}$$

that is

$$\begin{cases} \max c^\top x \\ Ax \leqq b \\ x \geqq [0] , \end{cases}$$

hence we have obtained the primal problem (P).

Now we verify property (*f*). We start from the primal problem in "standard form":

$$\begin{cases} \max c^\top x \\ Ax = b \\ x \geqq [0] , \end{cases}$$

that is

$$\begin{cases} \max c^\top x \\ Ax \leqq b \\ Ax \geqq b \\ x \geqq [0] , \end{cases}$$

that is

$$\begin{cases} \max c^\top x \\ Ax \leqq b \\ -Ax \leqq -b \\ x \geqq [0] , \end{cases} , \quad \text{that is} \quad \begin{cases} \max c^\top x \\ \begin{bmatrix} A \\ -A \end{bmatrix} x \leqq \begin{bmatrix} b \\ -b \end{bmatrix} \\ x \geqq [0] . \end{cases}$$

Now we write the dual of the last formulation of the problem:

$$\begin{cases} \min((y^1)^\top b - (y^2)^\top b) \\ ((y^1)^\top A - (y^2)^\top A) \geqq c^\top \\ y^1 \geqq [0] ,\ y^2 \geqq [0] , \end{cases}$$

that is, by putting $y = y^1 - y^2$, we obtain

$$\begin{cases} \min y^\top b \\ y^\top A \geqq c^\top , \end{cases}$$

as the vector $y = y^1 - y^2$ has no sign constraint, being the difference of two nonnegative vectors (obviously this difference may give rise to a vector with elements of any sign).

Example 5.4.1 Write the dual of the problem

$$\begin{cases} \min c^\top x \\ Ax \geqq b \\ x \geqq [0] . \end{cases}$$

First we write the primal problem in its usual canonical form (of a maximization problem):

$$\begin{cases} \max(-c^\top x) \\ (-A)x \leqq -b \\ x \geqq [0] . \end{cases}$$

Now we write the dual problem, on the grounds of the above given rules,

$$\begin{cases} \min y^\top(-b) \\ y^\top(-A) \geqq -c^\top \\ y \geqq [0] , \end{cases}$$

that is

$$\begin{cases} \max y^\top b \\ y^\top A \leqq c^\top \\ y \geqq [0] . \end{cases}$$

For the reader's convenience we give a general version of the primal problem (P) and of its dual problem (D). Here (P) is given as a minimization problem and hence (D) as a maximization problem. The reader is invited to write the primal as a maximization problem and to build its associated dual.

$$(P) : \quad \begin{cases} \min((c^1)^\top x^1 + (c^2)^\top x^2 + (c^3)^\top x^3) \\ A_{11}x^1 + A_{12}x^2 + A_{13}x^3 \geqq b^1 \\ A_{21}x^1 + A_{22}x^2 + A_{23}x^3 = b^2 \\ A_{31}x^1 + A_{32}x^2 + A_{33}x^3 \leqq b^3 \\ x^1 \geqq [0] ,\ x^2 \text{ unconstrained},\ x^3 \leqq [0] \end{cases}$$

$$(D) : \quad \begin{cases} \max((y^1)^\top b^1 + (y^2)^\top b^2 + (y^3)^\top b^3) \\ (y^1)^\top A_{11} + (y^2)^\top A_{21} + (y^3)^\top A_{31} \leqq (c^1)^\top \\ (y^1)^\top A_{12} + (y^2)^\top A_{22} + (y^3)^\top A_{32} = (c^2)^\top \\ (y^1)^\top A_{13} + (y^2)^\top A_{23} + (y^3)^\top A_{33} \geqq (c^3)^\top \\ y^1 \geqq [0] ,\ y^2 \text{ unconstrained},\ y^3 \leqq [0] . \end{cases}$$

From the previous formulation it is easy to obtain several special cases, for example the following ones.

$$(P): \begin{cases} \min c^\top x \\ Ax \geqq b \ \left[= b; \ \leqq b\right] \\ x \geqq [0] \end{cases} \quad (D): \begin{cases} \max y^\top b \\ y^\top A \leqq c^\top \\ y \geqq [0] \ \left[y \text{ unconstrained}; \ y \leqq [0]\right]. \end{cases}$$

$$(P): \begin{cases} \min c^\top x \\ Ax \geqq b \ \left[= b; \ \leqq b\right] \\ x \text{ unconstrained} \end{cases} \quad (D): \begin{cases} \max y^\top b \\ y^\top A = c^\top \\ y \geqq [0] \ \left[y \text{ unconstrained}; \ y \leqq [0]\right]. \end{cases}$$

$$(P): \begin{cases} \min((c^1)^\top x^1 + (c^2)^\top x^2) \\ A_{11}x^1 + A_{12}x^2 \geqq b \ \left[= b; \ \leqq b\right] \\ x^1 \geqq [0], \ x^2 \text{ unconstrained} \end{cases}$$

$$(D): \begin{cases} \max y^\top b \\ y^\top A_{11} \leqq (c^1)^\top \\ y^\top A_{12} = (c^2)^\top \\ y \geqq [0] \ \left[y \text{ unconstrained}; \ y \leqq [0]\right]. \end{cases}$$

$$(P): \begin{cases} \min c^\top x \\ A_{11}x \geqq b^1 \\ A_{21}x = b^2 \\ A_{31}x \leqq b^3 \\ x \geqq [0] \ \ [x \text{ unconstrained}] \end{cases}$$

$$(D): \begin{cases} \max((y^1)^\top b^1 + (y^2)^\top b^2 + (y^3)^\top b^3) \\ (y^1)^\top A_{11} + (y^2)^\top A_{21} + (y^3)^\top A_{31} \leqq c^\top \ \left[= c^\top\right] \\ y^1 \geqq [0], \ y^2 \text{ unconstrained}, \ y^3 \leqq [0]. \end{cases}$$

$$(P): \begin{cases} \max c^\top x \\ Ax \leqq b \ \left[= b; \ \leqq b\right] \\ x \text{ unconstrained} \end{cases} \quad (D): \begin{cases} \min y^\top b \\ y^\top A = c^\top \\ y \geqq [0] \ \left[y \text{ unconstrained}; \ y \leqq [0]\right]. \end{cases}$$

$$(P): \begin{cases} \max((c^1)^\top x^1 + (c^2)^\top x^2) \\ A_{11}x^1 + A_{12}x^2 \leqq b \\ x^1 \geqq [0], \ x^2 \text{ unconstrained} \end{cases}$$

$$D):\quad \begin{cases} \min y^\top b \\ y^\top A_{11} \geqq (c^1)^\top \\ y^\top A_{12} = (c^2)^\top \\ y \geqq [0] \quad \left[y \text{ unconstrained; } y \leqq [0]\right]. \end{cases}$$

$$(P):\quad \begin{cases} \max c^\top x \\ A_{11}x \leqq b^1 \\ A_{21}x = b^2 \\ A_{31}x \geqq b^3 \\ x \geqq [0]\,, \quad [x \text{ unconstrained}] \end{cases}$$

$$(D):\quad \begin{cases} \min((y^1)^\top b^1 + (y^2)^\top b^2 + (y^3)^\top b^3) \\ (y^1)^\top A_{11} + (y^2)^\top A_{21} + (y^3)^\top A_{31} \geqq c^\top \ \left[= c^\top\right] \\ y^1 \geqq [0]\,,\ y^2 \text{ unconstrained, } y^3 \leqq [0]\,. \end{cases}$$

For example, if we have the primal problem

$$\begin{cases} \min 2x_1 + 3x_2 + 4x_3 + x_4 \\ x_1 - 5x_3 + 2x_4 \geqq 7 \\ 2x_1 + 4x_2 - 6x_3 \geqq 9, \end{cases}$$

its associated dual is

$$\begin{cases} \max 7y_1 + 9y_2 \\ y_1 + 2y_2 = 2 \\ 4y_2 = 3 \\ -5y_1 - 6y_2 = 4 \\ 2y_1 = 1 \\ y_1 \geqq 0,\ y_2 \geqq 0. \end{cases}$$

If we have the primal problem

$$\begin{cases} \max 4x_1 + 3x_2 + 2x_3 \\ x_1 + 2x_2 + 3x_3 \leqq 8 \\ 2x_1 - x_3 \leqq 7 \\ 3x_1 + 4x_2 - x_3 \leqq 5 \\ x_2 + x_3 \leqq 6 \\ x_2 \geqq 0, \end{cases}$$

its associated dual is

$$\begin{cases} \min 8y_1 + 7y_2 + 5y_3 + 6y_4 \\ y_1 + 2y_2 + 3y_3 = 4 \\ 2y_1 + 4y_3 + y_4 \geqq 3 \\ 3y_1 - y_2 - y_3 + y_4 = 2 \\ y_1 \geqq 0,\ y_2 \geqq 0,\ y_3 \geqq 0,\ y_4 \geqq 0. \end{cases}$$

If we have the primal problem

$$\begin{cases} \min 2x_1 - 3x_2 + x_3 \\ 3x_1 + x_2 + 5x_3 \geqq 7 \\ x_1 + x_2 - 6x_3 \leqq 9 \\ 4x_1 - x_2 - 2x_3 = 8 \\ x_1 \geqq 0,\ x_2 \geqq 0 \end{cases}$$

its associated dual is

$$\begin{cases} \max 7y_1 - 9y_2 + 8y_3 \\ 3y_1 + y_2 + 4y_3 \leqq 2 \\ y_1 - y_2 - y_3 \leqq -3 \\ 5y_1 - 6y_2 - 2y_3 = 1 \\ y_1 \geqq 0,\ y_2 \geqq 0. \end{cases}$$

The relationships between the two L. P. problems, primal and dual, are very strict and interesting, also from a computational point of view and also for economic interpretations; but above all, these problems give rise to several results, very useful under theoretical aspects. We examine the problems (P) and (D) considered at the beginning of the present section (problems in their canonical form) and we write for the same the respective Lagrangian functions, denoted by $\mathcal{L}$ and $\mathcal{M}$:

$$\mathcal{L}(x, \lambda) = c^\top x - \lambda^\top (Ax - b) = c^\top x + \lambda^\top (b - Ax) = c^\top x + \lambda^\top b - \lambda^\top Ax;$$

$$\mathcal{M}(y, \mu) = y^\top b + (c^\top - y^\top A)\mu = y^\top b + c^\top \mu - y^\top A\mu.$$

It is immediately seen that, by putting in $\mathcal{L}$, $\lambda = y$, and by putting in $\mathcal{M}$, $\mu = x$, the two Lagrangian functions become equal, that is

$$\mathcal{L}(x, y) = c^\top x + y^\top b - y^\top Ax = \mathcal{M}(y, x) = y^\top b + c^\top x - y^\top Ax.$$

(The variables vector of a problem is the multipliers vector of the other one). There is something in the wind! Now we write the Karush-Kuhn-Tucker conditions for the two problems and recall that, owing to the linearity of the functions involved, these conditions are both necessary and sufficient for the optimality of the related points (moreover, the constraint qualifications are not needed). We have that $\hat{x}$ is a solution of (P) if and only if there exists a vector $\hat{y}$ such that

$$\begin{aligned} \nabla_x \mathcal{L}(\hat{x}, \hat{y}) = c^\top - \hat{y}^\top A &\leqq [0]\,; \\ (c^\top - \hat{y}^\top A)\hat{x} &= 0; \\ b - A\hat{x} &\geqq [0]\,; \\ \hat{y}^\top (b - A\hat{x}) &= 0; \\ \hat{x} \geqq [0]\,;\ \ \hat{y} &\geqq [0]\,. \end{aligned}$$

Similarly, $\hat{y}$ is a solution of (D) if and only if there exists a vector $\hat{x}$ such that

$$\begin{aligned} \nabla_y \mathcal{M}(\hat{y}, \hat{x}) = b - A\hat{x} &\geqq [0]\,; \\ \hat{y}^\top (b - A\hat{x}) &= 0; \\ c^\top - \hat{y}A &\leqq [0]\,; \\ (c^\top - y^\top A)\hat{x} &= 0; \\ \hat{y} \geqq [0]\,;\ \hat{x} &\geqq [0]\,. \end{aligned}$$

The Karush-Kuhn-Tucker conditions are therefore the same for the two problems: if $\hat{x}$ is a solution of (P), there exists a vector $\hat{y}$ which satisfies the above conditions. Vice-versa, if $\hat{y}$ is a solution of (D), there exists a vector $\hat{x}$ which satisfies the above conditions. This remark is very important for further developments. We begin with the basic "Existence Theorem".

Theorem 5.4.2 (Existence Theorem) *A necessary and sufficient condition such that a L. P. problem admits a solution is that its feasible set and the feasible set of its dual are nonempty. In other words: let K and Y be, respectively, the feasible sets of the primal (P) and of the dual (D). Then*

(a) The primal (P) has a solution if and only if $K \neq \emptyset$ and $Y \neq \emptyset$.
(b) The dual (D) has a solution if and only if $Y \neq \emptyset$ and $K \neq \emptyset$.

In order to prove this theorem we need some preliminary results. First of all we make the following obvious remark: The theorem (or lemma) of the alternative of Farkas (Theorem 3.4.6) can be rewritten in the form

- The following two systems are in alternative

$$S: \quad \begin{cases} Ax \geqq [0] \\ b^\top x < 0 \end{cases}; \qquad S^*: \quad \begin{cases} A^\top y = b \\ y \geqq [0]\,. \end{cases}$$

Now we prove the following two lemmas, the first one being in turn a corollary of the theorem of Farkas.

Lemma 5.4.3 *The following two systems are in alternative*

$$S_1: \quad \begin{cases} Ax \geqq [0] \\ b^\top x < 0 \\ x \geqq [0] \end{cases} ; \quad S_1^*: \quad \begin{cases} A^\top y \leqq b \\ y \geqq [0] . \end{cases}$$

Proof

(*i*) If S_1^* admits a solution, then S_1 is inconsistent, as from $y^\top Ax \leqq b^\top x < 0$ we deduce that we cannot have $y \geqq [0]$.

(*ii*) If S_1^* is inconsistent, then S_1 admits a solution. Indeed, as S_1^* can be rewritten as $A^\top y + z = b$, with $z \geqq [0]$, we have that the system

$$\left[A^\top, I\right]\begin{bmatrix} y \\ z \end{bmatrix} = b$$

has no solution $[y, z]^\top \geqq [0]$. From the theorem of Farkas it follows that there exists a vector x such that

$$\begin{bmatrix} A \\ I \end{bmatrix} x \geqq [0], \quad b^\top x < 0,$$

that is S_1 admits a solution.

□

The next result shows that if in a L. P. maximization problem, the objective function is *upper bounded*, then the maximum is indeed attained. Similarly, if in a L. P. minimization problem, the objective function is *lower bounded*, then the minimum is indeed attained. We prove this result with reference to the dual problem (D), that is with reference to a L. P. minimization problem. (The case of a maximization problem is proved in a quite similar way).

Lemma 5.4.4 *Let be in* (D)

$$\inf_{y \in Y} b^\top y = M > -\infty,$$

where $Y = \left\{y : y^\top A \geqq c^\top, \ y \geqq [0]\right\}$. *Then there exists* $y^* \in Y$ *such that*

$$b^\top y^* = \min_{y \in Y} b^\top y.$$

Proof Absurdly suppose that the said vector $y^* \in Y$ does not exist. Then the system

$$\begin{cases} A^\top y \geqq c \\ b^\top y \leqq M \\ y \geqq [0] \end{cases} \tag{5.2}$$

admits no solution. We put

$$B = \begin{bmatrix} -A^\top \\ b \end{bmatrix}, \quad d = \begin{bmatrix} -c \\ M \end{bmatrix}.$$

Now we rewrite system (5.2) in the following form

$$\begin{cases} By \leqq d \\ y \geqq [0]. \end{cases}$$

As this system admits no solution, by Lemma 5.4.3, the following system

$$\begin{cases} B^\top u \geqq [0] \\ d^\top u < 0 \\ u \geqq [0] \end{cases}$$

admits a solution. Now, if we give an increment $\varepsilon > 0$, sufficiently small, to the number M which appears in the second inequality of the last system, here below rewritten,

$$-\sum_{i=1}^{n} c_i u_i + M u_{n+1} < 0,$$

then the system

$$\begin{cases} B^\top u \geqq [0] \\ -\sum_{i=1}^{n} c_i u_i + (M+\varepsilon) u_{n+1} < 0 \\ u \geqq [0] \end{cases}$$

remains solvable. Always by Lemma 5.4.3, it is therefore inconsistent the system

$$By \leqq d_\varepsilon$$
$$y \geqq [0]$$

where

$$d_\varepsilon = \begin{bmatrix} -c \\ M+\varepsilon \end{bmatrix}.$$

In the old notation this system becomes

$$\begin{cases} A^\top y \geqq c \\ b^\top y \leqq M+\varepsilon \\ y \geqq [0]. \end{cases}$$

However, the assertion that this system is inconsistent contradicts the definition of the number M, as the lower bound of the function $b^\top y$ over Y. □

Proof of Theorem 5.4.2 Let us denote $v(x) = c^\top x$ $(v : \mathbb{R}^n \longrightarrow \mathbb{R})$ and $z(y) = y^\top b$ $(z : \mathbb{R}^n \longrightarrow \mathbb{R})$ the objective function, respectively, of the primal problem and of the dual problem. The functional constraints of the two problems are

$$Ax \leqq b; \quad y^\top A \geqq c^\top .$$

We multiply the first inequality by $y^\top$ (y feasible):

$$y^\top Ax \leqq y^\top b \equiv z(y).$$

Now we multiply the second inequality by x (x feasible):

$$y^\top Ax \geqq c^\top x \equiv v(x).$$

Therefore $v(x) \leqq z(y)$ for every feasible x and y: the value of the objective function of the primal problem cannot exceed the value of the objective function of the dual problem (many authors call this result "weak duality theorem"). If the two feasible sets are nonempty, that is there exist two feasible vectors $\hat{x}$ and $\hat{y}$, (feasible for the respective problems), we have therefore

$$v(x) \leqq z(\hat{y}), \text{ for every feasible } x \text{ for } (P),$$
$$v(\hat{x}) \leqq z(y), \text{ for every feasible } y \text{ for } (D).$$

In other words: $v(x)$ is upper bounded and $z(y)$ is lower bounded. By Lemma 5.4.4 we conclude that v attains its maximum and z attains its minimum. We have thus shown that if the two feasible sets are nonempty, then both problems admit a solution. Now we prove the vice-versa. Let $\hat{x}$ a solution of the primal. The Karush-Kuhn-Tucker conditions assure in particular the existence of a vector $\hat{y}$ such that

$$\hat{y}^\top A \geqq c^\top, \quad \hat{y} \geqq [0],$$

that is $\hat{y}$ is feasible for the dual problem. Being $\hat{x}$ a solution of the primal problem, a fortiori it is a feasible point for the same problem, and hence the two feasible sets are both nonempty. □

Remark 5.4.5 On the grounds of the Existence Theorem, denoting as usual, by K and Y the feasible sets, respectively of the primal problem (P) and of the dual problem (D), we can have the following situations:

(1) $\{K \neq \emptyset;\ Y \neq \emptyset\} \iff$ There exists a solution of (P) and (D).
(2) $K = \emptyset \implies$ either $Y = \emptyset$ or $Y \neq \emptyset$, but in this last case (D) has an objective function unbounded over Y.

(3) $Y = \emptyset \implies$ either $K = \emptyset$ or $K \neq \emptyset$, but in this last case (P) has an objective function unbounded over K.

In other words: either both problems admit a solution or both problems admit no solution, in opposition to what happens for the theorems of the alternative!

From the proof of Theorem 5.4.2 it results an intermediate property, known as *Weak Duality Theorem.* For the reader's convenience we report this theorem, as an autonomous result.

Theorem 5.4.6 (Weak Duality Theorem) *Let $v(x) = c^\top x$ be the objective function of the primal problem (P) and let $z(y) = y^\top b$ be the objective function of the dual problem (D). Then we have*

$$v(x) \leqq z(y), \quad \forall x \in K, \ \forall y \in Y.$$

Other results related to the primal and dual problems are the following ones.

Theorem 5.4.7 (Strong Duality Theorem) *Let $\hat{x}$ and $\hat{y}$ be feasible, respectively for (P) and (D). Then $\hat{x}$ is a solution of (P) and $\hat{y}$ is a solution of (D) if and only if $c^\top \hat{x} = \hat{y}^\top b$.*

Proof Being $\hat{x}$ a solution of (P) and $\hat{y}$ a solution of (D), the Karush-Kuhn-Tucker conditions hold for both problems, wit the related complementary slackness conditions:

$$(c^\top - \hat{y}^\top A)\hat{x} = 0;$$
$$\hat{y}^\top (b - A\hat{x}) = 0.$$

From these conditions it follows

$$\hat{y}^\top b - \hat{y}^\top A\hat{x} = c^\top \hat{x} - \hat{y}^\top A\hat{x} = 0,$$

that is $\hat{y}^\top b = c^\top \hat{x}$.

Vice-versa, let $\hat{x}$ and $\hat{y}$ be feasible, respectively, for (P) and (D), and let be $c^\top \hat{x} = \hat{y}^\top b$. By the weak duality theorem we have

$$c^\top x \leqq \hat{y}^\top b, \quad \forall x \in K.$$

We have also

$$c^\top x \leqq c^\top \hat{x} (= \hat{y}^\top b), \quad \forall x \in K.$$

In other words, $\hat{x}$ is a solution of (P). The proof that it holds $\hat{y}^\top b \leqq y^\top b$, $\forall y \in Y$, i.e. that $\hat{y}$ is a solution of (D), is quite similar. □

Theorem 5.4.8 (Equilibrium Theorem or Complementary Slackness Theorem) *A necessary and sufficient condition such that the two vectors $\hat{x}$ and $\hat{y}$, feasible* *respectively for (P) and (D), are solutions of (P) and (D), respectively, is that the following conditions hold*

$$(c^\top - \hat{y}^\top A)\hat{x} = 0;$$
$$\hat{y}^\top (b - A\hat{x}) = 0.$$

Proof Obviously these conditions are necessary optimality conditions, as they are part of the Karush-Kuhn-Tucker conditions for (P) and (D). For what concerns sufficiency, let us remark that, if $\hat{x}$ and $\hat{y}$ satisfy the above complementary slackness conditions, then it holds

$$c^\top \hat{x} - \hat{y}^\top A\hat{x} = 0 = \hat{y}^\top b - \hat{y}^\top A\hat{x}.$$

From this relation we have $v(\hat{x}) = c^\top \hat{x} = \hat{y}^\top b = z(\hat{y})$ and hence, by the strong duality theorem, it results that $\hat{x}$ is a solution of (P) and $\hat{y}$ is a solution of (D). □

For completeness, we give also a result which puts into relationship the primal (P) and the dual (D) with the saddle points of the Lagrangian function $\mathcal{L}(x, y)$. This result is due to [12] and it can be easily obtained from the previous statements.

Theorem 5.4.9 *The primal problem (P) and the dual problem (D) have the respective solutions $\hat{x}$ and $\hat{y}$ if and only if the pair $(\hat{x}, \hat{y})$ is a saddle point of the Lagrangian function $\mathcal{L}(x, y) = c^\top x + y^\top (b - Ax)$:*

$$\mathcal{L}(x, \hat{y}) \leqq \mathcal{L}(\hat{x}, \hat{y}) \leqq \mathcal{L}(\hat{x}, y), \quad \forall x \geqq [0], \ \forall y \geqq [0].$$

Proof (1) Necessity. If $(\hat{x}, \hat{y})$ is a pair of optimal solutions of (P) and (D), we have

$$c^\top \hat{x} = \hat{y}^\top b$$
$$\hat{x} \geqq [0], \ \hat{y} \geqq [0]$$
$$b - A\hat{x} \geqq [0]$$
$$c^\top - \hat{y}^\top A \leqq [0]$$
$$\hat{y}^\top (b - A\hat{x}) = (c^\top - \hat{y}^\top A)\hat{x} = 0.$$

Whence

$$\mathcal{L}(\hat{x}, \hat{y}) = \hat{y}^\top b = \hat{y}^\top b + (c^\top - \hat{y}^\top A)\hat{x}$$
$$\geqq \hat{y}^\top b + (c^\top - \hat{y}^\top A)x = \mathcal{L}(x, \hat{y}), \ \forall x \geqq [0],$$

proving necessity.

(2) Sufficiency. If $(\hat{x}, \hat{y})$ is a saddle point of $\mathcal{L}(\cdot, \cdot)$, then we have

$$y^\top (b - A\hat{x}) \geqq 0, \ \forall y \geqq [0],$$
$$(c^\top - \hat{y}^\top A)x \leqq 0, \ \forall x \geqq [0],$$

and these relations give rise to $A\hat{x} \leqq b$, $\hat{x} \geqq [0]$, $\hat{y}^\top A \geqq c^\top$, $\hat{y} \geqq [0]$. Hence $\hat{x}$ is feasible for (P) and $\hat{y}$ is feasible for (D). On the other hand, the definition of saddle point for $\mathcal{L}(\cdot, \cdot)$, when $x = [0]$ and $y = [0]$, implies $\hat{y}^\top b = \mathcal{L}([0], \hat{y}) \leqq \mathcal{L}(\hat{x}, \hat{y}) \leqq \mathcal{L}(\hat{x}, [0]) = c^\top \hat{x}$, which proves the optimality of the pair $(\hat{x}, \hat{y})$. Therefore the pair $(\hat{x}, \hat{y})$ solves (P) and (D), respectively.

□

We take again into consideration Theorem 5.4.8 and its related complementary slackness conditions

$$\hat{y}^\top (b - A\hat{x}) = 0$$
$$(c^\top - \hat{y}^\top A)\hat{x} = 0.$$

These conditions may be interpreted as follows:

(a) $\hat{y}_i > 0 \implies A_i\hat{x} = b_i$;
(b) $A_i\hat{x} < b_i \implies \hat{y}_i = 0$;
(c) $\hat{x}_j > 0 \implies \hat{y}^\top A^j = c_j$;
(d) $\hat{y}^\top A^j > c^j \implies \hat{x}_j = 0$.

As a comment to the above remark, consider the following "economic" example.

Example 5.4.10 We have an industry (or production system) which produces n goods by means of m resources (with n not necessary equal to m). The technology is described (similarly for what happens in the Leontief models and in the Activity Analysis of Koopmans) by a matrix $A \geq [0]$, of order (m, n), where a_{ij} measures the quantity of the i-th factor necessary to produce a unity of the j-th good. The vector $x \in \mathbb{R}^n_+$ is the *program vector,* $b \in \mathbb{R}^m_+$ is the vector of the total availability of the production factors. The vector $c \in \mathbb{R}^n_+$ is the vector of the *unitary profits.* We have to determine the optimal production plan, taking into account the constraints of the production processes, and in which the total profits are maximized. The related primal problem is given in the following canonical form

$$(P): \qquad \begin{cases} \max c^\top x \\ Ax \leqq b \\ x \geqq [0]. \end{cases}$$

For this problem and for its dual (D) the above complementary slackness conditions hold. These condition say that:

- If the dual variable is positive, the corresponding resource is fully used (relation (a) above).
- If a resource is not fully used, the corresponding dual variable is zero (relation (b) above).
- If a production is activated at a positive level, the corresponding constraint of the dual problem is active (relation (c) above).
- If the previous situation does not hold, the corresponding production is not activated (relation (d) above).

Example 5.4.11 Consider the dual of the problem of Example 5.3.4, that is

$$\begin{cases} \min(120y_1 + 420y_2 + 40y_3) \\ \quad y_1 + 5y_2 + 0.8y_3 \geqq 10 \\ \quad 3y_1 + 6y_2 + 0.4y_3 \geqq 15 \\ \quad y_1 \geqq 0, \; y_2 \geqq 0, \; y_3 \geqq 0. \end{cases}$$

By means of some computations (e.g. by means of the simplex method) we find that the vertices of the feasible set are:

$(4, 0, \frac{15}{2})$, $(\frac{15}{9}, \frac{15}{9}, 0)$, $(10, 0, 0)$, $(0, 0, \frac{75}{2})$, $(0, \frac{5}{2}, 0)$.

The values of the objective function at the said vertices are, respectively, 780, 900, 1200, 1500, 1050. The minimizer is therefore

$$(y^o)^\top = \left[4, 0, \frac{15}{2}\right].$$

Note that:

(i) The minimum of the objective function of the dual problem coincides with the maximum of the objective function of the primal problem (780).

(ii) At the minimum point of the dual problem, the second variable is zero and the other ones are positive. In the primal problem, the second constraint is not active and the other constraints are active.

The values of the dual variables give useful information on the primal problem. As previously remarked, the second variable y_2 is zero at the optimal solution: we take no advantage in increasing the hours of labour in the assembling department. The first and the third variable of the dual problem are positive and point out that an increasing of the available hours of labour in the departments of preparation and testing induces an increasing of the profit. As $y_3 = \frac{15}{2} > 4 = y_2$, the effect is higher in the testing department: it is more convenient to allocate a greater availability of hour of labour, if any, to the testing department.

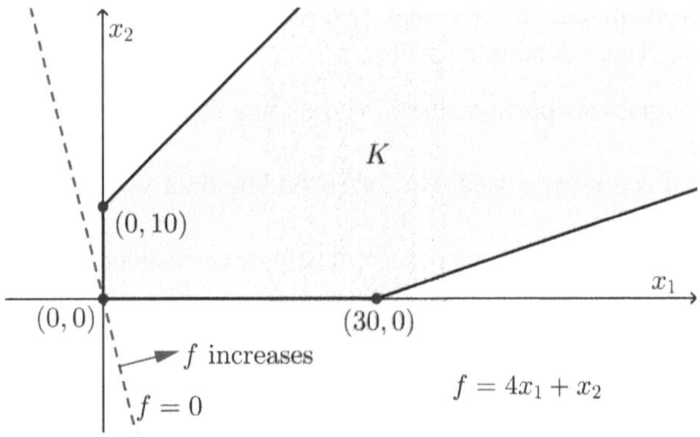

Fig. 5.6 Feasible set of Example 5.4.12. The objective function $f = 4x_1 + x_2$ is unbounded on it

Example 5.4.12 Consider the L. P. problem

$$\begin{cases} \max(4x_1 + x_2) \\ x_1 - 3x_2 \leqq 30 \\ -x_1 + x_2 \leqq 10 \\ x_1 \geqq 0, \ x_2 \geqq 0. \end{cases}$$

Its feasible set is represented in Fig. 5.6.

The three vertices are (0, 0), (30, 0), (0, 10). Without difficulty it is seen that the objective function is upper unbounded on the feasible set. The dual problem is

$$\begin{cases} \min(30y_1 + 10y_2) \\ y_1 - y_2 \geqq 4 \\ -3y_1 + y_2 \geqq 1 \\ y_1 \geqq 0, \ y_2 \geqq 0. \end{cases}$$

As was to be expected, its feasible set is empty.

Example 5.4.13 We give some suggestions on possible links between Linear Programming and Leontief models. The classical "open" Leontief model, concisely described in Chap. 1, Sect. 1.9, can be fitted into Linear Programming schemes, which put into evidence some interesting optimality features. We adopt the notations and conventions used in Chap. 1, Sect. 1.9. The vector $x \in \mathbb{R}^n$ of total productions must satisfy the relation

$$x = Ax + c,$$

i.e.

$$(I - A)x = c,$$

where A is the input-output matrix and $c \in \mathbb{R}^n$ is the vector representing the consumptions basket. In this classical version of the Leontief open model, the vector of labour requirements does not appear in an explicit way (contrary to what happens in the models of P. Sraffa). If we denote, as usual, by $\ell_i > 0$, $i = 1, \ldots, n$, the quantity of labour necessary to produce a unity of the i-th good, the total quantity of labour required to produce x, is given by $\ell^\top x$.

If we want to minimize the total quantity of labour, in order to get a net production greater or equal to the given vector c, we can write the following L. P. problem

$$\begin{cases} \min \ell^\top x \\ (I - A)x \geqq c \\ x \geqq [0], \end{cases}$$

whose dual is

$$\begin{cases} \max p^\top c \\ p^\top (I - A) \leqq \ell^\top \\ p \geqq [0], \end{cases}$$

which puts into evidence the problem of maximizing the total revenue ($p^\top c$), under fulfillment of the related constraints. See. e.g., [10, 14].

5.5 Final Remarks

Also for L. P. problems there is a quite abundant literature regarding results of *sensitivity, stability,* and in general of *post-optimal analysis* (and this fact is quite obvious, as the structure of L. P. problems is simpler than the structure of nonlinear programming problems). We only remark that the solution vector $\hat{y}$ of the dual problem (D) is nothing but the multiplier vector of the Karush-Kuhn-Tucker conditions of the primal problem (P). If the dual (D) admits a unique solution $\hat{y}$, this solution will measure the "sensitivity" of the optimal value function of (P), with respect to variations of the right-hand sided vector b of the constraints. The related economic interpretation is therefore the same of nonlinear programming problems: $\hat{y}$ describes the so-called "shadow prices". More formally, denoting by $F(b)$ the optimal value function of the primal problem, here rewritten,

$$(P): \quad \begin{cases} \max c^\top x \\ Ax \leqq b \\ x \geqq [0], \end{cases}$$

whose dual is

$$(D): \quad \begin{cases} \min y^\top b \\ y^\top A \geqq c \\ y \geqq [0], \end{cases}$$

we can state the following results.

Theorem 5.5.1 *The function $F(b)$ is continuous and concave over its (convex) domain $B = \{z \in \mathbb{R}^m : Ax \leqq z, \ x \geqq [0]\}$.*

Theorem 5.5.2 *If the dual problem (D) has a unique solution $\hat{y}$, then $F(b)$ is differentiable at the corresponding point b and it holds*

$$\nabla F(b) = \hat{y}.$$

If the dual problem (D) has not a unique solution, the situation becomes a bit more complicate, $F(b)$ is no longer differentiable, however it can be proved that $F(b)$ has right-hand and left-hand partial derivatives in the interior of B and that it holds

$$\left(\frac{\partial F}{\partial b_i}\right)^- \geqq \hat{y}_i \geqq \left(\frac{\partial F}{\partial b_i}\right)^+, \ i = 1, \ldots, n.$$

A final remark: the dual of a L. P. problem may sometimes be useful to find, by means of level curves, the solution of the primal problem, whenever this problem is not directly manageable in a geometrical way. Consider the following examples.

Example 5.5.3 Solve the following L. P. problem

$$\begin{cases} \min(175.000x_1 + 100.000x + 22.500x_3 + 15.000x_4) \\ 7x_1 + 3x_2 + x_3 \geqq 15 \\ 5x_1 + 6x_2 + x_4 \geqq 12,5 \\ x_1 \geqq 0, \ x_2 \geqq 0, \ x_3 \geqq 0, \ x_4 \geqq 0. \end{cases}$$

Consider its dual (in two variables!)

$$\begin{cases} \max(15y_1 + 12,5y_2) \\ 7y_1 + 5y_2 \leqq 175.000 \\ 3y_1 + 6y_2 \leqq 100.000 \\ y_1 \leqq 22.500 \\ y_2 \leqq 15.000 \\ y_1 \geqq 0, \ y_2 \geqq 0. \end{cases}$$

This last problem can be solved by means of level lines; it is seen that it admits the unique solution $y^0 = [550.000/27, \ 175.000/27]^\top$ and it results $(y^0)^\top b =$

10.437.500/27; moreover, the third and fourth constraint are not active at y^0. Hence the primal problem admits an optimal solution x^0 such that it results $x_3^0 = x_4^0 = 0$ and both constraints are active (otherwise it would hold $y_1^0 = y_2^0 = 0$, which is not verified). We have therefore a pair of equations which yield the unique nonnegative solution x^0, with $x_1^0 = 17, 5/9$, $x_2^0 = 12, 5/27$, $x_3^0 = 0$, $x_4^0 = 0$. Then we have $c^\top x^0 = 10.437.500/27 = (y^0)^\top b$. Hence x^0 is the unique solution of the primal problem.

Example 5.5.4 Find the solutions, if any, of the problem

$$\begin{cases} \max(3x_1 - 16x_2 - 8x_4) \\ x_1 + 2x_2 - 2x_3 - x_4 \leqq -3 \\ x_1 - x_2 - 3x_3 + x_4 \leqq 1 \\ x_1 \geqq 0,\ x_2 \geqq 0,\ x_3 \geqq 0,\ x_4 \geqq 0. \end{cases}.$$

With "pencil and paper" (and without the use of the simplex method) we are not able to study in a geometrical way the said problem.

Its dual is

$$\begin{cases} \min(-3y_1 + y_2) \\ y_1 + y_2 \geqq -3 \\ 2y_1 - y_2 \geqq -16 \\ -2y_1 - 3y_2 \geqq 0 \\ -y_1 + y_2 \geqq -8 \\ y_1 \geqq 0,\ y_2 \geqq 0. \end{cases}$$

It is easily seen that the feasible set of the dual is empty. So, the primal problem admits no solution.

References

1. S. Achmanov, *Programmation Linéaire* (Editions MIR, Moscou, 1984)
2. M.S. Bazaraa, J.J. Jarvis, H.D. Sherali, *Linear Programming and Network Flows* (Wiley, New York, 1990)
3. M.S. Bazaraa, H.D. Sherali, C.M. Shetty, *Nonlinear programming: Theory and Algorithms* (Wiley, New York, 1993)
4. M. Bertocchi, S. Stefani, G. Zambruno, *Matematica per l'Economia e la Finanza* (McGraw-Hill, Milano, 1995)
5. D. Bertsimas, J.N. Tsitsiklis, *Introduction to Linear Optimization* (Athena Scientific, Belmont, 1997)
6. R.W. Cottle, M.N. Thapa, *Linear and Nonlinear Optimization* (Springer, New York, 2017)
7. G.B. Dantzig, *Linear Programming and Extensions* (Princeton Univ, Press, Princeton, 1963)
8. G.B. Dantzig, M.N. Thapa, *Linear Programming I: Introduction* (Springer, New York, 1997)
9. G.B. Dantzig, M.N. Thapa, *Linear Programming II: Theory and Extensions* (Springer, New York, 2003)

10. R. Dorfman, P.A. Samuelson, R. Solow, *Linear Programming and Economic Analysis* (McGraw-Hill, New York, 1958)
11. G. Giorgi, B. Jiménez, V. Novo, *Basic Mathematical Programming Theory* (Springer, Berlin, 2023)
12. A.J. Goldman, A.W. Tucker, *Theory of Linear Programming: in Linear Inequalities and Related Systems* (Princeton University Press, Princeton, pp 53–98, 1957)
13. G. Hadley, *Linear Programming* (Addison-Wesley Publishing, Reading, 1962)
14. M.D. Intriligator, *Mathematical Optimization and Economic Theory* (Prentice-Hall, Englewood Cliffs, 1971)
15. D.G. Luenberger, Y. Ye, *Linear and Nonlinear Programming*, 3rd edn. (Springer, New York, 2008)
16. M. Simmonard, *Programmation Linéaire*, vols. I and II (Dunod, Paris, 1972, 1973)
17. R.J. Vanderbei, *Linear Programming: Foundations and Extensions* (Kluwer Acad. Publ., Boston, 1996)
18. R.V. Vohra, *Advanced Mathematical Economics* (Routledge, London, 2005)

Chapter 6
Continuous Dynamical Systems

For the reader's convenience, in the present chapter and in the next chapter the vector quantities will be written in bold; matrices, scalar quantities and the other "mathematical objects" will be written in the usual "normal body". The zero vector of $\mathbb{R}^n$ and the zero matrix will be denoted, as usual, by $[0]$.

6.1 Dynamical Systems

Dynamical systems are mathematical models which describe, explain and govern the *time evolution* of a set of variables. We speak in general of "systems", as we do not want or we are not able to consider separately (that is in a no joint way) the single variables; we speak of "dynamical systems" as we are interested in considering the evolution law, along time, of the variables. It is natural to represent the running of time through a scalar variable t, which assumes its values in an interval of the real axis, for instance from the time t_0 to the time t_1 : $[t_0, t_1]$ or also from the time t_0 onwards: $[t_0, +\infty)$. Usually $t_0 = 0$. On many occasions we have phenomena which take place occasionally, for instance the compilation of the balance sheet of a firm, the writing of the temperature in a clinical folder, etc. In these cases we have not to take into consideration all infinite values of an interval of $\mathbb{R}$, but it is sufficient to consider a sequence of times $t_0, t_1, t_2, t_3, \ldots$ on the real axis. We observe that also those phenomena which occur in a continuous way with respect to time (for example the evolution of the temperature in a certain spot), can be described, in an alternative way, by means of a periodic monitoring of time, and hence treated as they would be discrete phenomena.

Let us denote by T a set of time variables, considered as a subset of $\mathbb{R}$. We may have $T = [t_0, t_1]$ or also $T = [t_0, +\infty)$, in the continuous case, or also $T = \{t_0, t_1, t_2, \ldots t_s, \ldots\}$, that is a sequence, finite or not, of various times, in an increasing order. Therefore we have the following basic classification.

G. Giorgi et al., *Lectures on Mathematics for Economic and Financial Analysis*,
https://doi.org/10.1007/978-3-031-83339-7_6

- *CONTINUOUS DYNAMICAL SYSTEMS:* the time variable t runs over an interval of the real axis, usually identified with $T = [t_0,\ t_1]$ or with $T = [0,\ t_1]$ or with $T = [t_0,\ +\infty)$.
- *DISCRETE DYNAMICAL SYSTEMS:* the time variables are elements of a numerical sequence $t_0, t_1, t_2, \ldots, t_s, \ldots$, that is the phenomenon is considered as described by discrete instants, for example a vector whose elements are the components of a balance sheet at the end of the first year, of a certain firm, at the end of the second year, at the end of the third year, etc.

In both cases, we denote by $\mathbf{x} \in \mathbb{R}^n$ the vector whose elements are the n variables of the system, hence we can think that a dynamical system (continuous or discrete) is an application

$$\mathbf{x} : T \longrightarrow \mathbb{R}^n,$$

that is a vector function $\mathbf{x}(t) \longrightarrow \mathbb{R}^n$, where, if the system is continuous, $t \in [t_0,\ t_1]$ or also $t \in [t_0,\ +\infty)$ or also T any interval of $\mathbb{R}$ (usually $t_0 \geqq 0$). We have $t \in T = \{t_0,\ t_1, t_2, \ldots t_s, \ldots\}$, if the system is discrete. The time unit in discrete models can be one year, one month, etc. In many cases the differences $t_i - t_{i-1}$ are equal for all i.

The vector

$$\mathbf{x}(t) = \begin{bmatrix} x_1(t) \\ x_2(t) \\ \vdots \\ x_n(t) \end{bmatrix}$$

shows therefore the position of the dynamical system at the time $t \in T$. The components of $\mathbf{x}$ are usually called *state variables* of the system (they are the dependent variables, while time is the independent variable, considered also as a parameter). The number n of the state variables is the *dimension* of the system.

In the present chapter we shall be concerned with *continuous dynamical systems* and in the following chapter we shall take into consideration *discrete dynamical systems.* Suppose, for example, that the components of a vector $\mathbf{x} = \mathbf{x}(t) \in \mathbb{R}^n$ describe the sizes of some amounts which characterize the situation of a given economic system at a certain time t (we have previously considered the example of $\mathbf{x}(t)$ which describes the various components of a balance sheet of a firm). We propose to build a scheme which describes the time evolution of $\mathbf{x}(t)$. Therefore consider two consecutive instants t and $t + h$, and the related state vectors $\mathbf{x}(t)$ and $\mathbf{x}(t + h)$. The difference $\mathbf{x}(t + h) - \mathbf{x}(t)$ points out the "movement direction" of the system. We may think to a relation of the type

$$\mathbf{x}(t + h) - \mathbf{x}(t) = \mathbf{g}(\mathbf{x}(t), t, h).$$

If $\mathbf{g}$ is a sufficiently regular function (in terms of h) it is possible to write a relation of the type

$$\mathbf{x}(t + h) - \mathbf{x}(t) = \mathbf{f}(\mathbf{x}(t), t)h + \mathbf{o}(h). \tag{6.1}$$

Now we divide relation (6.1) by h and take the limit for $h \longrightarrow 0$, obtaining

$$\mathbf{x}'(t) = \mathbf{f}(\mathbf{x}(t), t),$$

which is an example of a *system of differential equations* which has to be satisfied by every vector $\mathbf{x}(t)$ which belongs to the scheme built above.

Definition 6.1.1 Let $\mathbf{f} : \mathbb{R}^n \times \mathbb{R} \longrightarrow \mathbb{R}^n$ be defined over $X \times T$, where $X \subset \mathbb{R}^n$ and $T \subset \mathbb{R}$ is a nonempty interval. Let $\mathbf{x} : \mathbb{R} \longrightarrow \mathbb{R}^n$ be defined (at least) over T. The relation

$$\mathbf{x}'(t) = \mathbf{f}(\mathbf{x}(t), t) \tag{6.2}$$

describes a *system of ordinary differential equations of the first order in a normal (or canonical) form.* When $n = 1$ we speak of an *ordinary differential equation of the first order in a normal form.*

The integer n is the *dimension of the system;* in an extended form relation (6.2) can be written in the following way

$$\begin{cases} x_1'(t) = f_1(x_1(t), x_2(t), \ldots, x_n(t), t) \\ x_2'(t) = f_2(x_1(t), x_2(t), \ldots, x_n(t), t) \\ \quad\cdots\cdots\cdots\cdots\cdots \\ x_n'(t) = f_n(x_1(t), x_2(t), \ldots, x_n(t), t). \end{cases}$$

For example, the following system

$$\begin{cases} x_1'(t) = 3x_2(t) - 2t \\ x_2'(t) = x_1(t) - tx_2(t) \\ x_3'(t) = \log|x_1(t)/x_2(t)| + 3t^4 x_3(t) \end{cases}$$

is a system of three ordinary differential equations of the first order (in a normal form). The expression "in a normal form" points out the fact that $\mathbf{x}'(t)$ is considered as an *explicit function* of $\mathbf{x}(t)$ and t. Indeed, another form of systems of ordinary differential equations of the first order is the following one

$$\mathbf{\Phi}(\mathbf{x}'(t), \mathbf{x}(t), t) = [0],$$

with $\mathbf{\Phi} : \mathbb{R}^n \times \mathbb{R}^n \times \mathbb{R} \longrightarrow \mathbb{R}^n$, but this *general form* is less easy to study than the normal form. The expression "of the first order" points out the fact that in the equation we have only derivatives of the first order. We have previously observed that when $n = 1$ we have *ordinary differential equations of the first order in a normal form:*

$$x'(t) = f(x(t), t), \quad \text{with } x(t) \in \mathbb{R}. \tag{6.3}$$

The expressions "ordinary differential equations" or system of "ordinary differential equations" point out the fact that the functions involved are functions of one variable (i.e. the time t). If the functions involved are functions of several variables, then we have *partial differential equations.* These equations are rather important, also in financial analysis (e.g. in the Black and Scholes models), but we shall not be concerned with this type of differential equations.

Consider again Eq. (6.3). We can also suppose that, besides the first order derivative, also higher order derivatives are involved. In this case we speak of an ordinary differential equation *of order n* ($n \geqq 1$). More precisely, we have the following definition.

Definition 6.1.2 An *ordinary differential equation of order n* is a relation of the type

$$F(x^{(n)}(t), \ldots, x'(t), x(t), t) = 0, \tag{6.4}$$

where F is a function of $(n + 2)$ real variables. When relation (6.4) is written in a *normal form*, we have

$$x^{(n)}(t) = \varphi(x^{(n-1)}(t), \ldots, x'(t), x(t), t),$$

where φ is a function of $(n + 1)$ real variables.

We remember that x and its various derivatives are function of time.

For example, the two following relations

$$x'(t) = 3t + 1;$$

$$x'(t) = 3x(t)$$

are two ordinary first order differential equations in a normal form; the expression

$$x''(t) = 3x'(t) - x(t) + 2t + 1$$

is a second order differential equation in a normal form.

It is always possible to transform a differential equation of order n into an equivalent system of the first order and vice-versa. More precisely, a differential equation of the type

$$x^{(n)}(t) = \varphi(x^{(n-1)}(t), \ldots, x'(t), x(t), t) \tag{6.5}$$

can be transformed into a system of the type

$$\mathbf{x}'(t) = \mathbf{f}(\mathbf{x}(t), t),$$

where $\mathbf{x}(t) \in \mathbb{R}^n$ and $\mathbf{f} = [f_1, f_2, \ldots, f_n]^\top$. It is sufficient to put

$$\begin{aligned} x &= x_1 \\ x' &= x_2 \text{ and hence } x_2 = x_1' \\ x'' &= x_3 \text{ and hence } x_3 = x_2' \\ &\cdots\cdots\cdots\cdots \\ x^{(n-1)} &= x_n \text{ and hence } x_n = x_{n-1}' \end{aligned}$$

and to observe that (6.5) is equivalent to the system

$$\begin{cases} x_n' = \varphi(x_n, \ldots, x_1, t) \\ x_{n-1}' = x_n \\ \vdots \\ x_2' = x_3 \\ x_1' = x_2, \end{cases} \tag{6.6}$$

in the sense that a solution $x(t)$ of (6.5) generates a vector $(x(t), x'(t), \ldots, x^{(n-1)}(t))^\top$ which is a solution of (6.6), and vice-versa, if $(x_1(t), \ldots, x_n(t))^\top$ is a solution of (6.6), then $x_1(t)$ is a solution of (6.5).

Example 6.1.3 Consider the following second order differential equation (in a normal form)

$$x''(t) = x'(t) - x(t).$$

Put $x_1(t) = x(t)$ and $x_2(t) = x'(t)$, so that

$$x_1'(t) = x'(t) = x_2(t)$$

and

$$x_2'(t) = x''(t).$$

By substituting, we have the system

$$\begin{cases} x_1'(t) = x_2(t) \\ x_2'(t) = x_2(t) - x_1(t), \end{cases}$$

that is

$$\begin{bmatrix} x_1'(t) \\ x_2'(t) \end{bmatrix} = \begin{bmatrix} 0x_1(t) + x_2(t) \\ -x_1(t) + x_2(t) \end{bmatrix},$$

that is

$$\mathbf{x}'(t) = \begin{bmatrix} 0 & 1 \\ -1 & 1 \end{bmatrix} \mathbf{x}(t).$$

It is easy to obtain the "vice-versa": from the second equation of the above system we have $x_2''(t) = -x_1'(t) + x_2'(t)$. We insert in this equation the first equation of the system:

$$x_2''(t) = -x_2(t) + x_2'(t)$$

i.e.

$$x''(t) = x'(t) - x(t).$$

Therefore we can limit ourselves to consider *systems of equations of the first order*.

Remark 6.1.4 We can also consider, as a generalization of the systems of the first order, *systems of ordinary differential equations of order k* $(k \geqq 1)$ *in a normal form*, that is systems of the type

$$\mathbf{x}^{(k)}(t) = \mathbf{F}(\mathbf{x}^{(k-1)}(t), \dots, \mathbf{x}'(t), \mathbf{x}(t), t) \tag{6.7}$$

with $\mathbf{F} : \underbrace{\mathbb{R}^n \times \mathbb{R}^n \times \cdots \times \mathbb{R}^n}_{k \text{ times}} \times \mathbb{R} \longrightarrow \mathbb{R}^n$.

Also for this case it is possible to transform the above equation into an equivalent system of the first order and of dimension nk.

We come back to the usual first order system

$$\mathbf{x}'(t) = \mathbf{f}(\mathbf{x}(t), t), \quad \mathbf{x}(t) \in \mathbb{R}^n. \tag{6.8}$$

If $\mathbf{f}(\mathbf{x}(t), t) = \mathbf{f}(\mathbf{x}(t))$, that is $\mathbf{f}$ *does not depend explicitly on* t, the system is said to be *autonomous;* otherwise, the system is said to be *non autonomous.* The same classification holds for differential equations of order n and for systems of type (6.7).

A function $x(t) : \mathbb{R} \longrightarrow \mathbb{R}$ which verifies (6.4) or (6.5), in the sense that the first member is equal to the second member for all values of t of a certain interval of $\mathbb{R}$, is said to be a *solution* or also (quite improperly) an *integral* of (6.4) or (6.5). The same definition applies to a vector function $\mathbf{x}(t) : \mathbb{R} \longrightarrow \mathbb{R}^n$, with reference to (6.8). For example, $x(t) = e^{-4t}$ is a solution of the second order differential equation

$$x''(t) + 3x'(t) - 4x(t) = 0,$$

being $x'(t) = -4e^{-4t}$ and $x''(t) = 16e^{-4t}$. Therefore

$$x''(t) + 3x'(t) - 4x(t) = 16e^{-4t} - 12e^{-4t} - 4e^{-4t}$$
$$= e^{-4t}(16 - 12 - 4) = 0, \quad \forall t \in \mathbb{R}.$$

Whereas, for example, the function $x(t) = 3+t^2$ is *not* a solution of the following third order differential equation

$$tx'''(t) + x''(t) + x'(t) - x(t) = 0.$$

Indeed, we have $x'(t) = 2t$, $x''(t) = 2$, $x'''(t) = 0$ and hence, by substituting,

$$0 + 2 + 2t - (3 + t^2) = -(t^2 - 2t + 1) = -(t - 1)^2 = 0,$$

which is verified only for $t = 1$.

Example 6.1.5 Find the values of the real parameter $k > 0$ for which the function

$$x(t) = 3t^k - 2t^{k-1}$$

is a solution of the following second order differential equation (in a general form):

$$t^2x''(t) - 2tx'(t) + 2x(t) = 0.$$

We have $x'(t) = 3kt^{k-1} - 2(k - 1)t^{k-2}$; $x''(t) = 3k(k - 1)t^{k-2} - 2(k - 1)(k - 2)t^{k-3}$. By substituting:

$$t^2(3k(k - 1)t^{k-2} - 2(k - 1)(k - 2)t^{k-3}) - 2t(3kt^{k-1} - 2(k - 1)t^{k-2})+$$
$$+ 2(3t^k - 2t^{k-1}) = 0.$$

We have

$$3t^k(k^2 - 3k + 2) - 2t^{k-1}(k^2 - 5k + 6) = 0.$$

Hence we must have

$$\begin{cases} k^2 - 3k + 2 = 0 \\ k^2 - 5k + 6 = 0. \end{cases}$$

We find $k = 2$.

Perhaps the simplest problem which involves a differential equation is the classical problem of finding the *primitives* (or *antiderivatives*) of a given function, that is find all functions $F(x)$ (in our notation $x(t)$) that in every point of an interval (a, b) have as first-order derivative a (continuous) function $f(x)$ ($f(t)$ in our notation), that is find all functions $F(x)$ such that $F'(x) = f(x), \forall x \in (a, b)$. In our notation:

$$x'(t) = f(t). \tag{6.9}$$

We know that (6.9) has infinitely many solutions, represented by

$$x(t) = \int_a^t f(z)dz + c, \tag{6.10}$$

where c is an *arbitrary constant.* We take the occasion to remark that if we have a differential equation of order n or a system of first-order differential equations of dimension n, then the related solution will usually contain n arbitrary constants.

Now we wish to find all functions $x(t)$ which coincide with their (first-order) derivative, that is we wish to solve the problem

$$x'(t) = x(t). \tag{6.11}$$

Obviously, this is another simple example of differential equations. In order to solve (6.11), we multiply both members of this relation by the nonzero factor e^{-t} (this factor is called "integrating factor", as it transforms a certain function into the derivative of another function). We obtain

$$e^{-t}(x'(t) - x(t)) = 0,$$

that is $D(e^{-t} \cdot x(t)) = 0$ and hence, by relation (6.10), $e^{-t}x(t) = c$, from which

$$x(t) = ce^t,$$

which represents the general solution of our problem. This can be seen as one of the reasons why the Euler number e is a sort of "parsely" in Mathematics.

Now we wish to solve the second-order differential equation

$$x''(t) = f(t), \tag{6.12}$$

being $f(t)$ a given function ($f : \mathbb{R} \longrightarrow \mathbb{R}$). From (6.12) it follows

$$x'(t) = c_1 + \int_a^t f(z)dz$$

which is an equation of type (6.9). If we put

$$\varphi(t) = \int_a^t f(z)dz,$$

always from (6.12) we get

$$x(t) = c_2 + c_1 t + \int_a^t \varphi(z)dz$$

where c_1 and c_2 are arbitrary constants.

In the previous examples the differential equations examined have infinitely many solutions; more precisely, Eqs. (6.9) and (6.11) have infinitely many solutions, depending on one arbitrary constant c, whereas the second-order equation (6.12) has infinitely many solutions depending on two arbitrary constants c_1 and c_2.

To *solve* or to *integrate* a differential equation of the type (6.4) or (6.5) means to find its *general solution* or its *general integral,* that is the most general expression formed by the family of *all* solutions which satisfy relations (6.4) or (6.5). Then, by giving particular values to the arbitrary constants which appear in the general solution, we find a *particular solution* or a *particular integral.* Similar considerations hold true for systems of differential equations of the first order, of the type (6.8): the general solution usually will contain n arbitrary constants $c_1, c_2, \ldots, c_n$, that is we have $\mathbf{x}(t) = \mathbf{x}(t, \mathbf{c})$. We may have also solutions of (6.4) or (6.5) or of (6.7) or (6.8), not deducible from the general solution, by means of particular values of the arbitrary constants which appear in the said solution; these types of solutions are called *singular solutions.*

Example 6.1.6 (Economic-Financial Example) Let $x(t) > 0$ be the principal and interest (i.e. the final sum) at time t of an investment under continuous capitalization scheme; if the interests produced at the instant rate of interest $\delta > 0$ in the time interval $(t, t + h)$ by the final capital, are immediately reinvested, we must have

$$x(t + h) = x(t) + x(t)\delta h + o(h)$$

that is

$$\frac{x(t + h) - x(t)}{h} = x(t)\delta + \frac{o(h)}{h}.$$

By letting $h \longrightarrow 0$ in both members of the last relation, we obtain

$$x'(t) = \delta x(t),$$

that is a first-order differential equation, quite similar to (6.11). Instead of using a suitable "integrating factor", we can divide both members by $x(t)$, as this quantity is nonzero by assumption. We obtain

$$\frac{x'(t)}{x(t)} = \delta \;\; (\delta > 0).$$

And now:

$$\int \frac{x'(t)}{x(t)} dt = \int \delta dt$$

from which

$$\log(x(t)) = \delta t + c \implies x(t) = e^{\delta t + c} = e^{\delta t} \cdot e^{c} = k e^{\delta t}.$$

Hence we have obtained the general solution of our differential equation. If we consider the initial time $t = 0$, with the initial capital $x(0) = C$, we can obtain a *particular solution* (or *particular integral*), that is

$$x(0) = e^{\delta \cdot 0} \cdot k = C$$

from which $k = C$ and hence $x(t) = Ce^{\delta t}$, which is the particular solution of the problem ("initial value problem" or "Cauchy problem")

$$\begin{cases} x'(t) = \delta x(t) \\ x(0) = C. \end{cases}$$

If we have a differential equation of type (6.5) or a system of differential equations of type (6.8), in many cases we are less interested in finding all solutions of the same, than in finding those solutions which verify some *fixed conditions.* We call *initial value problem* or *Cauchy problem* for Eq. (6.5), the problem consisting in finding the solutions $x(t)$ of (6.5) that for $t = t_0$ verify the *initial conditions*

$$x(t_0) = x_0,\ x'(t_0) = x_1, \dots,\ x^{(n-1)}(t_0) = x_{n-1}$$

where $x_0, x_1, \dots, x_{n-1}$ are *fixed quantities.*

Similarly, with reference to system (6.8), the above problem consists in finding a solution of (6.8) which satisfies an initial condition $\mathbf{x}(t_0) = \mathbf{x}^0$, with $\mathbf{x}^0 \in \mathbb{R}^n$. As previously said, we shall be mainly concerned with systems of this type, that is of systems of differential equations of the first order in a normal form:

$$\mathbf{x}'(t) = \mathbf{f}(\mathbf{x}(t), t),\ \ \mathbf{x}(t) \in \mathbb{R}^n,\ t \in \mathbb{R},\ \mathbf{f} = [f_1, f_2, \dots, f_n]^\top.$$

Special and important cases of the above systems are the following ones:

(1)

$$\mathbf{x}'(t) = A(t)\mathbf{x}(t) + \mathbf{b}(t),$$

where A is a real square matrix of order n, whose elements $a_{ij}(t)$ are functions of t, and $\mathbf{b}(t) \in \mathbb{R}^n$. These systems are said *systems of linear differential equations (of the first order).* The matrix A is the *coefficients matrix* of the system, whereas $\mathbf{b(t)}$ is usually called *forcing term* or also *right-hand side term.*

(2) *Homogeneous systems of linear differential equations:* in the above formulation we have $\mathbf{b}(t) = [0]$:

$$\mathbf{x}'(t) = A(t)\mathbf{x}(t).$$

(3) Systems of linear differential equations with *constant coefficients:*

$$\mathbf{x}'(t) = A\mathbf{x}(t) + \mathbf{b}(t)$$

(note that the forcing vector is not required to have constant components).

(4) Homogeneous systems of linear differential equations with constant coefficients:

$$\mathbf{x}'(t) = A\mathbf{x}(t).$$

In what follows we shall be concerned with some basic results on the existence and uniqueness of the solution $\mathbf{x}(t)$ of systems of type (6.8), solution which satisfies an initial condition of the type $\mathbf{x}(t_0) = \mathbf{x}^0 \in \mathbb{R}^n$. With reference to linear systems, with constant coefficients, we shall give the methods to obtain their general solution. In any case, the solution will be a vector function of the type $\mathbf{x} = \mathbf{x}(t, \mathbf{c})$, with $\mathbf{c} \in \mathbb{R}^n$.

6.2 Existence and Uniqueness of Solutions

We begin with a result which ensures the *local existence* of at least one solution $\mathbf{x}(t)$ for a system of type (6.8), with a given initial value condition.

Theorem 6.2.1 (Peano Theorem for Systems of First-Order Differential Equations in a Normal Form) *Consider the system*

$$\mathbf{x}'(t) = \mathbf{f}(\mathbf{x}(t), t)$$

where $f_i(x_1, x_2, \ldots, x_n, t), i = 1, \ldots, n$, are continuous functions on an open set $A \subset \mathbb{R}^{n+1}$. If $(\mathbf{x}^0, t_0)$ is an element of A, it is possible to determine a number $\delta > 0$ such that in the interval $(t_0 - \delta, t_0 + \delta)$ there exists at least one solution of the initial value problem (or Cauchy problem)

$$\begin{cases} \mathbf{x}'(t) = \mathbf{f}(\mathbf{x}(t), t), \\ \mathbf{x}(t_0) = \mathbf{x}^0. \end{cases}$$

For the proof of the previous theorem see, e.g., [10], [13]. It is possible to give an estimation of the width of the interval where the solution exists. Indeed, let $B_a(\mathbf{x}^0)$ and $B_b(t_0)$ be two neighbourhoods, respectively, of $\mathbf{x}^0$ and t_0, such that $K = \text{cl}(B_a(\mathbf{x}^0) \times B_b(t_0)) \subset A$, and let $M = \max_{(\mathbf{x},t)\in K} \|\mathbf{f}(\mathbf{x}(t), t)\|$. Then it holds $\delta = \min\left(a, \frac{b}{M}\right)$.

It is quite easy to check that the continuity assumption on $\mathbf{f}$ is not sufficient to ensure the *uniqueness* of the solution of the previous Cauchy problem. Consider the following classical counterexample:

$$x'(t) = 3\sqrt[3]{x^2(t)}$$

with initial condition $x(0) = 0$. It is evident that the constant $x(t) \equiv 0$ is a solution: $x'(t) = 0$. Then consider the function $x(t) = t^3$, from which $t = \sqrt[3]{x(t)}$. By deriving, we get

$$x'(t) = 3t^2 = 3\sqrt[3]{x^2(t)}$$

and hence also this function solves the differential equation, under the initial condition $x(0) = 0$.

It is possible, under further assumptions on $\mathbf{f}$, to obtain local existence and *uniqueness* of the solution of the initial value problem for the above systems.

Theorem 6.2.2 (Cauchy Theorem for Systems of First-Order Differential Equations in a Normal Form) *Let* $\mathbf{f}$ *satisfy the assumptions of Theorem 6.2.1 and moreover, assume that all partial derivatives of the first order of* $f_i, i = 1, \ldots, n$, *with respect to* $x_i, i = 1, \ldots, n$, *exist in the set* A *(that is every* $f_i, i = 1, \ldots, n$, *is of class* C^1 *on* A, *with respect to the variables* $x_i, i = 1, \ldots, n$*). Then there exists* $\delta > 0$ *such that the initial value problem*

$$\begin{cases} \mathbf{x}'(t) = \mathbf{f}(\mathbf{x}(t), t), \\ \mathbf{x}(t_0) = \mathbf{x}^0 \end{cases}$$

has a unique solution defined on the interval $(t_0 - \delta, t_0 + \delta)$.

Remark 6.2.3 The previous counterexample does not respect all assumptions of Theorem 6.2.2. Indeed, we have

$$\frac{\partial f}{\partial x} = 3 \cdot \frac{2}{3} x^{\frac{2}{3}-1} = 2 \cdot x^{-\frac{1}{3}} = \frac{2}{\sqrt[3]{x}},$$

which is not continuous at $x = 0$.

It is possible to prove that the (local) uniqueness of the solution of the Cauchy problem is ensured under weaker assumptions than the ones of Theorem 6.2.2, more precisely the function $\mathbf{f}$ must be *Lipschitz with respect to* $\mathbf{x}$, *uniformly with respect to*

t, on the open set $A \subset \mathbb{R}^{n+1}$, that is, there must exist a constant $K \geqq 0$ ("Lipschitz constant") such that

$$\left\| \mathbf{f}(\mathbf{x}^2, t) - \mathbf{f}(\mathbf{x}^1, t) \right\| \leqq K \left\| \mathbf{x}^2 - \mathbf{x}^1 \right\|, \quad \forall (\mathbf{x}^1, t),\ (\mathbf{x}^2, t) \in A.$$

In other words, the function $\mathbf{f}$ must have the following kind of difference quotient

$$\frac{\left\| \mathbf{f}(\mathbf{x}^2, t) - \mathbf{f}(\mathbf{x}^1, t) \right\|}{\left\| \mathbf{x}^2 - \mathbf{x}^1 \right\|}, \quad \mathbf{x}^2 \neq \mathbf{x}^1,$$

bounded on A by the constant K. (Rudolf Otto Lipschitz, 1832–1903, German mathematician). Many modern textbooks on Analysis or on Differential Equations follow this approach, but in the present introductory book we follow a more traditional approach.

Both Peano theorem and Cauchy theorem are so-called "local theorems", in the sense that they ensure the existence of a solution of the initial value problem for a system of type (6.8) in a neighbourhood of the given point $\mathbf{x}^0$. It is possible to obtain also "global theorems", where the initial value problem has a solution on a whole fixed interval; obviously this is possible under additional assumptions.

Theorem 6.2.4 (Global Existence and Uniqueness Theorem on Initial Value Conditions) *Let* $\mathbf{f}$ *be a vector function defined on the "stripe"* $S \times [a, b]$, *where it is continuous, together with all its partial derivatives* $\frac{\partial f_i}{\partial x_j}$. *If* $\mathbf{f}$ *or if every partial derivative* $\frac{\partial f_i}{\partial x_j}$ *is bounded on the whole stripe* S, *then the solution of the initial value problem*

$$\begin{cases} \mathbf{x}'(t) = \mathbf{f}(\mathbf{x}(t), t), \\ \mathbf{x}(t_0) = \mathbf{x}^0 \end{cases}$$

exists and is unique on the whole interval $[a, b]$.

In the *one-dimensional case,* if we consider the t variable on the abscissa axis of the point $P(t, x(t))$, where $x(t)$ is the solution of the differential equation $x'(t) = f(x(t), t)$, this point, by varying t, describes on the plane tOx a curve, said *integral line* or *integral curve.* Then the differential equation $x'(t) = f(x(t), t)$ says that at every point P of the said integral line, the tangent at P to the integral line, must have a slope given by $f(P)$. Hence the equation $x'(t) = f(x(t), t)$ specifies a set of "directions" in the various points of the plane tOx; it is also said that the equation specifies a "directions field". Always with reference to the one-dimensional case, the initial value problem consists therefore in imposing that at least one integral line passes through the given point $P(t_0, x_0)$.

We confirm once more that the results of the previous Theorems 6.2.1 and 6.2.2 are of a local nature, even if the assumptions of the theorems are satisfied on the whole $A = \mathbb{R}^{n+1}$. Indeed, consider the following example.

$$\begin{cases} x'(t) = x^2(t) \\ x(0) = 1. \end{cases}$$

The function $f(x) = x^2$ is continuously differentiable on the whole plane; moreover, being $x(0) = 1$, surely $x(t) \neq 0$ in a neighbourhood of the origin. Then we can divide both members by $x^2(t)$:

$$\frac{x'(t)}{x^2(t)} = 1,$$

and integrating, we have

$$\int \frac{x'(t)}{x^2(t)} dt = \int dt, \quad \text{that is} \quad -\frac{1}{x(t)} = t + c.$$

Therefore the general solution is

$$x(t) = -\frac{1}{t+c}, \quad c \in \mathbb{R}.$$

The condition $x(0) = 1$ is satisfied with $c = -1$ and hence the particular solution is

$$x(t) = \frac{1}{1-t},$$

which is not defined on the whole $\mathbb{R}$, but only on the interval $(-\infty, 1)$, which contains the initial value $t_0 = 0$.

6.3 Some First-Order Differential Equations

We consider in the present section some first-order differential equations which are often encountered in some applications and that can be "solvable", in the sense that we are able to express their solutions by means of suitable integration operations. However we observe that the are several efficient numerical methods for finding the solutions of differential equations, also of order $n > 1$ (and of systems of differential equations).

- **First-order linear equations.** First of all, in general, given the differential equation

$$x^{(n)}(t) = \varphi(x^{(n-1)}(t), \ldots, x'(t), x(t), t),$$

the same is *linear* if φ is a *first degree polynomial* with respect to $x, x', \ldots, x^{(n-1)}$. For instance,

$$x' = tx + 1; \;\; 3x'' - t^3x' - x = 1; \;\; x''' = (\log t)x - 3t^2 + 1$$

are all linear differential equations, whereas

$$x' = x^2; \;\; x'' = \log x; \;\; x''' = \sqrt{x'' + 4x' + 1}$$

are not linear differential equations.

A first-order linear differential equation has the form

$$x'(t) + a(t)x(t) = b(t),$$

where $a(t)$ and $b(t)$ are continuous functions on an interval $I \subset \mathbb{R}$. If $b(t) \equiv 0$, the equation is said to be *homogeneous,* otherwise it is said to be *non homogeneous* or *complete.* We begin with the first-order linear differential equations with *constant coefficients* and with forcing term constant:

$$x'(t) + ax(t) = b,$$

with $x', x, a, b \in \mathbb{R}$. If $a = 0$, we have the simple equation

$$x'(t) = b$$

whose obvious solution is

$$x(t) = bt + c, \quad c \in \mathbb{R}.$$

If $a \neq 0$, by multiplying both members of the equation by e^{at} ("integrating factor" for this kind of equations), we obtain

$$e^{at}x'(t) + ae^{at}x(t) = be^{at}.$$

Then we observe that the first member of the above equality is the derivative of $e^{at}x(t)$, that is

$$D\left[e^{at}x(t)\right] = be^{at},$$

and by integrating both members, we obtain

$$e^{at}x(t) = \int be^{at}dt.$$

that is

$$e^{at}x(t) = b\int e^{at}\,dt$$

from which

$$e^{at}x(t) = \frac{b}{a}e^{at} + c.$$

Hence the general solution is

$$x(t) = \frac{b}{a} + ce^{-at}, \quad c \in \mathbb{R}.$$

Example 6.3.1 Solve the equation

$$x'(t) - 3x(t) = 2.$$

Being $a = -3$, $b = 2$, we have

$$x(t) = -\frac{2}{3} + ce^{3t}, \quad c \in \mathbb{R}.$$

Note that if $b = 0$, we obtain the general solution

$$x(t) = ce^{-at}, \quad c \in \mathbb{R}.$$

Obviously, if instead of considering

$$x'(t) + ax(t) = 0,$$

we write the same equation in a normal form, that is

$$x'(t) = ax(t)$$

we obtain the solution (previously already obtained)

$$x(t) = ce^{at}, \; c \in \mathbb{R}.$$

Now we come back to our starting point, i.e. we take into consideration the first-order *linear differential equation with variable coefficients:*

$$x'(t) + a(t)x(t) = b(t),$$

where $a(t)$ and $b(t)$ are continuous functions on an interval $[a, b]$. Multiply both members by the following integrating factor:

$$I(t) = e^{\int a(t)dt}.$$

We obtain

$$x'(t)I(t) + a(t)x(t)I(t) = b(t)I(t). \tag{6.13}$$

We observe that it holds

$$I'(t) = e^{\int a(t)dt} \cdot a(t) = I(t) \cdot a(t);$$

$$D[x(t)I(t)] = x'(t)I(t) + x(t)I'(t) = x'(t)I(t) + x(t)I(t)a(t),$$

and that this last expression is the first member of (6.13). We have therefore

$$D[x(t)I(t)] = b(t)I(t)$$

from which, by integrating both members,

$$x(t)I(t) = \int b(t)I(t)dt.$$

We obtain

$$x(t) = [I(t)]^{-1} \int I(t)b(t)dt,$$

from which we obtain the general solution

$$x(t) = e^{-\int a(t)dt} \left(\int e^{\int a(t)dt} \cdot b(t)dt \right).$$

Example 6.3.2 Solve the equation

$$x'(t) + 2tx(t) = 5t.$$

Being $a(t) = 2t, b(t) = 5t$, it results $I(t) = e^{\int 2tdt} = e^{t^2}$, by omitting the arbitrary constant (which will appear in the general solution). Hence

$$\begin{aligned} x(t) &= e^{-t^2} \left(\int 5te^{t^2} dt \right) = e^{-t^2} \left(\frac{5}{2} \int 2te^{t^2} dt \right) \\ &= e^{-t^2} \left(\frac{5}{2} e^{t^2} + c \right) = \frac{5}{2} + ce^{-t^2}, \quad c \in \mathbb{R}. \end{aligned}$$

If we have an initial value problem $x(t_0) = x_0, t_0 \in [a, b]$, it is possible to determine the constant c such that the initial condition is satisfied, otherwise to make reference to the following formula:

$$x(t) = e^{-\int_{t_0}^{t} a(s)ds}\left(x_0 + \int_{t_0}^{t} b(u)e^{\int_{t_0}^{u} a(s)ds} du\right)$$

which may be useful when we are not able to compute the indefinite integrals. Moreover, note that, being $a(t)$ and $b(t)$ continuous on $[a, b]$, the initial value problem $x(t_0) = x_0, t_0 \in [a, b]$, has a unique solution on the whole interval $[a, b]$.

Example 6.3.3 Solve the initial value problem

$$\begin{cases} x'(t) = \frac{1}{t}x(t) + t, \ t > 0 \\ x(1) = 0. \end{cases}$$

From $x'(t) - \frac{1}{t}x(t) = t$, we have

$$x(t) = e^{-\int -\frac{1}{t}dt}\left(\int te^{\int -\frac{1}{t}} dt\right),$$

that is

$$x(t) = e^{\int \frac{1}{t}dt}\left(\int te^{-\log t} dt\right).$$

We note that $-\log t = \log 1 - \log t = \log \frac{1}{t}$, hence

$$x(t) = e^{\log t}\left(\int t \cdot \frac{1}{t} dt\right),$$

that is

$$x(t) = t(t + c) = t^2 + ct.$$

From $x(1) = 0$, we have $1 + c = 0$, that is $c = -1$, therefore the solution of the initial value problem considered, is $x(t) = t^2 - t$.

Note that if $b(t) = 0$, the general solution of the equation into consideration is

$$x(t) = ce^{-\int a(t)dt},$$

being of course $\int 0dt = c$. If, in the present case, $a(t) = a$ (constant) we find again the solution

$$x(t) = ce^{-at}, \ c \in \mathbb{R}.$$

- **Equations with separable variables.**

These equations have the form

$$x'(t) = b(t) \cdot a(x)$$

where the second member is the product (or also the quotient) of two continuous functions, one depending only on t and the other one depending only on x. Without pretensions to be rigorous from a logical-mathematical point of view, we adopt the following resolution procedure.

(A) We rewrite the equation in the form

$$\frac{dx}{dt} = b(t) \cdot a(x).$$

(B) We "separate" the variables, by assuming $a(x) \neq 0, \forall x$:

$$\frac{dx}{a(x)} = b(t)dt.$$

(C) We integrate both members, the first one with respect to x, the second one with respect to t. In this way we obtain the solution of the equation.

(D) If $a(x)$ vanishes at $x = a$, we have to add to the previous solutions, also the solution $x(t) = a$.

Example 6.3.4

(a) Solve the differential equation

$$x'(t) = \frac{2t}{x^2}.$$

(b) Solve the differential equation

$$x'(t) = -2tx^2.$$

(c) Solve the differential equation

$$x'(t) = t\log(1 + t^2).$$

(a)

$$\frac{dx}{dt} = 2t \cdot \frac{1}{x^2} \quad (x \neq 0).$$

Hence $b(t) = 2t$, $a(x) = \frac{1}{x^2}$, therefore $a(x) \neq 0$, $\forall x \neq 0$. Then we have

$$x^2 dx = 2t dt,$$

that is

$$\frac{x^3}{3} = t^2 + c \iff x^3 = 3t^2 + c$$

(obviously it makes no sense to write $3c$, being c arbitrary). The general solution is therefore

$$x(t) = \sqrt[3]{3t^2 + c}.$$

(b)

$$\frac{dx}{dt} = -2t \cdot x^2.$$

With $x^2 \neq 0$, i.e. $x \neq 0$, we have

$$\frac{1}{x^2} dx = -2t dt$$

and, by integrating both members,

$$\int \frac{1}{x^2} dx = -2 \int t dt,$$

that is

$$\frac{1}{x} = t^2 + c,$$

that is

$$x(t) = \frac{1}{t^2 + c}.$$

But as x^2 vanishes at $x = 0$, the general solution is given by

$$x(t) = \begin{cases} \frac{1}{t^2+c}, & \text{for } x \neq 0, \\ 0, & \text{for } x = 0. \end{cases}$$

(c)

$$x'(t) = t \log(1 + t^2).$$

As the function at the second member of the equation does not contain the variable x, this is a problem of determination of the primitives of this function.

$$\int x'(t)dt = \int t \log(1 + t^2)dt.$$

We have, by putting $1 + t^2 = z$, $dz = 2tdt$, and hence

$$\int t \log(1 + t^2)dt = \frac{1}{2} \int \log z dz$$

and then

$$\frac{1}{2} \int \log z dz = \frac{1}{2}(z \log z - z) + c.$$

Therefore

$$x(t) = \frac{1}{2} \left[(1 + t^2) \log(1 + t^2) - 1 - t^2 \right] + c,$$

that is

$$x(t) = \frac{1}{2} \left[(1 + t^2)(\log(1 + t^2) - 1) \right] + c.$$

Sometimes it is not possible to express $x(t)$ as an explicit function of t.

Example 6.3.5 Solve the differential equation

$$x'(t) = \frac{t^3}{x^6 + 1}.$$

It results $(1/(x^6 + 1)) \neq 0$, $\forall x$. We have

$$\frac{dx}{dt} = \frac{t^3}{x^6 + 1}.$$

Now we separate the variables:

$$(x^6 + 1)dx = t^3 dt.$$

By integrating both members, we get

$$\int (x^6 + 1)dx = \int t^3 dt$$

from which

$$\frac{1}{7}x^7 + x = \frac{1}{4}t^4 + c.$$

We can only say that the solutions of the differential equation are all functions $x(t)$ which satisfy the last written relation.

Now we propose, as an application of differential equations with separable variables, the continuous *logistic model* of Verhulst. This demographic model was proposed by P. F. Verhulst (1804–1849) in 1845. It is a model which studies the growth of an isolated population, where $x(t)$ denotes the number of individuals present at instant t.

Malthus in 1798 had proposed a simpler model concerning the dynamics of a population; this rather rough model was constructed under the assumption that the growth rate is proportional to the consistency of the population, hence with the assumption that space and food are available in an unlimited quantity:

$$x'(t) = ax(t).$$

The Malthus model is therefore expressed by a linear differential equation, whose solution, as previously seen, describes an exponential growth: $x(t) = x(0)e^{at}$. In a more realistic way, it is natural to assume that the more a population is numerous, the more the resources become scarce, and this entails a reduction of the growth rate. In the model proposed by Verhulst, space and food are limited and the population cannot overcome a saturation level denoted by M.

The Verhulst model is described by the following differential equations with separable variables:

$$x'(t) = ax(t)\left[M - x(t)\right],$$

with $a > 0, M > 0, (M - x(t)) > 0, x(t) > 0, \forall t$. We rewrite this differential equation in the form

$$\frac{dx}{dt} = ax(t)(M - x(t)),$$

that is, by separating the variables,

$$\frac{dx}{x(M - x)} = adt,$$

from which

$$\int \frac{dx}{x(M-x)} = \int a dt$$

and hence

$$\frac{1}{M} \log \frac{x}{M-x} = at + c,$$

from which we have

$$x(t) = \frac{kMe^{Mat}}{1+ke^{Mat}}, \quad \forall k \in \mathbb{R}.$$

For the reader's convenience we explicit the various mathematical steps. The fraction

$$\frac{1}{x(M-x)}$$

can be decomposed into the sum of two fractions of the type

$$\frac{A}{x} + \frac{B}{M-x},$$

with A and B constants to be determined. We impose that identically it must be

$$\frac{A}{x} + \frac{B}{M-x} = \frac{1}{x(M-x)},$$

that is

$$A(M-x) + Bx = 1,$$

that is

$$AM + (-A+B)x = 1,$$

from which

$$\begin{cases} -A+B=0 \\ AM=1. \end{cases}$$

We obtain $A = \frac{1}{M}$, $B = \frac{1}{M}$. Hence

$$\int \frac{dx}{x(M-x)} = \int \frac{dx}{Mx} + \int \frac{dx}{M(M-x)}$$
$$= \frac{1}{M}\log x - \frac{1}{M}\log(M-x) = \frac{1}{M}\log\left(\frac{x}{M-x}\right) + c.$$

And now:

$$\frac{1}{M}\log\frac{x}{M-x} = at + c$$

from which

$$\frac{x}{M-x} = e^{Mat+k}, \quad \text{with } k = Mc,$$

that is

$$\frac{x}{M-x} = Ke^{Mat}, \quad \text{with } K = e^k.$$

Finally we have

$$x(t) = \frac{KMe^{Mat}}{1+Ke^{Mat}}.$$

By multiplying numerator and denominator by e^{-Mat},

$$x(t) = \frac{KM}{e^{-Mat}+K}$$

and it holds

$$\lim_{t\longrightarrow+\infty} x(t) = M,$$

that is the population does not exceed the maximum given level. Figure 6.1 shows the classical "logistic function", which is a typical "S-shaped curve".

- **Bernoulli equations.**

These equations are quite important in several economic growth models, such as a model due to R. Solow (see. e.g., [11], [14] and Example 6.3.6,b) and are of the type

$$x'(t) + p(t)x = q(t)x^\alpha, \quad \alpha \neq 0, \ \alpha \neq 1,$$

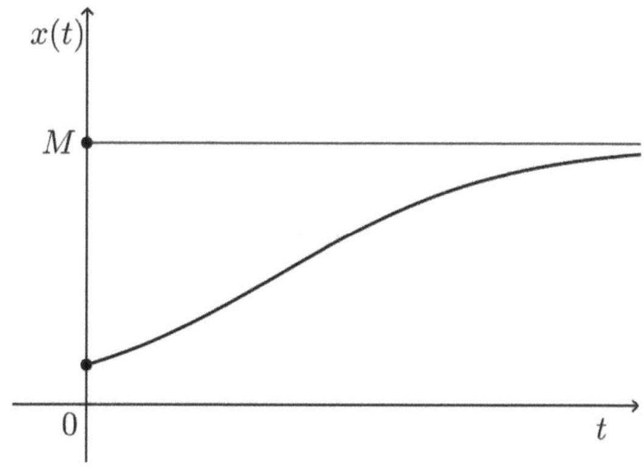

Fig. 6.1 Graph of the logistic function

where $p(t)$ and $q(t)$ are continuous functions on an interval $I \subset \mathbb{R}$ and α is different from one and zero, otherwise we come back to the case of linear equations. Bernoulli equations can be transformed into a linear differential equation; first of all, we remark that if $\alpha > 0$, the function $x(t) \equiv 0$ is a particular solution of the equation. For the moment we exclude this case and divide both members of the equation by x^{α}:

$$x^{-\alpha}x' + p(t)x^{1-\alpha} = q(t).$$

Then we put

$$u = x^{1-\alpha},$$

and being

$$u' = (1-\alpha)x^{-\alpha}x',$$

we have

$$u' + (1-\alpha)p(t)u = q(t)(1-\alpha)$$

which is a linear differential equation. If $u(t)$ is the general solution of this equation, then

$$x(t) = [u(t)]^{\frac{1}{1-\alpha}}$$

is the general solution of the Bernoulli equation. The solution $x(t) = [u(t)]^{\frac{1}{1-\alpha}}$ is to be considered in an algebraic sense; if, for example, $\alpha = -1$, we find $x = \pm\sqrt{u}$. If $\alpha > 0$, we have to take into consideration also the solution $x = 0$.

Example 6.3.6

(a) Solve the equation of Bernoulli

$$x' + tx = t^3 x^3.$$

We divide both members by $x^3 \neq 0$:

$$x^{-3}x' + tx^{-2} = t^3.$$

We put $u = x^{-2}$ and we obtain $u' = -2x^{-3}x'$. By substituting, we obtain

$$u' - 2tu = -2t^3$$

whose solution is

$$u = t^2 + 1 + ce^{t^2}.$$

The general solution of the given equation is therefore

$$x^{-2} = t^2 + 1 + ce^{t^2}, \text{ that is } x = \frac{1}{\sqrt{t^2 + 1 + ce^{t^2}}},$$

to which we have to add the solution $x = 0$, being $\alpha > 0$.

(b) (Sketch of the Solow model). Let $Y = F(K, L)$ be a production function with constant returns to scale, where the production factors are the capital K and the labour L (supplied exclusively by workers). Suppose that $K'(t) = sY(t)$, $L'(t) = \lambda L(t)$, with $s, \lambda > 0$. Consider the per capita production and the per capita capital, $y = Y/L$ and $k = K/L$. Taking into account the assumptions on the production function, we have

$$y = \frac{Y}{L} = \frac{F(K, L)}{L} = F(k, 1) = f(k).$$

Moreover,

$$k'(t) = \frac{K'(t)L(t) - K(L)L'(t)}{L^2(t)} =$$

$$= \frac{K'(t)}{L(t)} - \frac{K(t)}{L(t)}\frac{L'(t)}{L(t)} = sf(k(t)) - \lambda k(t).$$

If Y is a Cobb-Douglas function of the type $Y = AK^{1-\alpha}L^{\alpha}$, with $0 < \alpha < 1$, we obtain

$$y = f(k) = \frac{Y}{L} = \frac{AK^{1-\alpha}L^{\alpha}}{L} = Ak^{1-\alpha}$$

and from the differential equation $k'(t) = sf(k(t)) - \lambda k(t)$ we obtain a Bernoulli equation:

$$k'(t) = -\lambda k(t) + sAk^{1-\alpha}, \text{ with } 0 < (\alpha - 1) < 1,$$

which has the solution

$$k(t) = \left(ce^{-\lambda\alpha t} + \frac{sA}{\lambda}\right)^{\frac{1}{\alpha}} \vee k(t) = 0.$$

- **Riccati equations.**

  The *Riccati equations* have the form

$$x'(t) = p(t)x(t) + q(t)x^2(t) + r(t)$$

where p, q and r are continuous functions on an interval $I \subset \mathbb{R}$, and where $r(t) \neq 0$, otherwise we get a Bernoulli equation. If $t_0 \in I$ we have existence and uniqueness of the Cauchy problem. *There are no general methods* to solve a Riccati equation; however, if we know a particular solution $\phi(t)$ of the same, it is possible to make the following change of the variable:

$$x(t) = \phi(t) + \frac{1}{z(t)}$$

from which

$$x'(t) = \phi'(t) - \frac{z'(t)}{z^2(t)}.$$

By substituting into the original equation we get

$$\phi'(t) - \frac{z'(t)}{z^2(t)} = p(t)\left[\phi(t) + \frac{1}{z(t)}\right] + q(t)\left[\phi(t) + \frac{1}{z(t)}\right]^2 + r(t).$$

As $\phi(t)$ is a particular solution, we get

$$-\frac{z'(t)}{z^2(t)} = p(t)\frac{1}{z(t)} + q(t)\frac{1}{z^2(t)} + 2q(t)\frac{\phi(t)}{z(t)}$$

from which we obtain the linear equation in $z(t)$:

$$z'(t) = [-p(t) + 2q(t)\phi(t)]\, z(t) - q(t).$$

From the last equation it is possible to obtain the solution $x(t)$ of the original equation.

Example 6.3.7 Consider the Riccati equation

$$x'(t) = -2(t+1)x^2(t) - 2(2t+3)x(t) - 2t - 4.$$

It is seen that $\phi(t) = -1$ is a particular solution. Then we put

$$x(t) = -1 + \frac{1}{z(t)},$$

from which

$$x'(t) = -\frac{1}{z^2(t)} z'(t).$$

By substituting into the equation we have therefore

$$-\frac{1}{z^2(t)} z'(t) = -2(t+1)\left[-1 + \frac{1}{z(t)}\right]^2 - 2(2t+3)\left[-1 + \frac{1}{z(t)}\right] - 2t - 4.$$

Then we obtain

$$-\frac{1}{z^2(t)} z'(t) = [-2(t+1) + 2(2t+3) - 2t - 4] - 2(t+1)\cdot$$

$$\cdot\left[\frac{1}{z^2(t)} - \frac{2}{z(t)}\right] - \frac{2(2t+3)}{z(t)}.$$

As $\phi(t) = -1$ is a particular solution, the expression within the first brackets is equal to zero. Then, by multiplying by $z^2(t)$ and changing the sign, we arrive to the original equation in $z(t)$:

$$z'(t) = 2z(t) + 2(t+1)$$

whose general solution is

$$z(t) = ce^{2t} - \frac{3}{2}t - 1.$$

Finally, we have

$$x(t) = -1 + \frac{1}{ce^{2t} - \frac{3}{2}t - 1}.$$

6.4 First-Order Linear Systems of Differential Equations

As previously said, *first-order linear systems of differential equations* are systems of the type

$$\mathbf{x}'(t) = A(t)\mathbf{x}(t) + \mathbf{b}(t), \tag{6.14}$$

where $\mathbf{x}'(t), \mathbf{x}(t), \mathbf{b}(t) \in \mathbb{R}^n$, and $A(t)$ is a square matrix of order n. The matrix $A(t)$ is called the *coefficient matrix of the system,* whereas $\mathbf{b}(t)$ is the *forcing term* or *forcing vector.*

Example 6.4.1 The system

$$\begin{cases} x_1'(t) = 3tx_1(t) - x_2(t) + e^{-2t} \\ x_2'(t) = 4t^2 x_1(t) + 5x_2(t) + 4te^{-2t} \end{cases}$$

is a linear system (of differential equations), where

$$A(t) = \begin{bmatrix} 3t & -1 \\ 4t^2 & 5 \end{bmatrix}; \quad \mathbf{b}(t) = \begin{bmatrix} e^{-2t} \\ 4te^{-2t} \end{bmatrix}.$$

If $\mathbf{b}(t) = [0]$, the system is *homogeneous,* otherwise it is called *non homogeneous* (or *complete*). If $A(t) = A$, the system is with *constant coefficients.*

The particular form of systems described in (6.14) generates interesting and important results for what concerns existence, uniqueness, and structure of solutions. For what concerns existence and uniqueness, we can state the following result on the "global" validity of a related initial value problem (or Cauchy problem), similarly to what asserted for first-order differential linear equations.

Theorem 6.4.2 *If in* (6.14) *the matrix $A(t)$ and the vector $\mathbf{b}(t)$ are continuous on $[a, b]$ (that is, every element $a_{ij}(t)$ of $A(t)$ and every element $b_j(t)$ of $\mathbf{b}(t)$ is a continuous function of the variable t), then there exists on the whole interval $[a, b]$ a unique solution of* (6.14) *which verifies the initial value condition $\mathbf{x}(t_0) = \mathbf{x}^0$, for each choice of $t_0 \in [a, b]$.*

Proof The result is a quite immediate consequence of Theorem 6.2.4, being $\mathbf{f}(\mathbf{x},t) = A(t)\mathbf{x}(t) + \mathbf{b}(t)$ and hence

$$f_i(\mathbf{x}, t) = \sum_{j=1}^{n} a_{ij}(t)x_j(t).$$

We have therefore $\frac{\partial f_i}{\partial x_j} = a_{ij}(t)$. Therefore the assumptions of the said theorem hold on the "stripe" $\mathbb{R}^n \times [a, b]$ and the thesis follows. □

It is worth remarking that the previous result holds also if $A(t)$ and $\mathbf{b}(t)$ are continuous on any interval $I \subset \mathbb{R}$, not necessarily compact. In particular it holds (everywhere) when a_{ij} and b_j are *constant.*

For what concerns the structure of the solutions of (6.14), we have the important *superposition principle*, wholly similar to the one described in Chap. 1, with reference to linear systems (systems of linear equations).

Theorem 6.4.3 (Superposition Principle) *The general solution of system* (6.14) *is given by*

$$\mathbf{x}(t) = \mathbf{z}(t) + \mathbf{x}^0(t),$$

where $\mathbf{z}(t)$ *is the general solution of the homogeneous system associated with* (6.14), *i.e.*

$$\mathbf{z}'(t) = A(t)\mathbf{z}(t),$$

and where $\mathbf{x}^0(t)$ *is any particular solution of* (6.14), *usually obtained from an initial value problem.*

Proof

(1) We begin to prove that any function of the type

$$\mathbf{x}(t) = \hat{\mathbf{x}}(t) + \mathbf{z}(t),$$

where $\hat{\mathbf{x}}(t)$ is a particular solution of (6.14) and $\mathbf{z}(t)$ is the general solution of the homogeneous associated system, is a solution of (6.14). By substituting, we get:

(A) $\left[\hat{\mathbf{x}}(t) + \mathbf{z}(t)\right]' = \hat{\mathbf{x}}(t)' + \mathbf{z}(t)'$;
(B) $A(t)\left[\hat{\mathbf{x}}(t) + \mathbf{z}(t)\right] + \mathbf{b}(t) = A(t)\hat{\mathbf{x}}(t) + A(t)\mathbf{z}(t) + \mathbf{b}(t)$.

But $\mathbf{z}'(t) = A(t)\mathbf{z}(t)$ and hence (A) becomes

$$\left[\hat{\mathbf{x}}(t) + \mathbf{z}(t)\right]' = \hat{\mathbf{x}}(t)' + A(t)\mathbf{z}(t).$$

It results then (A) = (B), because

$$\hat{\mathbf{x}}(t)' + A(t)\mathbf{z}(t) = A(t)\hat{\mathbf{x}}(t) + A(t)\mathbf{z}(t) + \mathbf{b}(t),$$

that is

$$\hat{\mathbf{x}}'(t) = A(t)\hat{\mathbf{x}}(t) + \mathbf{b}(t),$$

where the equality is verified, being $\hat{\mathbf{x}}(t)$ a particular solution of (6.14).

(2) Vice-versa, we now prove that *all* solutions of (6.14) are of the type

$$\mathbf{x}(t) = \mathbf{z}(t) + \mathbf{x}^0(t).$$

It is sufficient to show that the difference of any two solutions of the said type is a solution of the associated homogeneous system

$$\mathbf{z}'(t) = A(t)\mathbf{z}(t).$$

Therefore consider two solutions $\hat{\mathbf{x}}(t)$ and $\bar{\mathbf{x}}(t)$ of system (6.14):

$$\begin{cases} \hat{\mathbf{x}}'(t) = A\mathbf{x}(t)\hat{} + b(t) \\ \bar{\mathbf{x}}'(t) = A\bar{\mathbf{x}}(t) + \mathbf{b}(t). \end{cases}$$

By subtracting member to member we have

$$\left[\hat{\mathbf{x}}(t) - \bar{\mathbf{x}}(t)\right]' = A\left[\hat{\mathbf{x}}(t) - \bar{\mathbf{x}}(t)\right].$$

By putting $\mathbf{z}(t) = \hat{\mathbf{x}}(t) - \bar{\mathbf{x}}(t)$, we have the conclusion. □

Let us verify the previous result by an example. Consider the following one-dimensional system (i.e. a differential equation)

$$x'(t) + ax(t) = b, \; a, b \neq 0.$$

We have seen that its general solution is

$$x(t) = \frac{b}{a} + ce^{-at}, \; c \in \mathbb{R},$$

and that if $b = 0$ (homogeneous case), we have

$$x(t) = ce^{-at}, \quad c \in \mathbb{R}.$$

We rewrite the equation in its normal form

$$x'(t) = ax(t) + b, \quad a, b \neq 0. \tag{6.15}$$

Its general solution is

$$x(t) = -\frac{b}{a} + ce^{at}, \; c \in \mathbb{R},$$

and if $b = 0$ (homogeneous case), we have $x(t) = ce^{at}$. Now we verify the above results by means of the superposition principle. First of all, we try with a constant solution of (6.15):

$$x_0(t) = k, \; k \in \mathbb{R}.$$

We must have $(x_0(t))' = 0$, from which

$$0 = ak + b, \text{ that is } k = -\frac{b}{a}.$$

We have therefore found a particular solution of (6.15), i.e.

$$x_0(t) = -\frac{b}{a}.$$

The associated homogeneous equation is

$$z'(t) = az(t),$$

whose general solution is

$$z(t) = ce^{at}.$$

Applying the superposition principle we find again the general solution of (6.15):

$$x(t) = -\frac{b}{a} + ce^{at}, \; c \in \mathbb{R}.$$

The simple result concerning the superposition principle suggests a "line of attack" for the determination of the general solution of the system

$$\mathbf{x}'(t) = A\mathbf{x}(t) + \mathbf{b}(t).$$

We could proceed as follows.

- To look for all solutions of the associated homogeneous system $\mathbf{x}'(t) = A(t)\mathbf{x}(t)$.
- To look for a particular solution of the system $\mathbf{x}'(t) = A(t)\mathbf{x}(t) + \mathbf{b}(t)$, by trial-and-error or by means of the method of "variations of the arbitrary constants" (see further).

  Thus we have determined the general solution of the system $\mathbf{x}'(t) = A(t)\mathbf{x}(t) + \mathbf{b}(t)$. However we have to note that, except the case of particular systems (e.g. the case $A(t) = A$, with A a constant matrix), we encounter several difficulties, mainly in the second one of the two above steps. Therefore we pay therefore our attention in the problem of identifying all solutions of the associated homogeneous system

$$\mathbf{x}'(t) = A(t)\mathbf{x}(t).$$

First of all, we remark that the concepts of linear independence and linear dependence, recalled in Chap. 1 with reference to vectors of $\mathbb{R}^n$, can be extended to elements of any linear space S. For example, if we have n vector functions $\mathbf{x}^1(t), \mathbf{x}^2(t), \ldots, \mathbf{x}^n(t)$, defined on an interval $I \subset \mathbb{R}$ and taking values in $\mathbb{R}^n$, they are *linearly dependent on I* if there exist multipliers $c_i, i = 1, \ldots, n$, *not all zero,* such that

$$\sum_{i=1}^{n} c_i \mathbf{x}^i(t) = [0], \ \forall t \in I.$$

Otherwise the said functions are *linearly independent on I.*

The following result is fundamental.

Theorem 6.4.4 *Consider the following homogeneous system of first order n equations with n variables:*

$$\mathbf{x}'(t) = A(t)\mathbf{x}(t), \tag{6.16}$$

where the elements of $A(t)$ are continuous functions on an interval $I \subset \mathbb{R}$. Then the set S of all its solutions (i.e. its general solution) is a linear space of dimension n.

Proof It is clear that if c_1 and c_2 are two arbitrary constants and $\mathbf{x}^1(t), \mathbf{x}^2(t)$ are two solutions of the system $\mathbf{x}'(t) = A(t)\mathbf{x}(t)$, also $c_1\mathbf{x}^1(t) + c_2\mathbf{x}^2(t)$ is a solution of the system, therefore the set S of all solutions is a linear space. Now let us prove that the dimension of S is just n, i.e. there exist in this space n and only n linearly independent solutions. Let us consider the n standard vectors of $\mathbb{R}^n$: $\mathbf{e}^1, \mathbf{e}^2, \ldots, \mathbf{e}^n$ (we recall that they are linearly independent and form a basis for $\mathbb{R}^n$). Let us consider $\mathbf{x}^1(t), \mathbf{x}^2(t), \ldots, \mathbf{x}^n(t)$, solutions of the Cauchy problems $\mathbf{x}^1(t_0) = \mathbf{e}^1, \mathbf{x}^2(t_0) = \mathbf{e}^2, \ldots, \mathbf{x}^n(t_0) = \mathbf{e}^n$, with $t_0 \in I = [a, b]$. Since such solutions are linearly independent at $t = t_0$, they are linearly independent on the whole I. Indeed, if they were dependent on I, there would exist n scalars $c_1, c_2, \ldots, c_n$, not all zero, such that

$$c_1\mathbf{x}^1(t) + c_2\mathbf{x}^2(t) + \cdots + c_n\mathbf{x}^n(t) = [0]$$

in I, and hence also at $t = t_0$, in contradiction with the assumptions. The dimension of S is therefore greater or equal to n. In order to prove that it is exactly n, we show that every other solution $\mathbf{x}(t)$ of the system can be expressed as a linear combination of the previous considered solutions $\mathbf{x}^1(t), \mathbf{x}^2(t), \ldots, \mathbf{x}^n(t)$. Indeed, let $\mathbf{x}(t)$ be a solution, with $\mathbf{x}(t_0) = \mathbf{x}^0$. Since the vectors $\mathbf{e}^1, \mathbf{e}^2, \ldots, \mathbf{e}^n$ are a basis for $\mathbb{R}^n$, it is possible to write

$$\mathbf{x}^0 = \lambda_1\mathbf{e}^1 + \lambda_2\mathbf{e}^2 + \cdots + \lambda_n\mathbf{e}^n,$$

with $\lambda_1, \lambda_2, \ldots, \lambda_n$ suitable scalars. Define now

$$\boldsymbol{\psi}(t) = \sum_{j=1}^{n} \lambda_j \mathbf{x}^j(t).$$

Since $\mathbf{x}(t_0) = \mathbf{x}^0 = \boldsymbol{\psi}(t_0)$, the vector functions $\mathbf{x}(t)$ and $\boldsymbol{\psi}(t)$ are solutions of the same Cauchy problem and therefore they must coincide, thanks to the uniqueness asserted in Theorem 6.4.2. But then we have

$$\mathbf{x}(t) = \sum_{j=1}^{n} \lambda_j \mathbf{x}^j(t).$$

□

From Theorem 6.4.4 we have the following important consequence.

Corollary 6.4.5 *Let* $\mathbf{x}^1(t), \mathbf{x}^2(t), \ldots, \mathbf{x}^n(t)$ *be* n *linearly independent solutions of the homogeneous system* (6.16). *Then the general solution of* (6.16) *is given by*

$$\mathbf{x}(t) = \sum_{i=1}^{n} c_i \mathbf{x}^i(t), \ c_i \in \mathbb{R},$$

where $c_1, c_2, \ldots, c_n$ *are arbitrary real constants (or also arbitrary complex constants, if we wish to make this choice).*

If we have n linearly independent solutions of system (6.16), we say that they form a *fundamental system of solutions.* Note that the zero vector $[0]$ is a solution of (6.16), but obviously it is not an element of the fundamental system. In general, it is not a trivial problem the one of finding a fundamental system of solutions for (6.16), however we shall show that in the particular case of A constant, this problem can be solved.

Now we see how it is possible to establish if n solutions $\mathbf{x}^1(t), \mathbf{x}^2(t), \ldots, \mathbf{x}^n(t)$ of (6.16) are or not linearly independent on the interval $I \subset \mathbb{R}$. We introduce the matrix

$$W(t) = \left[\mathbf{x}^1(t), \mathbf{x}^2(t), \ldots, \mathbf{x}^n(t)\right]$$

obtained putting together the column vectors $\mathbf{x}^i(t), i = 1, \ldots, n$. This matrix is called *Wronskian matrix* (from the Polish mathematician J. Wronski). The determinant of the Wronskian matrix is simply called *Wronskian*. The following useful proposition can be proved.

- If $\det(W(t_0)) \neq 0$ at a point $t_0 \in I$, then the vectors $\mathbf{x}^1(t), \mathbf{x}^2(t), \ldots, \mathbf{x}^n(t)$ are linearly independent on I and hence they are a fundamental system of solutions for (6.16). Moreover, it holds $\det(W(t_0)) \neq 0$ on the whole I.

Hence, under our assumptions, it results that either $\det(W(t)) = 0$ on the whole interval I or $\det(W(t)) \neq 0$ on the whole interval I. In this last case the Wronskian matrix is also said to be *fundamental* for system (6.16). Finally, note that, if we put

$$\mathbf{c} = \begin{bmatrix} c_1 \\ c_2 \\ \vdots \\ c_2 \end{bmatrix},$$

the general solution of the homogeneous system (6.16), with $\det(W(t)) \neq 0$, can be written in the compact form

$$\mathbf{x}(t) = W(t) \cdot \mathbf{c}, \ \ \mathbf{c} \in \mathbb{R}^n.$$

If we have a fundamental system of solutions of the homogeneous system (6.16), by means of the superposition principle, it is possible to obtain the general solution of system (6.14). Another form to express the general solution of (6.14) is given by the so-called "method of variations of parameters", due to J. L. Lagrange. We have observed that if the set $\mathbf{x}^1(t), \mathbf{x}^2(t), \ldots, \mathbf{x}^n(t)$ is a fundamental system for the homogeneous system (6.16), then the general solution of the same system can be written in the form

$$\mathbf{x}(t) = W(t) \cdot \mathbf{c},$$

with $\mathbf{c} = [c_1, c_2, \ldots, c_n]^\top$. Now we are looking for the general solution of (6.14) in the form

$$\mathbf{x}(t) = W(t) \cdot \mathbf{c}(t),$$

with $\mathbf{c}(t)$ differentiable vector function, to be determined. By substituting this expression into the original system, we have

$$W'(t)\mathbf{c}(t) + W(t)\mathbf{c}'(t) = A(t)W(t)\mathbf{c}(t) + \mathbf{b}(t).$$

But, being $W'(t) = A(t)W(t)$, we have

$$A(t)W(t)\mathbf{c}(t) + W(t)\mathbf{c}'(t) = A(t)W(t)\mathbf{c}(t) + \mathbf{b}(t),$$

that is

$$W(t)\mathbf{c}'(t) = \mathbf{b}(t)$$

from which

$$\mathbf{c}'(t) = W^{-1}(t)\mathbf{b}(t)$$

and hence

$$\mathbf{c}(t) = \int W^{-1}(t)\mathbf{b}(t)dt,$$

where the indefinite integral is determined by means of the antiderivatives of each component of the vector function $W^{-1}(t)\mathbf{b}(t)$. We conclude that the general solution of (6.14) is given by

$$\mathbf{x}(t) = W(t)\int W^{-1}(t)\mathbf{b}(t)dt.$$

If we have an initial value problem (or Cauchy problem) of the type

$$\begin{cases} \mathbf{x}'(t) = A(t)\mathbf{x}(t) + \mathbf{b}(t), \\ \mathbf{x}(t_0) = \mathbf{x}^0, \ t_0 \in I \subset \mathbb{R}, \end{cases}$$

we can write the expression of $\mathbf{c}(t)$ in the equivalent form

$$\mathbf{c}(t) = \mathbf{c} + \int_{t_0}^{t} W^{-1}(s)\mathbf{b}(s)ds,$$

with $\mathbf{c} \in \mathbb{R}^n$ arbitrary. The general solution of (6.14) which satisfies the above initial value condition, becomes then

$$\mathbf{x}(t) = W(t)\mathbf{c} + W(t)\int_{t_0}^{t} W^{-1}(s)\mathbf{b}(s)ds.$$

Putting $t = t_0$, we have

$$\mathbf{x}(t_0) = W(t_0)\mathbf{c}$$

and $\mathbf{c}$ is determined by solving the system

$$W(t_0)\mathbf{c} = \mathbf{x}^0.$$

Example 6.4.6 Consider the system

$$\mathbf{x}'(t) = \begin{bmatrix} 2 & -2 \\ -2 & 2 \end{bmatrix}\mathbf{x}(t) + \begin{pmatrix} \frac{1}{\sqrt{1-t^2}} + \frac{e^{4t}}{1+t^2} \\ \frac{1}{\sqrt{1-t^2}} - \frac{e^{4t}}{1+t^2} \end{pmatrix}.$$

A fundamental system of solutions of the homogeneous system is (see the next section)

$$\mathbf{w}^1(t) = \begin{bmatrix} 1 \\ 1 \end{bmatrix}; \ \mathbf{w}^2(t) = e^{4t} \begin{bmatrix} 1 \\ -1 \end{bmatrix}.$$

We have therefore

$$W(t) = \begin{bmatrix} 1 & e^{4t} \\ 1 & -e^{4t} \end{bmatrix}; \ W^{-1}(t) = \frac{1}{2} \begin{bmatrix} 1 & 1 \\ e^{-4t} & -e^{-4t} \end{bmatrix}.$$

It follows that

$$W^{-1}(t)\mathbf{b}(t) = \begin{pmatrix} \frac{1}{\sqrt{1-t^2}} \\ \frac{1}{1+t^2} \end{pmatrix}$$

and hence

$$\int W^{-1}(t)\mathbf{b}(t)dt = \begin{pmatrix} \arcsin t + c_1 \\ \arctan t + c_2 \end{pmatrix}.$$

The general solution of the system is therefore given by

$$\mathbf{x}(t) = \begin{pmatrix} \arcsin t + c_1 + e^{4t}(\arctan t + c_2) \\ \arcsin t + c_1 - e^{4t}(\arctan t + c_2) \end{pmatrix}.$$

6.5 Homogeneous Linear Systems with Constant Coefficients

First of all we remark that for systems of the type

$$\mathbf{x}'(t) = A\mathbf{x}(t) + \mathbf{b}(t)$$

(A square matrix of order n and $\mathbf{b}(t)$ continuous vector on $I \subset \mathbb{R}^n$) the theorems described in the previous section a fortiori hold true also for the present case: it holds the "superposition principle" and the solution of an initial value problem (or Cauchy problem) $\mathbf{x}(t_0) = \mathbf{x}^0$ is unique and global. First we consider the linear *homogeneous systems with constant coefficients,* that is

$$\mathbf{x}'(t) = A\mathbf{x}(t), \tag{6.17}$$

with A square matrix of order n; for these systems we establish the procedures to deduce their general solution ("general integral"). We recall that the set S of all solutions of (6.17) is a linear space of dimension n. Therefore, if we are able to find n linearly independent solutions of the system, the set of all solutions (general solution, general integral) is given by the linear combination, with arbitrary constants, of the said linearly independent solutions.

We have previously seen that with $n = 1$, that is with $A = a \in \mathbb{R}$, the general solution of the equation $x'(t) = ax(t)$ is

$$x(t) = ce^{at}, \quad c \in \mathbb{R}.$$

The structure of this solution suggests that also system (6.17) has a general solution with a similar structure. First we note that if λ_j is an eigenvalue of A, then the function

$$\mathbf{x}^j(t) = e^{\lambda_j t}\mathbf{v}^j, \tag{6.18}$$

where $\mathbf{v}^j$ is an eigenvector associated with λ_j, is a solution of (6.17). Indeed, we have

$$(\mathbf{x}^j(t))' = e^{\lambda_j t} \cdot \lambda_j \mathbf{v}^j$$

and, being $A\mathbf{v}^j = \lambda_j \mathbf{v}^j$, we obtain

$$(\mathbf{x}^j(t))' = e^{\lambda_j t} A\mathbf{v}^j,$$

that is

$$(\mathbf{x}^j(t))' = A\mathbf{x}^j(t).$$

Hence (6.18) is a solution of system (6.17).

(A) **Case of A diagonalizable.**

A first result concerns the case of A diagonalizable, that is A has n linearly independent eigenvectors, or, equivalently, every eigenvalue is *regular* (algebraic multiplicity equal to geometric multiplicity).

Theorem 6.5.1 *If A has n linearly independent eigenvectors $\mathbf{v}^1, \mathbf{v}^2, \ldots, \mathbf{v}^n$, and if $\lambda_1, \lambda_2, \ldots, \lambda_n$ are the related eigenvalues (not necessarily distinct!), the general solution of* (6.17) *is given by*

$$\mathbf{x}(t) = c_1\mathbf{v}^1 e^{\lambda_1 t} + c_2\mathbf{v}^2 e^{\lambda_2 t} + \ldots . + c_n\mathbf{v}^n e^{\lambda_n t}$$

with $c_1, c_2, \ldots, c_n$ arbitrary constants. If we have an initial value problem $\mathbf{x}(t_0) = \mathbf{x}^0$, then the constants c_i are uniquely determined.

Proof It is sufficient to note that the Wronskian formed by the above solutions is given by

$$\det\left[\mathbf{v}^1 e^{\lambda_1 t},\ \mathbf{v}^2 e^{\lambda_2 t}, \ldots, \mathbf{v}^n e^{\lambda_n t}\right].$$

By a well known property of determinants, the previous quantity is given by

$$e^{(\lambda_1+\lambda_2+\cdots+\lambda_n)t}\det\left[\mathbf{v}^1,\ \mathbf{v}^2, \ldots, \mathbf{v}^n\right].$$

This Wronskian is different from zero, as the vectors $\mathbf{v}^1; \mathbf{v}^2, \ldots, \mathbf{v}^n$ are, by assumption, linearly independent. We have therefore found a fundamental system of solutions for (6.17). □

The previous result can be also obtained in another way: we know that if A is diagonalizable, then it is possible to find a nonsingular matrix P such that

$$D = P^{-1}AP,$$

where D is a diagonal matrix with the eigenvalues of A on its main diagonal. Multiply both members of (6.17) by P^{-1}:

$$P^{-1}\mathbf{x}'(t) = P^{-1}A\mathbf{x}(t),$$

from which, being $PP^{-1} = I$,

$$(P^{-1}\mathbf{x}(t))' = P^{-1}APP^{-1}\mathbf{x}(t)$$

and hence

$$(P^{-1}\mathbf{x}(t))' = D(P^{-1}\mathbf{x}(t)).$$

By putting $\mathbf{z}(t) = P^{-1}\mathbf{x}(t)$, we obtain the system

$$\mathbf{z}'(t) = D\mathbf{z}(t)$$

whose components

$$z_i'(t) = \lambda_i z_i(t), \quad i = 1, \ldots, n,$$

have the solution

$$z_i(t) = c_i e^{\lambda_i t}$$

with c_i arbitrary constant ($i = 1, \ldots, n$). In a vector notation:

$$\mathbf{z}(t) = L\mathbf{c},$$

being

$$L = \begin{bmatrix} e^{\lambda_1 t} & 0 & \cdots & 0 \\ 0 & e^{\lambda_2 t} & \cdots & 0 \\ \vdots & \vdots & \ddots & \vdots \\ 0 & 0 & \cdots & e^{\lambda_n t} \end{bmatrix}$$

and $\mathbf{c} = [c_1,\ c_2, \ldots, c_n]^\top$.

Being $\mathbf{x}(t) = P\mathbf{z}(t)$, we obtain the general solution of (6.17), as the columns of P are the n linearly independent eigenvectors of A, i.e. the vectors $\mathbf{v}^i$, $i = 1, \ldots, n$, associated with the eigenvalues λ_i, $i = 1, \ldots, n$:

$$\mathbf{x}(t) = c_1\mathbf{v}^1 e^{\lambda_1 t} + c_2\mathbf{v}^2 e^{\lambda_2 t} + \ldots. + c_n\mathbf{v}^n e^{\lambda_n t}.$$

Note that if we choose $c_1 = c_2 = \cdots = c_n = 0$, we find the zero solution $\mathbf{x}(t) = [0]$. We recall that if A has n *distinct* eigenvalues, then A is diagonalizable; the same is true if A is *symmetric*. If the eigenvalues are all real, we have also real eigenvectors and the general solution of (6.17) will be real, of course by choosing all real constants.

Example 6.5.2 Let be given the system

$$\begin{cases} x_1'(t) = x_2(t) \\ x_2'(t) = -6x_1(t) + 5x_2(t). \end{cases}$$

It results

$$A = \begin{bmatrix} 0 & 1 \\ -6 & 5 \end{bmatrix}; \quad |A - \lambda I| = \begin{vmatrix} -\lambda & 1 \\ -6 & 5-\lambda \end{vmatrix}.$$

The eigenvalues of A are $\lambda_1 = 2, \lambda_2 = 3$. The corresponding eigenvectors (linearly independent) are

$$\mathbf{v}^1 = c_1\begin{bmatrix} 1 \\ 2 \end{bmatrix}, \ c_1 \neq 0; \ \mathbf{v}^2 = c_2\begin{bmatrix} 1 \\ 3 \end{bmatrix}, \ c_2 \neq 0.$$

The general solution is therefore

$$\mathbf{x}(t) = c_1\begin{bmatrix} 1 \\ 2 \end{bmatrix} e^{2t} + c_2\begin{bmatrix} 1 \\ 3 \end{bmatrix} e^{3t}, \quad c_1, c_2 \in \mathbb{R}.$$

Example 6.5.3 Solve the system

$$\mathbf{x}'(t) = \begin{bmatrix} 0.6 & 0.1 \\ 0.4 & 0.9 \end{bmatrix} \mathbf{x}(t).$$

It results $|A - \lambda I| = \lambda^2 - 1.5\lambda + 0.5$. Hence $\lambda_1 = 1, \lambda_2 = 0.5$. The corresponding eigenvectors are

$$\mathbf{v}^1 = \alpha \begin{bmatrix} \frac{1}{4} \\ 1 \end{bmatrix}, \alpha \neq 0; \quad \mathbf{v}^2 = \beta \begin{bmatrix} -1 \\ 1 \end{bmatrix}, \quad \beta \neq 0.$$

Hence the general solution is

$$\mathbf{x}(t) = c_1 \begin{bmatrix} \frac{1}{4} \\ 1 \end{bmatrix} e^t + c_2 \begin{bmatrix} -1 \\ 1 \end{bmatrix} e^{\frac{1}{2}t}$$

and a particular solution is determined by imposing an initial condition. By imposing, for instance,

$$\mathbf{x}(0) = \begin{bmatrix} 500 \\ 3000 \end{bmatrix} = \mathbf{x}^0,$$

we obtain the system

$$c_1 \begin{bmatrix} \frac{1}{4} \\ 1 \end{bmatrix} + c_2 \begin{bmatrix} -1 \\ 1 \end{bmatrix} = \begin{bmatrix} 500 \\ 3000 \end{bmatrix},$$

that is

$$\begin{cases} \frac{1}{4}c_1 - c_2 = 500 \\ c_1 + c_2 = 3000, \end{cases}$$

from which $c_1 = 2800, c_2 = 200$. The particular solution generated by the initial value problem is therefore

$$\mathbf{x}^0(t) = 2800 \begin{bmatrix} \frac{1}{4} \\ 1 \end{bmatrix} e^t + 200 \begin{bmatrix} -1 \\ 1 \end{bmatrix} e^{\frac{1}{2}t}.$$

We have seen how to obtain a fundamental system of solutions for (6.17), when we have n linearly independent eigenvectors, that is, when the matrix A of the system is diagonalizable. Now we wish to treat the case of complex (conjugate) eigenvalues; this case is not excluded in Theorem 6.5.1, however we show that it is always possible to obtain, in this case, an equivalent general solution, expressed only by real functions, which it may be more interesting and meaningful in economic and financial applications.

Let us suppose, just to fix ideas, to have one complex eigenvalue $\lambda_1 = a + ib$, together with its conjugate $\bar{\lambda}_1 = a - ib$. It is always possible to associate with these eigenvalues two conjugate eigenvectors of $\mathbb{C}^n$, say

$$\mathbf{v}^1 = \mathbf{u}+i\mathbf{w};\ \ \mathbf{v}^2 = \mathbf{u}-i\mathbf{w},$$

being $\mathbf{u}, \mathbf{w} \in \mathbb{R}^n$. In the general solution of (6.17) it appears therefore a sum of the following type (by neglecting the arbitrary constants)

$$(\mathbf{u}+i\mathbf{w})e^{(a+ib)t} + (\mathbf{u}-i\mathbf{w})e^{(a-ib)t}.$$

We have, from this last expression,

$$\begin{aligned}\mathbf{u}e^{at+ibt} + i\mathbf{w}e^{at+ibt} + \mathbf{u}e^{at-ibt} - i\mathbf{w}e^{at-ibt} \\ = \mathbf{u}(e^{at+ibt} + e^{at-ibt}) + i\mathbf{w}(e^{at+ibt} - e^{at-ibt}).\end{aligned}$$

Now we recall the "Euler formulas" for the cosine and the sine (Sect. 1.5):

$$\cos t = \frac{e^{it} + e^{-it}}{2};\ \sin t = \frac{e^{it} - e^{-it}}{2i}.$$

We multiply and divide by 2 the first addendum of the last expression above, and multiply and divide by $2i$ the second addendum. We obtain

$$2\mathbf{u}e^{at}\left(\frac{e^{ibt} + e^{-ibt}}{2}\right) + 2i \cdot i\mathbf{w}e^{at}\left(\frac{e^{ibt} - e^{-ibt}}{2i}\right).$$

Being $i \cdot i = i^2 = -1$, we have

$$\begin{aligned}e^{at}\left(2\mathbf{u}\frac{e^{ibt} + e^{-ibt}}{2} - 2\mathbf{w}\frac{e^{ibt} - e^{-ibt}}{2i}\right) &= e^{at}(2\mathbf{u}\cos bt - 2\mathbf{w}\sin bt) \\ &= e^{at}(\mathbf{k}^1\cos bt + \mathbf{k}^2\sin bt),\end{aligned}$$

being $\mathbf{k}^1 = 2\mathbf{u} \in \mathbb{R}^n$ and $\mathbf{k}^2 = -2\mathbf{w} \in \mathbb{R}^n$.

Therefore, we have transformed the contribution to the solution of (6.17) of two complex (conjugate) eigenvalues and eigenvectors into real quantities of $\mathbb{R}^n$. Basically, when we have a pair of complex conjugate eigenvalues $a \pm ib$, we consider the expression

$$e^{at}(\mathbf{k}^1\cos bt + \mathbf{k}^2\sin bt),$$

with $\mathbf{k}^1 \in \mathbb{R}^n$ and $\mathbf{k}^2 \in \mathbb{R}^n$, to be determined (overall they will contain two arbitrary constants), in such a way that the system is satisfied. If we have more than two complex eigenvalues, the procedure is similar. An example will clarify the matter.

Example 6.5.4 Solve the system

$$\mathbf{x}'(t) = \begin{bmatrix} 0 & 1 \\ -1 & 0 \end{bmatrix} \mathbf{x}(t).$$

We have

$$|A - \lambda I| = \begin{vmatrix} -\lambda & 1 \\ -1 & -\lambda \end{vmatrix} = \lambda^2 + 1,$$

hence $\lambda_1 = i, \lambda_2 = -i$ $(a = 0, b = 1)$. We know that the solution has the following structure:

$$\mathbf{x}(t) = 1 \cdot (\mathbf{k}^1 \cos t + \mathbf{k}^2 \sin t).$$

In order to determine $\mathbf{k}^1$ and $\mathbf{k}^2$, we insert this solution into the system:

$$\mathbf{x}(t) = \left\{ \begin{bmatrix} a \\ b \end{bmatrix} \cos t + \begin{bmatrix} c \\ d \end{bmatrix} \sin t \right\};$$

$$\mathbf{x}'(t) = \left\{ - \begin{bmatrix} a \\ b \end{bmatrix} \sin t + \begin{bmatrix} c \\ d \end{bmatrix} \cos t \right\}.$$

Consequently we impose

$$\begin{bmatrix} -a \\ -b \end{bmatrix} \sin t + \begin{bmatrix} c \\ d \end{bmatrix} \cos t = \begin{bmatrix} 0 & 1 \\ -1 & 0 \end{bmatrix} \left\{ \begin{bmatrix} a \\ b \end{bmatrix} \cos t + \begin{bmatrix} c \\ d \end{bmatrix} \sin t \right\},$$

that is

$$\begin{bmatrix} -a \sin t + c \cos t \\ -b \sin t + d \cos t \end{bmatrix} = \begin{bmatrix} 0 & 1 \\ -1 & 0 \end{bmatrix} \begin{bmatrix} a \cos t + c \sin t \\ b \cos t + d \sin t \end{bmatrix}.$$

We have

$$\begin{bmatrix} -a \sin t + c \cos t \\ -b \sin t + d \cos t \end{bmatrix} = \begin{bmatrix} b \cos t + d \sin t \\ -a \cos t - c \sin t \end{bmatrix}.$$

Consequently, it must hold

$$\begin{cases} c = b \\ -a = d \\ -b = -c \\ d = -a \end{cases} \quad \text{from which} \quad \begin{cases} c = b \\ d = -a. \end{cases}$$

Therefore in $\mathbf{k}^1$ and $\mathbf{k}^2$ (real vectors of $\mathbb{R}^2$) we have 4 constants: a and b are arbitrary, whereas c and d must respect the above conditions. We have therefore

$$\mathbf{x}(t) = \begin{bmatrix} a \\ b \end{bmatrix} \cos t + \begin{bmatrix} b \\ -a \end{bmatrix} \sin t, \quad a, b \in \mathbb{R}.$$

This solution may be rewritten as follows

$$\mathbf{x}(t) = a \begin{bmatrix} \cos t \\ -\sin t \end{bmatrix} + b \begin{bmatrix} \sin t \\ \cos t \end{bmatrix},$$

that is

$$\mathbf{x}(t) = c_1 \begin{bmatrix} \cos t \\ -\sin t \end{bmatrix} + c_2 \begin{bmatrix} \sin t \\ \cos t \end{bmatrix}, \quad c_1, c_2 \in \mathbb{R}.$$

(B) **Case of A non diagonalizable (multiple eigenvalues, with geometric multiplicity less that the respective algebraic multiplicity).**

Let us suppose, just to fix ideas, to have one real eigenvalue λ_1 with algebraic multiplicity $m^A = 2$ ("double root" of the characteristic equation), *but not regular*, i.e. with geometric multiplicity $m^G = 1$. Hence we have not two linearly independent eigenvectors associated with λ_1 and Theorem 6.5.1 is not directly applicable. In any case, by Theorem 6.4.4, we must going looking for two linearly independent solutions, in order to build a "fundamental system". We begin by considering a simple example.

Example 6.5.5 Let be given the system

$$\mathbf{x}'(t) = \begin{bmatrix} 1 & 2 \\ 0 & 1 \end{bmatrix} \mathbf{x}(t).$$

We have

$$\begin{vmatrix} 1 - \lambda & 2 \\ 0 & 1 - \lambda \end{vmatrix} = (1 - \lambda)^2.$$

Hence $\lambda = 1$, with $m_\lambda^A = 2$. Then we have

$$A - I = \begin{bmatrix} 0 & 2 \\ 0 & 0 \end{bmatrix}$$

and hence $m_\lambda^G = 2 - \text{rank}(A - I) = 2 - 1 = 1$. The associated eigenvector is

$$\mathbf{v}^1 = \begin{bmatrix} x_1 \\ 0 \end{bmatrix}, \quad x_1 \neq 0.$$

However, we need two linearly independent solutions, in order to build a fundamental system. We rewrite the system:

$$\begin{cases} x_1'(t) = x_1(t) + 2x_2(t) \\ x_2'(t) = x_2(t). \end{cases}$$

The second equation of the system has the solution $x_2(t) = c_2 e^t$. By substituting this solution into the first equation, we get

$$x_1'(t) = x_1(t) + 2c_2 e^t.$$

This last equation is a first-order linear differential equation with variable coefficients; its solution is $x_1(t) = c_1 e^t + c_2 t e^t$. Hence the general solution of our system is

$$\mathbf{x}(t) = \begin{bmatrix} c_1 e^t + c_2 t e^t \\ c_2 e^t \end{bmatrix} = c_1 e^t \begin{bmatrix} 1 \\ 0 \end{bmatrix} + c_2 \left(e^t \begin{bmatrix} 0 \\ 1 \end{bmatrix} + t e^t \begin{bmatrix} 1 \\ 0 \end{bmatrix} \right)$$

$$= k_1 e^t \begin{bmatrix} 1 \\ 1 \end{bmatrix} + k_2 t e^t \begin{bmatrix} 1 \\ 0 \end{bmatrix} = (\mathbf{a} + \mathbf{b}t)e^t, \quad \mathbf{a}, \mathbf{b} \in \mathbb{R}^2.$$

If we put

$$\mathbf{x}^1(t) = e^t \begin{bmatrix} 1 \\ 0 \end{bmatrix}; \quad \mathbf{x}^2(t) = \left(e^t \begin{bmatrix} 0 \\ 1 \end{bmatrix} + t e^t \begin{bmatrix} 1 \\ 0 \end{bmatrix} \right),$$

the Wronskian matrix $W(t) = \left[\mathbf{x}^1(t); \mathbf{x}^2(t)\right]$, evaluated at $t_0 = 0$, is

$$W(0) = \begin{bmatrix} 1 & 0 \\ 0 & 1 \end{bmatrix}$$

and hence $\det(W(0)) = 1$, which confirms that $\mathbf{x}^1(t)$ and $\mathbf{x}^2(t)$ form a fundamental system of solutions.

Therefore, coming back to the case described at the beginning of this section, it results that in the general solution of (6.17), instead of the term

$$c_1 \mathbf{v}^1 e^{\lambda_1 t} + c_2 \mathbf{v}^2 e^{\lambda_1 t},$$

it appears a term of the type

$$(\mathbf{a} + \mathbf{b}t)e^{\lambda_1 t}$$

with $\mathbf{a}, \mathbf{b}$ vectors of $\mathbb{R}^n$ *to be determined* and containing overall two arbitrary constants.

If λ_1 has algebraic multiplicity equal to 3 and is not regular, we shall have an addendum of the type

$$(\mathbf{a} + \mathbf{b}t + \mathbf{c}t^2)e^{\lambda_1 t}$$

with $\mathbf{a}, \mathbf{b}$ and $\mathbf{c}$ vectors of $\mathbb{R}^n$ *to be determined* and containing overall three arbitrary constants.

Basically, we can follow the procedure previously seen for the complex roots, i.e. we substitute the "modified solution" into the system, in order to determine the structure of vectors $\mathbf{a}, \mathbf{b}, \mathbf{c}$, etc.

Example 6.5.6

(*a*) Solve the system

$$\begin{cases} x_1'(t) = 2x_1(t) \\ x_2'(t) = x_1(t) + 2x_2(t), \end{cases}$$

that is

$$\mathbf{x}'(t) = \begin{bmatrix} 2 & 0 \\ 1 & 2 \end{bmatrix} \mathbf{x}(t).$$

It results

$$\begin{vmatrix} 2-\lambda & 0 \\ 1 & 2-\lambda \end{vmatrix} = (2-\lambda)^2.$$

Hence $\lambda = 2$ is an eigenvalue with algebraic multiplicity equal to 2. Then we have

$$|A - 2I| = \begin{bmatrix} 0 & 0 \\ 1 & 0 \end{bmatrix}$$

and $2 - \text{rank}(A - 2I) = 2 - 1 = 1$, so λ *is not regular.* Now we compute $\mathbf{x}'(t)$, being $\mathbf{x}(t) = (\mathbf{a} + \mathbf{b}t)e^{2t}$. We have

$$\mathbf{x}'(t) = \mathbf{b}e^{2t} + 2(\mathbf{a} + \mathbf{b}t)e^{2t} = e^{2t}(\mathbf{b}+2\mathbf{a}+2\mathbf{b}t).$$

Therefore

$$\mathbf{x}'(t) = \left(\begin{bmatrix} 2a_1 + b_1 \\ 2a_2 + b_2 \end{bmatrix} + t\begin{bmatrix} 2b_1 \\ 2b_2 \end{bmatrix}\right) e^{2t}.$$

Then we impose the equality

$$\left(\begin{bmatrix} 2a_1 + b_1 \\ 2a_2 + b_2 \end{bmatrix} + t\begin{bmatrix} 2b_1 \\ 2b_2 \end{bmatrix}\right) e^{2t} = \begin{bmatrix} 2 & 0 \\ 1 & 2 \end{bmatrix}\left(\begin{bmatrix} a_1 \\ a_2 \end{bmatrix} + t\begin{bmatrix} b_1 \\ b_2 \end{bmatrix}\right) e^{2t}.$$

By dividing both members by $e^{2t} \neq 0$, we obtain

$$\begin{cases} \begin{bmatrix} 2a_1 + b_1 \\ 2a_2 + b_2 \end{bmatrix} = \begin{bmatrix} 2a_1 \\ a_1 + 2a_2 \end{bmatrix} \\ \begin{bmatrix} 2b_1 \\ 2b_2 \end{bmatrix} = \begin{bmatrix} 2b_1 \\ b_1 + 2b_2 \end{bmatrix} \end{cases},$$

that is

$$\begin{cases} 2a_1 + b_1 = 2a_1 \\ 2a_2 + b_2 = a_1 + 2a_2 \\ 2b_1 = 2b_1 \text{ (useless and trivial relation)} \\ 2b_2 = b_1 + 2b_2 \Longrightarrow b_1 = 0. \end{cases}$$

Therefore, being $b_1 = 0$, also the first equation of the system becomes useless and trivial, whereas from the second equation we get $a_1 = b_2$.

Therefore a_1 and a_2 are arbitrary: $a_1 = \alpha$ and $a_2 = \beta$, whereas $b_1 = 0$ and $b_2 = a_1 = \alpha$. Therefore the general solution of the system is

$$\mathbf{x}(t) = \left(\begin{bmatrix} \alpha \\ \beta \end{bmatrix} + \begin{bmatrix} 0 \\ \alpha \end{bmatrix} t\right) e^{2t},$$

with α, β arbitrary constants.

(*b*) Consider the system

$$\mathbf{x}'(t) = A\mathbf{x}(\mathbf{t}), \text{ with } A = \begin{bmatrix} 1 & 1 & 0 \\ 0 & 0 & 1 \\ -1 & -1 & 3 \end{bmatrix}.$$

It results $\lambda_1 = 0, \lambda_2 = 2, \lambda_3 = 2$. Therefore $m^A(2) = 2$. We have $m^G(2) = 3 - \text{rank}(A - 2I)$.

$$A - 2I = \begin{bmatrix} -1 & 1 & 0 \\ 0 & -2 & 1 \\ -1 & -1 & 1 \end{bmatrix}; \quad \text{rank}(A - 2I) = 2.$$

Therefore $m^G(2) = 3 - 2 = 1$. The general solution will have the form

$$\mathbf{x}(t) = \mathbf{a}e^{0t} + \mathbf{b}e^{2t} + \mathbf{c}te^{2t} = \mathbf{a} + (\mathbf{b} + \mathbf{c}t)e^{2t},$$

with $\mathbf{a}, \mathbf{b},$ and $\mathbf{c}$ vectors of $\mathbb{R}^3$ to be determined. These vectors will contain overall the arbitrary constants c_1, c_2 and c_3. By inserting the above solution into the system, we obtain

$$[0] + 2\mathbf{b}e^{2t} + (1 + 2t)e^{2t}\mathbf{c} = A\mathbf{a}+e^{2t}A\mathbf{b}+te^{2t}A\mathbf{c}.$$

By equating the coefficients of the corresponding terms, we have

$$\begin{cases} A\mathbf{a} = [0] \\ A\mathbf{b} = 2\mathbf{b} + \mathbf{c} \\ A\mathbf{c} = 2\mathbf{c}. \end{cases}$$

Note that $\mathbf{a}$ is the eigenvector associated with $\lambda_1 = 0$, $\mathbf{c}$ is the eigenvector associated with $\lambda_2 = 2$, but $\mathbf{b}$ is *not* an eigenvector of A. By solving the system we find

$$\mathbf{a} = c_1 \begin{pmatrix} 1 \\ -1 \\ 0 \end{pmatrix}; \ \mathbf{b} = c_2 \begin{pmatrix} 1 \\ 1 \\ 2 \end{pmatrix} + c_3 \begin{pmatrix} -1 \\ 0 \\ 1 \end{pmatrix}; \mathbf{c} = c_3 \begin{pmatrix} 1 \\ 1 \\ 2 \end{pmatrix}.$$

The general solution is therefore

$$\mathbf{x}(t) = c_1 \begin{pmatrix} 1 \\ -1 \\ 0 \end{pmatrix} + \left[c_2 \begin{pmatrix} 1 \\ 1 \\ 2 \end{pmatrix} + c_3 \begin{pmatrix} -1 \\ 0 \\ 1 \end{pmatrix} + c_3 t \begin{pmatrix} 1 \\ 1 \\ 2 \end{pmatrix} \right] e^{2t}.$$

It is worth remarking that not always multiple eigenvalues are non regular. For instance, consider the system

$$\begin{cases} x_1'(t) = x_1(t) \\ x_2'(t) = x_2(t) \end{cases},$$

that is

$$\mathbf{x}'(t) = \begin{bmatrix} 1 & 0 \\ 0 & 1 \end{bmatrix} \mathbf{x}(t).$$

We have the eigenvalue $\lambda = 1$, with algebraic multiplicity equal to 2, however regular (the matrix of the system is a diagonal matrix!). We have therefore two linearly independent eigenvectors associated with this eigenvalue:

$$\mathbf{v}^1 = \alpha \begin{bmatrix} 1 \\ 0 \end{bmatrix}, \quad \alpha \neq 0; \quad \mathbf{v}^2 = \beta \begin{bmatrix} 0 \\ 1 \end{bmatrix}, \quad \beta \neq 0.$$

The general solution is therefore given by

$$\mathbf{x}(t) = c_1 \mathbf{v}^1 e^t + c_2 \mathbf{v}^2 e^t = e^t (c_1 \mathbf{v}^1 + c_2 \mathbf{v}^2) = \begin{bmatrix} c_1 \\ c_2 \end{bmatrix} e^t, \quad c_1, c_2 \in \mathbb{R}.$$

Another example concerning the present case is the following one:

$$\mathbf{x}'(t) = \begin{bmatrix} 1 & 0 & 2 \\ 0 & 1 & 1 \\ 0 & 0 & 3 \end{bmatrix} \mathbf{x}(t).$$

The eigenvalues of A are $\lambda_1 = \lambda_2 = 1$ and $\lambda_3 = 3$. We have

$$\operatorname{rank}(A - 1I) = \operatorname{rank} \begin{pmatrix} 0 & 0 & 2 \\ 0 & 0 & 1 \\ 0 & 0 & 2 \end{pmatrix} = 1$$

and hence $n - \operatorname{rank}(A - I) = 3 - 1 = 2$. Therefore the eigenvalue $\lambda = 1$ is regular and it follows that it is possible to associate with this eigenvalue two linearly independent eigenvectors, in order to have in total three linearly independent eigenvectors. Indeed, it results, with $\lambda = 1$,

$$\mathbf{v}^1 = \alpha \begin{bmatrix} 1 \\ 0 \\ 0 \end{bmatrix}, \quad \alpha \neq 0; \quad \mathbf{v}^2 = \beta \begin{bmatrix} 0 \\ 1 \\ 0 \end{bmatrix}, \quad \beta \neq 0.$$

With $\lambda_3 = 3$ we associate the eigenvectors

$$\mathbf{v}^3 = \gamma \begin{bmatrix} 1 \\ \frac{1}{2} \\ 1 \end{bmatrix}, \quad \gamma \neq 0.$$

It follows that the general solution of the system is

$$\mathbf{x}(t) = c_1 \begin{bmatrix} 1 \\ 0 \\ 0 \end{bmatrix} e^t + c_2 \begin{bmatrix} 0 \\ 1 \\ 0 \end{bmatrix} e^t + c_3 \begin{bmatrix} 1 \\ \frac{1}{2} \\ 1 \end{bmatrix} e^{3t}, \ c_1, c_2, c_3 \in \mathbb{R}.$$

Moreover, we know that if A is a *symmetric* matrix of order n, it is always possible to have n linearly independent eigenvectors, even in correspondence of multiple eigenvalues.

Some other theoretical remarks are appropriate. If we make reference, in order to justify the rules given for the case of multiple eigenvalues, regular or not, to the Schur Theorem on the "triangularization" of a square matrix A (Theorem 1.6.21), it is possible to get the following considerations. If A, of order n, has k distinct eigenvalues $\lambda_1, \lambda_2, \ldots, \lambda_k$, with respective algebraic multiplicity $m_1, m_2, \ldots, m_k$ $(m_1 + m_2 + \cdots + m_k = n)$, the general solution of (6.17) can be expressed in the following way:

$$\mathbf{x}(t) = \sum_{s=1}^{k} \mathbf{p}^s(t) e^{\lambda_s t},$$

where $\mathbf{p}^s(t)$ is a suitable vector of polynomials in t, of degree less than or equal to $m_s - 1$, and containing as many arbitrary constants as the degree of algebraic multiplicity of λ_s. Furthermore, it can be shown that all the solutions appearing in the above formula are linearly independent, i.e. they form a *fundamental system,* i.e. the said formula gives just the general solution.

Indeed, the *Schur Theorem* (Theorem 1.6.21) says that there exists always a nonsingular matrix S of order n such that $A = SBS^{-1}$, B being, for instance, a lower triangular matrix, having therefore the eigenvalues of A on its main diagonal. The homogeneous system (6.17) becomes therefore

$$\mathbf{x}'(t) = SBS^{-1}\mathbf{x}(t),$$

and, with $\mathbf{y}(t) = S^{-1}\mathbf{x}(t)$,

$$\mathbf{y}'(t) = B\mathbf{y}(t),$$

where

$$B = \begin{bmatrix} \lambda_1 & 0 & 0 & \cdots & 0 \\ b_{21} & \lambda_2 & 0 & \cdots & 0 \\ b_{31} & b_{32} & \lambda_3 & \cdots & 0 \\ \vdots & \vdots & \vdots & \ddots & \vdots \\ b_{n1} & b_{n2} & b_{n3} & \cdots & \lambda_n \end{bmatrix}.$$

We have therefore

$$
\begin{aligned}
y_1'(t) &= \lambda_1 y_1(t) \\
y_2'(t) &= b_{21} y_1(t) + \lambda_2 y_2(t) \\
y_3'(t) &= b_{31} y_1(t) + b_{32} y_2(t) + \lambda_3 y_3(t) \\
& \cdots\cdots\cdots\cdots\cdots\cdots\cdots\cdots \\
y_n'(t) &= b_{n1} y_1(t) + b_{n2} y_2(t) + b_{n3} y_3(t) + \cdots + \lambda_n y_n(t).
\end{aligned}
$$

Let us suppose that, in case of repeated eigenvalues, these ones appear in a consecutive position on the main diagonal of B. If there are complex eigenvalues, it is always possible, as previously seen, to consider real solutions. The advantage of the "triangularization" is that the above system can be solved in a recursive way, beginning from the first equation:

$$y_1(t) = c_1 e^{\lambda_1 t}, \quad c_1 \text{ arbitrary.}$$

Now consider two cases: $\lambda_2 \neq \lambda_1$ and $\lambda_2 = \lambda_1$.

(i) If $\lambda_2 \neq \lambda_1$ (hence λ_1 is a simple root of the characteristic equation of A), we have

$$y_2'(t) = \lambda_2 y_2(t) + b_{21} c_1 e^{\lambda_1 t}.$$

We have obtained a linear differential equation (of the first order) whose solution is

$$y_2(t) = c_2 e^{\lambda_2 t} + \frac{b_{21} c_1}{\lambda_1 - \lambda_2} e^{\lambda_1 t}, \quad c_2 \text{ arbitrary.}$$

(ii) If $\lambda_2 = \lambda_1$, the second equation becomes

$$y_2'(t) = \lambda_1 y_2(t) + b_{21} c_1 e^{\lambda_1 t},$$

from which we have

$$y_2(t) = (c_2 + b_{21} c_1 t) e^{\lambda_1 t}, \quad c_2 \text{ arbitrary.}$$

Going on in this procedure, we see that the general solution of $\mathbf{y}'(t) = B\mathbf{y}(t)$, when there are repeated eigenvalues, can be expressed as

$$\mathbf{y}(t) = \sum_{s=1}^{k} \mathbf{p}^s(t) e^{\lambda_s t},$$

where $\lambda_1, \lambda_2, \ldots, \lambda_k$ are the *distinct* eigenvalues of A and $\mathbf{p}^s(t)$ is a vector of polynomials in t of degree less than or equal to the algebraic multiplicity of λ_s, *minus one.*

As for what concerns the original system $\mathbf{x}'(t) = A\mathbf{x}(t)$, being $\mathbf{x}(t) = S\mathbf{y}(t)$, its general solution has the same structure of the solution of $\mathbf{y}'(t) = B\mathbf{y}(t)$. For a complete and more formal proof of what asserted, see. e.g., [1] and [12]. Further detailed information on the structure of the vector of polynomials $\mathbf{p}^s(t)$ can be obtained by means of the *Jordan canonical form* of A (Sect. 1.6), but this is beyond the aims of the present book. We remark that, from a practical point of view, the presence of repeated and non regular eigenvalues of A is not worrying, as *any* small perturbation in the coefficients a_{ij} will yield a matrix with distinct roots. More precisely, we have the following result, due to [7]. Let $\|A\|$ be any *matrix norm* of the square real matrix A, of order n. For example, $\|A\| = \max_j \sum_i |a_{ij}|$ or $\|A\| = \max_i \sum_j |a_{ij}|$ or $\|A\| = \sqrt{\sum_i \sum_j |a_{ij}|^2}$, etc.

Theorem 6.5.7 *Given a real square matrix A of order n, a square matrix B of the same order can always be found, with distinct eigenvalues and such that*

$$\|A - B\| \leqq \varepsilon,$$

where $\varepsilon > 0$ can be chosen arbitrarily small.

Summing Up
When we have n eigenvalues, someone real and distinct, someone real multiple and non regular, someone complex and distinct, someone complex multiple and non regular, all previous expressions must be considered. Hence the general solution of a system of the type

$$\mathbf{x}'(t) = A\mathbf{x}(t)$$

can be obtained by summing functions from $\mathbb{R}$ to $\mathbb{R}^n$ of the following types:

(a)

$$c_s \mathbf{v}^s e^{\lambda_s t}, \quad \mathbf{v}^s \in \mathbb{R}^n,$$

if λ_s is real and a simple root of the characteristic equation of A, and where $\mathbf{v}^s$ is an eigenvector associated with λ_s (c_s is an arbitrary constant).

(b)

$$e^{at}\left[\mathbf{k}^1 \cos bt + \mathbf{k}^2 \sin bt\right], \quad \mathbf{k}^1, \mathbf{k}^2 \in \mathbb{R}^n,$$

if $\lambda_s = a + ib$ (and then A will have also the conjugate eigenvalue $\bar{\lambda}_s = a - ib$) and is a simple root of the characteristic equation of A.

(c)

$$e^{\lambda_s t}\mathbf{p}^s(t),$$

if λ_s is a real multiple root, with algebraic multiplicity k, being $\mathbf{p}^s(t)$ a suitable vector of polynomials, of degree less than or equal to $k-1$, and containing k arbitrary constants.

(d)

$$e^{at}\left[\mathbf{p}^s(t)\cos bt + \mathbf{q}^s(t)\sin bt\right],$$

if $\lambda_s = a+ib$ has algebraic multiplicity equal to k (and then A will have also the conjugate eigenvalue $\bar{\lambda}_s = a - ib$, with the same multiplicity), and regarding $\mathbf{p}^s(t)$ and $\mathbf{q}^s(t)$, the same considerations of the case (c) hold true.

Finally, we observe that another formal way to express the general solution of a system of type (6.17), is to use the so-called *exponential* of the matrix A. Given $A = \left[a_{ij}\right], i, j = 1, \dots, n$, the *exponential of* A, denoted by e^A or $\exp(A)$, is defined as the series

$$e^A = \sum_{k=0}^{+\infty} \frac{1}{k!}(A)^k = I + A + \frac{1}{2}(A)^2 + \cdots + \frac{1}{k!}(A)^k + \dots$$

It is possible to prove that this series is always convergent, for any square matrix A. It is also possible to verify that an alternative definition of $\exp(A)$ is

$$e^A = \lim_{k \longrightarrow +\infty}\left(I + \frac{1}{k}A\right)^k.$$

In general, it is not an easy task to compute $\exp(A)$, unless A is diagonalizable. However, the above definition is useful, since the related series converges very quickly. Now, let us consider system (6.17). It is quite easy to prove that its general solution can be expressed as

$$\mathbf{x}(t) = e^{tA}\cdot\mathbf{c},$$

where $\mathbf{c} \in \mathbb{R}^n$ is arbitrary. Indeed, it results, just as for scalar quantities,

$$\frac{d}{dt}e^{tA} = Ae^{tA}, \text{ for every } t \in \mathbb{R}.$$

Moreover, the columns of e^{tA} form a fundamental system of solutions of (6.17), i.e. the above formula gives the general solution of (6.17). If we have a Cauchy problem of the type $\mathbf{x}(t_0) = \mathbf{x}^0$, then the related solution is expressed as

$$\mathbf{x}(t) = e^{(t-t_0)A}\cdot\mathbf{x}^0.$$

If A is diagonalizable, it is not difficult to obtain the general solution of (6.17) by means of $\exp(A)$, solution having the same form as the one previously obtained. Indeed, in this case, there will exist a nonsingular matrix S (whose columns are the linearly independent eigenvectors of A) such that $A = SDS^{-1}$.

We recall that

$$e^A = I + A + \frac{1}{2}(A)^2 + \dots$$

and that

$$(A)^n = (SDS^{-1})^n = SDS^{-1} \cdot SDS^{-1} \cdot \dots \cdot SDS^{-1} = S(D)^n S^{-1}.$$

Thus

$$e^A = SS^{-1} + SDS^{-1} + \frac{1}{2}S(D)^2 S^{-1} + \dots$$

Factoring out S and S^{-1} on the left and right gives

$$e^A = S(I + D + \frac{1}{2}(D)^2 + \dots)S^{-1} = Se^D S^{-1}.$$

Therefore

$$\begin{aligned} e^A &= Se^D S^{-1} = S\left(\sum_{k=0}^{+\infty} \frac{1}{k!}(D)^k\right) S^{-1} \\ &= S\left(\sum_{k=0}^{+\infty} \operatorname{diag}\left(\frac{\lambda_1^k}{k!}, \frac{\lambda_2^k}{k!}, \dots, \frac{\lambda_n^k}{k!}\right)\right) S^{-1} = S \cdot \operatorname{diag}\left(e^{\lambda_1}, e^{\lambda_2}, \dots, e^{\lambda_n}\right) S^{-1} \end{aligned}$$

(since $\sum_{k=0}^{+\infty} \frac{\lambda_i^k}{k!} = e^{\lambda_i}, i = 1, \dots, n$). Thus

$$e^{tA} = S \cdot \operatorname{diag}(e^{\lambda_i t}) S^{-1}.$$

Observe now that if $\mathbf{k} \in \mathbb{R}$ is arbitrary, then also $\mathbf{c} = S^{-1}\mathbf{k}$ is a vector of arbitrary constants, so that in the expression of the general solution we can avoid to make reference to S^{-1} and write simply

$$\mathbf{x}(t) = S \cdot \operatorname{diag}(e^{\lambda_i t}) \cdot \mathbf{c}.$$

Expanding the product we get

$$\mathbf{x}(t) = c_1 \mathbf{v}^1 e^{\lambda_1 t} + c_2 \mathbf{v}^2 e^{\lambda_2 t} + \dots + c_n \mathbf{v}^n e^{\lambda_n t},$$

i.e. the expression previously given, of the general solution of (6.17) for the case considered.

If A is not diagonalizable, the computation of $\exp(A)$ is more difficult and usually the Jordan canonical form of A must be taken into consideration. See, e.g., [12] and, at a more elementary level [15].

6.6 Non Homogeneous Linear Systems with Constant Coefficients

Consider the system

$$\mathbf{x}'(t) = A\mathbf{x}(t) + \mathbf{b}(t) \tag{6.19}$$

of dimension n, and with $\mathbf{b}(t) \in \mathbb{R}^n$ a continuous vector, not identically equal to zero, with $t \in I \subset \mathbb{R}$. On the grounds of the general result of Theorem 6.4.3, ("superposition principle"), the knowing of a *particular solution* of (6.19), say $\mathbf{x}^0(t)$, allows to know all solutions of the system ("general solution" or "general integral"), hence we have

$$\mathbf{x}(t) = \mathbf{z}(t) + \mathbf{x}^0(t),$$

where $\mathbf{z}(t)$ is the general solution of the associated homogeneous system. Therefore we are requested to obtain a particular solution of (6.19) and then to add the same to the general solution $\mathbf{z}(t)$ of

$$\mathbf{x}'(t) = A\mathbf{x}(t).$$

A particular solution of (6.19) can be often obtained by means of the so-called "method of undetermined coefficients", called also sometimes "similarity principle". This method can be synthesized as follows: a particular solution of system (6.19) is a function $\mathbf{x}^0(t) : \mathbb{R} \longrightarrow \mathbb{R}^n$, probably of the same "structure" of $\mathbf{b}(t) : \mathbb{R} \longrightarrow \mathbb{R}^n$. For instance, if $\mathbf{b}(t)$ is a vector of polynomials, we impose that also $\mathbf{x}^0(t)$ has this structure; if $\mathbf{b}(t)$ is a vector whose elements are trigonometric functions, we impose that $\mathbf{x}^0(t)$ has this structure, etc. If $\mathbf{b}(t) = \mathbf{b}$, a constant vector, therefore we can to try to impose that a particular solution of the non homogeneous system is a constant vector:

$$\bar{\mathbf{x}}(t) = \bar{\mathbf{x}},$$

whose components will be determined by substituting this vector into the original system

$$\mathbf{x}'(t) = A\mathbf{x}(t) + \mathbf{b}.$$

Thus we obtain

$$[0] = A\mathbf{x} \mathbin{\bar{+}} b$$

from which

$$\mathbf{x} \mathrel{\bar{=}} -A^{-1}\mathbf{b},$$

if $|A| \neq 0$, that is if 0 is not an eigenvalue of A. If $|A| = 0$, we have to look for a particular solution of the type $\bar{\mathbf{x}}(t) = \mathbf{a}^1 + \mathbf{a}^2 t$, with $\mathbf{a}^1$ and $\mathbf{a}^2$ vectors to be determined.

Example 6.6.1 Solve the system

$$\begin{cases} x_1'(t) = x_1(t) + 2 \\ x_2'(t) = x_1(t) + 2x_2(t) + 3. \end{cases}$$

Hence $\mathbf{b}(t) = \mathbf{b}$, a constant vector. The matrix of the homogeneous system is

$$A = \begin{bmatrix} 1 & 0 \\ 1 & 2 \end{bmatrix}$$

with eigenvalues $\lambda_1 = 1$ and $\lambda_2 = 2$. The corresponding associated eigenvectors are

$$\mathbf{v}^1 = \begin{bmatrix} \alpha \\ -\alpha \end{bmatrix}, \quad \alpha \in \mathbb{R}, \alpha \neq 0; \quad \mathbf{v}^2 = \begin{bmatrix} 0 \\ \beta \end{bmatrix}, \quad \beta \in \mathbb{R}, \beta \neq 0.$$

It follows that the general solution of the associated homogeneous system $\mathbf{x}'(t) = A\mathbf{x}(t)$ is

$$\mathbf{x}(t) = c_1 \begin{bmatrix} 1 \\ -1 \end{bmatrix} e^t + c_2 \begin{bmatrix} 0 \\ 1 \end{bmatrix} e^{2t}, \quad c_1, c_2 \in \mathbb{R}.$$

The matrix A is nonsingular and it holds

$$A^{-1} = \begin{bmatrix} 1 & 0 \\ -\frac{1}{2} & \frac{1}{2} \end{bmatrix}; \quad -A^{-1}\mathbf{b} = \begin{bmatrix} -1 & 0 \\ \frac{1}{2} & -\frac{1}{2} \end{bmatrix} \begin{bmatrix} 2 \\ 3 \end{bmatrix} = \begin{bmatrix} -2 \\ -\frac{1}{2} \end{bmatrix}.$$

The general solution of the original system is therefore

$$\mathbf{x}(t) = c_1 \begin{bmatrix} 1 \\ -1 \end{bmatrix} e^t + c_2 \begin{bmatrix} 0 \\ 1 \end{bmatrix} e^{2t} - \begin{bmatrix} 2 \\ \frac{1}{2} \end{bmatrix}.$$

Example 6.6.2 Consider the system

$$\mathbf{x}'(t) = \begin{bmatrix} 0.6 & 0.1 \\ 0.4 & 0.9 \end{bmatrix} \mathbf{x}(t) + \begin{bmatrix} 3 \\ 2t \end{bmatrix}.$$

Consider the associated homogeneous system and the characteristic equation of matrix A:

$$\begin{vmatrix} 0.6-\lambda & 0.1 \\ 0.4 & 0.9-\lambda \end{vmatrix} = 0 \implies \lambda^2 - 1.5\lambda + 0.5 = 0.$$

It follows that we have two real eigenvalues $\lambda_1 = 1$ and $\lambda_2 = \frac{1}{2}$. The corresponding eigenvectors are

$$\mathbf{v}^1 = \alpha \begin{bmatrix} \frac{1}{4} \\ 1 \end{bmatrix}, \quad \alpha \neq 0; \quad \mathbf{v}^2 = \beta \begin{bmatrix} -1 \\ 1 \end{bmatrix}, \quad \beta \neq 0.$$

Therefore, the general solution of the associated homogeneous system is

$$\mathbf{z}(t) = c_1 \begin{bmatrix} \frac{1}{4} \\ 1 \end{bmatrix} e^t + c_2 \begin{bmatrix} -1 \\ 1 \end{bmatrix} e^{\frac{1}{2}t}.$$

Now we look for a particular solution $\mathbf{x}^0(t)$ of the original non homogeneous system. As the forcing term is a vector of polynomials of degree less than or equal to 1, we try a solution of the following type

$$\mathbf{x}^0(t) = \begin{bmatrix} a_1 + b_1 t \\ a_2 + b_2 t \end{bmatrix},$$

with $a_1, b_1, a_2, b_2 \in \mathbb{R}$, to be determined. We have

$$(\mathbf{x}^0)'(t) = \begin{bmatrix} b_1 \\ b_2 \end{bmatrix}.$$

By substituting into the system, we obtain

$$\begin{bmatrix} b_1 \\ b_2 \end{bmatrix} = \begin{bmatrix} 0.6 & 0.1 \\ 0.4 & 0.9 \end{bmatrix} \begin{bmatrix} a_1 + b_1 t \\ a_2 + b_2 t \end{bmatrix} + \begin{bmatrix} 3 \\ 2t \end{bmatrix}$$

that is

$$\begin{cases} 0.6(a_1 + b_1 t) + 0.1(a_2 + b_2 t) + 3 = b_1 \\ 0.4(a_1 + b_1 t) + 0.9(a_2 + b_2 t) + 2t = b_2 \end{cases}$$

that is

$$\begin{cases} 0.6a_1 + 0.1a_2 - b_1 + (0.6b_1 + 0.1b_2)t = -3 \\ 0.4a_1 + 0.9a_2 - b_2 + (0.4b_1 + 0.9b_2)t = -2t. \end{cases}$$

Therefore, owing to the identity principle of polynomials, we must have

$$\begin{cases} 0.6a_1 + 0.1a_2 - b_1 = -3 \\ 0.4a_1 + 0.9a_2 - b_2 = 0 \\ 0.6b_1 + 0.1b_2 = 0 \\ 0.4b_1 + 0.9b_2 = -2. \end{cases}$$

From the third equation we have $b_2 = -6b_1$. By substituting into the fourth equation, we have $0.4b_1 + 0.9(-6b_1) = -2$, from which $b_1 = \frac{2}{5}$ and $b_2 = -6\left(\frac{2}{5}\right) = -\frac{12}{5}$. From the first equation we get

$$0.6a_1 + 0.1a_2 = -\frac{13}{5}.$$

From the second equation we get

$$0.4a_1 + 0.9a_2 = -\frac{12}{5}.$$

Now consider the system formed by the last two equation; its solution is given by $a_1 = -\frac{21}{5}$, $a_2 = -\frac{4}{5}$. Hence

$$\mathbf{x}^0(t) = \begin{bmatrix} -\frac{21}{5} + \frac{2}{5}t \\ -\frac{4}{5} - \frac{12}{5}t \end{bmatrix} = -\begin{bmatrix} \frac{21}{5} \\ \frac{4}{5} \end{bmatrix} + \begin{bmatrix} \frac{2}{5} \\ -\frac{12}{5} \end{bmatrix} t.$$

Therefore the general solution of the original system is

$$\mathbf{x}(t) = c_1 \begin{bmatrix} \frac{1}{4} \\ 1 \end{bmatrix} e^t + c_2 \begin{bmatrix} -1 \\ 1 \end{bmatrix} e^{\frac{1}{2}t} - \begin{bmatrix} \frac{21}{5} \\ \frac{4}{5} \end{bmatrix} + \begin{bmatrix} \frac{2}{5} \\ -\frac{12}{5} \end{bmatrix} t.$$

Example 6.6.3 Consider a case where the matrix A is singular (i.e. $|A| = 0$):

$$\mathbf{x}'(t) = \begin{bmatrix} 2 & 4 \\ 4 & 8 \end{bmatrix} \mathbf{x}(t) + \begin{bmatrix} 3 \\ 2 \end{bmatrix}.$$

The coefficient matrix of the system is singular and hence it has a zero eigenvalue (its characteristic equation is $\lambda^2 - 10\lambda = 0$). The general solution of the associated homogeneous system is

$$\mathbf{z}(t) = c_1 \begin{bmatrix} 2 \\ -1 \end{bmatrix} + c_2 \begin{bmatrix} 1 \\ 2 \end{bmatrix} e^{10t}, \quad c_1, c_2 \in \mathbb{R}.$$

As vector $\mathbf{b}$ is a constant vector, we could try to find a particular solution of the original system having the same structure, i.e. a constant particular solution. Hence we try with the vector

$$\mathbf{x}^0(t) = \begin{bmatrix} u_1 \\ u_2 \end{bmatrix}$$

and therefore

$$(\mathbf{x}^0(t))' = \begin{bmatrix} 0 \\ 0 \end{bmatrix}.$$

By substituting into the system we have

$$\begin{bmatrix} 2 & 4 \\ 4 & 8 \end{bmatrix} \begin{bmatrix} u_1 \\ u_2 \end{bmatrix} = \begin{bmatrix} -3 \\ -2 \end{bmatrix}.$$

Being $\text{rank}(A) = 1$ and $\text{rank}(A \mid \mathbf{b}) = 2$, by the Rouché-Capelli theorem, the above system is inconsistent. The presence of the zero eigenvalue rules out the possibility to find a particular solution of the original system, made of constants. We can try with a vector of polynomials of degree one:

$$\mathbf{x}^0(t) = \begin{bmatrix} u_1 + w_1 t \\ u_2 + w_2 t \end{bmatrix}.$$

Then we have

$$(\mathbf{x}^0(t))' = \begin{bmatrix} w_1 \\ w_2 \end{bmatrix}.$$

By substituting, as usual, into the system, we obtain

$$\begin{bmatrix} w_1 \\ w_2 \end{bmatrix} = \begin{bmatrix} 2 & 4 \\ 4 & 8 \end{bmatrix} \begin{bmatrix} u_1 + w_1 t \\ u_2 + w_2 t \end{bmatrix} + \begin{bmatrix} 3 \\ 2 \end{bmatrix}$$

from which

$$\begin{cases} w_1 = 2u_1 + 2tw_1 + 4u_2 + 4tw_2 + 3 \\ w_2 = 4u_1 + 4tw_1 + 8u_2 + 8tw_2 + 2. \end{cases}$$

Then, by the same procedure used in the previous example, we obtain the following system

$$\begin{cases} w_1 - 2u_1 - 4u_2 = 3 \\ 2w_1 + 4w_2 = 0 \\ w_2 - 4u_1 - 8u_2 = 2 \\ 4w_1 + 8w_2 = 0. \end{cases}$$

The fourth equation is equivalent to the second equation and hence it is eliminated. From the second equation we deduce $w_1 = -2w_2$, and by substituting this relation into the first equation, we obtain the system

$$\begin{cases} 2u_1 + 4u_2 + 2w_2 = -3 \\ 4u_1 + 8u_2 - w_2 = -2. \end{cases}$$

We consider u_1 as a parameter and solve this system by the Cramer rule:

$$u_2 = -\frac{7 + 10u_1}{20}, \quad w_2 = -\frac{4}{5}, \quad u_1 \in \mathbb{R}.$$

Hence $w_1 = \frac{8}{5}$. For simplicity we can choose $u_1 = 0$ and hence $u_2 = -\frac{7}{20}$. Then we obtain the following particular solution of the original system

$$\mathbf{x}^0(t) = \begin{bmatrix} \frac{8}{5}t \\ -\frac{7}{20} - \frac{4}{5}t \end{bmatrix}.$$

Therefore the general solution of the non homogeneous original system is

$$\mathbf{x}(t) = c_1 \begin{bmatrix} 2 \\ -1 \end{bmatrix} + c_2 \begin{bmatrix} 1 \\ 2 \end{bmatrix} e^{10t} + \begin{bmatrix} \frac{8}{5}t \\ -\frac{7}{20} - \frac{4}{5}t \end{bmatrix}, \quad c_1, c_2 \in \mathbb{R}.$$

Another method useful to find a particular solution of system (6.19) is the so-called "method of variations of parameters", previously described, with reference to systems with variable coefficients. Consider again the system of Example 6.4.6:

$$\mathbf{x}'(t) = \begin{bmatrix} 2 & -2 \\ -2 & 2 \end{bmatrix} \mathbf{x}(t) + \begin{pmatrix} \frac{1}{\sqrt{1-t^2}} + \frac{e^{4t}}{1+t^2} \\ \frac{1}{\sqrt{1-t^2}} - \frac{e^{4t}}{1+t^2} \end{pmatrix}.$$

Now we are able to find a fundamental system for the associated homogeneous system. We have

$$\begin{vmatrix} 2-\lambda & -2 \\ -2 & 2-\lambda \end{vmatrix} = (2-\lambda)^2 - 4 = \lambda^2 - 4\lambda + 4 - 4 = \lambda(\lambda - 4).$$

Then we have two distinct real eigenvalues, $\lambda_1 = 0$ and $\lambda_2 = 4$, with associated eigenvectors

$$\mathbf{v}^1 = \alpha \begin{bmatrix} 1 \\ 1 \end{bmatrix}, \ \alpha \neq 0; \ \ \mathbf{v}^2 = \beta \begin{bmatrix} 1 \\ -1 \end{bmatrix}, \ \ \beta \neq 0.$$

The general solution of the homogeneous system is therefore

$$\mathbf{z}(t) = c_1 \begin{bmatrix} 1 \\ 1 \end{bmatrix} + c_2 \begin{bmatrix} 1 \\ -1 \end{bmatrix} e^{4t}.$$

For the conclusion of the exercise see Example 6.4.6.

6.7 Differential Linear Equations of Order n with Constant Coefficients

In the present section we are concerned with equations of the type

$$x^{(n)}(t) + a_1 x^{(n-1)}(t) + \cdots + a_{n-1} x'(t) + a_n x(t) = b(t) \tag{6.20}$$

where $n \geqq 2$, $a_1, a_2, \ldots a_n$ are real constants and $b(t)$ is a continuous function on an interval $I \subset \mathbb{R}$. If $b(t) \equiv 0$, the equation is said to be *homogeneous,* otherwise *non homogeneous* or *complete.* We have previously remarked that it is always possible to transform these equations into a first-order system (of dimension n) and vice-versa. Indeed, if we put

$$\begin{cases} x_1(t) = x(t) \\ x_2(t) = x'(t) = x_1'(t) \\ x_3(t) = x''(t) = x_2'(t) \\ \cdots\cdots\cdots\cdots\cdots\cdots \\ x_n(t) = x^{(n-1)}(t) \implies x_n'(t) = x^{(n)}(t) \end{cases}$$

it is quite easy to see that (6.20) is equivalent to the first-order system

$$\begin{cases} x_1'(t) = x_2(t) \\ x_2'(t) = x_3(t) \\ x_3'(t) = x_4(t) \\ \dots\dots\dots\dots \\ x_n'(t) = -a_1 x_n(t) - \cdots - a_{n-1} x_2(t) - a_n x_1(t) + b(t) \end{cases}$$

and hence Eq. (6.20) can be equivalently written in the form

$$\mathbf{x}'(t) = A\mathbf{x}(t) + \mathbf{b}(t)$$

where $\mathbf{x}(t) = [x_1(t), x_2(t), \dots, x_n(t)]^\top$,

$$A = \begin{bmatrix} 0 & 1 & 0 & \cdots & 0 & 0 \\ 0 & 0 & 1 & \cdots & 0 & 0 \\ \vdots & \vdots & \vdots & \ddots & \vdots & \vdots \\ 0 & 0 & 0 & \cdots & 0 & 1 \\ -a_n & -a_{n-1} & \cdots & \cdots & -a_2 & -a_1 \end{bmatrix}, \quad \mathbf{b}(t) = \begin{bmatrix} 0 \\ 0 \\ \vdots \\ 0 \\ b(t) \end{bmatrix}.$$

The above matrix A is said "companion matrix"; the first component of $\mathbf{x}(t)$ is the solution of Eq. (6.20). From the said equivalence, it is immediate to deduce the following basic results.

(i) The set S_0 of all solutions of the associated homogeneous equation

$$x^{(n)}(t) + a_1 x^{(n-1)}(t) + \cdots + a_{n-1} x'(t) + a_n x(t) = 0 \tag{6.21}$$

is a linear space of dimension n.

(ii) The set S_b of all solutions of (6.20) is the affine space $S_b = S_0 + x_0$, where x_0 is any particular solution of (6.20).

(iii) The method of undetermined coefficients continues to hold for equations of type (6.20).

In many cases it is convenient to treat directly Eq. (6.20), instead of the equivalent first-order system previously described. Therefore consider the associated homogeneous equation (6.21). Obviously, it admits always the zero solution $x(t) = 0$. We are looking for nonzero solutions of the type

$$x(t) = e^{\lambda t}.$$

Now we compute the subsequent derivatives of order n of this function. We get

$$x'(t) = \lambda e^{\lambda t}; \;\; x''(t) = \lambda^2 e^{\lambda t}; \dots ; x^{(n-1)}(t) = \lambda^{n-1} e^{\lambda t}; \;\; x^{(n)}(t) = \lambda^n e^{\lambda t}.$$

Hence, if we suppose that $e^{\lambda t}$ is a solution of the homogeneous equation (6.21), by substituting the above derivatives, we obtain

$$\lambda^n e^{\lambda t} + a_1 \lambda^{n-1} e^{\lambda t} + \cdots + a_{n-1} \lambda e^{\lambda t} + a_n e^{\lambda t} = 0.$$

Now we divide both members of the previous equation by $e^{\lambda t}$ (quantity different from zero, for all t):

$$\lambda^n + a_1 \lambda^{n-1} + \cdots + a_{n-1} \lambda + a_n = 0. \tag{6.22}$$

Hence, the homogeneous equation (6.21) has a solution of the form $e^{\lambda t}$ only if λ is a root of Eq. (6.22); this algebraic equation, of degree n, is said the *characteristic equation* of (6.21) and its roots are nothing but the eigenvalues of the companion matrix A previously described. Then, it is possible to prove what follows.

(*i*) If the n roots $\lambda_1, \lambda_2, \ldots, \lambda_n$ of (6.22) are real and distinct, the general solution of (6.21) is given by

$$x(t) = c_1 e^{\lambda_1 t} + c_2 e^{\lambda_2 t} + \cdots + c_n e^{\lambda_n t},$$

with $c_1, c_2, \ldots, c_n$ arbitrary constants.

(*ii*) If in (6.22) there are p roots all equal to λ_1, instead of $c_1 e^{\lambda_1 t}$ in the expression of the general solution of (6.22), it will appear the sum

$$c_1 e^{\lambda_1 t} + c_2 t e^{\lambda_1 t} + \cdots + c_{p-1} t^{p-1} e^{\lambda_1 t},$$

with $c_1, c_2, \ldots, c_{p-1}$ arbitrary constants.

(*iii*) if in (6.22) there are complex (conjugate) roots, for example the roots $\alpha + i\beta$ and $\alpha - i\beta$, it is always possible to transform the complex solutions into equivalent real solutions; the "contribution" to the general solution of (6.21) of the said two complex roots, expressed in real terms, is given by

$$c_1 e^{\alpha t} \cos \beta t + c_2 e^{\alpha t} \sin \beta t,$$

that is

$$e^{\alpha t}(c_1 \cos \beta t + c_2 \sin \beta t),$$

being c_1 and c_2 arbitrary constants.

Example 6.7.1

(1) Solve the differential equation

$$x''(t) + 3x'(t) + 2x(t) = 0.$$

From $\lambda^2 + 3\lambda + 2 = 0$ we obtain $\lambda_1 = -1$ and $\lambda_2 = -2$, from which

$$x(t) = c_1 e^{-t} + c_2 e^{-2t}, \quad c_1, c_2 \in \mathbb{R}.$$

(2) Solve the equation

$$x''(t) - 4x'(t) + 4x(t) = 0.$$

From $\lambda^2 - 4\lambda + 4 = 0 = (\lambda - 2)^2 = 0$, we obtain $\lambda_1 = \lambda_2 = 2$, from which

$$x(t) = c_1 e^{2t} + c_2 t e^{2t} = (c_1 + c_2 t)e^{2t}, \quad c_1, c_2 \in \mathbb{R}.$$

(3) Solve the equation (of order four)

$$x^{(4)}(t) - x(t) = 0.$$

Its characteristic equation $\lambda^4 - 1 = 0$ has the following distinct roots

$$\begin{cases} \lambda_1 = 1 \\ \lambda_2 = -1 \\ \lambda_3 = i \\ \lambda_4 = -i. \end{cases}$$

Its (real) general solution is therefore

$$x(t) = c_1 e^t + c_2 e^{-t} + c_3 \cos t + c_4 \sin t.$$

(4) Solve the equation

$$x^{(4)}(t) + 7x^{(3)}(t) + 9x''(t) - 27x'(t) - 54x(t) = 0.$$

The associated characteristic equation is

$$\lambda^4 + 7\lambda^3 + 9\lambda^2 - 27\lambda - 54 = 0,$$

which has the solutions $\lambda_1 = 2$ (simple root), $\lambda_2 = \lambda_3 = \lambda_4 = -3$. Therefore the general solution of the equation is

$$x(t) = c_1 e^{2t} + (c_2 + c_3 t + c_4 t^2)e^{-3t}, \quad c_1, c_2, c_3, c_4 \in \mathbb{R}.$$

(5) Solve the following Cauchy problem:

$$\begin{cases} x^{(5)}(t) + 8x^{(3)}(t) + 16x'(t) = 0, \\ x(0) = x'(0) = x''(0) = x'''(0) = 0, \\ x^{(4)}(0) = 1. \end{cases}$$

By proceeding as in the previous examples, we find that the general solution is

$$c_1 + (c_2 t + c_3)\sin 2t + (c_4 t + c_5)\cos 2t.$$

Then, by imposing the initial conditions, we obtain the system

$$\begin{cases} c_1 + c_5 = 0 \\ 2c_3 + c_4 = 0 \\ 3c_2 - 2c_5 = 0 \\ 2c_3 + 3c_4 = 0 \\ -40c_2 + 16c_5 = 1. \end{cases}$$

This system has the solutions $c_1 = \frac{3}{32}$, $c_2 = -\frac{1}{16}$, $c_3 = c_4 = 0$, $c_5 = -\frac{3}{32}$.

(6) Find the family of functions $x : \mathbb{R} \longrightarrow \mathbb{R}$ such that

$$x''(t) = x(t).$$

From $x''(t) - x(t) = 0$ we have $\lambda^2 - 1 = 0$, which has the roots $\lambda_1 = 1$ and $\lambda_2 = -1$. Hence the family of functions satisfying the given request is

$$x(t) = c_1 e^t + c_2 e^{-t}, \quad c_1, c_2 \in \mathbb{R}.$$

Example 6.7.2 Find the solution of the equation

$$x'''(t) - 4x''(t) + 3x'(t) = 0$$

and verify the solution by means of the companion matrix.

We have

$$\lambda^3 - 4\lambda^2 + 3\lambda = 0$$

which gives $\lambda_1 = 0$, $\lambda_2 = 1$, $\lambda_3 = 3$. Hence

$$x(t) = c_1 + c_2 e^t + c_3 e^{3t}.$$

The companion matrix is

$$A = \begin{bmatrix} 0 & 1 & 0 \\ 0 & 0 & 1 \\ 0 & -3 & 4 \end{bmatrix};$$

we have $|A - \lambda I| = -\lambda^3 + 4\lambda^2 - 3\lambda = 0$, which gives $\lambda_1 = 0$, $\lambda_2 = 1$, $\lambda_3 = 3$ (as before). Then we have, with $\lambda_1 = 0$,

$$\mathbf{v}^1 = c_1 \begin{bmatrix} 1 \\ 0 \\ 0 \end{bmatrix}, \quad c_1 \neq 0;$$

with $\lambda_2 = 1$,

$$\mathbf{v}^2 = c_2 \begin{bmatrix} 1 \\ 1 \\ 1 \end{bmatrix}, \quad c_2 \neq 0;$$

with $\lambda_3 = 3$,

$$\mathbf{v}^3 = c_3 \begin{bmatrix} 1 \\ 3 \\ 9 \end{bmatrix}, \quad c_3 \neq 0.$$

By recalling that we have to take into consideration only the *first* component of $\mathbf{x}(t)$, we obtain the same solution as before.

For what concerns the *non homogeneous* linear differential equations of order n with constant coefficients

$$x^{(n)}(t) + a_1 x^{(n-1)}(t) + \cdots + a_{n-1}x'(t) + a_n x(t) = b(t), \tag{6.23}$$

what said with reference to the non homogeneous systems, continues to hold for the (scalar) equations; for example, it holds the superposition principle and the method of undetermined coefficients (or similarity principle). In particular, with reference to the similarity principle, the following rules are useful:

- If $b(t)$ is a *polynomial* of degree r, when $\lambda = 0$ is not a root of the characteristic equation of the associated homogeneous equation, then there exists a particular solution of (6.23) given by $x_0(t) = q(t)$, where $q(t)$ is a polynomial of degree r. When $\lambda = 0$ is a root of multiplicity m of the characteristic equation of the associated homogeneous equation, then a particular solution of (6.23) is given by $x_0 = t^m q(t)$.
- Let $b(t) = R(t)e^{\alpha t}$, where $R(t)$ is a polynomial of degree r. If α is not a solution of the characteristic equation of the associated homogeneous equation, there exists a particular solution of (6.23) given by $x_0(t) = Q(t)e^{\alpha t}$, where $Q(t)$ is a polynomial of degree r. If α is a solution of the characteristic equation of the associated homogeneous equation, with multiplicity m, then (6.23) has a particular solution of the form $x_0(t) = t^m Q(t)e^{\alpha t}$.
- Let $b(t) = R(t)e^{\alpha t} \sin \beta t$ (or $b(t) = R(t)e^{\alpha t} \cos \beta t$), where $R(t)$ is a polynomial of degree r. If $\alpha \pm i\beta$ is not a solution of the characteristic equation of the associated homogeneous equation, a particular solution of (6.23) has the form

$$x_0(t) = e^{\alpha t} \left[Q(t) \sin \beta t + S(t) \cos \beta t \right],$$

where $Q(t)$ and $S(t)$ are polynomials of degree r. If $\alpha \pm i\beta$ is a solution of multiplicity m of the characteristic equation, then a particular solution of (6.23) has the form

$$x_0(t) = t^m e^{\alpha t} \left[Q(t) \sin \beta t + S(t) \cos \beta t \right],$$

where $Q(t)$ and $S(t)$ are polynomials of degree r.

Obviously, also the method of variations of parameters can be applied to the case of non homogeneous linear differential equations of order n, with constant coefficients, but usually the similarity principle, with the rules described above, results a more convenient method.

6.8 Equilibrium Solutions. Stability of Equilibrium Solutions

Consider a first-order *autonomous* dynamical system of the form

$$\mathbf{x}'(t) = \mathbf{f}(\mathbf{x}(t)), \tag{6.24}$$

where $\mathbf{x}, \mathbf{x}' : I \subset \mathbb{R} \longrightarrow \mathbb{R}^n$ and $\mathbf{f} : \mathbb{R}^n \longrightarrow \mathbb{R}^n$. We are interested in the existence of a *constant solution* $\mathbf{x}^*$, that is a vector $\mathbf{x}^* \in \mathbb{R}^n$ such that if at a time t_0 the system (6.24) assumes the values described by $\mathbf{x}^*$, then the trajectory of the dynamical system becomes equal to the constant vector $\mathbf{x}^*$, for the remaining time $t \geqq t_0$. Such a point $\mathbf{x}^*$ is called an *equilibrium point* or *equilibrium solution* or also *steady point* for (6.24). We give the following basic definition.

Definition 6.8.1 A *constant solution* of (6.24) $\mathbf{x}^*(t) = \mathbf{x}^*, \forall t \in I \subset \mathbb{R}$, is called an *equilibrium solution* or *equilibrium point* for (6.24).

How it is possible to realize the existence of equilibrium points for (6.24) and to locate these points? We are not obliged to solve the system: it is sufficient to note that $\mathbf{x}^*(t) = \mathbf{x}^*$, a constant vector, if and only if all derivatives are zero. Therefore the vector equation which identifies the equilibrium points (if any) for (6.24) is, since $(\mathbf{x}^*)' = [0]$,

$$\mathbf{f}(\mathbf{x}^*) = [0].$$

In other words, the equilibrium solutions (if any) are the zero points of $\mathbf{f}$.

Example 6.8.2

(1) Consider the n-dimensional non homogeneous system

$$\mathbf{x}'(t) = A\mathbf{x}(t) + \mathbf{b}.$$

The condition which defines the equilibrium points of the above system is

$$A\mathbf{x}^* + \mathbf{b} = [0] \implies A\mathbf{x}^* = -\mathbf{b},$$

which is a linear system of n algebraic equations with n unknowns (the components of the equilibrium solutions). If $\det(A) \neq 0 \iff \text{rank}(A) = n$, we have a Cramerian system and hence a unique equilibrium point; if $\det(A) = 0 \iff \text{rank}(A) < n$, we have to verify the validity or not of the Rouché-Capelli theorem:

$$\text{rank}(A) = \text{rank}(A \mid \mathbf{b}) \iff \text{there are } \infty^{n-r} \text{ equilibrium solutions;}$$

$$\text{rank}(A) < \text{rank}(A \mid \mathbf{b}) \iff \text{there are no equilibrium solutions } \mathbf{x}^*.$$

Consider the following system

$$\mathbf{x}'(t) = \begin{bmatrix} 1 & 2 \\ 2 & \alpha \end{bmatrix} \mathbf{x}(t) + \begin{bmatrix} 0 \\ 1 \end{bmatrix}, \quad \alpha \in \mathbb{R}.$$

The system which produces the equilibrium points (if any) is

$$\begin{bmatrix} 1 & 2 \\ 2 & \alpha \end{bmatrix} \begin{bmatrix} x_1^* \\ x_2^* \end{bmatrix} = \begin{bmatrix} 0 \\ -1 \end{bmatrix}.$$

If $\alpha \neq 4$, the coefficients matrix is nonsingular, hence

$$\begin{bmatrix} x_1^* \\ x_2^* \end{bmatrix} = \begin{bmatrix} 1 & 2 \\ 2 & \alpha \end{bmatrix}^{-1} \begin{bmatrix} 0 \\ -1 \end{bmatrix} = \frac{1}{\alpha - 4} \begin{bmatrix} 2 \\ 1 \end{bmatrix}.$$

If $\alpha = 4$, the system has no equilibrium solutions.

(2) Suppose that the linear system of the previous point is homogeneous:

$$\mathbf{x}'(t) = A\mathbf{x}(t).$$

In this case there exists always an equilibrium point, that is $\mathbf{x}^* = [0]$. If $\det(A) \neq 0$, there exists a *unique* equilibrium solution, i.e. $\mathbf{x}^* = [0]$; if $\det(A) = 0$, there exist ∞^{n-r} equilibrium solutions, being $r = \text{rank}(A)$. In any case the equilibrium solutions are all those vectors belonging to $\ker(\mathbf{f})$, with $\mathbf{f} = A\mathbf{x}$. Consider the previous example, with $\mathbf{b} = [0]$. We find that:

- If $\alpha \neq 4$, $\mathbf{x}^* = [0]$ is the unique equilibrium point.
- If $\alpha = 4$, all vectors of the type

$$\begin{bmatrix} \alpha \\ -\frac{\alpha}{2} \end{bmatrix}, \quad \text{with } \alpha \in \mathbb{R},$$

are equilibrium solutions.

(3) Consider the following (nonlinear) system

$$\begin{cases} x_1'(t) = (x_1 - 5)t(1 + (x_2)^2)(x_2 - 3) \\ x_2'(t) = (x_1 + 5)t(x_2 + 1)(x_2 - 3). \end{cases}$$

The vector

$$\mathbf{x}^* = \begin{bmatrix} 5 \\ -1 \end{bmatrix}$$

is an equilibrium solution of the system. Also all points of the line of equation $x_2 = 3$ are equilibrium solutions.

(4) The process of price adjustment ("tâtonnement" in the terminology of L. Walras) in an economic model of Walrasian competitive market of pure exchange is described (in the formulation due to P. A. Samuelson) by the system of differential equations

$$\mathbf{p}'(t) = \mathbf{E}(\mathbf{p}(t)),$$

where $\mathbf{p}$ is the (positive) vector of prices and E_i $(i = 1, \ldots, n)$ is the *excess demand function* for the i-th good exchanged in the market:

$$E_i(\mathbf{p}) = D_i(\mathbf{p}) - S_i(\mathbf{p}),$$

being $D_i(\mathbf{p})$ the demand function for the i-th good and $S_i(\mathbf{p})$ the corresponding supply function. The vector $\mathbf{p}^*$ is an equilibrium vector if

$$\mathbf{E}(\mathbf{p}^*) = [0],$$

that is, for every i,

$$D_i(\mathbf{p}^*) = S_i(\mathbf{p}^*).$$

Obviously, it is also possible to give a definition of an equilibrium solution for non autonomous systems of the type

$$\mathbf{x}'(t) = \mathbf{f}(\mathbf{x}(t), t),$$

but for simplicity we shall consider only autonomous systems.

Equilibrium solutions of a dynamical system are interesting if their existence help us to forecast the behaviour of the other solutions of the system (maybe related to an initial value problem), with respect to the equilibrium solution we have found. Consider, for instance, the case of a portrait hanged to a wall; the portrait is "in equilibrium" and if we give a rotating stroke to the portrait, the same, after some oscillations, comes back to its original equilibrium state (like a pendulum). In this case we say that the equilibrium is *stable:* by modifying this "quiet position", we give rise to a dynamical law that tends to bring again the system to its original position. Again: if we put a bottle on a table, the bottle is obviously in an equilibrium position. If we give a little stroke to the bottle, the same, after some totterings, comes back to its original position, but if we give a strong stroke, the bottle falls on the table (and perhaps on the floor). The equilibrium is in this case only *locally stable,* where the term "locally" means that the displacement from the equilibrium position must not be too wide, if we want to maintain the original equilibrium state.

It is time to give the formal definitions. Consider system (6.24), that is

$$\mathbf{x}'(t) = \mathbf{f}(\mathbf{x}(t)),$$

with $\mathbf{f}$ of class C^1 on an open set $\Omega \subset \mathbb{R}^n$ and consider its solution $\mathbf{x}(t)$, together with the initial value problem (Cauchy problem) $\mathbf{x}(t_0) = \mathbf{x}^0$, being $t \in I = [0, +\infty)$.

Definition 6.8.3 An equilibrium point $\mathbf{x}^*$ of (6.24) is said to be *stable in the sense of Lyapunov* or *L-stable* or *neutrally stable* for $t \geqq t_0$, if for every $\varepsilon > 0$ there exists $\delta > 0$ such that, if $\left\|\mathbf{x}^0 - \mathbf{x}^*\right\| < \delta$, then the solution $\mathbf{x}(t)$ with initial value $\mathbf{x}(t_0) = \mathbf{x}^0$, for every $t \geqq t_0$ satisfies the condition

$$\left\|\mathbf{x}(t) - \mathbf{x}^*\right\| < \varepsilon.$$

In other words, $\mathbf{x}^*$ is stable in the sense of Lyapunov (A. M. Lyapunov, 1857–1918, Russian mathematician) if all solutions of the system that start from $\mathbf{x}^*$ with a distance lower than δ with respect to $\mathbf{x}^*$, are well defined for all $t \geqq t_0$ and never go away from $\mathbf{x}^*$ of a distance greater than ε. For the one-dimensional case, that is for the autonomous differential equation $x'(t) = f(x(t)), x \in \mathbb{R}, f \in \mathbb{R}$, we can visualize the previous situation by means of Fig. 6.2 ($t_0 = 0$).

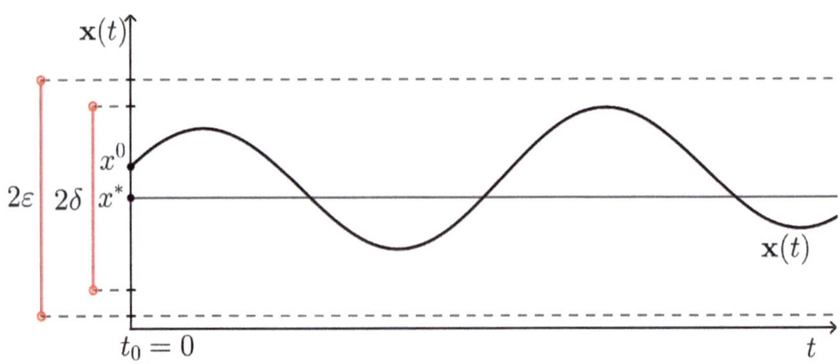

Fig. 6.2 An equilibrium point $\mathbf{x}^*$, it is stable in the sense of Lyapunov

We remark again that the concept introduced above is a "local" concept, as it makes reference to the behaviour of the solutions in a neighbourhood of $\mathbf{x}^*$, and δ may be quite small. Moreover, the definition does not require that $\mathbf{x}(t)$ must converge to $\mathbf{x}^*$, when $t \longrightarrow +\infty$, but essentially it requires that if $\mathbf{x}^0$ is sufficiently close to $\mathbf{x}^*$, then $\mathbf{x}(t)$, for $t \geqq t_0$ remains indefinitely close to $\mathbf{x}^*$.

Now we give the formal definition of "local stability" for an equilibrium point $\mathbf{x}^*$.

Definition 6.8.4 An equilibrium point $\mathbf{x}^*$ of (6.24) is said to be *locally (asymptotically) stable* if it is stable in the sense of Lyapunov and there exists $\delta_1 > 0$ such that for every solution $\mathbf{x}(t)$ with initial value condition $\mathbf{x}(t_0) = \mathbf{x}^0$, for which it holds

$$\left\| \mathbf{x}^0 - \mathbf{x}^* \right\| < \delta_1,$$

we have

$$\lim_{t \longrightarrow +\infty} (\mathbf{x}(t) - \mathbf{x}^*) = [0],$$

that is

$$\lim_{t \longrightarrow +\infty} \mathbf{x}(t) = \mathbf{x}^*.$$

In other words, the equilibrium solution $\mathbf{x}^*$ is locally asymptotically stable if it is L-stable and there exists a neighbourhood of $\mathbf{x}^0$ such that every solution path starting within the said neighbourhood, converges towards $\mathbf{x}^*$.

In the one-dimensional case we may depict the above situation by means of Fig. 6.3 ($t_0 = 0$).

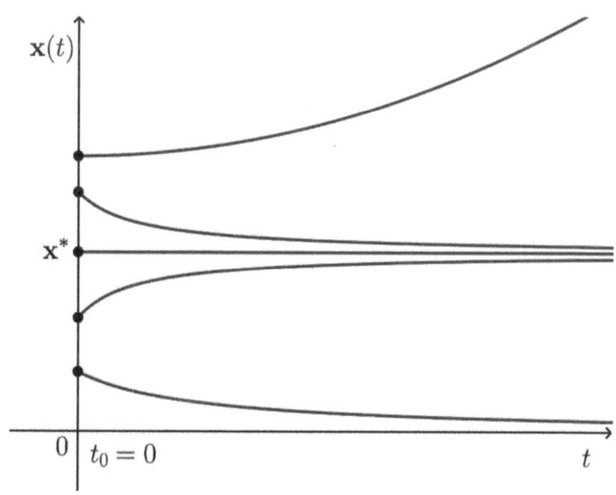

Fig. 6.3 $\mathbf{x}^*$ is locally asymptotically stable

When we have asymptotic stability, it is interesting to know how wide is the set containing the initial points whose related trajectories converge to the equilibrium solution. This leads to the following definition.

Definition 6.8.5 Let be given a dynamical system of type (6.24), with $\mathbf{f}$ of class $\mathcal{C}^1$ on an open set $\Omega \subset \mathbb{R}^n$. Let us consider an equilibrium point $\mathbf{x}^*$; the *basin of attraction of* $\mathbf{x}^*$ is the set

$$A(\mathbf{x}^*) = \left\{ \mathbf{x}^0 \in \Omega \subset \mathbb{R}^n : \lim_{t \longrightarrow +\infty} \mathbf{x}(t) = \mathbf{x}^* \right\},$$

that is the set of all solutions starting from $\mathbf{x}^0 = \mathbf{x}(t_0)$ and converging to the equilibrium solution $\mathbf{x}^*$.

If the attraction basin coincides with the domain Ω of $\mathbf{f}$, that is $A(\mathbf{x}^*) = \Omega$, then the equilibrium point $\mathbf{x}^*$ is *globally asymptotically stable* (and in this case necessarily it is the unique equilibrium point of the system). We point out the obvious implication

$$\mathbf{x}^* \text{ globally asymptotically stable} \implies \mathbf{x}^* \text{ locally asymptotically stable.}$$

In the one-dimensional case we may depict the above situation by means of Fig. 6.4 ($t_0 = 0$)

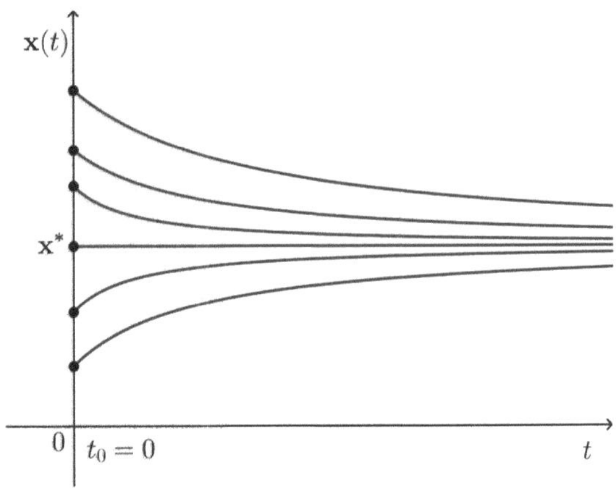

Fig. 6.4 $\mathbf{x}^*$ is globally asymptotically stable

6.9 Stability of the Equilibrium Solutions of Homogeneous Linear Systems with Constant Coefficients and of Homogeneous Linear Equations of Order *n* with Constant Coefficients

Consider the system

$$\mathbf{x}'(t) = A\mathbf{x}(t). \tag{6.25}$$

As previously remarked, the zero vector $[0] \in \mathbb{R}^n$ is an equilibrium solution of (6.25), being obviously $A\,[0] = [0]$. We have the following basic result.

Theorem 6.9.1 *Consider system* (6.25).

(i) *All eigenvalues of A have a negative real part, i.e.* $\mathrm{Re}(\lambda) < 0$ *for every* λ, *if and only if the equilibrium solution* $\mathbf{x}^* = [0]$ *is globally asymptotically stable (and hence stable in the sense of Lyapunov).*

(ii) *The equilibrium solution* $\mathbf{x}^* = [0]$ *is stable in the sense of Lyapunov if and only if all eigenvalues of A have a non positive real part, i.e.* $\mathrm{Re}(\lambda) \leqq 0$ *for every* λ, *and the eigenvalues with zero real part have an algebraic multiplicity equal to the respective geometric multiplicity (i.e. they are regular).*

(iii) *The equilibrium solution* $\mathbf{x}^* = [0]$ *is unstable in all other cases, that is if there exists an eigenvalue with positive real part or with zero real part, but with algebraic multiplicity greater than the corresponding geometric multiplicity.*

Proof We prove only point (i) which requires an easy proof. One becomes convinced of the thesis, by recalling the structure of the solutions of system (6.25). Let λ_s be a real eigenvalue of A; we know that the "contribution" of the eigenvalues of this type to the solution $\mathbf{x}(t)$ is given by the sum of addenda of the type $\mathbf{p}^s(t)e^{\lambda_s t}$, where $\mathbf{p}^s(t)$ is a suitable vector of real polynomials of degree less than or equal to $m_s^A - 1$, where m_s^A is the algebraic multiplicity of λ_s. If $\lambda_s = \alpha_s \pm \beta_s i$, we have addenda of the type

$$e^{\alpha_s t}\left[\mathbf{p}^s(t)\cos(\beta_s t) + \mathbf{q}^s(t)\sin(\beta_s t)\right],$$

if we want to obtain only real solutions. In any case, by considering that the functions $\sin x$ and $\cos x$ are *bounded* on the whole $\mathbb{R}$, if $\mathrm{Re}(\lambda_s) < 0$ for every complex λ_s, then $\mathbf{x}(t) \longrightarrow [0]$ for $t \longrightarrow +\infty$ and hence $[0]$ is an equilibrium asymptotically stable, with basin of attraction coinciding with the whole $\mathbb{R}^n$. Vice-versa, if $[0]$ is an equilibrium asymptotically stable, then we cannot have eigenvalues with a nonnegative real part. Indeed, in this case there will exist solutions which do not converge to the zero vector, when $t \longrightarrow +\infty$. □

Example 6.9.2

(1) The system

$$\mathbf{x}'(t) = \begin{bmatrix} -2 & 2 & 0 \\ 0 & 1 & 1 \\ 0 & -5 & -3 \end{bmatrix} \mathbf{x}(t)$$

has the eigenvalues $\lambda_1 = -2, \lambda_2 = -1 + i, \lambda_3 = -1 - i$. It follows that the equilibrium solution $\mathbf{x}^* = [0]$ is globally asymptotically stable.

(2) The system

$$\mathbf{x}'(t) = \begin{bmatrix} 1 & 0 & -8 \\ 2 & 1 & 8 \\ 3 & 1 & -8 \end{bmatrix} \mathbf{x}(t)$$

has the eigenvalues $\lambda_1 = 0, \lambda_2 = \lambda_3 = -3$. Therefore $\mathbf{x}^* = [0]$ is not a globally asymptotically stable solution, but being $\lambda_1 = 0$ a simple root of the characteristic equation, we can assert that $\mathbf{x}^* = [0]$ is stable in the sense of Lyapunov.

(3) The system

$$\mathbf{x}'(t) = \begin{bmatrix} 1 & 2 \\ -2 & 0 \end{bmatrix} \mathbf{x}(t)$$

has the eigenvalues $\lambda_1 = \frac{1}{2} + i\frac{\sqrt{15}}{2}, \lambda_2 = \frac{1}{2} - i\frac{\sqrt{15}}{2}$. Therefore the equilibrium solution $\mathbf{x}^* = [0]$ is unstable.

Definition 6.9.3 A square real matrix A of order n whose all eigenvalues have a negative real part, i.e. $\text{Re}(\lambda_j) < 0, \forall j$, is said a *stable matrix.*

In the specialized literature we encounter also the notion of *positive stable matrices:* A is positive stable if all its eigenvalues have *positive* real part. Obviously, A is stable if and only if $-A$ is positive stable. We note that if A is stable (or positive stable) then A is nonsingular and A^{-1} is again a stable (or positive stable) matrix. If, in addition, S is a nonsingular matrix, then SAS^{-1} is a stable (or positive stable) matrix as well.

The most known necessary and sufficient conditions for the stability of a square matrix are the following ones:

(a) *Routh-Hurwitz conditions.* We put

$$k_i = \begin{cases} 1, \text{ for } i = 0; \\ (-1)^i \cdot \text{ trace of order } i \text{ of the matrix } A, \; i = 1, \ldots, n; \\ 0, \text{ for } i < 0 \text{ or for } i > n. \end{cases}$$

We remark that k_i, $i = 1, \ldots, n$, is the coefficient of the term λ^{n-i} of the characteristic equation of A, when this equation is written in the form

$$\lambda^n + k_1\lambda^{n-1} + k_2\lambda^{n-2} + \cdots + k_{n-1}\lambda + k_n = 0.$$

We recall that the *trace of order i of the matrix A* is given by the sum of all $\binom{n}{i}$ principal minors of A of order i, with $i = 1, 2, \ldots, n$. The trace of order 1 (or, shortly, trace of A) is therefore given by $\sum_{i=1}^{n} a_{ii}$ and usually denoted by $\mathrm{tr}(A)$.

Now we build the following determinants

$$H_j = \begin{vmatrix} k_1 & k_3 & k_5 & \cdots\cdots \\ k_0 = 1 & k_2 & k_4 & \cdots\cdots \\ k_{-1} = 0 & k_1 & k_3 & \cdots\cdots \\ \vdots & \vdots & \vdots & \vdots \quad \vdots \\ 0 & 0 & \cdots\cdots & k_j \end{vmatrix},$$

$j = 1, \ldots, n$. A straightforward way to build the matrix H (of order n) is: we write the main diagonal of H with the elements k_i in the i-th position; then we move to the right of k_i, by increasing of two units the index of k_i at every step (that is, writing k_{i+2}, k_{i+4}, etc.), then we move to the left of k_i, by decreasing of two units the index of k_i at every step (that is, writing k_{i-2}, k_{i-4}, etc.). For example, with $n = 5$ we have

$$H_5 = \begin{vmatrix} k_1 & k_3 & k_5 & 0 & 0 \\ 1 & k_2 & k_4 & 0 & 0 \\ 0 & k_1 & k_3 & k_5 & 0 \\ 0 & 1 & k_2 & k_4 & 0 \\ 0 & 0 & k_1 & k_3 & k_5 \end{vmatrix}.$$

It can be proved that A is stable if and only if

$$H_1 > 0,\ H_2 > 0, \ldots, H_n > 0.$$

A necessary (but not sufficient) condition for the stability of A is

$$k_1 > 0,\ k_2 > 0, \ldots, k_n > 0.$$

(*b*) *Lyapunov conditions.* A is stable if and only if there exists a *symmetric positive definite* matrix B such that the symmetric matrix

$$BA + A^\top B$$

is negative definite.

For $n = 2$ the Routh-Hurwitz conditions become

$$\operatorname{tr}(A) < 0, \ \det(A) > 0.$$

These conditions can be easily proved as follows (obviously the proof of the Routh-Hurwitz conditions for the general case is much more elaborate). We know that $\operatorname{tr}(A) = a_{11} + a_{22} = \lambda_1 + \lambda_2$ and that $\det(A) = \lambda_1\lambda_2$. Therefore we have the conditions

$$\begin{cases} \lambda_1 + \lambda_2 < 0 \\ \lambda_1\lambda_2 > 0. \end{cases}$$

(1) If $\lambda_1, \lambda_2 \in \mathbb{R}$, the condition $\lambda_1\lambda_2 > 0$ implies that either $\lambda_1 > 0, \lambda_2 > 0$, *but this is excluded from the first relation of the system,* or $\lambda_1 < 0, \lambda_2 < 0$ and hence the solution $x^* = [0]$ is globally asymptotically stable.
(2) If $\lambda_1, \lambda_2 \in \mathbb{C}$, then $\lambda_1 = (a + ib), \lambda_2 = (a - ib)$. It results $\lambda_1 + \lambda_2 < 0 \iff 2a < 0 \iff a < 0$, that is $\operatorname{Re}(\lambda_1) = \operatorname{Re}(\lambda_2) < 0$. In this case the condition $\lambda_1\lambda_2 > 0$ is automatically satisfied, as $(a + ib)(a - ib) = a^2 + b^2$.

In the two dimensional case we can be more accurate about the classification of the equilibrium points. Let us consider, with $n = 2$, the characteristic equation of A:

$$P(\lambda) = \lambda^2 - \operatorname{tr}(A)\lambda + \det(A) = 0.$$

We suppose $\det(A) \neq 0$, i.e. the origin is the unique equilibrium point and A has no zero eigenvalues. If we want to plot the trajectories around the equilibrium point we have the following possibilities. We recall that the eigenvalues are given by

$$\lambda = \frac{\operatorname{tr}(A) \pm \sqrt{(\operatorname{tr}(A))^2 - 4\det(A)}}{2}$$

and that the "discriminant" Δ is

$$\Delta = (\operatorname{tr}(A))^2 - 4\det(A).$$

1. $\Delta = 0$, i.e. there is one real eigenvalue λ with algebraic multiplicity 2. In this case we have the following possibilities.

   (*a*) $\lambda = \lambda_1 = \lambda_2 < 0$, λ regular. In this case the origin is stable.
   (*b*) $\lambda = \lambda_1 = \lambda_2 > 0$, λ regular. In this case the origin is unstable.
   In both cases the origin is also called a "star node".

(*c*) $\lambda = \lambda_1 = \lambda_2, \lambda$ not regular. The origin is called an "improper node" or "Jordan node" and is asymptotically stable if $\lambda < 0$, unstable if $\lambda > 0$.

2. $\Delta > 0$ (i.e. A has two eigenvalues, real and distinct, λ_1 and λ_2). We have the following possibilities.

   (*a*) $\lambda_1 < 0, \lambda_2 < 0$. In this case the origin is stable.
   (*b*) $\lambda_1 > 0, \lambda_2 > 0$. In this case the origin is unstable.
   In both cases the origin is also called a "proper node".
   (*c*) The eigenvalues λ_1 and λ_2 have opposite sign, for instance $\lambda_1 > 0, \lambda_2 < 0$. The origin is unstable and is also called a "saddle point".

3. $\Delta < 0$; the matrix has two complex conjugate eigenvalues λ and $\bar{\lambda} \in \mathbb{C}$. In this case we have the following possibilities.

   (*a*) $\text{Re}(\lambda) < 0, \text{Re}(\bar{\lambda}) < 0$. In this case the origin is stable.
   (*b*) $\text{Re}(\lambda) > 0, \text{Re}(\bar{\lambda}) > 0$. In this case the origin is unstable.
   In both cases the origin is also called a "focus" or "vortex" or "spiral".
   (*c*) $\text{Re}(\lambda) = 0, \text{Re}(\bar{\lambda}) = 0$. In this case the origin is stable, but only in the sense of Lyapunov (hence it is not asymptotically stable). In this case the origin is called a "centre".

There are also several sufficient conditions for the stability of a matrix, some of them important for economic applications. We mention the following ones.

(*i*) If A is *symmetric,* then A is stable if and only if A is negative definite. In this case, indeed, the matrix A has all its negative eigenvalues.
(*ii*) If A is *negative quasidefinite* (that is the symmetric matrix $\frac{1}{2}(A + A^\top)$ is negative definite), then A is stable [31].
(*iii*) If in A it results $a_{ij} \geqq 0, \forall i \neq j$, that is A is a "Metzlerian matrix" (from the American economist L. Metzler), then A is stable if and only if all its leading principal minors (or North-West principal minors) alternate in sign, beginning with the negative sign.
(*iv*) The square matrix A has a *dominant diagonal in the sense of McKenzie* (Lionel McKenzie, 1919–2010, American economist), if there exist positive numbers $d_1, d_2, \ldots, d_n$, such that ("row dominance")

$$d_i \left|a_{ii}\right| > \sum_{j \neq i} d_j \left|a_{ij}\right|, \quad i = 1, 2, \ldots, n$$

or, *equivalently,* ("column dominance")

$$d_j \left|a_{jj}\right| > \sum_{i \neq j} d_i \left|a_{ij}\right|, \quad j = 1, 2, \ldots, n.$$

If in the previous inequalities, it results $d_1 = d_2 = \cdots = d_n = 1$, we have the special case of matrices with a *dominant diagonal in the sense of Hadamard:*

$$|a_{ii}| > \sum_{j \neq i} \left|a_{ij}\right|, \; i = 1, 2, \ldots, n$$

("row dominance in the sense of Hadamard");

$$\left|a_{jj}\right| > \sum_{i \neq j} \left|a_{ij}\right|, \; j = 1, 2, \ldots, n$$

("column dominance in the sense of Hadamard").

Example 6.9.4 The matrix

$$A = \begin{bmatrix} -4 & 2 & -1 \\ -1 & -3 & 1 \\ 1 & 2 & 5 \end{bmatrix}$$

has a row dominance in the sense of Hadamard:

$|-4| > |2| + |-1|$;
$|-3| > |-1| + |1|$;
$|5| > |1| + |2|$.

However, the same matrix has not a column dominance in the sense of Hadamard:

$|4| > |-1| + |1|$;
$|-3| < |2| + |2|$;
$|5| > |-1| + |1|$.

The definition of McKenzie is obviously more general of the one of Hadamard: not only the definition of McKenzie contains the one of Hadamard, but it can be proved that if a matrix has a row dominance in the sense of McKenzie, then it has a column dominance in the sense of McKenzie, and vice-versa. As seen in the previous example, this is no longer true for the definition of Hadamard. For example, the matrix

$$A = \begin{bmatrix} -3 & 6 & -5 \\ 1 & 5 & -4 \\ 2 & -1 & 20 \end{bmatrix}$$

has no row dominance, nor column dominance in the sense of Hadamard, however it satisfies the definition of McKenzie. For instance, with reference to the rows, with $d_1 = 35, d_2 = 12, d_3 = 6$, we have

$35\,|-3| > 12\,|6| + 6\,|-5|$, that is $105 > 102$.
$12\,|5| > 35\,|1| + 6\,|-4|$, that is $60 > 59$.
$6\,|20| > 35\,|2| + 12\,|-1|$, that is $120 > 82$.

Therefore the row dominance (in the sense of McKenzie) is verified.

With reference to the columns, with $d_1 = 20, d_2 = 28, d_3 = 15$, we have

$20\,|-3| > 28\,|1| + 15\,|2|$, that is $60 > 58$.

$28\,|5| > 20\,|6| + 15\,|-1|$, that is $140 > 135$.

$15\,|20| > 20\,|-5| + 28\,|-4|$, that is $300 > 212$.

Therefore the column dominance (in the sense of McKenzie) is verified.

By means of the definition of McKenzie we can thus speak of "matrices with dominant diagonal" (in the sense of McKenzie), without specification if the property is referred to rows or columns. It is used also the term "matrices with quasi-dominant diagonal".

Theorem 6.9.5 (McKenzie) *If the matrix A, of order n, has a dominant diagonal (in the sense of McKenzie), then it is nonsingular.*

Finally, we give the sufficient stability condition of the present point iv).

Theorem 6.9.6 (McKenzie) *If A, of order n, has a negative dominant diagonal (in the sense of McKenzie), that is A has a dominant diagonal and moreover, $a_{ii} < 0, i = 1, \ldots, n$, then A is stable.*

We have seen that a necessary condition for the stability of a square matrix A is that the quantities k_i which appear in the definition of the Routh-Hurwitz determinants, are all positive, Hence the condition $\mathrm{tr}(A) = \sum_{i=1}^{n} a_{ii} < 0$ is a necessary condition for the stability of A. Therefore, if $\mathrm{tr}(A) \geqq 0$, surely A is not stable. Another necessary condition for the stability of A is: $(-1)^n \det(A) > 0$.

Another important notion of stability for square matrices has been considered within the studies concerning the tâtonnement process of Walrasian markets of pure exchange: it is the notion of D-stability (see, e.g., K. J. Arrow and M. McManus, *A note on dynamic stability,* Econometrica, vol. 26, 1958, 448–454).

Definition 6.9.7 A square matrix A of order n, is said to be *D-stable,* if DA is stable for *every* diagonal matrix D having positive diagonal elements: $d_i > 0, \forall i = 1, \ldots, n$.

The previous definition draws its utility from the fact that the various "adjustment processes" concerning demand and supply functions of n goods exchanged in a market, have usually different "speeds", which a priori are not known. Therefore it is desirable that DA (where the diagonal matrix D gathers the various "adjustment speeds") is stable, for any diagonal matrix D, with a positive diagonal. It is obvious that if A is D-stable, then it is also stable (it is sufficient to choose $D = I$), but the vice-versa does not hold (Arrow and McManus give a counterexample, with $n = 2$). The following theorem gives some sufficient conditions for the D-stability of a square matrix.

Theorem 6.9.8 *The square matrix A, of order n, is D-stable if any one of the following conditions holds true:*

(1) There exists a diagonal matrix C, with a positive diagonal, such that the symmetric matrix $AC + A^{\top}C$ is negative definite (Arrow and McManus).

(2) *A is a Metzlerian matrix,* $(a_{ij} \geqq 0, \forall i \neq j)$ *and its leading principal minors (or North-West principal minors) alternate in sign, beginning with the negative sign:* $(-1)^i \Delta_i > 0, \forall i = 1, \ldots, n.$
(3) *A is a Metzlerian matrix and there exists a vector* $h \geq [0]$ *such that* $Ah < [0]$.
(4) *A has a negative dominant diagonal, in the sense of McKenzie.*
(5) *A is negative quasidefinite.*
(6) *A is symmetric and negative definite.*

We have to note that, at present, no general conditions on the elements of A, which are necessary and sufficient for A to be D-stable, are known (it is indeed one of the most difficult questions born within Mathematical Economics, concerning the stability of matrices!). However, it can be proved (see, e.g., [20]) the following result.

- If A, of order n, is *Metzlerian,* that is $a_{ij} \geqq 0, \forall i \neq j$, the following conditions are equivalent:

  (a) A is stable.
  (b) A is D-stable.
  (c) A is *totally stable,* i.e. every principal submatrix of A is D-stable.
  (d) A is *Hicksian,* i.e. every principal minor of A of order r has the sign of $(-1)^r, r = 1, \ldots, n.$
  (e) A has all its leading principal minors (or North-West principal minors) alternating in sign, by beginning with the negative sign.
  (f) A has a negative dominant diagonal, in the sense of McKenzie.

For what concerns the stability of the equilibrium solutions of *non homogeneous linear systems,* with a constant forcing term, i.e. of the systems

$$\mathbf{x}'(t) = A\mathbf{x}(t) + \mathbf{b},$$

there is no need of further results, in order to study the stability of these systems. Indeed, if A has no zero eigenvalues (i.e. A is nonsingular), the above system has a unique equilibrium solution:

$$\mathbf{x}^* = -A^{-1}\mathbf{b}.$$

In this case the general solution of the system is

$$\mathbf{x}(t) = \mathbf{x}^* + \mathbf{z}(t),$$

where $\mathbf{z}(t)$ is the general solution of the associated homogeneous system. But then, we have, when $t \longrightarrow +\infty$,

$$\lim_{t \longrightarrow +\infty} \mathbf{x}(t) = \lim_{t \longrightarrow +\infty} (\mathbf{x}^* + \mathbf{z}(t)) = \mathbf{x}^* + \lim_{t \longrightarrow +\infty} \mathbf{z}(t) = \mathbf{x}^*$$

if and only if

$$\lim_{t\longrightarrow+\infty} \mathbf{z}(t) = [0] \,.$$

Hence, the equilibrium vector $\mathbf{x}^* = -A^{-1}\mathbf{b}$ of the system $\mathbf{x}'(t) = A\mathbf{x}(t) + \mathbf{b}$, when $\det(A) \neq 0$, is globally asymptotically stable if and only if the zero vector $[0]$ is globally asymptotically stable for the associated homogeneous system

$$\mathbf{x}'(t) = A\mathbf{x}(t).$$

Note that $\det(A) = 0$ if and only if at least one eigenvalue of A is zero, but in this case A cannot be a stable matrix.

We consider now the stability of the zero solution of a homogeneous linear differential equation, with constant coefficients, of order n:

$$x^{(n)}(t) + a_1 x^{(n-1)}(t) + a_2 x^{(n-2)}(t) + \cdots + a_n x(t) = 0. \tag{6.26}$$

Also for this case the stability of the zero solution depends on the negativity of the real parts of all roots of the characteristic equation of (6.26). More precisely:

- $x^* = 0$ is an equilibrium solution globally asymptotically stable for (6.26) if and only if all roots of the characteristic polynomial $P(\lambda)$ have a negative real part.
- $x^* = 0$ is an equilibrium solution stable in the sense of Lyapunov for (6.26), if and only if all roots of the characteristic polynomial $P(\lambda)$ have non positive real part and the roots with zero real part are simple roots.

Also for (6.26) the Routh-Hurwitz theorem gives necessary and sufficient conditions for the real part of the roots of $P(\lambda)$ to be negative, by analyzing the coefficients of $P(\lambda)$. Therefore consider the characteristic polynomial in the form

$$P(\lambda) = \lambda^n + a_1\lambda^{n-1} + a_2\lambda^{n-2} + \cdots + a_n$$

(note that we have assumed. without loss of generality, $a_0 = 1$). Then we build the matrix H ("Hurwitz matrix"), square of order n, on the grounds of the following rules.

- The first row of H contains the coefficients of the polynomial $P(\lambda)$ with an *odd* index $(a_1, a_3, a_5, \ldots)$, as long as there are, and then complete the row with zeros.
- The second row contains $a_0 = 1$ and then all coefficients of the polynomial $P(\lambda)$ with an even index, as long as there are, and then complete the row with zeros.
- The third row repeats the first row, but with a zero in the first position (first column); the other elements of the first row are therefore shifted to right.
- The fourth row repeats the second row, but with a shifting to right.

And so on.

$$H = \begin{bmatrix} a_1 & a_3 & a_5 & a_7 & \cdots & 0 & \cdots & 0 \\ 1 & a_2 & a_4 & a_6 & \cdots & 0 & \cdots & 0 \\ 0 & a_1 & a_3 & a_5 & \cdots & 0 & \cdots & 0 \\ 0 & 1 & a_2 & a_4 & \cdots & 0 & \cdots & 0 \\ \vdots & \vdots & \vdots & \vdots & \vdots & \vdots & \vdots & \vdots \\ 0 & 0 & \cdots & \cdots & \cdots & \cdots & \cdots & a_n \end{bmatrix}.$$

We have the following result.

- A necessary and sufficient condition such that all roots of $P(\lambda) = 0$ have a negative real part, is that all leading principal minors (North-West principal minors) of H are *positive.*

For the case $n = 2$, we have

$$H = \begin{bmatrix} a_1 & 0 \\ 1 & a_2 \end{bmatrix},$$

hence the Routh-Hurwitz conditions become $a_1 > 0, a_1 a_2 > 0$, that is $a_1 > 0, a_2 > 0$.

For the case $n = 3$, we have the following Hurwitz matrix

$$H = \begin{bmatrix} a_1 & a_3 & 0 \\ 1 & a_2 & 0 \\ 0 & a_1 & a_3 \end{bmatrix}.$$

The Routh-Hurwitz conditions become: $a_1 > 0, a_2 > 0, a_3 > 0, a_1 a_2 - a_3 > 0$.

Example 6.9.9 Consider the following characteristic equation

$$\lambda^4 + 5\lambda^3 + 8\lambda^2 + 7\lambda + 2 = 0.$$

The corresponding Hurwitz matrix is

$$H = \begin{bmatrix} 5 & 7 & 0 & 0 \\ 1 & 8 & 2 & 0 \\ 0 & 5 & 7 & 0 \\ 0 & 1 & 8 & 2 \end{bmatrix}.$$

We have $\Delta_1 = 5 > 0$,

$$\Delta_2 = \begin{vmatrix} 5 & 7 \\ 1 & 8 \end{vmatrix} = 33 > 0; \;\; \Delta_3 = \begin{vmatrix} 5 & 7 & 0 \\ 1 & 8 & 2 \\ 0 & 5 & 7 \end{vmatrix} = 181 > 0; \;\; \Delta_4 = |H| = 362 > 0.$$

Hence all roots of $P(\lambda) = 0$ have negative real parts.

6.10 Stability of the Equilibrium Solutions of First-Order Nonlinear Autonomous Systems

We have seen that the study of the stability of linear systems can be quite easily carried out, with reference to the stability of the zero equilibrium solution. Now we give some basic notions on the stability of the equilibrium solutions of nonlinear systems of the type

$$\mathbf{x}'(t) = \mathbf{f}(\mathbf{x}(t)).$$

Let us suppose that this system has (at least) one equilibrium solution $\mathbf{x}^*$. We shall give some hints on the so-called "linearization method" and on the so-called "second method of Lyapunov". The first method is suitable for the study of local stability, whereas the method of Lyapunov is suitable also for the study of global stability and, for this reason, it has a great importance in the analysis of stability of the equilibrium solutions of Walrasian economic models. We point out the basic works of Arrow, Hurwicz, Uzawa, McKenzie, Nikaido, Morishima and others on this important questions of economic analysis.

Reconsider the system

$$\mathbf{x}'(t) = \mathbf{f}(\mathbf{x}(t)) \tag{6.27}$$

and suppose that the vector function $\mathbf{f}$ is of class C^1 on an open set of $\mathbb{R}^n$; moreover, suppose that the system has (at least) one equilibrium point $\mathbf{x}^* \in \mathbb{R}^n$, so that by Taylor's formula, we can write

$$\mathbf{f}(\mathbf{x}) = \mathbf{f}(\mathbf{x}^*) + J\mathbf{f}(\mathbf{x}^*)(\mathbf{x} - \mathbf{x}^*) + \mathbf{o}(\|\mathbf{x} - \mathbf{x}^*\|).$$

Therefore we can approximate (locally) system (6.27) by the above linear system generated by Taylor's formula. We write the following "approximating" system

$$\mathbf{x}'(t) = \mathbf{f}(\mathbf{x}^*) + J\mathbf{f}(\mathbf{x}^*)(\mathbf{x} - \mathbf{x}^*) + \mathbf{o}(\|\mathbf{x} - \mathbf{x}^*\|),$$

and, recalling that $\mathbf{f}(\mathbf{x}^*) = [0]$ and that $(\mathbf{x}^*)' = [0]$:

$$(\mathbf{x}(t) - \mathbf{x}^*)' = J\mathbf{f}(\mathbf{x}^*)(\mathbf{x} - \mathbf{x}^*) + \mathbf{o}(\|\mathbf{x} - \mathbf{x}^*\|),$$

from which, neglecting the term $\mathbf{o}(\|\mathbf{x} - \mathbf{x}^*\|)$ and putting $\mathbf{z}(t) = \mathbf{x}(t) - \mathbf{x}^*$, we have

$$\mathbf{z}'(t) = J\mathbf{f}(\mathbf{x}^*)\mathbf{z}(t),$$

which is an homogeneous linear system with constant coefficients (the coefficients matrix is obviously the Jacobian matrix J of $\mathbf{f}$, evaluated at the equilibrium point $\mathbf{x}^*$). In other words, we have operated a "linearization" of the original system (6.27).

With reference to this type of systems, we give now some conditions which allow to obtain from the behaviour of the above linearized system, some useful information (unfortunately only "local information") concerning the behaviour of the original system, that is of

$$\mathbf{x}'(t) = \mathbf{f}(\mathbf{x}(t)).$$

We recall that $\mathbf{x}^*$ is an equilibrium point or equilibrium solution of the above system, if $\mathbf{f}(\mathbf{x}^*) = [0]$. By substituting the function $\mathbf{z}(t) = \mathbf{x}(t) - \mathbf{x}^*$ to the function $\mathbf{x}(t)$, the zero vector $[0]$ becomes an equilibrium point of the new system. Therefore we can always assume that $\mathbf{x}^* = [0]$. We have the following fundamental result.

Theorem 6.10.1 (Theorem of Lyapunov-Poincaré or Theorem of Transfer of the Information) *Consider a system of the type*

$$\mathbf{x}'(t) = \mathbf{f}(\mathbf{x}(t)). \tag{6.28}$$

Let $\mathbf{x}^* = [0]$ *be an equilibrium solution of the system (i.e.* $\mathbf{f}([0]) = [0]$*) and let* $\mathbf{f}$ *be of class* C^1 *on a neighbourhood of* $[0]$*, with Jacobian matrix* $J\mathbf{f}([0]) = A$*. Let us suppose* $\det(A) \neq 0$*, which ensures that in a neighbourhood of* $[0]$ *there are no other equilibrium points. If* $\mathbf{x}^* = [0]$ *is globally asymptotically stable for the linearized system* $\mathbf{x}'(t) = A\mathbf{x}(t)$*, then* $\mathbf{x}^* = [0]$ *is locally asymptotically stable for the original system* (6.28)*. If* A *has an eigenvalue with a positive real part, then* $\mathbf{x}^* = [0]$ *is unstable for system* (6.28)*.*

For the proof see, e.g., [13]. In other words: if $\mathbf{x}^* = [0]$ is (globally) asymptotically stable (resp. unstable) for the associated linear system, then $\mathbf{x}^* = [0]$ is locally asymptotically stable (resp. unstable) for the original nonlinear system. Moreover, we remark that Theorem 6.10.1 gives only *sufficient conditions:* it may happen that an equilibrium point is locally or globally stable in the original system and is *not* stable in its associated linear approximation system. This is due to the fact that the "approximation error" may act favourably for stability. Note that if the Jacobian matrix $J\mathbf{f}([0]) = A$ has a zero eigenvalue, then it is a singular matrix and we have no longer a unique equilibrium solution $\mathbf{x}^* = [0]$. However, as in Theorem 6.5.7, in this case a small perturbation in the coefficients of A can restore the matrix to have a full rank.

Example 6.10.2

(a) Let be given the system

$$\begin{cases} x_1'(t) = -2x_1(t) + (x_2)^2(t) \\ x_2'(t) = -2(x_1)^3(t) - x_2(t) \end{cases}$$

for which the zero solution $\mathbf{x}^* = [0]$ is clearly an equilibrium point. The Jacobian matrix of $\mathbf{f}$ is

$$J\mathbf{f}(\mathbf{x}) = \begin{bmatrix} -2 & 2x_2 \\ -6(x_1)^2 & -1 \end{bmatrix}$$

and hence

$$J\mathbf{f}([0]) = \begin{bmatrix} -2 & 0 \\ 0 & -1 \end{bmatrix}.$$

This (diagonal) matrix has two real negative eigenvalues, therefore the zero vector is locally asymptotically stable for the original nonlinear system.

(b) Let be given the system, with $\alpha \in \mathbb{R}$,

$$\begin{cases} x_1'(t) = -x_1(t) + \alpha x_2(t) - (x_1)^2(t) + (x_2)^3(t) \\ x_2'(t) = e^{x_1(t)} - x_2(t) - 1. \end{cases}$$

The zero vector $\mathbf{x}^* = [0]$ is an equilibrium solution, for any $\alpha \in \mathbb{R}$. The Jacobian matrix is

$$J\mathbf{f}(\mathbf{x}) = \begin{bmatrix} -1 - 2x_1 & \alpha + 3(x_2)^2 \\ e^{x_1} & -1 \end{bmatrix}.$$

Hence we have

$$A \equiv J\mathbf{f}([0]) = \begin{bmatrix} -1 & \alpha \\ 1 & -1 \end{bmatrix},$$

$$|A - \lambda I| = \lambda^2 + 2\lambda + 1 - \alpha.$$

We have the following two eigenvalues

$$\lambda_1 = -1 - \sqrt{\alpha}; \quad \lambda_2 = -1 + \sqrt{\alpha}.$$

(i) If $\alpha < 1$, then $\mathbf{x}^* = [0]$ is locally asymptotically stable for the original system.

(ii) If $\alpha = 1$, we have $\lambda_1 = -2, \lambda_2 = 0$. The Jacobian matrix becomes singular and hence we exclude this case.

(iii) If $\alpha > 1$, then $\mathrm{Re}(\lambda_1) < 0$, $\mathrm{Re}(\lambda_2) > 0$ and hence $\mathbf{x}^* = [0]$ is an unstable equilibrium point for the original system.

Unfortunately the theorem of Lyapunov-Poincaré allows to have information on the possible *local* stability of an equilibrium point for system (6.28), thus remaining unsolved the question on the possible global stability of the same. A device rather useful and which often allows to investigate the solutions of system (6.28), with reference also to their global stability, is the so-called "second method of Lyapunov" , called also "direct method of Lyapunov", while the "indirect method" or "first method" is the linearization method described above. However, there is not uniformity of denomination in the literature, notably in the economic literature. The "direct method" leaves out of consideration the research of the solutions of system (6.28) and makes use of the so-called "Lyapunov functions". We give only some basic notions on this method, always with reference to systems of the type

$$\mathbf{x}'(t) = \mathbf{f}(\mathbf{x}(t)), \tag{6.29}$$

where $\mathbf{f} : X \longrightarrow \mathbb{R}^n$, $X \subset \mathbb{R}^n$, and for which $\mathbf{x}^* = [0]$ is an equilibrium point (i.e. $\mathbf{f}([0]) = [0]$). In order to prove that every solution of the above system converges to $[0]$, when $t \longrightarrow +\infty$, we may try to associate with $\mathbf{x}(t)$ a scalar function $V(\mathbf{x}(t))$, which in a sense that will be better precised, measures some sort of "distance " of $\mathbf{x}(t)$ from the origin. If, for every solution of the system, we are able to prove that

$$v(t) = V(\mathbf{x}(t))$$

converges to zero, when $t \longrightarrow +\infty$, it is reasonable to expect that we have also

$$\lim_{t \longrightarrow +\infty} \mathbf{x}(t) = [0] .$$

The decreasing behaviour of $v(t)$ can be sometimes obtained, without knowing $\mathbf{x}(t)$, as it results

$$v'(t) = V'(\mathbf{x}(t))\mathbf{x}'(t)$$

and hence

$$v'(t) = V'(\mathbf{x}(t))\mathbf{f}(\mathbf{x}(t)),$$

an expression where $\mathbf{x}'(t)$ does not longer appear, and whose sign is determinable, at least in several cases coming out from applications.

Definition 6.10.3 Let be given system (6.29) for which $\mathbf{x}^* = [0] \in X \subset \mathbb{R}^n$ is the unique equilibrium point, with X open set (maybe $X = \mathbb{R}^n$). A *Lyapunov function* for system (6.29) is a function $V : \mathbb{R}^n \longrightarrow \mathbb{R}$ such that:

(i) V is defined at least on X and is of class C^1 on this set.
(ii) $V(\mathbf{x}) > 0, \forall \mathbf{x} \neq [0], \mathbf{x} \in U([0]), V([0]) = 0.$

(iii)

$$\frac{dV(\mathbf{x}(t))}{dt} \leqq 0, \ \forall \mathbf{x}(t) \neq [0], \ \ \mathbf{x}(t) \in U([0]),$$

that is

$$\sum_{i=1}^{n} \frac{\partial V(\mathbf{x})}{\partial x_i} \frac{dx_i}{dt} = \sum_{i=1}^{n} \frac{\partial V(\mathbf{x})}{\partial x_i} f_i(\mathbf{x}) \leqq 0, \ \ \forall \mathbf{x}(t) \neq [0], \ \ \mathbf{x}(t) \in U([0]).$$

Condition (ii) of the above definition is also expressed by saying that $V(\mathbf{x})$ is "positive definite" on $U([0])$. Relation (iii) is also called "orbital derivative" or "derivative of V along the trajectories of the system", and it is often denoted by $\dot{V}$.

We have the following basic result.

Theorem 6.10.4 *Let be given system* (6.29) *with* $\mathbf{x}^* = [0]$ *as its unique equilibrium point.*

(a) *If there exists a Lyapunov function defined in a neighbourhood* $U([0])$, *then* $\mathbf{x}^* = [0]$ *is locally stable in the sense of Lyapunov.*
(b) *If the previous condition* (a) *holds, and moreover it holds*

$$\frac{dV(\mathbf{x}(t))}{dt} < 0 \text{ in } U([0]) \setminus [0],$$

then $\mathbf{x}^* = [0]$ *is locally asymptotically stable.*
(c) *If* (a) *and* (b) *hold on the whole set* X *and, moreover,* $V(\mathbf{x}) \longrightarrow +\infty$, *when* $\|\mathbf{x}\| \longrightarrow +\infty$, *then* $\mathbf{x}^* = [0]$ *is globally asymptotically stable.*

The second method of Lyapunov is appealing, above all in economic applications, as it does not require the research of the explicit solution of the related system. The problem lies in the fact that the method does not show how to find, every time, the appropriate Lyapunov function. The literature on the second method of Lyapunov is now rather abundant and often rather technical and advanced. See, for instance, [9], [30], [22], [32], [37], [38]. For economic applications, see the basic works of [2], [14], [17], [25]. A useful reference on problems of economic stability is also [35].

Example 6.10.5 Consider the following linear system, with constant coefficients,

$$\mathbf{x}'(t) = A\mathbf{x}(t)$$

and suppose that A is *symmetric and negative definite.* The vector $\mathbf{x}^* = [0]$ is the unique equilibrium point. By means of the previous theorem we obtain a result we already know. We choose the following Lyapunov function: $V(\mathbf{x}) = \|\mathbf{x}\|^2 = \mathbf{x}^\top \mathbf{x}$; this function is indeed a Lyapunov function, as:

(i) $V(\mathbf{x}) > 0, \forall \mathbf{x} \in \mathbb{R}^n, \mathbf{x} \neq [0]$;
(ii) $V([0]) = 0$;
(iii) $V'(t) = 2\mathbf{x}^\top \mathbf{x}'(t) = 2\mathbf{x}^\top A\mathbf{x} < 0, \forall \mathbf{x} \in \mathbb{R}^n, \mathbf{x} \neq [0]$.

It follows that $\mathbf{x}^* = [0]$ is a globally asymptotically stable equilibrium point for the given system; obviously this result does not require the Lyapunov method, since we know that a (symmetric) negative definite matrix has all its (real) eigenvalues less than zero.

Example 6.10.6

(a) Consider the system

$$\begin{cases} x' = -y - 3x^3 \\ y' = x^5 - 2y^3. \end{cases}$$

From $x' = 0$, $y' = 0$, we have

$$\begin{cases} y = -3x^3 \\ x^5 - 2(-27x^9) = 0, \end{cases}$$

that is

$$\begin{cases} y = -3x^3 \\ x^5(1 + 54x^4) = 0, \end{cases}$$

from which we deduce that $(x^*, y^*) = (0, 0)$ is the unique equilibrium point. We consider the following Lyapunov function:

$$V(x, y) = x^6 + 3y^2.$$

It results $x^6 + 3y^2 > 0, \forall (x, y) \neq [0]$, hence $V(x, y)$ is "positive definite" on $\mathbb{R}^2$ and, moreover, $V(x, y) \longrightarrow +\infty$ when $x^2 + y^2 \longrightarrow +\infty$. Furthermore,

$$\begin{aligned} \frac{dV}{dt} &= \frac{\partial V}{\partial x}(x(t), y(t)) f_1(x(t), y(t)) + \frac{\partial V}{\partial y}(x(t), y(t)) f_2(x(t), y(t)) \\ &= -6yx^5 - 18x^8 + 6x^5y - 12y^4 = -18x^8 - 12y^4 < 0, \ \forall (x, y) \neq [0]. \end{aligned}$$

It follows that $(x^*, y^*) = [0]$ is an equilibrium solution globally asymptotically stable for the proposed system.

(b) Consider the system

$$\begin{cases} x'(t) = y(t) + x^3(t) - 3x(t) \\ y'(t) = -x(t). \end{cases}$$

The point $(x^*, y^*) = (0, 0)$ is an equilibrium point. The function $V(x, y) = \frac{1}{2}x^2 + \frac{1}{2}y^2$ is positive definite on $\mathbb{R}^2$ and we have

$$\frac{dV(x, y)}{dt} = x\left[y + (x^3 - 3x)\right] + y(-x) = x^4 - 3x^2 = x^2(x^2 - 3).$$

Hence $\frac{dV}{dt} < 0$ on the "stripe" $(x, y) \in \mathbb{R}^2$, with

$$-\sqrt{3} < x < \sqrt{3}, \; y \in \mathbb{R}.$$

It follows that $(0, 0)$ is locally asymptotically stable.

Note that for Exercise sub (a) the "linearization method" does not work. Indeed, we have

$$\frac{\partial f_1}{\partial x} = -9x^2; \; \frac{\partial f_1}{\partial y} = -1; \quad \frac{\partial f_2}{\partial x} = 5x^4; \; \frac{\partial f_2}{\partial y} = -6y^2.$$

It follows that the Jacobian matrix of the system, evaluated at $[0]$, is

$$A = \begin{bmatrix} 0 & -1 \\ 0 & 0 \end{bmatrix}$$

which has the eigenvalues $\lambda_1 = \lambda_2 = 0$.

(c) Consider the system

$$\begin{cases} x'(t) = x(t)(10 - 2y(t) - x(t)) \\ y'(t) = y(t)(2x(t) - 2y(t) - 4). \end{cases}$$

The points $(0, 0)$ and $(\frac{38}{7}, \frac{16}{7})$ are equilibria. Let us study the stability of $(\frac{38}{7}, \frac{16}{7})$. The function $V(x, y) = 2(x - \frac{38}{7}\log x) + 2(y - \frac{16}{7}\log y)$, with $x, y > 0$, is a Lyapunov function for the system: it is positive definite and $V(x, y) \longrightarrow +\infty$ when $(x^2 + y^2) \longrightarrow +\infty$. Moreover,

$$\begin{aligned} \dot{V} &= 2\left(x - \frac{38}{7}\right)(10 - 2y - x) + 2\left(y - \frac{16}{7}\right)(2x - 3y - 4) \\ &= -2\left(x - \frac{38}{7}\right)^2 - 6\left(y - \frac{16}{7}\right)^2 < 0, \; \forall (x, y) \neq \left(\frac{38}{7}, \frac{16}{7}\right) \end{aligned}$$

and $\left(\frac{38}{7}, \frac{16}{7}\right)$ is a global minimum for V in the positive quadrant of $\mathbb{R}^2$. Thus $\left(\frac{38}{7}, \frac{16}{7}\right)$ is locally asymptotically stable and the positive quadrant of $\mathbb{R}^2$ is its basin of attraction.

Example 6.10.7 (Stability of the Walrasian Equilibrium) The subject is rather complex and would require a long discussion. The interested reader can see the classical book of [2]; also useful are the works of [14] and [35]. Here we give only a short presentation of the problem. Consider a Walrasian market of pure exchange where $E_i(\mathbf{p}), i = 1, 2, \ldots, n$, denotes the *excess demand function* for the good i, whereas $\mathbf{p} > [0]$ is the vector of prices. An *equilibrium price vector* $\mathbf{p}^* > [0]$ exists if $E_i(\mathbf{p}^*) = 0, i = 1, \ldots, n$. Moreover, we suppose that the so-called *weak axiom of revealed preference* of Samuelson holds in this economic system, according to which it holds $(\mathbf{p}^*)'\mathbf{E}(\mathbf{p}) > 0$, for every $\mathbf{p} \neq \lambda \mathbf{p}^*, \lambda > 0$. This result states that in any disequilibrium situation, the sum of the excess demands weighted by the equilibrium prices is always positive. The price adjustment rule (Walras' tâtonnement mechanism) is the following one:

$$\frac{dp_i(t)}{dt} = k_i E_i(\mathbf{p}), \; i = 1, \ldots, n,$$

where $k_i > 0$ measures the "speed of adjustment" of the i-th market. Let $\mathbf{p}^*$ be the vector of equilibrium prices of the market ($\mathbf{E}(\mathbf{p}^*) = [0]$) and assume that it holds the so-called "Walras law", i.e. $\mathbf{p}^\top \mathbf{E}(\mathbf{p}) = 0$, that is the total value of the excess demand vectors, evaluated at any price vector, is equal to zero. We wish to prove that all price paths which are solutions of the above adjustment rule, converge to the equilibrium price vector $\mathbf{p}^*$, in other words the equilibrium vector $\mathbf{p}^*$ is globally stable. We choose the following Lyapunov function (here the equilibrium is not the zero vector [0]):

$$V(\mathbf{p}) = \sum_{i=1}^{n} \frac{(p_i - p_i^*)}{k_i},$$

which is positive for $p_i \neq p_i^*$. We have

$$\begin{aligned} \frac{dV(\mathbf{p})}{dt} &= (2 \sum_i (p_i - p_i^*) p_i')/k_i = (2 \sum_i (p_i - p_i^*) E_i(p) k_i)/k_i \\ &= 2(\mathbf{p} - \mathbf{p}^*)^\top \mathbf{E}(\mathbf{p}). \end{aligned}$$

As the Walras law holds, we have

$$\frac{dV(\mathbf{p})}{dt} = -2(\mathbf{p}^*)^\top \mathbf{E}(\mathbf{p}) < 0,$$

being $(\mathbf{p}^*)^\top \mathbf{E}(\mathbf{p}) > 0$, by the weak axiom of revealed preferences. Moreover, $V(\mathbf{p})$ is equal to zero only at the equilibrium point. Hence, with the assumptions adopted, the equilibrium vector $\mathbf{p}^*$ is globally asymptotically stable.

Example 6.10.8 A simple microeconomic model of market dynamics.

We suppose that the demanded and supplied quantities of a certain good exchanged in a market, depend on the price of the same good. Suppose that p denotes the market price of the said good and that the associated demand and supply functions are linear affine, of the type

$$D(p) = a - bp; \quad S(p) = -c + dp$$

with a, b, c, d positive parameters.

Note that the demand function $D(p)$ is *decreasing* if the price increases and that the supply $S(p)$ is *increasing* if the price increases. The market is in equilibrium if

$$D(p) = S(p), \text{ that is } E(p) = D(p) - S(p) = 0,$$

that is

$$a - bp = -c + dp$$

from which we get the equilibrium price p^*:

$$p^* = \frac{a + c}{b + d}.$$

Consider now $p(t)$, that is the price as a function of time. If at $t_0 = 0$ we have $p(0) = p^*$, the market is in a position of equilibrium. If $p(0) \neq p^*$, it is desirable an adjustment process (tâtonnement) which stimulates the price $p(t)$ to approach to the equilibrium price p^*. Hence we put $p = p(t)$ and suppose that its dynamic law makes the related variation rate proportional to the excess demand $E(p) = D(p) - S(p)$:

$$p'(t) = kE(p), \ k > 0.$$

Then we have

$$p'(t) = k\left[a - bp(t) + c - dp(t)\right],$$

that is

$$p'(t) + k(b + d)p(t) = k(a + c).$$

If we put $g = k(b+d)$ and $h = k(a+c)$, we obtain a linear differential equation with constant coefficients:

$$p'(t) + gp(t) = h.$$

We recall that the differential equation

$$x'(t) + ax(t) = b, \quad a \neq 0,$$

has the general solution

$$x(t) = \frac{b}{a} + ce^{-at}.$$

Hence the general solution of our differential equation is

$$p(t) = \frac{h}{g} + ce^{-gt},$$

that is

$$p(t) = \frac{k(a+c)}{k(b+d)} + ce^{-k(b+d)t},$$

that is

$$p(t) = p^* + ce^{-k(b+d)t}.$$

Now suppose to have the initial value problem (Cauchy problem) $p(0) = p_0$. At $t_0 = 0$ then we have

$$p(0) = p_0 = p^* + ce^0$$

from which

$$c = p_0 - p^*.$$

Therefore the solution of the initial value problem is

$$p(t) = p^* + (p_0 - p^*)e^{-k(b+d)t}.$$

Being $k > 0, b > 0, d > 0$, we have

$$\lim_{t \longrightarrow +\infty} e^{-k(b+d)t} = 0$$

and hence, with $t \longrightarrow +\infty$,

$$p(t) \longrightarrow p^*,$$

in other words, $p(t)$ converges to the equilibrium price of the market: this equilibrium price is globally asymptotically stable.

6.11 Phase Diagrams

In this section we are concerned with differential equations of the type

$$x'(t) = f(x(t))$$

with $x, x', f : \mathbb{R} \longrightarrow \mathbb{R}$, namely with first-order autonomous differential equations. In order to study the qualitative behaviour of this type of equations, it is often customary to use, mainly in economic analysis, certain graphical techniques, known as "phase diagrams". In Economics these techniques were introduced by the English economist A. Marshall and subsequently used by J. Hicks and, above all, by J. K. Arrow and L. Hurwicz in their basic studies on the stability of Walrasian systems; see [3].

Phase diagrams are nothing but Cartesian diagrams which contain the diagram of the function $x' = f(x)$, on the Cartesian plane xOx'. Hence, we plot on the plane xOx' the diagram of the function f, as if it were a function of the variable x only. If, for instance, we have the differential equation

$$x'(t) = x(t) - x^2(t),$$

its diagram is the parabola, strictly concave, with vertex at $\left(\frac{1}{2}, \frac{1}{4}\right)$, and that intersects the horizontal axis at 0 and 1. See Fig. 6.5.

We remark at once, that, as the equilibrium points are the solutions of $x' = 0$, they are the points of intersection of the phase diagram with the horizontal axis, in our example the points of abscissa 0 and 1. As $f > 0$ points out that $x' > 0$, that is $x(t)$ is (strictly) increasing (the values of x increase), and, vice-versa, $f < 0$ points

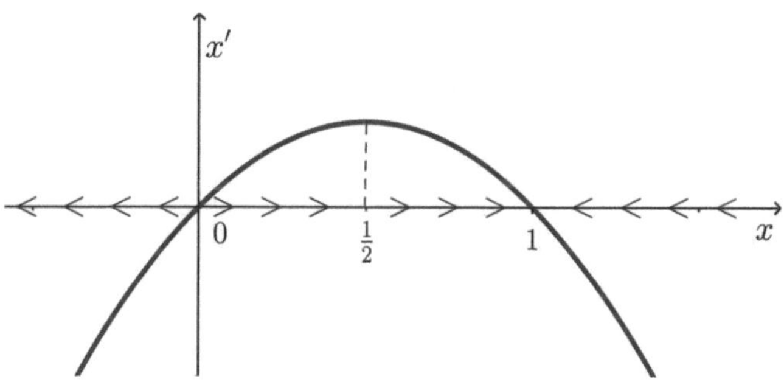

Fig. 6.5 Phase diagram of the equation $x'(t) = x(t) - x^2(t)$

out (strictly) decreasing values of x, it is possible to deduce from the diagram, by means of "oriented arrows", the stability or instability of the equilibrium points.

With reference to our example, if we consider the values of x for which the diagram lies above the horizontal axis (we have therefore $x' > 0$), if the system is in this position, with the flow of time, the value of x will increase and the point corresponding to the position of the system will shift towards right. We have an opposite behaviour in the case where the values of x correspond to negative values in the diagram. Always with reference to our example, we may distinguish on the horizontal axis three parts, that is the intervals $(-\infty, 0)$, $(0, 1)$, $(1, +\infty)$. In the first interval f is negative: therefore, if the system starts from the position $x(0) < 0$, it will move indefinitely towards left. In the second interval f is positive and hence, if the system starts from the initial position $x(0) \in (0, 1)$, the system tends to approach 1 from the left. If the system starts from $x(0) > 1$, the negativity of f pushes the same to the left, again towards 1. We can therefore conclude that the equilibrium point 0 is unstable, whereas 1 is locally asymptotically stable. Its basin of attraction is the interval $(0, +\infty)$, namely if $x(0) > 0$, then $x(t) \longrightarrow 1$ when $t \longrightarrow +\infty$.

One is easily persuaded that equilibrium points x^* that are at least locally stable, are characterized from the fact that f is strictly decreasing (it is positive on the left of x^* and negative on the right). A sufficient condition for this behaviour is obviously $f'(x^*) < 0$. More formally, we have the following result.

Theorem 6.11.1 *Consider the following autonomous first-order differential equation:*

$$x'(t) = f(x(t)),$$

with $f : \mathbb{R} \longrightarrow \mathbb{R}$. If f is continuous and strictly decreasing at an equilibrium point x^*, then x^* is locally asymptotically stable.

Proof We have $f(x) < 0$ on a right neighbourhood U^+ of x^*. Consider a solution x of the equation, with $x(t_0) \in U$. For $t > t_0$ it results that x is strictly decreasing and it holds $\lim_{t \longrightarrow +\infty} x(t) = x^*$. Indeed, if this would not be the case, with

$$\xi = \lim_{t \longrightarrow +\infty} x(t),$$

(being $x^* < \xi \leq x(t_0)$), we have then $\lim_{t \longrightarrow +\infty} x'(t) = f(\xi) < 0$, which is absurd, as in this case it would hold $\lim_{t \longrightarrow +\infty} x(t) = -\infty$. Similar reasoning for the case $x(t_0) < x^*$. □

Example 6.11.2 Consider the equation

$$x'(t) = x^2(t) - 3x(t) + 2.$$

Its two equilibrium points are $x = 1$ and $x = 2$. We have $f'(1) = -1$, $f'(2) = 1$. See Fig. 6.6.

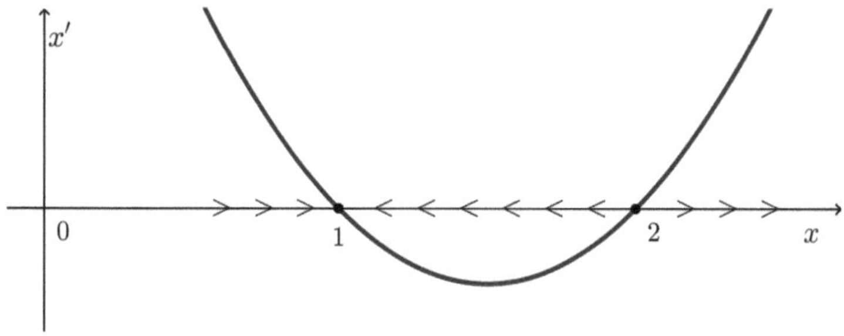

Fig. 6.6 Phase diagram of Example 6.11.2

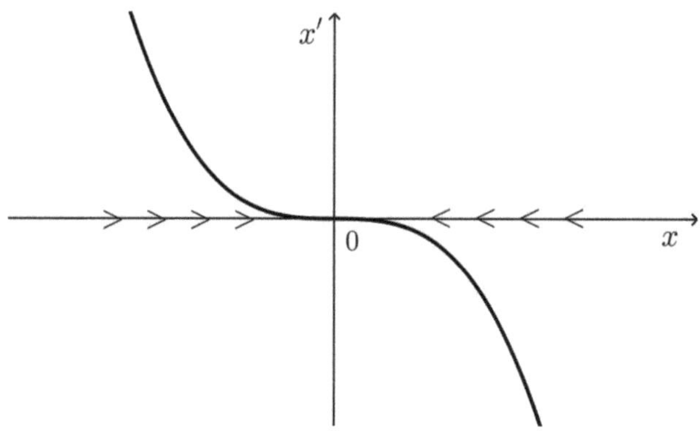

Fig. 6.7 Phase diagram of the equation $x'(t) = -x^3(t)$

Hence $x_1 = 1$ is locally asymptotically stable, $x_2 = 2$ is an unstable equilibrium solution.

In some cases the phase diagrams technique allows to locate even the globally stable equilibrium points. Consider, for instance, the equation $x'(t) = -x^3(t)$. See Fig. 6.7.

We note that the origin is its unique equilibrium point. If the system starts under 0, the positivity of f on the negative semiaxis, pushes the same towards 0; if the system starts on the right of zero, the negativity of f on the positive semiaxis, pushes the same again towards zero. If the system starts from zero, it remains in this position, and hence zero is the unique global asymptotically stable equilibrium point.

Phase diagrams are also used to study two-dimensional autonomous systems ($n = 2$). In this case we speak of "phase portraits". For these subjects the reader may make reference to [16] or [34].

Besides the works quoted along the chapter, we refer to the reader interested in the topics of this chapter to the references [4–6, 8, 18, 19, 21, 23, 24, 26–29, 33, 36].

References

1. H. Amann, *Ordinary Differential Equations. An Introduction to Nonlinear Analysis* (W. de Gruyter, Berlin, 1990)
2. K.J. Arrow, F.H. Hahn, *General Competitive Analysis* (North Holland, Amsterdam, 1971)
3. K.J. Arrow, L. Hurwicz, On the stability of the competitive equilibrium I. Econometrica **26**, 522–552 (1958)
4. K.J. Arrow, M.D. Intriligator (eds.), *Handbook of Mathematical Economics* (North Holland, Amsterdam, 1981)
5. K.J. Arrow, H.D. Block, L. Hurwicz, On the stability of the competitive equilibrium II. Econometrica **27**, 82–109 (1959)
6. R.E. Bellman, *Stability Theory of Differential Equations* (McGraw-Hill, New York, 1953)
7. R. Bellman, *Dynamic Programming* (Princeton Univ. Press, Princeton, 1960)
8. R. Bellman, K.L. Cooke, *Differential-Difference Equations* (Academic Press, New York, 1963)
9. N. Bhatia, G.P. Szegö, *Dynamical Systems: Stability Theory and Applications* (Springer, New York, 1967)
10. M. Braun, *Differential Equations: An Introduction to Applied Mathematics* (Springer, New York, 1992)
11. E. Burmeister, A.R. Dobell, *Mathematical Theories of Economic Growth* (The Macmillan Company, New York, 1970)
12. E.A. Coddington, R. Carlson, *Linear Ordinary Differential Equations* (SIAM, Philadelphia, 1997)
13. E.A. Coddington, N. Levinson, *Theory of Ordinary Differential Equations* (McGraw-Hill, New York, 1955)
14. G. Gandolfo, *Economic Dynamics* (Springer, Berlin, 1997)
15. G. Giorgi, C. Zuccotti, Forma canonica di Jordan e sistemi di equazioni differenziali: note didattiche, in *DEM Working Paper Series N. 167* (Department of Economics and management, University of Pavia (Italy), 2018), pp. 11–18 (http://economiaweb.unipv.it)
16. A. Guerraggio, S. Salsa, *Metodi Matematici per l'Economia e le Scienze Sociali* (Giappichelli, Torino, 1997)
17. F.H. Hahn, Stability, in *Handbook of Mathematical Economics*, ed. by K.J. Arrow, M.D. Intriligator, vol. II (North Holland, Amsterdam, 1982), pp. 756–793
18. M.W. Hirsch, S. Smale, R.L. Devaney, *Differential Equations, Dynamical Systems and an Introduction to Chaos* (Elsevier Academic Press, Amsterdam, 2004)
19. R.A. Holmgren, *A First Course in Discrete Dynamical Systems* (Springer, New York, 1994)
20. M.C. Kemp, Y. Kimura, *Introduction to Mathematical Economics* (Springer, New York, 1978)
21. K. Lancaster, *Mathematical Economics* (Macmillan, London, 1968)
22. J.P. Lasalle, S. Lefschetz, *Stability by Lyapunov's Direct Method with Applications* (Academic Press, New York, 1961)
23. H.M. Lorenz, *Nonlinear Dynamical Economics and Chaotic Motion* (Springer, Berlin, 2011)
24. L.W. McKenzie, Matrices with dominant diagonal and economic theory, in *Mathematical Methods in the Social Sciences*, ed. by K.J. Arrow, S. Karlin, P. Suppes (Stanford Univ. Press, Stanford, 1960), pp. 47–62
25. A. Medio, M. Lines, *Nonlinear Dynamics: A Primer* (Cambridge Univ. Press, Cambridge (UK), 2001)
26. P. Michel, *Cours de Mathématiques pour Economistes* (Economica, Paris, 1984)
27. P. Newman, Approaches to stability analysis. Economica **28**, 12–29 (1961)

28. L. Perko, *Differential Equations and Dynamical Systems* (Springer, Berlin, 1991)
29. R. Rammanathan, *Introduction to the Theory of Economic Growth* (Springer, Berlin, 1982)
30. N. Rouche, P. Habets, M. Laloy, *Stability Theory by Lyapunov's Direct Method* (Springer, Berlin, 1977)
31. P.A. Samuelson, *Foundations of Economic Analysis* (Harvard Univ. Press, Cambridge, 1947)
32. G. Sansone, R. Conti, *Non-linear Differential Equations* (Pergamon Press, London, 1964)
33. R. Shone, *Economic Dynamics: Phase Diagrams and Their Economic Applications* (Cambridge Univ. Press, Cambridge, 1997)
34. C.P. Simon, L. Blume, *Mathematics for Economists* (W.W. Norton, New York, 1994)
35. A. Takayama, *Mathematical Economics* (Cambridge Univ. Press, Cambridge, 1985)
36. A. Takayama, *Analytical Methods in Economics* (University of Michigan Press, Anna Arbor, 1993)
37. G. Teschl, *Ordinary Differential Equations and Dynamical Systems* (American Math. Society, Providence, 2012)
38. T. Yoshizawa, *Stability Theory by Liapunov's Second Method* (The Mathematical Society of Japan, Tokyo, 1966)

Chapter 7
Discrete Dynamical Systems

7.1 General Concepts

Also in this chapter vector quantities are written in bold.

We have already given in the previous chapter some hints on *discrete dynamical systems*. Several models of economic analysis and mathematics of finance are described by means of discrete variables which move along precise time moments and often the time intervals are equal. Think, for example, to the investment of a capital C : the amount at time $(t + 1)$ is usually given by the amount at time t, plus the accumulated interest, which is computed according to the financial law chosen. If we denote by i the constant interest rate and put $x(0) = C$, and if we choose the *simple interest law*, we get the equation

$$x(t + 1) = x(t) + ix(0).$$

If we choose the *compound interest law*, we get the equation

$$x(t + 1) = x(t) + ix(t).$$

The previous equations are examples of *first-order linear finite difference equations, with constant coefficients*. Their solutions are, respectively,

$$x(t) = x(0)(1 + it);$$
$$x(t) = x(0)(1 + i)^t,$$

as it is well known from the basic courses of Financial Mathematics.

In discrete one-dimensional models the *state variables* are described by a real sequence $x(t)$, defined for $t \geqq t_0$, $t_0 \in \mathbb{N}$ and usually with $t_0 = 0$, whereas the *evolution law* is a *finite difference equation of order* $n \geqq 1$, that is an equation

G. Giorgi et al., *Lectures on Mathematics for Economic and Financial Analysis*,
https://doi.org/10.1007/978-3-031-83339-7_7

containing an unknown $x(t)$, evaluated at the consecutive times $t, t+1, t+2, \ldots, t+n$. Its *general form* can be written as

$$F(x(t), x(t+1), x(t+2), \ldots, x(t+n), t) = 0, \tag{7.1}$$

where F is a real function of $(n+2)$ real variables. If F does not explicitly depend on t, the equation is said to be *autonomous*, otherwise it is said to be *non autonomous*. If F is a polynomial of the first degree in $x(t), x(t+1), \ldots, x(t+n)$, the equation is said to be *linear*. A finite difference equation of order n is said to be in a *normal form* if it is given by expressing the value of $x(t+n)$ in terms of the sequence $x(t)$, evaluated at the previous times:

$$x(t+n) = f(x(t), x(t+1), \ldots, x(t+n-1), t), \tag{7.2}$$

where f is a real function of $(n+1)$ real variables.

In a finite difference equation, the *order n* is given by the difference between the highest and the lowest indices of the various times, appearing in the sequence.

Example 7.1.1

(a) The equation

$$x(t+1) = 3x(t)$$

is a linear autonomous finite difference equation, of order one.

(b) The equation

$$x(t+2) = e^{3t}x(t)$$

is a linear non autonomous finite difference equation of order two.

(c) The equation

$$x(t+1) = \frac{1}{x(t)}$$

is a (nonlinear) autonomous difference equation of order one.

(d) The equation

$$x(t+1) = \sqrt{x(t+1) + 2t}$$

is a (nonlinear) non autonomous difference equation of order one.

(e) The equation

$$x(t+2) + 2x(t+1) - tx(t-1) + 1 = 0$$

is of order 3, since $t+2$ is the maximum time index, $t-1$ is the minimum, and the difference is 3.

A *solution* of (7.1) or (7.2) is a sequence $x(t)$ which satisfies the equation for every $t \geqq t_0$. In order to solve (7.1) or (7.2) one has to find *all* sequences that are solutions of (7.1) or (7.2). In this case we speak of "general solution".

The *Cauchy problem* or *initial value problem* for difference equations of order n, in a normal form, consists in finding a solution of the said equation, that in the consecutive times $t_0, (t_0 + 1), \ldots, (t_0 + n - 1)$ satisfies n *initial conditions*

$$x(t_0) = \alpha_0;\ x(t_0 + 1) = \alpha_1;\ \ldots;\ x(t_0 + n - 1) = \alpha_{n-1},$$

with $\alpha_0, \alpha_1, \ldots, \alpha_{n-1}$ fixed.

If we have, for instance, a difference equation of the first order, in a normal form:

$$x(t + 1) = f(x(t), t),$$

and its related Cauchy problem

$$\begin{cases} x(t + 1) = f(x(t), t) \\ \quad x(0) = \alpha, \end{cases}$$

where, without loss of generality, we have put $t_0 = 0$, then in general it is possible to assert the *existence and uniqueness* of the solution of the said problem. Indeed, let I be a real interval and suppose that the function f is defined on $I \times \mathbb{N}$ and that the same function assumes its values on the same interval I. If we fix an initial value $x(0) \in I$, we have

$$\begin{aligned} x(1) &= f(x(0), 0) \in I; \\ x(2) &= f(x(1), 1) \in I; \\ x(3) &= f(x(2), 2) \in I, \end{aligned}$$

and so on.

Hence, from the equation $x(t + 1) = f(x(t), t)$, it is possible, by a *recursive procedure* (or *iterative procedure*) to find all the elements of the solution starting from an initial value and operating step by step. The basic assumption of this procedure is that the range (or codomain) of f must be contained in I, when its variables are chosen in $I \times \mathbb{N}$. When this assumption is not satisfied, the iterative procedure could stop.

Consider e.g., the Cauchy problem

$$\begin{cases} x(t + 1) = \sqrt{x(t) - 1} \\ \quad x(0) = 5. \end{cases}$$

It results $f : [1, +\infty) \longrightarrow [0, +\infty)$. Starting from the initial condition $x(0) = 5$, we obtain $x(1) = \sqrt{4} = 2$; $x(2) = \sqrt{1} = 1$; $x(3) = \sqrt{0} = 0$, but it is not possible to compute $x(4)$, as $x(4) = \sqrt{-1}$, not real.

If I is a proper subinterval of $\mathbb{R}$, it is not always immediate to verify a priori that at every iteration the range of f is contained in I. Obviously, if I coincides with the whole $\mathbb{R}$, the above anomaly does not appear. Furthermore, we have to note that the iterative method exposed does not allow to know the value of the variable for a fixed time value t^*, for instance $t^* = 1.000$, without knowing the value at $t^* - 1 = 999$, etc. Furthermore, it does not allow to predict what happens when t is large enough or even when $t \longrightarrow +\infty$.

We remark that the iterative procedure can be, at least formally, applied to finite difference equations of order $k \geqq 1$. Indeed we have the following result.

Theorem 7.1.2 (Existence and Uniqueness of the Solution of the Cauchy Problem) *Let $I \subset \mathbb{R}$ be an interval; denote by I^k the Cartesian product*

$$I^k = I \times I \times \cdots \times I \text{ (k times).}$$

If f is such that $f : I^k \times \mathbb{N} \longrightarrow I$, then from the equation

$$x(t+k) = f(x(t), x(t+1), x(t+2), \ldots, x(t+k-1), t)$$

it is possible to obtain, by the iterative procedure, one and only one solution for every choice of the initial conditions

$$(x(0), x(1), x(2), \ldots, x(k-1)) \in I^k.$$

Taking into account what previously remarked (just before Theorem 7.1.2), the following basic question arises: it is possible to find the solution $x(t)$ as an *explicit* function of the time, or, equivalently, it is possible to find $x(t)$ in a *closed form?* The answer is in general negative; the answer is positive for some particular classes of difference equations, just as it happens for the case of differential equations. One of the said classes is given by *linear difference equations.* We end the present section with two "amusing" examples.

Example 7.1.3 The well known "Fibonacci sequence" is given by

$$1, 1, 2, 3, 5, 8, 13, \ldots$$

and is defined, in terms of a difference equation, by

$$x(t+2) = x(t+1) + x(t),$$

with initial conditions

$$x(0) = 1 \text{ and } x(1) = 1.$$

Therefore we have a *second order* difference equation. The Fibonacci sequence, i.e. the solution of the above difference equation, is a sequence of integers in which each element is given by the sum of the two preceding elements of the sequence.

Example 7.1.4 (The Hanoi Tower Game) . This is a classical example to describe a recursive procedure. The tower of Hanoi is a puzzle that consists of three vertical poles and n disks of different sizes. Each disk has a hole in the middle, so it can be moved between the poles. Initially the disks are situated on the middle pole, ordered by size, with the largest disk at the bottom. The disks are only allowed to be moved in such a way that a larger disk never sits on top of a smaller disk. Other than that disks are allowed to be moved freely between the poles. The puzzle is solved when every disk either sits all on the left pole or all in the right pole.

What is the *minimum number* of moves to solve the puzzle? Let $y(n)$ denote the minimum number of moves required to solve the puzzle with n disks. Clearly $y(0) = 0,\ y(1) = 1,\ y(2) = 3.$

What about the general case? In order to move the bottom disk to an empty pole, we must move the remaining $n - 1$ disks to a different empty pole. By definition, this requires $y(n-1)$ moves. Once the $(n-1)$ disks have been moved aside and the bottom disk moved to its final location, we need to move back the $(n - 1)$ disks on top of the bottom disk to solve the puzzle. This requires another $y(n - 1)$ moves.

Therefore in total we need to use

$$y(n) = 2y(n - 1) + 1$$

moves. This describes a first order difference equation. This equation has the solution (see the next section)

$$y(n) = c2^n - 1.$$

Our initial condition $y(0) = 0$ yields $0 = c2^0 - 1 \implies c = 1$. Hence the minimum number of moves required to solve the puzzle is $y(n) = 2^n - 1$.

7.2 Linear Difference Equations

Linear difference equations of order n are those difference equations where the first member of (7.1) is a linear affine function of the terms of the sequence. In other words, we have equations of the type

$$a_0(t)x(t+n) + a_1(t)x(t+n-1) + \cdots + a_n(t)x(t) = b(t), \tag{7.3}$$

where $a_0(t),\ a_1(t), \ldots, a_n(t)$ are the *coefficients* of the equation and $b(t)$ is the right-hand side term, called also the “forcing term”. When $b(t) = 0, \forall t$, the equation is said *homogeneous,* otherwise *non homogeneous.*

There are no "closed formulas" which solve a general linear difference equation of order n, however, just as for the continuous case, we have general principles on the "structure" of the solutions. Let us consider the *homogeneous* difference equation of the type

$$a_0(t)x(t+n) + a_1(t)x(t+n-1) + \cdots + a_n(t)x(t) = 0. \tag{7.4}$$

We have the following important results.

- The set of all solutions of a linear homogeneous difference equation of order n, is a *linear space* of dimension n.
  This fact entails that in the general solution of this kind of equations there will be n arbitrary constants (to be determined by an initial value problem).
  For what concerns the linear *non homogeneous* difference equations of order n, the *superposition principle* holds:
- All solutions of a linear non homogeneous difference equation of order n are given by the sum of the general solution (i.e. all solutions) of the associated homogeneous equation, and a *particular* solution of the non homogeneous given equation.
  In other words, if $x(t)$ is the general solution of the homogeneous Eq. (7.4) (i.e. we have infinite sequences which satisfy (7.4)), and if $\bar{x}(t)$ is a particular solution of the non homogeneous Eq. (7.3) (i.e. we have a unique sequence, originated by a cauchy problem), then the infinite sequences

  $$\{x(t) + \bar{x}(t)\}$$

  are the set of all solutions of the non homogeneous Eq. (7.3).

Now we give some information on the case of linear finite difference equations of the *first order*, for which there is the possibility to obtain a solution in a closed form. These equations are obtained from (7.3), with $n = 1$:

$$a_0(t)x(t+1) + a_1(t)x(t) = b(t).$$

By assuming, without loss of generality, $a_0(t) \neq 0$, we can rewrite this equation in the form

$$x(t+1) = a(t)x(t) + b(t), \tag{7.5}$$

i.e. in a *normal form*. We begin by considering the homogeneous equation, i.e. in (7.5) we have $b(t) = 0$:

$$x(t+1) = a(t)x(t), \tag{7.6}$$

with $a(t) \neq 0$, $\forall t$. We have, by the iterative procedure,

$$\begin{aligned}
x(1) &= a(0)x(0)\\
x(2) &= a(1)x(1) = a(1)a(0)x(0)\\
x(3) &= a(2)x(2) = a(2)a(1)x(1) = a(2)a(1)a(0)x(0)\\
&\vdots\\
x(t) &= a(t-1)x(t-1) = a(t-1)a(t-2)\ldots a(1)a(0)x(0).
\end{aligned}$$

Hence, the general solution of (7.6) is

$$x(t) = \left(\prod_{s=0}^{t-1} a(s)\right) c,$$

where $c = x(0)$.

Note that if the equation is *autonomous,* i.e. $a(t)$ is a constant, say $a(t) = a$, $a \neq 0$, then the general solution becomes

$$x(t) = a^t \cdot c,$$

where $c = x(0)$.

Example 7.2.1 Find the general solution of

$$x(t+1) = 3^t x(t),$$

such that $x(0) = 1$.

We have

$$x(t) = \left(\prod_{s=0}^{t-1} 3^s\right) c = c3^{\sum_{s=0}^{t-1} s} = c3^{\frac{t(t-1)}{2}}.$$

(We recall that the sum of the first $(t-1)$ natural numbers is $\frac{t(t-1)}{2}$). By putting $x(0) = 1$, we obtain $c = 1$.

With reference to the non homogeneous Eq. (7.5), we can proceed as before, by means of the iterative procedure, with a given $x(0)$ at $t = 0$. We have

$$\begin{aligned}
x(1) &= a(0)x(0) + b(0);\\
x(2) &= a(1)x(1) + b(1) = a(1)(a(0)x(0) + b(0)) + b(1)\\
&= a(1)a(0)x(0) + a(1)b(0) + b(1).
\end{aligned}$$

Then, omitting the details for the next two calculations, we have

$$\begin{aligned}x(3) &= a(2)a(1)a(0)x(0) + a(2)a(1)b(0) + a(2)b(1) + b(2)\\ x(4) &= a(3)a(2)a(1)a(0)x(0) + a(3)a(2)a(1)b(0)\\ &\quad + a(3)a(2)b(1) + a(3)b(2) + b(3).\end{aligned}$$

In fact, the general solution is given by the formula (which can be proved quite easily by induction on t):

$$x(t) = x(0)\prod_{s=0}^{t-1} a(s) + \left(\prod_{s=1}^{t-1} a(s)\right) b(0) + \left(\prod_{s=2}^{t-1} a(s)\right) b(1) + \cdots +$$
$$+ \left(\prod_{s=t-1}^{t-1} a(s)\right) b(t-2) + b(t-1)$$

which can be rewritten as

$$x(t) = x(0)\prod_{s=0}^{t-1} a(s) + \sum_{k=0}^{t-1}\left(\prod_{s=k+1}^{t-1} a(s)\right) b(k), \tag{7.7}$$

provided we agree that the product $\Pi_{s=t}^{t-1}$ of zero terms is 1.

Note that the first addendum of the last expression is the solution of the associated homogeneous Eq. (7.6); we deduce that the second addendum is a particular solution of the non homogeneous Eq. (7.5). In case in (7.5) the coefficient $a(t)$ is *constant,* i.e. $a(t) = a$, $\forall t$, $a \neq 0$, relation (7.7) becomes

$$x(t) = x(0)a^t + \sum_{k=0}^{t-1} b(k)a^{t-k-1},$$

and, if also the forcing term $b(t)$ is constant, i.e. $b(t) = b$, $\forall b$, then we have

$$\begin{cases} x(t) = x(0)a^t + b\frac{a^t-1}{a-1}, & \text{if } a \neq 1;\\ x(t) = x(0) + tb, & \text{if } a = 1.\end{cases}$$

Example 7.2.2

(a) Find the solution of

$$x(t+1) = -x(t) + 3,$$

with the initial condition $x(0) = 3$.

We have

$$x(t) = 3(-1)^t - \frac{3}{2}((-1)^t - 1) = 3(-1)^t + \frac{3}{2} - \frac{3}{2}(-1)^t$$
$$= \frac{3}{2}(-1)^t + \frac{3}{2}.$$

(*b*) Find the solution of

$$x(t+1) - 5x(t) = 1,$$

with the initial condition $x(0) = \frac{7}{4}$.

We rewrite the equation in the normal form

$$x(t+1) = 5x(t) + 1.$$

Then we have

$$x(t) = \frac{7}{4} \cdot 5^t + \frac{5^t - 1}{4} = \frac{7}{4} \cdot 5^t + \frac{5^t}{4} - \frac{1}{4} = 2 \cdot 5^t - \frac{1}{4}.$$

7.3 Systems of First Order Difference Equations: The Case of Constant Coefficients

We now discuss briefly the *systems* of finite difference equations of the *first order* and of *dimension* n, i.e. of the type

$$\begin{cases} x_1(t+1) = g_1(x_1(t), x_2(t), \ldots, x_n(t), t) \\ x_2(t+1) = g_2(x_1(t), x_2(t), \ldots, x_n(t), t) \\ \vdots \\ x_n(t+1) = g_n(x_1(t), x_2(t), \ldots, x_n(t), t) \end{cases}$$

or, in a vector notation,

$$\mathbf{x}(t+1) = \mathbf{g}(\mathbf{x}(t), t),$$

where $\mathbf{g} : D \longrightarrow \mathbb{R}^n$, $D \subset \mathbb{R}^{n+1}$. If we put $I^n = I \times I \times I \times \cdots \times I$, n times, if the function $\mathbf{g}$ is defined on $I^n \times \mathbb{N}$ and has its values on I^n (obviously this always occurs if $\mathbf{g} : \mathbb{R}^n \times \mathbb{N} \longrightarrow \mathbb{R}^n$), the recursive procedure, used in the previous section for first order difference equations, allows to find one and only one solution of the Cauchy problem (or initial value problem) $\mathbf{x}(0) = \mathbf{x}^0 \in \mathbb{R}^n$. An explicit formula for the solution of this type of systems is not in general available, unless for particular cases, such as the linear autonomous systems, i.e. systems of the type

$$\mathbf{x}(t+1) = A\mathbf{x}(t) + \mathbf{b},$$

where A is a square matrix of order n and $\mathbf{b} \in \mathbb{R}^n$.

If, in our initial formulation, we have

$$\mathbf{x}(t+1) = \mathbf{g}(\mathbf{x}(t)),$$

i.e. the right-side member does not depend explicitly on t, then we speak of *autonomous systems.* We recall that for "dimension" of a system of difference equations we mean the number n of equations which form the system. We have the following basic result.

Theorem 7.3.1 *Every difference equation of order n of the type*

$$x(t+n) = f\left[x(t+n-1), x(t+n-2), \dots, x(t+1), x(t), t\right]$$

is equivalent to a first order system of difference equations having dimension n :

$$\mathbf{x}(t+1) = \mathbf{g}\left[\mathbf{x}(t), t\right],$$

where $\mathbf{g} : \mathbb{R}^{n+1} \longrightarrow \mathbb{R}^n$ and the state vector $\mathbf{x}(t) \in \mathbb{R}^n$ is

$$\begin{bmatrix} x(t) \\ x(t+1) \\ \vdots \\ x(t+n-1) \end{bmatrix}.$$

Example 7.3.2 Consider the system

$$\begin{cases} x_1(t+1) = x_1(t) + 2x_2(t) \\ x_2(t+1) = -x_1(t) - x_2(t), \end{cases}$$

which may be rewritten in a matrix form

$$\begin{bmatrix} x_1(t+1) \\ x_2(t+1) \end{bmatrix} = \begin{bmatrix} 1 & 2 \\ -1 & -1 \end{bmatrix} \begin{bmatrix} x_1(t) \\ x_2(t) \end{bmatrix}.$$

We show that this system is equivalent to a second order difference equation. We rewrite the second equation at $(t+2)$:

$$x_2(t+2) = -x_1(t+1) - x_2(t+1).$$

We substitute in this equation the first equation of the system

$$x_2(t+2) = -\left[x_1(t) + 2x_2(t)\right] - x_2(t+1)$$

and then we get $x_1(t)$ from the second equation

$$x_1(t) = -x_2(t+1) - x_2(t).$$

By substituting in the second member of the expression giving $x_2(t+2)$, we obtain finally

$$x_2(t+2) = -\left[-x_2(t+1) - x_2(t) + 2x_2(t)\right] - x_2(t+1)$$

and hence

$$x_2(t+2) = -x_2(t),$$

which is a second order difference equation. The discussion of the other implication of the equivalence is left to the reader.

Now we give some information on systems of n *linear* first order difference equations with n state variables, i.e. systems of the form

$$\mathbf{x}(t+1) = A(t)\mathbf{x}(t) + \mathbf{b}(t), \tag{7.8}$$

where $A(t)$ is a matrix of order n, with general element $a_{ij}(t)$ and $\mathbf{b}(t) \in \mathbb{R}^n$ is the so-called *forcing term.* If $\mathbf{b}(t) = [0]$, $\forall t$, then we have a *linear homogeneous* difference equations system:

$$\mathbf{x}(t+1) = A(t)\mathbf{x}(t). \tag{7.9}$$

For example, the system of difference equations

$$\begin{cases} x_1(t+1) = 3tx_1(t) - x_2(t) + e^{-2t} \\ x_2(t+1) = 4t^2x_1(t) + 5x_2(t) + 4te^{-2t} \end{cases}$$

is linear. We have

$$A(t) = \begin{bmatrix} 3t & -1 \\ 4t^2 & 5 \end{bmatrix}; \quad \mathbf{b}(t) = \begin{bmatrix} e^{-2t} \\ 4e^{-2t} \end{bmatrix}.$$

We have the following basic result.

Theorem 7.3.3 *The set of all solutions ("general solution") of the homogeneous system* (7.9) *is a linear space of dimension* n.

On the grounds of the previous theorem, the general solution of (7.9) will contain n arbitrary constants and will be given by the linear combination of n linearly independent solutions, i.e. of a "fundamental system of solutions".

Theorem 7.3.4 (Superposition Principle) *The general solution of system* (7.8) *is given by the sum of a particular solution of* (7.8) *with the general solution of the associated homogeneous system* (7.9).

In order to find a *fundamental system of solutions* of (7.9), which, by linear combination, gives the linear space (of dimension n) of all solutions of (7.9), it is important the following result.

Theorem 7.3.5 *Let* $\mathbf{x}^1(t)$, $\mathbf{x}^2(t), \ldots, \mathbf{x}^n(t)$ *be solutions of the homogeneous system* (7.9). *These solutions are a fundamental system for* (7.9), *i.e. they are linearly independent, if and only if*

$$\det\left[\mathbf{x}^1(t), \mathbf{x}^2(t), \ldots, \mathbf{x}^n(t)\right] \neq 0. \tag{7.10}$$

The determinant (7.10) is also called *Casorati determinant* or, briefly, *Casoratian* (Felice Casorati, 1835–1890, mathematician of Pavia University). Hence this determinant has, in the theory of linear systems of difference equations, the role that the *Wronskian* (see the previous section) has in the theory of continuous linear dynamical systems. From Theorem 7.3.3 it appears that the general solution of system (7.9) is given by the linear combination of n linearly independent solutions $\mathbf{x}^1(t)$, $\mathbf{x}^2(t), \ldots, \mathbf{x}^n(t)$:

$$\mathbf{x}(t) = c_1\mathbf{x}^1(t) + c_2\mathbf{x}^2(t) + \cdots + c_n\mathbf{x}^n(t),$$

where $c_1, c_2, \ldots, c_n$ are arbitrary constants, which will be determined by an initial value problem (or Cauchy problem). Note that with $c_1 = c_2 = \cdots = c_n = 0$ we find the zero solution of (7.9) $\mathbf{x}(t) = [0]$, which always exists.

For what concerns the initial value problem for system (7.8), we can assert the existence and uniqueness of solutions, as soon as we assume that the elements of $A(t)$ and the elements of $\mathbf{b}(t)$ are defined for all t.

Now we discuss the case of linear difference equations systems with *constant coefficients,* i.e. with $A(t) = A$, $\forall t$:

$$\mathbf{x}(t + 1) = A\mathbf{x}(t) + \mathbf{b}(t), \tag{7.11}$$

with A square matrix of order n. Obviously for these systems a fortiori Theorems 7.3.3, 7.3.4 and 7.3.5 continue to hold true. First we discuss the case of *homogeneous systems,* i.e. in (7.11) we have $\mathbf{b}(t) = [0]$, $\forall t$. Therefore we consider the system

$$\mathbf{x}(t + 1) = A\mathbf{x}(t). \tag{7.12}$$

An approach to solve (7.12) is a generalization of the recursive procedure, previously used for the case $n = 1$. If we know that $\mathbf{x}(0) = \mathbf{x}^0$, we obtain, subsequently,

$$\begin{aligned}
\mathbf{x}(1) &= A\mathbf{x}^0;\\
\mathbf{x}(2) &= A\mathbf{x}(1) = A \cdot A\mathbf{x}^0 = (A)^2\mathbf{x}^0;\\
\mathbf{x}(3) &= A\mathbf{x}(2) = A(A)^2\mathbf{x}^0 = (A)^3\mathbf{x}^0;\\
&\vdots\\
\mathbf{x}(t) &= (A)^t\mathbf{x}^0.
\end{aligned}$$

This procedure, theoretically simple, really presents some difficulties, as the computation of the power of a matrix is in general not easy, when the order of the matrix is quite important. However, in certain cases, the solution can be obtained in an easier way.

(a) Case of A *diagonalizable*, i.e. A has n linearly independent eigenvectors, i.e. for each eigenvalue it holds $m^A = m^G$ (algebraic multiplicity = geometric multiplicity). In this case we know that there exists a nonsingular matrix V such that $A = VDV^{-1}$, where D is a diagonal matrix, with the eigenvalues of A on its main diagonal and V is the *polar matrix*, whose columns are the (linearly independent) n eigenvectors of A :

$$V = \left[\mathbf{v}^1, \mathbf{v}^2, \ldots, \mathbf{v}^n\right].$$

Now we put $\mathbf{z}(t) = V^{-1}\mathbf{x}(t)$ and obtain the diagonal system ("uncoupled system")

$$\mathbf{z}(t+1) = D\mathbf{z}(t).$$

Indeed, from $\mathbf{x}(t+1) = A\mathbf{x}(t)$, we have $\mathbf{x}(t+1) = VDV^{-1}\mathbf{x}(t)$, i.e. $\mathbf{x}(t+1) = VD\mathbf{z}(t)$, i.e. $V^{-1}\mathbf{x}(t+1) = D\mathbf{z}(t)$, i.e. $\mathbf{z}(t+1) = D\mathbf{z}(t)$. This last system admits the solution

$$\mathbf{z}(t) = \begin{bmatrix} c_1\lambda_1^t \\ c_2\lambda_2^t \\ \vdots \\ c_n\lambda_n^t \end{bmatrix},$$

where $c_1, c_2, \ldots, c_n$ are arbitrary constants, real if the corresponding eigenvector is real, or also complex if the corresponding eigenvector is complex.

Now we come back to the original variables and obtain the *general solution* of the homogeneous system (7.12):

$$\mathbf{x}(t) = V\mathbf{z}(\mathbf{t}) = c_1\mathbf{v}^1\lambda_1^t + c_2\mathbf{v}^2\lambda_2^t + \cdots + c_n\mathbf{v}^n\lambda_n^t.$$

Note that, being $\mathbf{v}^1, \mathbf{v}^2, \ldots, \mathbf{v}^n$ linearly independent, the Casoratian is different from zero. Note that with $c_1 = c_2 = \cdots = c_n = 0$, we obtain the zero solution $[0]$ which always exists. We recall that if all eigenvalues of A are *distinct* (algebraic multiplicity = 1), then A is diagonalizable.

Example 7.3.6 Consider the system

$$\begin{cases} x_1(t+1) = x_1(t) \\ x_2(t+1) = x_1(t) + 2x_2(t). \end{cases}$$

Hence we have

$$A = \begin{bmatrix} 1 & 0 \\ 1 & 2 \end{bmatrix}$$

and the eigenvalues are $\lambda_1 = 1, \lambda_2 = 2$. The corresponding eigenvectors are

$$\mathbf{v}^1 = \begin{bmatrix} \alpha \\ -\alpha \end{bmatrix}, \ \alpha \in \mathbb{R}, \alpha \neq 0; \ \ \mathbf{v}^2 = \begin{bmatrix} 0 \\ \beta \end{bmatrix}, \ \beta \in \mathbb{R}, \beta \neq 0.$$

Therefore the general solution is

$$\begin{bmatrix} x_1(t) \\ x_2(t) \end{bmatrix} = \begin{bmatrix} \alpha 1^t \\ -\alpha 1^t + \beta 2^t \end{bmatrix},$$

with $\alpha, \beta \in \mathbb{R}$.

If we have the initial value problem

$$\mathbf{x}(0) = \begin{bmatrix} 1 \\ 1 \end{bmatrix},$$

we have

$$\begin{bmatrix} \alpha \\ -\alpha + \beta 2^0 \end{bmatrix} = \begin{bmatrix} 1 \\ 1 \end{bmatrix},$$

that is

$$\begin{cases} \alpha = 1 \\ \beta = 2, \end{cases}$$

and the particular solution, identified by the above initial value problem, is

$$\begin{cases} x_1(t) = 1 \\ x_2(t) = -1 + 2^{t+1}. \end{cases}$$

Example 7.3.7 Consider the system

$$\begin{cases} x(t+1) = 3x(t) - y(t) \\ y(t+1) = -x(t) + 2y(t) - z(t) \\ z(t+1) = -y(t) + 3z(t). \end{cases}$$

The coefficients matrix is

$$A = \begin{bmatrix} 3 & -1 & 0 \\ -1 & 2 & -1 \\ 0 & -1 & 3 \end{bmatrix}.$$

The associated characteristic equation is

$$-\lambda^3 + 8\lambda^2 - 19\lambda + 12 = 0$$

and the eigenvalues are $\lambda_1 = 3$, $\lambda_2 = 4$, $\lambda_3 = 1$. Hence, A is diagonalizable; the associated eigenvectors are, respectively,

$$\mathbf{v}^1 = \alpha \begin{bmatrix} 1 \\ 0 \\ -1 \end{bmatrix}; \ \mathbf{v}^2 = \beta \begin{bmatrix} 1 \\ -1 \\ 1 \end{bmatrix}; \ \mathbf{v}^3 = \gamma \begin{bmatrix} 1 \\ 2 \\ 1 \end{bmatrix}, \ \alpha, \beta, \gamma \neq 0.$$

The general solution is therefore

$$\mathbf{x}(t) = c_1 \begin{bmatrix} 1 \\ 0 \\ -1 \end{bmatrix} 3^t + c_2 \begin{bmatrix} 1 \\ -1 \\ 1 \end{bmatrix} 4^t + c_3 \begin{bmatrix} 1 \\ 2 \\ 1 \end{bmatrix}.$$

In the above examples there are *real* eigenvalues and real eigenvectors. The previous expression of the general solution of (7.12) holds also for the complex case. However, in this case (as for continuous dynamical systems) it is possible, by means of the "conjugation operation", to transform complex solutions into real solutions. Let us consider, for simplicity, two complex (conjugate) eigenvalues $\lambda_1 = \alpha + i\beta$ and $\lambda_2 = \bar{\lambda}_1 = \alpha - i\beta$. Let us denote by $\mathbf{v}^1 = \mathbf{w}^1 + i\mathbf{w}^2$ and $\mathbf{v}^2 = \bar{\mathbf{v}}^1 = \mathbf{w}^1 - i\mathbf{w}^2$ the respective associated complex (conjugate) eigenvectors. The contribution of these quantities to the general solution of the homogeneous system (7.12) is

$$C_1\lambda_1^t\mathbf{v}^1 + C_2\lambda_2^t\mathbf{v}^2, \quad C_1, C_2 \in \mathbb{C},$$

i.e.

$$C_1(\alpha + i\beta)^t(\mathbf{w}^1 + i\mathbf{w}^2) + C_2(\alpha - i\beta)^t(\mathbf{w}^1 - i\mathbf{w}^2).$$

In order to obtain the *real* solutions, we can operate as follows. Recall that the sum of a complex number $\alpha + i\beta$ and its conjugate $\alpha - i\beta$ is the *real* number 2α. If we choose C_1 to be any complex constant $c_1 + ic_2$ and then choose C_2 to be its complex conjugate $c_1 - ic_2$, then the first term in the last expression will be the conjugate of the second term and the sum of these two terms will be a real solution. We rewrite the said expression as follows

$$(c_1 + ic_2)(\alpha + i\beta)^t(\mathbf{w}^1 + i\mathbf{w}^2) + (c_1 - ic_2)(\alpha - i\beta)^t(\mathbf{w}^1 - i\mathbf{w}^2)$$
$$= 2\,\mathrm{Re}\left\{(c_1 + ic_2)(\alpha + i\beta)^t(\mathbf{w}^1 + i\mathbf{w}^2)\right\}.$$

We recall the so-called *De Moivre's formula:*

$$(\alpha + i\beta)^t = \rho^t(\cos(\theta t) + i\sin(\theta t)),$$

where $\rho = \sqrt{\alpha^2 + \beta^2}$ and $\theta = \arg(\alpha + i\beta)$ is the (principal) *argument* of $(\alpha + i\beta)$, i.e. the polar angle of $(\alpha + i\beta)$.

We apply De Moivre's formula to the last written expression and then perform the multiplications:

$$2\,\mathrm{Re}\left[(c_1 + ic_2)\rho^t(\cos(\theta t) + i\sin(\theta t))(\mathbf{w}^1 + i\mathbf{w}^2)\right]$$
$$= 2\rho^t\,\mathrm{Re}\left[((c_1\cos(\theta t) - c_2\sin(\theta t)) + i(c_2\cos(\theta t) + c_1\sin(\theta t)))(\mathbf{w}^1 + i\mathbf{w}^2)\right]$$
$$= 2\rho^t\left[(c_1\cos(\theta t) - c_2\sin(\theta t))\mathbf{w}^1 - (c_2\cos(\theta t) + c_1\sin(\theta t))\mathbf{w}^2\right]$$

and, incorporating 2 in the above constants, we obtain

$$\rho^t\left[(c_1\cos(\theta t) - c_2\sin(\theta t))\mathbf{w}^1 - (c_2\cos(\theta t) + c_1\sin(\theta t))\mathbf{w}^2\right],\quad c_1, c_2 \in \mathbb{R},$$

which are real solutions.

By setting

$$\mathbf{k}^1 = c_1\mathbf{w}^1 - c_2\mathbf{w}^2;\;\; \mathbf{k}^2 = -c_2\mathbf{w}^1 - c_1\mathbf{w}^2$$

we get

$$\rho^t\left[\mathbf{k}^1\cos(\theta t) + \mathbf{k}^2\sin(\theta t)\right],$$

$\mathbf{k}^1, \mathbf{k}^2 \in \mathbb{R}^n$, that implicitly depend on the two real quantities c_1 and c_2.

Example 7.3.8 Consider the system

$$\begin{bmatrix} x_1(t+1) \\ x_2(t+1) \end{bmatrix} = \begin{bmatrix} 0 & 1 \\ -1 & 0 \end{bmatrix} \begin{bmatrix} x_1(t) \\ x_2(t) \end{bmatrix}.$$

The characteristic equation of the coefficients matrix is

$$\begin{vmatrix} -\lambda & 1 \\ -1 & -\lambda \end{vmatrix} = \lambda^2 + 1 = 0,$$

which has roots $\lambda_1 = i = \lambda$ and $\lambda_2 = -i = \bar{\lambda}$. Hence $\rho = 1$; $\theta = \frac{\pi}{2}$. Now we have

$$\mathbf{v}^1 = \alpha \begin{bmatrix} 1 \\ i \end{bmatrix} = \left(\begin{bmatrix} 1 \\ 0 \end{bmatrix} + i \begin{bmatrix} 0 \\ 1 \end{bmatrix} \right), \quad \alpha \neq 0.$$

If we choose $\alpha = 1$, we get

$$\mathbf{w}^1 = \begin{bmatrix} 1 \\ 0 \end{bmatrix}; \ \mathbf{w}^2 = \begin{bmatrix} 0 \\ 1 \end{bmatrix}.$$

Finally, the general solution is

$$\begin{aligned} \mathbf{x}(t) &= 1^t \left[\left(c_1 \begin{bmatrix} 1 \\ 0 \end{bmatrix} - c_2 \begin{bmatrix} 0 \\ 1 \end{bmatrix} \right) \cos\left(\frac{\pi}{2} t\right) - \left(c_2 \begin{bmatrix} 1 \\ 0 \end{bmatrix} + c_1 \begin{bmatrix} 0 \\ 1 \end{bmatrix} \right) \sin\left(\frac{\pi}{2} t\right) \right] \\ &= \begin{bmatrix} c_1 \\ -c_2 \end{bmatrix} \cos\left(\frac{\pi}{2} t\right) - \begin{bmatrix} c_2 \\ c_1 \end{bmatrix} \sin\left(\frac{\pi}{2} t\right) \\ &= \begin{bmatrix} c_1 \cos\left(\frac{\pi}{2} t\right) - c_2 \sin\left(\frac{\pi}{2} t\right) \\ -c_2 \cos\left(\frac{\pi}{2} t\right) - c_1 \sin\left(\frac{\pi}{2} t\right) \end{bmatrix}, \quad c_1, c_2 \in \mathbb{R}. \end{aligned}$$

In order to find the particular solution for the initial value problem $\mathbf{x}(0) = \mathbf{x}^0$, it is sufficient to compute the solution at $t = 0$:

$$\begin{bmatrix} c_1 \\ -c_2 \end{bmatrix} \cos\left(\frac{\pi}{2} 0\right) - \begin{bmatrix} c_2 \\ c_1 \end{bmatrix} \sin\left(\frac{\pi}{2} 0\right) = \begin{bmatrix} c_1 \\ -c_2 \end{bmatrix}.$$

Hence we have

$$\mathbf{x}(t) = \begin{bmatrix} x_1^0 \\ x_2^0 \end{bmatrix} \cos\left(\frac{\pi}{2} t\right) + \begin{bmatrix} x_2^0 \\ -x_1^0 \end{bmatrix} \sin\left(\frac{\pi}{2} t\right).$$

(*b*) *Case of A non diagonalizable.* In this case we have some repeated (i.e. multiple) eigenvalues *non regular* (geometric multiplicity strictly less than algebraic multiplicity). We have no longer the existence of n linearly independent eigenvectors associated with the respective eigenvalues and this fact obliges us to find n linearly independent solutions (i.e. a "fundamental system"), in order to obtain the set of all solutions, set which is a linear space of dimension n. The general theory requires the application of properties of the *Jordan normal form* of A (see Sect. 1.6), which is behind the aims of the present book.

It is possible to prove (see, e.g., [18]) that if, for instance, λ is a real eigenvalue with algebraic multiplicity $m^A = 2$, but *not regular*, in the expression of the general solution of the homogeneous system, instead of the term $\mathbf{v}\lambda^t$, it appears an addendum of the type

$$(\mathbf{a} + \mathbf{b}t)\lambda^t,$$

with $\mathbf{a}, \mathbf{b} \in \mathbb{R}^n$, to be determined in a suitable way. If λ (real) has algebraic multiplicity $m^A = 3$ and is not regular, we have to find a solution of the type

$$(\mathbf{a} + \mathbf{b}t + \mathbf{c}t^2)\lambda^t,$$

with $\mathbf{a}, \mathbf{b}, \mathbf{c} \in \mathbb{R}^n$, to be determined in a suitable way, and so on. The following example hopefully will clarify the question.

Example 7.3.9 Consider the system

$$\begin{cases} x_1(t+1) = 2x_1(t) \\ x_2(t+1) = x_1(t) + 2x_2(t). \end{cases}$$

The coefficient matrix is

$$A = \begin{bmatrix} 2 & 0 \\ 1 & 2 \end{bmatrix};$$

we have $\lambda_{1,2} = 2$; $m^A = 2$;

$$m^G = 2 - \text{rank}(A - 2I) = 2 - \text{rank} \begin{bmatrix} 0 & 0 \\ 1 & 0 \end{bmatrix} = 2 - 1 = 1.$$

Hence the eigenvalue $\lambda = 2$ is *not regular.* We look for a solution of the type

$$\mathbf{x}(t) = \left(\begin{bmatrix} a_1 \\ a_2 \end{bmatrix} + t \begin{bmatrix} b_1 \\ b_2 \end{bmatrix} \right) 2^t,$$

with $a_1, a_2, b_1, b_2 \in \mathbb{R}$ to be determined, and containing two arbitrary constants.

We have, by inserting into the system the above solution,

$$\mathbf{x}(t+1) = \left(\begin{bmatrix} a_1 \\ a_2 \end{bmatrix} + (t+1) \begin{bmatrix} b_1 \\ b_2 \end{bmatrix} \right) 2^{t+1} = \begin{bmatrix} 2 & 0 \\ 1 & 2 \end{bmatrix} \left(\begin{bmatrix} a_1 \\ a_2 \end{bmatrix} + t \begin{bmatrix} b_1 \\ b_2 \end{bmatrix} \right) 2^t.$$

We divide both members by 2^t and obtain

$$\begin{bmatrix} 2a_1 + 2b_1 \\ 2a_2 + 2b_2 \end{bmatrix} + t \begin{bmatrix} 2b_1 \\ 2b_2 \end{bmatrix} = \begin{bmatrix} 2 & 0 \\ 1 & 2 \end{bmatrix} \left(\begin{bmatrix} a_1 \\ a_2 \end{bmatrix} + t \begin{bmatrix} b_1 \\ b_2 \end{bmatrix} \right),$$

that is

$$\begin{bmatrix} 2a_1 + 2b_1 \\ 2a_2 + 2b_2 \end{bmatrix} + t \begin{bmatrix} 2b_1 \\ 2b_2 \end{bmatrix} = \begin{bmatrix} 2a_1 \\ a_1 + 2a_2 \end{bmatrix} + t \begin{bmatrix} 2b_1 \\ b_1 + 2b_2 \end{bmatrix},$$

that is

$$\begin{bmatrix} 2a_1 + 2b_1 \\ 2a_2 + 2b_2 \end{bmatrix} = \begin{bmatrix} 2a_1 \\ a_1 + 2a_2 \end{bmatrix};$$

$$\begin{bmatrix} 2b_1 \\ 2b_2 \end{bmatrix} = \begin{bmatrix} 2b_1 \\ b_1 + 2b_2 \end{bmatrix}.$$

Finally, we obtain the system

$$\begin{cases} 2a_1 + 2b_1 = 2a_1 \\ 2a_2 + 2b_2 = a_1 + 2a_2 \\ 2b_1 = 2b_1 \\ 2b_2 = b_1 + 2b_2, \end{cases}$$

from which (first or fourth equation) $b_1 = 0$; $a_1 = 2b_2$ (second equation). Hence $a_1 = \alpha$, $a_2 = \beta$ and $b_2 = \frac{\alpha}{2}$. Therefore we can write the general solution:

$$\mathbf{x}(t) = \left(\begin{bmatrix} \alpha \\ \beta \end{bmatrix} + t \begin{bmatrix} 0 \\ \frac{\alpha}{2} \end{bmatrix} \right) 2^t, \quad \alpha, \beta \in \mathbb{R},$$

or also

$$\mathbf{x}(t) = \left(\begin{bmatrix} c_1 \\ c2 \end{bmatrix} + t \begin{bmatrix} 0 \\ \frac{c_1}{2} \end{bmatrix} \right) 2^t, \quad c_1, c_2 \in \mathbb{R},$$

or also

$$\mathbf{x}(t) = \begin{bmatrix} c_1 2^t \\ c_2 2^t + c_1 t 2^{t-1} \end{bmatrix}, \quad c_1, c_2 \in \mathbb{R}.$$

For what concerns the general solution of non homogeneous systems of difference equations of the type (7.11), we begin with the case of a *constant forcing term,* i.e. $\mathbf{b}(t) = \mathbf{b}$, $\forall t$, $\mathbf{b} \neq [0]$:

$$\mathbf{x}(t+1) = A\mathbf{x}(t) + \mathbf{b}.$$

Exploiting the superposition principle, we have to find a particular solution of the above system. We try a *constant solution* (said also *steady state solution* or *equilibrium solution*) $\mathbf{x}(t) = \mathbf{x}^*$; we have $\mathbf{x}^* = A\mathbf{x}^* + \mathbf{b}$, i.e. $(I - A)\mathbf{x}^* = \mathbf{b}$ and, if the matrix $(I - A)$ is invertible, i.e. $\lambda = 1$ is *not* an eigenvalue of A, we get $\mathbf{x}^* = (I - A)^{-1}\mathbf{b}$. Therefore we have the following result.

- The general solution of the system of difference equations

$$\mathbf{x}(t+1) = A\mathbf{x}(t) + \mathbf{b},$$

where A is diagonalizable and $(I - A)$ is nonsingular, is

$$\mathbf{x}(t) = c_1\mathbf{v}^1\lambda_1^t + c_2\mathbf{v}^2\lambda_2^t + \cdots + c_n\mathbf{v}^n\lambda_n^t + (I - A)^{-1}\mathbf{b},$$

with $c_1, c_2, \ldots, c_n \in \mathbb{R}$ if all eigenvalues are real, otherwise $c_1, c_2, \ldots, c_n \in \mathbb{C}$.

Example 7.3.10 Consider the system

$$\mathbf{x}(t+1) = A\mathbf{x}(t) + \mathbf{b},$$

where

$$A = \begin{bmatrix} 2 & 1 \\ 0 & 3 \end{bmatrix}; \quad \mathbf{b} = \begin{bmatrix} 1 \\ 2 \end{bmatrix}.$$

The eigenvalues of A are $\lambda_1 = 2$, $\lambda_2 = 3$. The associated eigenvectors are, respectively,

$$\mathbf{v}^1 = \alpha \begin{bmatrix} 1 \\ 0 \end{bmatrix}, \quad \alpha \in \mathbb{R}, \alpha \neq 0; \quad \mathbf{v}^2 = \beta \begin{bmatrix} 1 \\ 1 \end{bmatrix}, \quad \beta \in \mathbb{R}, \beta \neq 0.$$

The general solution of the homogeneous system is

$$\mathbf{x}(t) = c_1 \begin{bmatrix} 1 \\ 0 \end{bmatrix} 2^t + c_2 \begin{bmatrix} 1 \\ 1 \end{bmatrix} 3^t = \begin{bmatrix} c_1 2^t + c_2 3^t \\ c_2 3^t \end{bmatrix}, \quad c_1, c_2 \in \mathbb{R}.$$

Then we have

$$\mathbf{x}^* = \begin{bmatrix} 2 & 1 \\ 0 & 3 \end{bmatrix} \mathbf{x}^* + \begin{bmatrix} 1 \\ 2 \end{bmatrix}$$

from which

$$(I - A)\mathbf{x}^* = \begin{bmatrix} 1 \\ 2 \end{bmatrix}$$

and so

$$\mathbf{x}^* = (I - A)^{-1} \begin{bmatrix} 1 \\ 2 \end{bmatrix} = \begin{bmatrix} 0 \\ -1 \end{bmatrix}.$$

Therefore the general solution of the non homogeneous system is

$$\mathbf{x}(t) = \begin{bmatrix} c_1 2^t + c_2 3^t \\ c_2 3^t \end{bmatrix} + \begin{bmatrix} 0 \\ -1 \end{bmatrix} = \begin{bmatrix} c_1 2^t + c_2 3^t \\ c_2 3^t - 1 \end{bmatrix}, \quad c_1, c_2 \in \mathbb{R}.$$

What happens if $(I - A)$ is singular, i.e. $\lambda = 1$ is an eigenvalue of A? Two possibilities may occur:

(a) If $\text{rank}(I - A) = \text{rank}(I - A \mid \mathbf{b})$, the dynamical system has infinitely many constant solutions $\mathbf{x}^*$. Also in this case the general solution is

$$\mathbf{x}(t) = c_1 \mathbf{v}^1 \lambda_1^t + c_2 \mathbf{v}^2 \lambda_2^t + \cdots + c_n \mathbf{v}^n \lambda_n^t + \mathbf{x}^*,$$

where $\mathbf{x}^*$ is one of the constant solutions, no matter which of them is chosen.

(b) If $\text{rank}(I - A) < \text{rank}(I - A \mid \mathbf{b})$, the dynamical system has no steady state. In such a case we have to look for a *non constant* solution $\mathbf{x}(t)$. Usually, it is sufficient to look for polynomials of degree $n \geqq 1$. See, for further insights, the books on discrete dynamical systems quoted in the References.

If we have the non homogeneous system (7.11), with the forcing term depending on t, we can try again to exploit the superposition principle and the "method of undetermined coefficients", i.e. if, for instance, $\mathbf{b}(t)$ is a vector of polynomials, to look for a particular solution expressed by the same mathematical form. If $\mathbf{b}(t)$ is, for instance, a vector of exponential functions with a certain base, to look for a particular solution, given by a vector of exponential functions with the same base, and so on.

Example 7.3.11 Consider the system

$$\begin{cases} x_1(t+1) = x_1(t) - x_2(t) + 2^t \\ x_2(t+1) = x_1(t) + x_2(t) + 3^t. \end{cases}$$

We look for a particular solution of the type

$$\mathbf{y}(t) = \begin{bmatrix} a \\ b \end{bmatrix} 2^t + \begin{bmatrix} c \\ d \end{bmatrix} 3^t = \begin{bmatrix} a2^t + c3^t \\ b2^t + d3^t \end{bmatrix}.$$

By substituting into the system, we have

$$\begin{bmatrix} a2^{t+1} + c3^{t+1} \\ b2^{t+1} + d3^{t+1} \end{bmatrix} = \begin{bmatrix} 1 & -1 \\ 1 & 1 \end{bmatrix} \begin{bmatrix} a2^t + c3^t \\ b2^t + d3^t \end{bmatrix} + \begin{bmatrix} 2^t \\ 3^t \end{bmatrix},$$

that is

$$\begin{bmatrix} 2a2^t + 3c3^t \\ 2b2^t + 3d3^t \end{bmatrix} = \begin{bmatrix} (a-b+1)2^t + (c-d)3^t \\ (a+b)2^t + (c+d+1)3^t \end{bmatrix}.$$

From the previous equality we obtain the system

$$\begin{cases} 2a = a - b + 1 \\ 3c = c - d \\ 2b = a + b \\ 3d = c + d + 1, \end{cases}$$

from which

$$\begin{cases} a + b = 1 \\ 2c + d = 0 \\ a - b = 0 \\ c - 2d = -1. \end{cases}$$

The first and the third equation give $a = b = \frac{1}{2}$; the second and the fourth equation give $c = -\frac{1}{5}$, $d = \frac{2}{5}$. Therefore the particular solution we have found is

$$\mathbf{y}(t) = \begin{bmatrix} \frac{1}{2} - \frac{t}{5} \\ \frac{t}{2} + \frac{2t}{5} \end{bmatrix}.$$

We can apply to system (7.8), i.e.

$$\mathbf{x}(t+1) = A(t)\mathbf{x}(t) + \mathbf{b}(t),$$

also the "recursive procedure", even if sometimes the said procedure is not convenient, at least from a computational point of view. Therefore we consider the above system, with $\mathbf{x} : \mathbb{N} \longrightarrow \mathbb{R}^n$, $\mathbf{b}(t) \in \mathbb{R}^n$ and $A(t)$ square matrix of order n. We wish to find, by the recursive procedure, its general solution. We write the relation for $s = 0, 1, 2, \ldots, t$, and obtain

$$\begin{aligned} \mathbf{x}(1) &= A(0) + \mathbf{b}(0); \\ \mathbf{x}(2) &= A(1)\mathbf{x}(1) + \mathbf{b}(1) = A(1)\left[A(0)\mathbf{x}(\mathbf{0}) + \mathbf{b}(0)\right] + \mathbf{b}(1) \\ &= A(1)A(0)\mathbf{x}(0) + A(1)\mathbf{b}(\mathbf{0}) + \mathbf{b}(1); \\ \mathbf{x}(3) &= A(2)\mathbf{x}(2) + \mathbf{b}(2) \\ &= A(2)A(1)A(0)\mathbf{x}(0) + A(2)A(1)\mathbf{b}(0) + A(2)\mathbf{b}(1) + \mathbf{b}(2), \end{aligned}$$

and so on. In general it holds

$$\mathbf{x}(t) = \prod_{s=0}^{t-1} A(t-1-s)\mathbf{x}(0) + \sum_{s=0}^{t-1} \prod_{u=s+1}^{t-1} A(t+s-u)\mathbf{b}(s).$$

If the lower index value of the product is greater than the upper index value, the product is assumed to be equal to the identity matrix I. ($\Pi_{u=\alpha}^{\beta} A(u) = I$, if $\alpha > \beta$).

Some special cases of the previous formula (not too appealing) are the following ones.

1. Coefficients matrix made of constants: $A(t) = A$. We have

$$\mathbf{x}(t) = (A)^t \mathbf{x}(0) + \sum_{s=0}^{t-1} (A)^{t-s-1} \mathbf{b}(s).$$

2. Coefficients matrix made of constants and $\mathbf{b}(t) = \mathbf{b}$. We have

$$\mathbf{x}(t) = (A)^t \mathbf{x}(0) + \left(\sum_{s=0}^{t-1} (A)^s \right) \mathbf{b},$$

where $(A)^0 = I$. We can also provide the closed formula for $\sum_{s=0}^{t-1} (A)^s$. Consider the equality

$$S = I + A + (A)^2 + \cdots + (A)^{t-1} = \sum_{s=0}^{t-1} (A)^s.$$

By right multiplying both sides by A we get

$$SA = A + (A)^2 + (A)^3 + \cdots + (A)^{t-1} + (A)^t.$$

Hence

$$S - SA = I + A - A + \cdots + (A)^{t-1} - (A)^{t-1} - (A)^t = I - (A)^t,$$

that is

$$S(I - A) = I - (A)^t.$$

If A is nonsingular, we obtain

$$S = \sum_{s=0}^{t-1} (A)^s = (I - (A)^t)(I - A)^{-1}.$$

Moreover, it can be proved that $(I-A)^{-1}$ and $(I-(A)^t)$ commute, and hence we have also

$$\sum_{s=0}^{t-1}(A)^s = (I-(A)^t)(I-A)^{-1} = (I-A)^{-1}(I-(A)^t).$$

In conclusion, if A has no eigenvalue equal to one, the solution with the initial value problem $\mathbf{x}(0) = \mathbf{x}^0$, can be written as

$$\begin{aligned}\mathbf{x}(t) &= (A)^t\mathbf{x}^0 + (I-A)^{-1}(I-(A)^t)\mathbf{b} \\ &= (A)^t\mathbf{x}^0 + (I-A)^{-1}\mathbf{b} - (I-A)^{-1}(A)^t\mathbf{b} \\ &= (A)^t\left[\mathbf{x}^0 - (I-A)^{-1}\mathbf{b}\right] + (I-A)^{-1}\mathbf{b}.\end{aligned}$$

3. Scalar case. When the system is one-dimensional, the matrix $A(t)$ becomes the scalar $a(t)$, the multiplication between scalars is commutative and we obtain the general solution previously given in Sect. 7.2, i.e.

$$x(t) = \prod_{s=0}^{t-1} a(s)x(0) + \sum_{s=0}^{t-1}\prod_{u=s+1}^{t-1} a(u)b(s).$$

7.4 Linear Difference Equations of Order n with Constant Coefficients

In this section we take into consideration linear difference equations of order n of the following type, where $a_n, a_0 \neq 0$:

$$a_nx(t+n)+a_{n-1}x(t+n-1)+a_{n-2}x(t+n-2)+\cdots+a_1x(t+1)+a_0x(t) = b(t),$$

or also, being $a_n \neq 0$,

$$x(t+n)+a_{n-1}x(t+n-1)+a_{n-2}x(t+n-2)+\cdots+a_1x(t+1)+a_0x(t) = b(t).$$

If $b = 0$ we obtain the *homogeneous* case.

We have previously asserted that every difference equation of order n is *equivalent* to a system of n first order difference equations. Hence, also linear difference equations of order n, written in the form

$$x(t+n) = -a_{n-1}x(t+n-1) - a_{n-2}x(t+n-2) - \cdots - a_0x(t) + b(t),$$

are equivalent to a linear system of dimension n, of the type

$$\mathbf{x}(t+1) = A\mathbf{x}(t) + \mathbf{b}(t).$$

For example, the homogeneous equation of order n, written as

$$x(t+n)+a_{n-1}x(t+n-1)+a_{n-2}x(t+n-2)+\cdots+a_0x(t) = 0, \quad a_0 \neq 0, \tag{7.13}$$

is equivalent to

$$\mathbf{x}(t+1) = A\mathbf{x}(t),$$

where A is of order n and is of the form

$$A = \begin{bmatrix} 0 & 1 & 0 & 0 & \cdots & 0 \\ 0 & 0 & 1 & 0 & \cdots & 0 \\ 0 & 0 & 0 & 1 & \cdots & 0 \\ \vdots & \vdots & \vdots & \vdots & \cdots & \vdots \\ -a_0 & -a_1 & -a_2 & -a_3 & \cdots & -a_{n-1} \end{bmatrix}.$$

For example, if we have the equation

$$x(t+2) = 3x(t+1) - 2x(t),$$

i.e.

$$x(t+2) - 3x(t+1) + 2x(t) = 0,$$

by putting

$$\begin{cases} x_1(t) = x(t) \\ x_2(t) = x(t+1), \end{cases}$$

we get

$$\begin{cases} x_1(t+1) = x(t+1) = x_2(t) \\ x_2(t+1) = x(t+2) = 3x_2(t) - 2x_1(t), \end{cases}$$

i.e. the homogeneous linear system

$$\begin{bmatrix} x_1(t+1) \\ x_2(t+1) \end{bmatrix} = \begin{bmatrix} 0 & 1 \\ -2 & 3 \end{bmatrix} \begin{bmatrix} x_1(t) \\ x_2(t) \end{bmatrix}.$$

Let us consider the case of the homogeneous linear difference equation of order n of type (7.13). For this equations the general solution is a *linear space of dimension* n and if we consider the non homogeneous case (i.e. $b \neq 0$), the *superposition principle* continues to hold. Similarly to the case of linear difference equations with constant coefficients, of the first order, we can look for solutions of (7.13) of the type

$$x(t) = c\lambda^t, \; c \neq 0, \; \lambda \neq 0.$$

(Indeed, if $c = 0$, we have $x(t) = 0$, $\forall t$ and the equation is trivially satisfied; the same holds if $\lambda = 0$).

If the said sequence solves (7.13), we obtain, by substitution,

$$c\lambda^{t+n} + a_{n-1}c\lambda^{t+n-1} + \cdots + a_1 c\lambda^{t+1} + a_0 c\lambda^t = 0$$

and, dividing by $c\lambda^t$:

$$\lambda^n + a_{n-1}\lambda^{n-1} + \cdots + a_1\lambda + a_0 = 0,$$

which is an algebraic equation of degree n in λ, which is said "the characteristic equation" of (7.13). Nihil sub sole novum.

Now we give the solution rules for (7.13). The reader will note the analogy with the continuous case examined in the previous chapter.

(*a*) The characteristic equation

$$P(\lambda) = \lambda^n + a_{n-1}\lambda^{n-1} + \cdots + a_1\lambda + a_0 = 0$$

has n real distinct roots $\lambda_1, \lambda_2, \ldots, \lambda_n$. Then the general solution of (7.13) is

$$x(t) = c_1\lambda_1^t + c_2\lambda_2^t + \cdots + c_n\lambda_n^t,$$

where $c_1, c_2, \ldots, c_n$ are arbitrarily chosen real constants. The assumption $a_0 a_n \neq 0$ assures that the characteristic equation of (7.13) has no zero solutions. This is not a serious restriction, as whenever the above assumption is not satisfied, it is always possible to change the origin of the time sequence, and to obtain an equivalent equation with a characteristic equation with roots all different from zero. For example, if we have the equation

$$x(t+3) - x(t+2) = 0,$$

it is easy to see that this equation is equivalent to

$$x(t+1) - x(t) = 0,$$

whose characteristic equation has the root $\lambda = 1$.

(*b*) The characteristic equation $P(\lambda) = 0$ has at least one real root λ of multiplicity $r > 1$. Then the contribution of the root λ to the general solution of (7.13) is

$$(c_1 + c_2 t + \cdots + c_r t^{r-1})\lambda^t.$$

Consider the following examples.

Example 7.4.1

(*a*) Consider the equation

$$x(t+2) - 5x(t+1) + 6x(t) = 0.$$

Its characteristic equation is

$$\lambda^2 - 5\lambda + 6 = 0,$$

which has two real and distinct roots: $\lambda_1 = 2$, $\lambda_2 = 3$. It follows that the general solution is

$$x(t) = c_1 2^t + c_2 3^t, \quad c_1, c_2 \in \mathbb{R}.$$

(*b*) We take again into consideration the homogeneous second order linear difference equation

$$x(t+2) = x(t+1) + x(t)$$

of Example 7.1.3, equation which gives rise to the well known *Fibonacci sequence* (Fibonacci is the nickname of the Italian mathematician Leonardo Pisano, 1175–1240). Now we are able to solve this equation. Its characteristic equation is

$$\lambda^2 - \lambda - 1 = 0$$

which has two real distinct roots:

$$\lambda_1 = \frac{1-\sqrt{5}}{2}; \quad \lambda_2 = \frac{1+\sqrt{5}}{2}.$$

Therefore the general solution is

$$x(t) = c_1 \left(\frac{1-\sqrt{5}}{2}\right)^t + c_2 \left(\frac{1+\sqrt{5}}{2}\right)^t, \quad c_1, c_2 \in \mathbb{R}.$$

The Fibonacci sequence is obtained by imposing the initial value problem $x(0) = x(1) = 1$. We obtain the system

$$\begin{cases} c_1 + c_2 = 1 \\ c_1\left(\frac{1-\sqrt{5}}{2}\right) + c_2\left(\frac{1+\sqrt{5}}{2}\right) = 1 \end{cases}$$

which yields

$$c_1 = -\frac{1-\sqrt{5}}{2\sqrt{5}}, \quad c_2 = \frac{1+\sqrt{5}}{2\sqrt{5}}.$$

Therefore

$$x(t) = \frac{1}{\sqrt{5}}\left[\left(\frac{1+\sqrt{5}}{2}\right)^{t+1} - \left(\frac{1-\sqrt{5}}{2}\right)^{t+1}\right].$$

The first twelve terms of the Fibonacci sequence are:

$$1,\ 1,\ 2,\ 3,\ 5,\ 8,\ 13,\ 21,\ 34,\ 55,\ 89,\ 144.$$

It is amazing that the said sequence is obtained by a formula which contains irrational numbers. In particular, the irrational number

$$\frac{1+\sqrt{5}}{2} \approxeq 1.61803398875\ldots$$

which is the result of the limit

$$\lim_{t \longrightarrow +\infty} \frac{x(t+1)}{x(t)},$$

is the well known "golden ratio", which "has inspired thinkers of all disciplines like no other number in the history of mathematics" ([8]).

Example 7.4.2 Consider the equation

$$x(t+3) - 6x(t+2) + 12x(t+1) - 8x(t) = 0.$$

Its characteristic equation is

$$\lambda^3 - 6\lambda^2 + 12\lambda - 8 = 0,$$

that is

$$(\lambda - 2)^3 = 0.$$

Therefore, the equation has one root $\lambda = 2$, with algebraic multiplicity $r = 3$. Then, the general solution is

$$x(t) = (c_1 + c_2 t + c_3 t^2)2^t,$$

with $c_1, c_2, c_3 \in \mathbb{R}$.

We note that the general solution given sub a) holds also for the case of *complex (conjugate)* roots of $P(\lambda)$. However, in this case, it is possible to obtain real solutions.

(c) The characteristic equation $P(\lambda) = 0$ has two complex solutions $\lambda = a + ib$ and $\bar{\lambda} = a - ib$. Then, the (real) contribution of λ and $\bar{\lambda}$ to the general solution of (7.13) is

$$\rho^t \left[c_1 \cos(\theta t) + c_2 \sin(\theta t)\right],$$

where $\theta = \arg \lambda$ and $\rho = \sqrt{a^2 + b^2}$. Indeed, let us choose the original coefficients c_1 and c_2 as follows:

$$c_1 = \alpha + i\beta; \quad c_2 = \alpha - i\beta.$$

Then we have

$$c_1\lambda^t + c_2\bar{\lambda}^t = (\alpha + i\beta)\rho^t(\cos\theta t + i\sin\theta t) + (\alpha - i\beta)\rho^t(\cos\theta t - i\sin\theta t),$$

i.e.

$$c_1\lambda^t + c_2\bar{\lambda}^t = \rho^t(2\alpha\cos\theta t + 2\beta\sin\theta t) = \rho^t\left[c_1\cos(\theta t) + c_2\sin(\theta t)\right],$$

where now $c_1 = 2\alpha$, $c_2 = 2\beta$ are real arbitrary constants, being α and β real and arbitrary.

If there are complex solutions of multiplicity $r > 1$, the previous formula is modified as follows:

$$\rho^t \left[P_1(t)\cos(\theta t) + P_2(t)\sin(\theta t)\right],$$

where $P_1(t)$ and $P_2(t)$ are polynomials of degree $(r - 1)$.

Example 7.4.3 Consider the equation

$$x(t+3) - 2x(t+2) + 4x(t+1) - 8x(t) = 0.$$

Its characteristic equation is

$$P(\lambda) = \lambda^3 - 2\lambda^2 + 4\lambda - 8 = 0,$$

which can be factorized as follows:

$$(\lambda - 2)(\lambda^2 + 4) = 0.$$

It turns out that $P(\lambda)$ has the roots:

$$\begin{cases} \lambda_1 = 2i \\ \lambda_2 = \bar{\lambda}_1 = -2i \\ \lambda_3 = 2. \end{cases}$$

In this case we have $\rho = 2$; $\theta = \frac{\pi}{2}$, therefore the general solution is

$$x(t) = 2^t(c_1 \cos\frac{\pi}{2}t + c_2 \sin\frac{\pi}{2}t + c_3).$$

Example 7.4.4 Consider the equation

$$x(t+6) - x(t+4) - x(t+2) + x(t) = 0.$$

Its characteristic equation is

$$P(\lambda) = \lambda^6 - \lambda^4 - \lambda^2 + 1 = 0.$$

If we put $\mu = \lambda^2$, we obtain

$$\mu^3 - \mu^2 - \mu + 1 = 0,$$

which has $\mu = 1$ as a root; then, by means of the beloved *Ruffini rule,* we obtain

$$(\mu - 1)(\mu^2 - 1) = 0,$$

which has as its roots $\mu_1 = 1$, $\mu_2 = 1$, $\mu_3 = -1$. It turns out that the roots of $P(\lambda) = 0$ are

$$\begin{cases} \lambda_1 = 1 \\ \lambda_2 = 1 \\ \lambda_3 = -1 \\ \lambda_4 = -1 \\ \lambda_5 = i \\ \lambda_6 = -i. \end{cases}$$

We have, for the complex roots, $\rho = 1$ and $\theta = \frac{\pi}{2}$. Therefore, the general solution of the equation is

$$x(t) = c_1 + c_2 t + (c_3 + c_4 t)(-1)^t + c_5 \cos \frac{\pi}{2} t + c_6 \sin \frac{\pi}{2} t.$$

Example 7.4.5 Consider the problem

$$x(t+2) - 7x(t+1) + 12x(t) = 0,$$

$$x(0) = 0, \ x(1) = \alpha.$$

Determine α such that $x(5) = 20$.
The characteristic equation is

$$\lambda^2 - 7\lambda + 12 = 0,$$

hence

$$\lambda = \frac{7 \pm \sqrt{49 - 48}}{2} \Longrightarrow \lambda_1 = 3, \ \lambda_2 = 4.$$

Therefore the general solution is

$$x(t) = c_1 3^t + c_2 4^t, \quad c_1, c_2 \in \mathbb{R}.$$

Now we impose the conditions of the initial value problem. We find $c_1 = -\alpha$; $c_2 = \alpha$. Therefore

$$x(t) = -\alpha 3^t + \alpha 4^t.$$

Then we have

$$20 = -\alpha 3^5 + \alpha 4^5$$

from which

$$\alpha = \frac{20}{781}.$$

Now we give some insights on the non homogeneous case:

$$x(t+n) + a_{n-1}x(t+n-1) + \cdots + a_1 x(t+1) + a_0 x(t) = b(t). \tag{7.14}$$

As usual, we could exploit the "superposition principle" and the "method of undetermined coefficients". We give only some hints on the following cases (see, e.g., [13], [14]).

(a) If $b(t) = P(t)$, where $P(t)$ is an m-degree polynomial, with $m \geqq 0$ and $\lambda = 1$ is *not* a root of the characteristic equation, then (7.14) has a particular solution of the form

$$y(t) = Q(t),$$

where $Q(t)$ is an m-degree polynomial. If $\lambda = 1$ is a root of the characteristic equation, with multiplicity r, then (7.14) has a particular solution of the form

$$y(t) = t^r Q(t).$$

(*b*) If $b(t) = P(t)\sigma^t$, where $P(t)$ is a polynomial of degree $m \geqq 0$ and σ is *not* a root of the characteristic equation, then (7.14) has a particular solution of the form

$$y(t) = Q(t)\sigma^t,$$

where $Q(t)$ is an m-degree polynomial. If σ is a root of the characteristic equation, with multiplicity r, then (7.14) has a particular solution of the form

$$y(t) = t^r Q(t)\sigma^t.$$

Practically, we insert into the equation to be solved the supposed solution and then we find the related coefficients of the polynomial.

Example 7.4.6 Solve the difference equation

$$x(t+2) + 3x(t+1) - 10x(t) = 8.$$

The characteristic equation associated with the homogeneous equation is

$$\lambda^2 + 3\lambda - 10 = 0.$$

We have

$$\lambda = \frac{-3 \pm \sqrt{9+40}}{2} \Longrightarrow \lambda_1 = 2, \;\; \lambda_2 = -5.$$

The general solution of the homogeneous equation is

$$x(t) = c_1 2^t + c_2(-5)^t, \;\; c_1, c_2 \in \mathbb{R}.$$

As $b(t) = b$, a constant, we look for a constant particular solution. By substituting into the non homogeneous equation, we obtain

$$k + 3k - 10k = 8,$$

from which $k = -\frac{4}{3}$. Thus, the general solution of the proposed equation is

$$x(t) = c_1 2^t + c_2(-5)^t - \frac{4}{3}, \quad c_1, c_2 \in \mathbb{R}.$$

Example 7.4.7 Solve the equation

$$x(t+2) - 8x(t+1) + 16x(t) = 27.$$

The characteristic equation of the homogeneous equation is

$$\lambda^2 - 8\lambda + 16 = 0 \iff (\lambda - 4)^2 = 0.$$

We have the double real root $\lambda = 4$. The general solution of the homogeneous equation is

$$x(t) = c_1 4^t + c_2 t 4^t, \quad c_1, c_2 \in \mathbb{R}.$$

We look for a constant particular solution of the non homogeneous equation. By substituting, we obtain

$$k - 8k + 16k = 27,$$

from which we get $k = 3$. The general solution of the proposed equation is

$$x(t) = c_1 4^t + c_2 t 4^t + 3, \quad c_1, c_2 \in \mathbb{R}.$$

Example 7.4.8 Solve the equation

$$x(t+4) - 2x(t+2) + x(t) = t.$$

The characteristic equation of the homogeneous equation is

$$\lambda^4 - 2\lambda^2 + 1 = 0,$$

which has two (double) roots: $\lambda_1 = \lambda_2 = 1$ and $\lambda_3 = \lambda_4 = -1$. The general solution of the associated homogeneous equation is

$$x(t) = c_1 + c_2 t + (-1)^t (c_3 + c_4 t), \quad c_1, c_2.c_3, c_4 \in \mathbb{R}.$$

Since 1 is a double solution of the characteristic equation and $b(t)$ is a first degree polynomial, we look for a particular solution of the form

$$y(t) = t^2(at + b).$$

By substituting into the proposed equation, we obtain

$$(t+4)^2(at+4a+b) - 2(t+2)^2(at+2a+b) + t^2(at+b) = t,$$

whence

$$48a + 8b + 24at = t.$$

Equating the coefficients of t and of the zero degree terms, we get $a = \frac{1}{24}$ and $b = -\frac{1}{4}$. Hence, the general solution of the proposed equation is

$$x(t) = c_1 + c_2 t + (-1)^t(c_3 + c_4 t) + \frac{1}{24}t^3 - \frac{1}{4}t^2, \quad c_1, c_2.c_3, c_4 \in \mathbb{R}.$$

7.5 Equilibrium Solutions and Stability

Let us consider for simplicity an autonomous first order difference equations system

$$\mathbf{x}(t+1) = \mathbf{f}(\mathbf{x}(t)), \tag{7.15}$$

where $\mathbf{x} : T \subset \mathbb{N} \longrightarrow \mathbb{R}^n$ and $\mathbf{f} : \mathbb{R}^n \longrightarrow \mathbb{R}^n$. A *constant solution* $\mathbf{x}^*$ of (7.15), if there exists, i.e. a solution such that $\mathbf{x}(t+1) = \mathbf{x}(t) = \mathbf{x}^*$, $\forall t = 0, 1, \ldots,$ is called an *equilibrium solution* or a *steady state solution*. In other words, we must have

$$\mathbf{f}(\mathbf{x}^*) = \mathbf{x}^*. \tag{7.16}$$

On the grounds of the definition contained in Lemma 1.8.7, relation (7.16) says that the equilibrium solution $\mathbf{x}^*$ is nothing but a *fixed point* for $\mathbf{f}$. A dynamical system of the form (7.15) may have:

(i) No equilibrium solutions.
(ii) A finite number $k \geqq 1$ of equilibrium solutions.
(iii) Infinitely many equilibrium solutions.

Let us consider the dynamical linear system described by

$$\mathbf{x}(t+1) = A\mathbf{x}(t) + \mathbf{b},$$

where A is square of order n.

Then this system has one equilibrium solution

$$\mathbf{x}^* = A\mathbf{x}^* + \mathbf{b}$$

if and only if $\det(I - A) \neq 0$, i.e. $(I - A)$ is invertible. If $\det(I - A) = 0$ and $\operatorname{rank}(I - A) = \operatorname{rank}[(I - A); \mathbf{b}]$, there are surely infinite many equilibria. Finally, if $\det(I - A) = 0$ and $\operatorname{rank}(I - A) \neq \operatorname{rank}[(I - A); \mathbf{b}]$, there are no equilibrium solutions. If $\mathbf{b} = [0]$, the equilibrium solution $\mathbf{x}^* = [0]$ always exists.

The notions of stability of a trajectory with initial point $x(0) = \mathbf{x}^0$, with respect to an equilibrium solution $\mathbf{x}^*$, are formally the same already introduced in the previous chapter for continuous dynamical systems and we do not repeat here the formal definitions. Obviously, in the discrete case the state variables (or state vectors) are given by a sequence. We recall only that:

- The *Lyapunov stability* (or *neutral stability*) means that all the nearby solutions stay close to the equilibrium solution $\mathbf{x}^*$ for ever: the pointwise distance between them and the equilibrium can be "controlled" and is small than $\varepsilon > 0$ for every $t \geqq t_0$ (usually $t_0 = 0$).
- The *local asymptotic stability* requires that all nearby solutions stay close to the equilibrium solution (i.e. the solutions are stable in the sense of Lyapunov) and moreover, the said solutions converge to the equilibrium solution $\mathbf{x}^*$, as $t \longrightarrow +\infty$.
- The *global asymptotic stability* requires that the previous process holds for *any* initial value problem.
- In all other cases, not described in the previous points, we speak of *unstable solutions.*

Now we study the stability of the zero equilibrium solution $\mathbf{x}^* = [0]$ of the homogeneous linear system

$$\mathbf{x}(t+1) = A\mathbf{x}(t), \tag{7.17}$$

with A square of order n. For this system we note that:

1. All solutions are defined for every $t \in \mathbb{N}_+$.
2. $\mathbf{x}^* = [0]$ is always an equilibrium solution.
3. There are infinitely many equilibrium solutions if and only if

$$\det(I - A) = 0 \iff \lambda = 1 \text{ is an eigenvalue of } A.$$

We focus our attention on the stability of $\mathbf{x}^* = [0]$. A first result we do not prove, is the following one.

Theorem 7.5.1 *If* $\mathbf{x}^* = [0]$ *is a locally asymptotically stable solution of* (7.17), *then* $\mathbf{x}^* = [0]$ *is globally (asymptotically) stable.*

On the grounds of the previous result, we can therefore assert that $\mathbf{x}^* = [0]$ is, for system (7.17), either globally asymptotically stable or stable in the sense of Lyapunov or unstable. We have the following basic result.

Theorem 7.5.2 *Let be given the system described by* (7.17).

(i) *If and only if all eigenvalues* λ_i *of* A *are such that*

$$|\lambda_i| < 1, \ \forall i,$$

then $\mathbf{x}^* = [0]$ *is globally asymptotically stable for* (7.17).

(ii) *If for every eigenvalue* λ_i *of* A *one has*

$$|\lambda_i| \leqq 1, \ \forall i,$$

and the eigenvalues λ_i *for which* $|\lambda_i| = 1$ *are regular eigenvalues of* A *(i.e. their algebraic multiplicity is equal to the respective geometric multiplicity), then* $\mathbf{x}^* = [0]$ *is stable in the sense of Lyapunov (we recall that simple eigenvalues are always regular).*

(iii) *In all other cases* $\mathbf{x}^* = [0]$ *is unstable.*

The proof of point (i) is quite simple. We have seen that the general solution of (7.17), with the initial condition $\mathbf{x}(0) = \mathbf{x}^0$, is given by

$$\mathbf{x}(t) = (A)^t \mathbf{x}^0.$$

If we want that $\mathbf{x}(t) \longrightarrow [0]$ for $t \longrightarrow +\infty$ and for all $\mathbf{x}^0 \in \mathbb{R}^n$, we have to assume that for all $\mathbf{x}^0 \in \mathbb{R}^n$ it holds $\lim_{t \longrightarrow +\infty} (A)^t = [0]$. In other words, the matrix A must be *small* (or *convergent*). This occurs (see Theorem 1.6.26) if and only if

$$\rho_A < 1,$$

where ρ_A is the *spectral radius* of A, i.e.

$$\rho_A = \max_{1 \leqq i \leqq n} \{|\lambda_i|\},$$

i.e. for all eigenvalues λ_i of A we must have

$$|\lambda_i| < 1, \ \forall i.$$

On the other hand, if A is *diagonalizable,* we have the following form of the general solution of (7.17):

$$\mathbf{x}(t) = c_1 \mathbf{v}^1 \lambda_1^t + c_2 \mathbf{v}^2 \lambda_2^t + \cdots + c_n \mathbf{v}^n \lambda_n^t$$

where $\lambda_1, \ldots, \lambda_n$, are the eigenvalues of A, $\mathbf{v}^1, \ldots, \mathbf{v}^n$, are the associated eigenvectors and $c_1, \ldots, c_n$ are constants to be determined by an initial value problem. Obviously, we have $\mathbf{x}(t) \longrightarrow [0]$ for $t \longrightarrow +\infty$ if and only if $|\lambda_i| < 1$ for every i.

Example 7.5.3 Consider the system

$$\mathbf{x}(t+1) = A\mathbf{x}(t),$$

where

$$A = \begin{bmatrix} -1 & -1 \\ 1 & \frac{1}{2} \end{bmatrix}.$$

The characteristic equation of A is

$$\lambda^2 + \frac{1}{2}\lambda + \frac{1}{2} = 0.$$

We have two complex (conjugate) solutions

$$\lambda_1 = -\frac{1}{4} + \frac{1}{4}i\sqrt{7};\ \bar{\lambda}_1 = -\frac{1}{4} - \frac{1}{4}i\sqrt{7}.$$

Since

$$|\lambda_1| = |\bar{\lambda}_1| = \sqrt{\frac{1}{16} + \frac{7}{16}} = \sqrt{\frac{1}{2}} < 1,$$

we conclude that the equilibrium solution $\mathbf{x}^* = [0]$ is globally asymptotically stable.

Example 7.5.4 Consider system (7.17) with

$$A = \begin{bmatrix} -1 & 0 & 1 \\ 0 & -\frac{1}{2} & 0 \\ 0 & 0 & -\frac{1}{3} \end{bmatrix}.$$

Being A a triangular matrix, its eigenvalues are $\lambda_1 = -1$, $\lambda_2 = -\frac{1}{2}$, $\lambda_3 = -\frac{1}{3}$. Since

$$|\lambda_1| = 1,\ |\lambda_2| = \frac{1}{2},\ |\lambda_3| = \frac{1}{3},$$

and λ_1 is a simple eigenvalue, the equilibrium point $\mathbf{x}^* = [0]$ is *stable in the sense of Lyapunov* (or *neutrally stable*) for system (7.17).

Example 7.5.5 Consider system (7.17) with

$$A = \begin{bmatrix} -1 & 2 & 3 \\ 0 & 0 & 4 \\ 0 & 0 & -1 \end{bmatrix}.$$

The eigenvalues are $\lambda_1 = \lambda_2 = -1$, $\lambda_3 = 0$. Hence

$$|\lambda_1| = |\lambda_2| = 1;\ |\lambda_3| = 0.$$

However, the eigenvalue $\lambda_1 = \lambda_2 = -1$ is *not regular,* as $m_\lambda^A = 2$, but $3 - \text{rank}(A - \lambda I) = 3 - \text{rank}(A + I) = 3 - 2 = 1$, being

$$(A + I) = \begin{bmatrix} 0 & 2 & 3 \\ 0 & 1 & 4 \\ 0 & 0 & 0 \end{bmatrix}.$$

Therefore the equilibrium point $\mathbf{x}^* = [0]$ is unstable for system (7.17).

Now we consider the non homogeneous case of the form

$$\mathbf{x}(t+1) = A\mathbf{x}(t) + \mathbf{b}, \ (\mathbf{b} \neq [0]). \tag{7.18}$$

By the superposition principle, if system (7.18) has an equilibrium point $\mathbf{x}^*$, then its general solution is given by

$$\mathbf{x}(t) = \mathbf{z}(t) + \mathbf{x}^*,$$

where $\mathbf{z}(t)$ is the general solution of the associated homogeneous system

$$\mathbf{x}(t+1) = A\mathbf{x}(t).$$

Therefore

$$\lim_{t \longrightarrow +\infty} \mathbf{x}(t) = \mathbf{x}^* \iff \lim_{t \longrightarrow +\infty} \mathbf{z}(t) = [0] .$$

Hence an equilibrium point $\mathbf{x}^*$ of (7.18) is:

- Globally asymptotically stable if and only if $[0]$ is globally asymptotically stable for the associated homogeneous system (7.17).
- Stable in the sense of Lyapunov if and only if $[0]$ is stable in the sense of Lyapunov for the associated homogeneous system (7.17).
- Unstable if and only if $[0]$ is unstable for system (7.17).

We recall again that (7.18) has an equilibrium point $\mathbf{x}^*$ if $\mathbf{x}^*$ is a solution of the system

$$\mathbf{x}^* = A\mathbf{x}^* + \mathbf{b}$$

and this equilibrium point is unique if and only if $(I - A)$ is nonsingular. Then we have

$$\mathbf{x}^* = (I - A)^{-1}\mathbf{b}.$$

Similarly to what happens for the continuous case, we have criteria which ensure that the roots of the characteristic equation of A, i.e. its eigenvalues, have modulus less than one, *without solving* the said equation. In the continuous case we have the *Routh-Hurwitz conditions* (see Sect. 6.9); in the discrete case we have the *Cohn-Schur conditions.* Let us consider again the autonomous system (7.17):

$$\mathbf{x}(t+1) = A\mathbf{x}(t)$$

and let $P(\lambda)$ be its characteristic polynomial given in the form

$$P(\lambda) = \lambda^n + a_{n-1}\lambda^{n-1} + \cdots + a_1\lambda + a_0.$$

In order to describe the *Cohn-Schur conditions* (a bit intricate), we need build some auxiliary matrices by means of the coefficients $a_0, a_1, \ldots, a_{n-1}$ of $P(\lambda)$. First we consider the following square matrices

$$\underset{(n\times n)}{B} = \begin{bmatrix} 1 & 0 & 0 & \cdots & 0 \\ a_{n-1} & 1 & 0 & \cdots & 0 \\ a_{n-2} & a_{n-1} & 1 & \cdots & 0 \\ \vdots & \vdots & \vdots & \ddots & \vdots \\ a_1 & a_2 & a_3 & \cdots & 1 \end{bmatrix}; \quad \underset{(n\times n)}{C} = \begin{bmatrix} a_0 & a_1 & a_2 & \cdots & a_{n-1} \\ 0 & a_0 & a_1 & \cdots & a_{n-2} \\ 0 & 0 & a_0 & \cdots & a_{n-3} \\ \vdots & \vdots & \vdots & \ddots & \vdots \\ 0 & 0 & 0 & \cdots & a_0 \end{bmatrix}.$$

Let $B(k)$ and $C(k)$ be, respectively, the North-West principal submatrices of order k of B and C :

$$B(1) = [1], \; B(2) = \begin{bmatrix} 1 & 0 \\ a_{n-1} & 1 \end{bmatrix}, \ldots, B(n) = B.$$

$$C(1) = [a_0]; \; C(2) = \begin{bmatrix} a_0 & a_1 \\ 0 & a_0 \end{bmatrix}, \ldots, C(n) = C.$$

Now we consider the *Cohn-Schur matrices* $S(k)$ defined as follows:

$$S(k) = \begin{bmatrix} B(k) & C(k) \\ (C(k))^\top & (B(k))^\top \end{bmatrix}, \quad k = 1, \ldots, n.$$

Therefore we have

$$S(1) = \begin{bmatrix} 1 & a_0 \\ a_0 & 1 \end{bmatrix};$$

$$S(2) = \begin{bmatrix} 1 & 0 & a_0 & a_1 \\ a_{n-1} & 1 & 0 & a_0 \\ a_0 & 0 & 1 & a_{n-1} \\ a_1 & a_0 & 0 & 1 \end{bmatrix};$$

$$\cdots$$

$$S(n) = \begin{bmatrix} 1 & 0 & \cdots & 0 & a_0 & a_1 & \cdots & a_{n-1} \\ a_{n-1} & 1 & \cdots & 0 & 0 & a_0 & \cdots & a_{n-2} \\ \vdots & \vdots & \cdots & \vdots & \vdots & \vdots & & \vdots \\ a_1 & a_2 & \cdots & 1 & 0 & 0 & \cdots & a_0 \\ a_0 & 0 & \cdots & 0 & 1 & a_{n-1} & \cdots & a_1 \\ a_1 & a_0 & \cdots & 0 & 0 & 1 & \cdots & a_2 \\ \vdots & \vdots & & \vdots & \vdots & \vdots & & \vdots \\ a_{n-1} & a_{n-2} & \cdots & a_0 & 0 & 0 & \cdots & 1 \end{bmatrix}.$$

Finally we have the following result.

Theorem 7.5.6 (Theorem of Cohn-Schur) *Every solution of*

$$P(\lambda) = \lambda^n + a_{n-1}\lambda^{n-1} + \cdots + a_1\lambda + a_0 = 0$$

has modulus less than one if and only if the North-West principal minors of $S(k)$, $k = 1, \ldots, n$, *are all positive.*

There are also stability criteria involving directly the elements of the matrix A :

- If $n = 2$, i.e. A is of order 2, then $\mathbf{x}^* = [0]$ is globally asymptotically stable for system (7.17) if and only if

$$|\mathrm{tr}(A)| < 1 + \det(A) \;\text{ and }\; \det(A) < 1.$$

  Indeed, we recall that with

$$A = \begin{bmatrix} a_{11} & a_{12} \\ a_{21} & a_{22} \end{bmatrix},$$

the characteristic equation of A is given by

$$\lambda^2 - (a_{11} + a_{22})\lambda + (a_{11}a_{22} - a_{12}a_{21}) = \lambda^2 - \mathrm{tr}(A)\lambda + \det(A) = 0,$$

being, moreover, $\mathrm{tr}(A) = \lambda_1 + \lambda_2$ and $\det(A) = \lambda_1\lambda_2$.

- The following condition is *sufficient* so that $\mathbf{x}^* = [0]$ is globally asymptotically stable for (7.17), i.e. all eigenvalues of A have modulus less than one (see [6])):

$$\sum_{j=1}^{n} |a_{ij}| < 1, \ \forall i = 1, \ldots, n.$$

  Recalling that the transpose $A^\top$ has the same characteristic roots of A, we can also apply the above criterion to the sum of the absolute values of the elements of the columns of A.
- If $A \geqq [0]$, then a necessary and sufficient condition such that $|\lambda_i| < 1$ for all i, is that the matrix $(I - A)$ satisfies the *Hawkins-Simon conditions* (see Theorem 1.9.1):

$$(1 - a_{11}) > 0; \quad \begin{vmatrix} 1 - a_{11} & -a_{12} \\ -a_{21} & 1 - a_{22} \end{vmatrix} > 0; \ldots ; |I - A| > 0.$$

  For the related proof, see, e.g., [3], [11], [12], [20].

Example 7.5.7 If in system (7.17) the coefficient matrix is

$$A = \begin{bmatrix} \frac{1}{2} & \frac{1}{4} \\ \frac{1}{4} & \frac{1}{2} \end{bmatrix},$$

we have

$$(I - A) = \begin{bmatrix} \frac{1}{2} & -\frac{1}{4} \\ -\frac{1}{4} & \frac{1}{2} \end{bmatrix},$$

and $\det(I - A) = \frac{1}{4} - \frac{1}{16} = \frac{3}{16}$; therefore the Hawkins-Simon conditions are satisfied.

- If $A \geqq [0]$, a sufficient condition for stability of $\mathbf{x}^* = [0]$ in system (7.17) is that all sums of the columns (or of the rows) of A are less than one. If, moreover, A is *irreducible,* then it is sufficient that all sums of the columns (or of the rows) of A are less or equal than one and at least one sum is strictly less than one.
- (Samuelson [17]). If $(A^\top A - I)$ is negative definite, then A has all eigenvalues with modulus less than one.
- Each one of the following conditions is *necessary* so that $\mathbf{x}^* = [0]$ is globally asymptotically stable for (7.17):

  (i)

$$\left| \sum_{i=1}^{n} a_{ii} \right| < n.$$

(ii)

$$|\det(A)| < 1.$$

For other considerations on these issues the reader may consult [2] and [5].

For what concerns the stability of the zero solution of linear homogeneous difference equations of order n of the type

$$x(t+n) + a_{n-1}x(t+n-1) + \cdots + a_1 x(t+1) + a_0 x(t) = 0, \tag{7.19}$$

the same considerations made before hold also for this case. If we consider the related characteristic equation

$$P(\lambda) = \lambda^n + a_{n-1}\lambda^{n-1} + \cdots + a_1\lambda + a_0 = 0,$$

then we have:

(i) $x^* = 0$ is globally asymptotically stable for (7.19) if and only if $|\lambda_i| < 1$ for all $i = 1, \ldots, n$.
(ii) If $|\lambda_i| \leqq 1$ for all $i = 1, \ldots, n$ and every root such that $|\lambda_i| = 1$ is simple (i.e. not repeated), then $x^* = 0$ is stable for (7.19) in the sense of Lyapunov (or neutrally stable).
(iii) In all other cases $x^* = 0$ is unstable.

Obviously, the Cohn-Schur conditions can be applied to the characteristic equation of (7.19). If in (7.19) we have $n = 2$, it is not difficult to see that $x^* = 0$ is globally asymptotically stable for (7.19) if and only if, in the related characteristic equation

$$\lambda^2 + a_1\lambda + a_0 = 0,$$

we have $a_0 < 1$, $a_0 + a_1 + 1 > 0$, $a_0 - a_1 + 1 > 0$. It seems that these conditions were first obtained by [17], without the use of the Cohn-Schur Theorem (Theorem 7.5.6). We note that the same conditions can be given in the equivalent form

$$a_0 < 1, \;\; 1 + a_0 > |a_1|.$$

If we have an algebraic equation of the type

$$a_n x^n + a_{n-1}x^{n-1} + \cdots + a_0 = 0,$$

two sufficient conditions for the same to have all roots with modulus less than one (i.e. which lie within the unit circle in the complex plane) are:

- $\sum_{i=0}^{n-1} |a_i| < |a_n|$.
- (Kakeya Theorem) $a_n > a_{n-1} > \cdots > a_0 > 0$.
  See [10].

Finally, we point out that [17] has shown how to use the Routh-Hurwitz conditions, instead of the Cohn-Schur conditions, as necessary and sufficient conditions for an algebraic equation to have all roots with modulus less than one. See also Samuelson [16].

7.6 Phase Diagrams and Some Economic Applications

Also in the discrete case there is the possibility to use a graphical method to study the behaviour of the solution of a first order (or also of a second order) difference equation, with respect to an equilibrium point. We limit ourselves to a first order autonomous difference equation of the type

$$x(t+1) = f(x(t)), \tag{7.20}$$

with $f : \mathbb{R} \longrightarrow \mathbb{R}$. The graphical technique used for (7.20) is known as a "phase diagram", just as for the continuous case; however, for the discrete case, it is used also the name "stairstep diagram". This technique consists of the following procedure.

1. We consider a Cartesian plane with horizontal axis containing the values of $x(t)$ and with vertical axis containing the values $x(t+1)$.
2. We plot the diagram of the function f, defined by relation (7.20), and we plot the bisecting line of the first and third quadrant. The equilibrium points are given by the intersection between the diagram of f and the said bisecting line.
3. We consider an arbitrary initial value $x_0 \neq x^*$, with x^* equilibrium point. Then we determine the value $f(x_0)$.
4. From the point $(x_0, f(x_0))$, we trace a parallel line to the horizontal axis, in order to intersect the bisecting line. The abscissa of the intersection point is, say, $x_1 = f(x_0)$.
5. We repeat the previous operations, by starting directly from (x_1, x_1), determining in this way $f(x_1)$, and so on.

By means of the above procedure we construct a broken line, from which the name of "stairstep diagram", which can put into evidence the trajectory of $x(t)$, with initial value x_0, with respect to the equilibrium point x^*.

Some example will hopefully clarify the procedure. We confine our diagrams to the first quadrant, because economic variables are typically nonnegative.

Example 7.6.1 Consider the first order autonomous equation

$$x(t+1) = \sqrt{x(t)}.$$

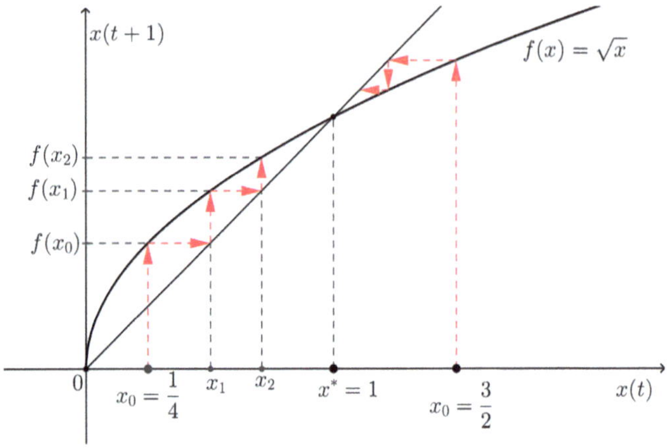

Fig. 7.1 Example 7.6.1. Graphical procedure starting at $x_0 = \frac{1}{4}$ and $x_0 = \frac{3}{2}$. $x(t)$ converges to $x^* = 1$

We have two equilibrium points at $x_1^* = 0$ and $x_2^* = 1$. See Fig. 7.1.

Consider an arbitrary initial value, for instance $x_0 = \frac{1}{4}$, and apply to this point the graphical procedure described above under the points 1)-5). We conclude that $x(t)$ appears to be convergent to $x^* = 1$, and this happens from any arbitrary starting initial value point x_0, with $x_0 \in (0, 1)$. If we investigate the motion of the solution to the right of the equilibrium point $x^* = 1$, for instance $x_0 = \frac{3}{2}$, we see that again $x(t)$ appears to be converging to the equilibrium point $x^* = 1$. We conclude that from any starting point $x_0 > 0$ the path $x(t)$ converges to $x^* = 1$, making therefore $x^* = 1$ a stable equilibrium, with basin of attraction $I = (0, +\infty)$.

On the other hand, the point $x^* = 0$ is an unstable equilibrium, because for $x(t) > 0$, $x(t)$ diverges from zero.

Example 7.6.2 Consider the linear equation

$$x(t+1) = \frac{1}{2}x(t),$$

which has an equilibrium point $x^* = 0$. See Fig. 7.2.

Starting, for instance, from $x_0 = 5$, we see that the stairstep moves towards the unique equilibrium point $x^* = 0$. The same happens if we start from a negative initial point, e.g., $x_0 = -7$. We conclude that the origin is globally asymptotically stable (its basin of attraction is the whole interval $(-\infty, +\infty)$).

Example 7.6.3 Consider the equation

$$x(t+1) = 2x(t).$$

Also in this case the unique equilibrium is $x^* = 0$; see Fig. 7.3.

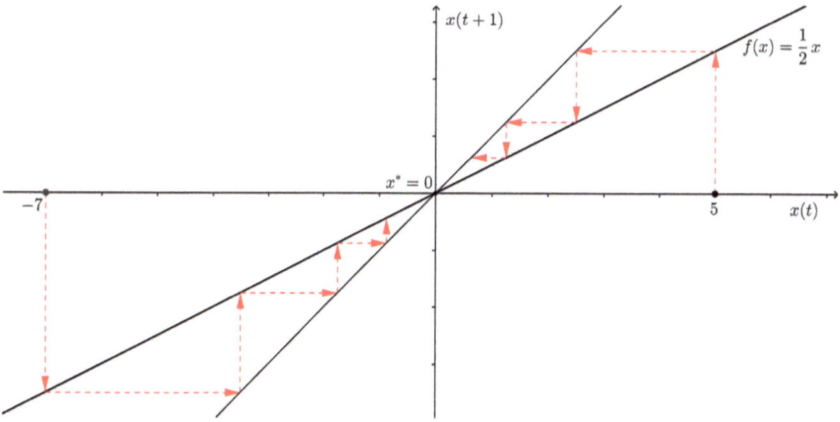

Fig. 7.2 Example 7.6.2. Graphical procedure, $x(t)$ converges to the unique equilibrium point $x^* = 0$

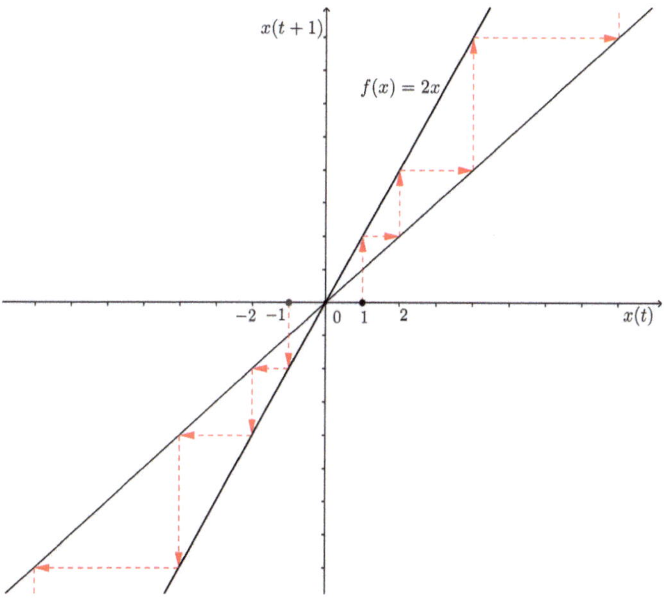

Fig. 7.3 Example 7.6.3. Graphical procedure, there is a unique equilibrium point $x^* = 0$, which is unstable

The stairstep diagram shows that if we start from the right of the origin, the corresponding trajectory diverges to $+\infty$. If we start from the left of the origin, the corresponding trajectory diverges to $-\infty$. We conclude that the origin is an unstable equilibrium point.

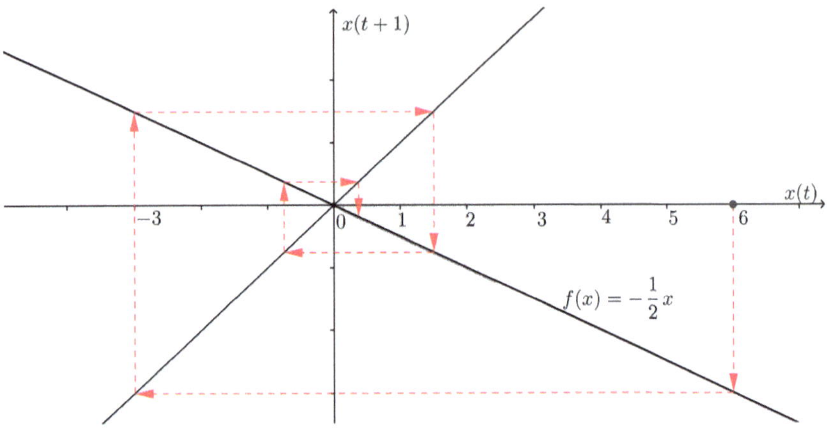

Fig. 7.4 Example 7.6.4. Graphical procedure. In this case we obtain a cobweb and a unique equilibrium point $x^* = 0$, which is globally asymptotically stable

Example 7.6.4 Consider the equation

$$x(t+1) = -\frac{1}{2}x(t).$$

The unique equilibrium point is obviously $x^* = 0$. See Fig. 7.4.

Starting from a positive initial value, say $x_0 = 6$, instead of a "stairstep", we obtain a sort of cobweb, wrapping the equilibrium point closer and closer. In this case the equilibrium point is globally asymptotically stable. Note that if the equation is, for example,

$$x(t+1) = -2x(t),$$

even starting from a neighbourhood of the origin, we obtain an "explosive" cobweb.

Example 7.6.5 Consider the following discrete logistic equation

$$x(t+1) - x(t) = ax(t)\,[M - x(t)]$$

with $a = \frac{1}{2}$ and $M = 1$, that is

$$x(t+1) = \frac{3}{2}x(t) - \frac{1}{2}x^2(t).$$

This equation has two equilibrium points: $x^* = 0$ and $x^{**} = 1$ (see Fig. 7.5).

- If $x_0 < 0$, it is easily seen that $x(t) \longrightarrow -\infty$.
- If $0 < x_0 < 3$, we find that $x(t) \longrightarrow 1$ for $t \longrightarrow +\infty$.
- If $x_0 > 3$, we find that $x(t) \longrightarrow +\infty$.

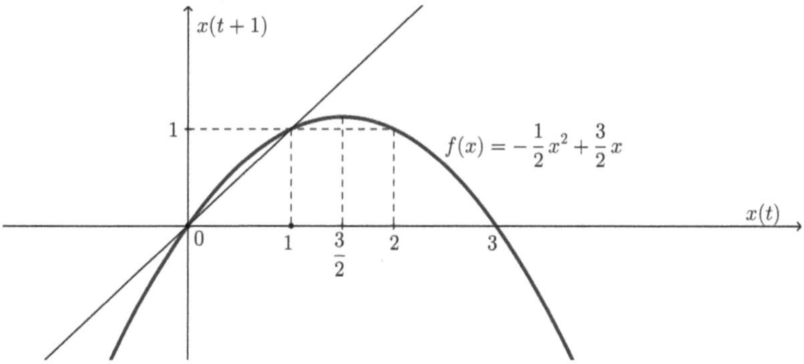

Fig. 7.5 Example 7.6.5 has two equilibrium points: $x^* = 0$ is unstable and $x^{**} = 1$ is locally asymptotically stable

Hence we conclude that $x^* = 0$ is an unstable equilibrium point, whereas $x^{**} = 1$ is a locally asymptotically stable equilibrium point.

The reader is invited to study the stability of the equilibrium points of

$$x(t + 1) = x^3(t) - 5x^2(t) + 7x(t).$$

He will find three equilibrium points: 0, 2, 3, and will find that for $x_0 > 3$ the equilibrium point $x^* = 3$ is unstable, for $2 < x_0 < 3$, the equilibrium point $x^* = 2$ is locally stable; for $0 < x_0 < 2$, the equilibrium point $x^* = 2$ is locally stable, ...

For simple autonomous difference equations of type (7.20) it is possible to give a general result.

Theorem 7.6.6 *Let be given a first order autonomous difference equation of the following type*

$$x(t + 1) = f(x(t)),$$

with $f : \mathbb{R} \longrightarrow \mathbb{R}$ *increasing. Let* x^* *be an equilibrium point.*

(*i*) *If in a neighbourhood of* x^* *we have*

$$f(x) > x, \text{ if } x < x^*;$$
$$f(x) < x, \text{ if } x > x^*,$$

then x^* *is locally asymptotically stable.*

(*ii*) *If the previous inequalities hold for every* $x \in \operatorname{dom}(f)$, *then* x^* *is globally asymptotically stable.*

See Figs. 7.5, 7.6 and 7.7, where the equilibrium point x^* is, respectively, locally asymptotically stable (point 1), globally asymptotically stable and unstable.

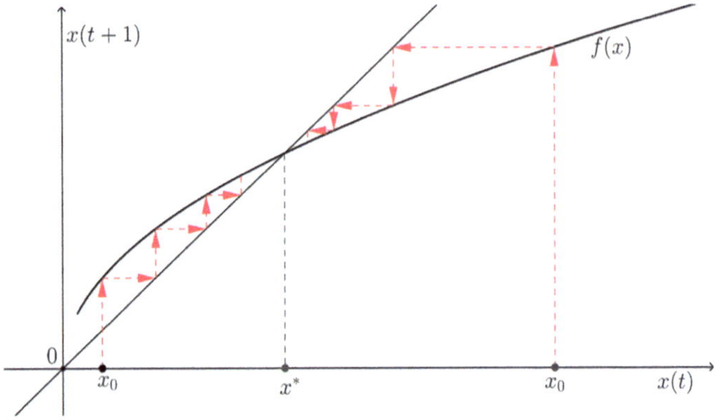

Fig. 7.6 Illustration of Theorem 7.6.6(i). x^* is globally asymptotically stable

Fig. 7.7 The pont x^* is unstable

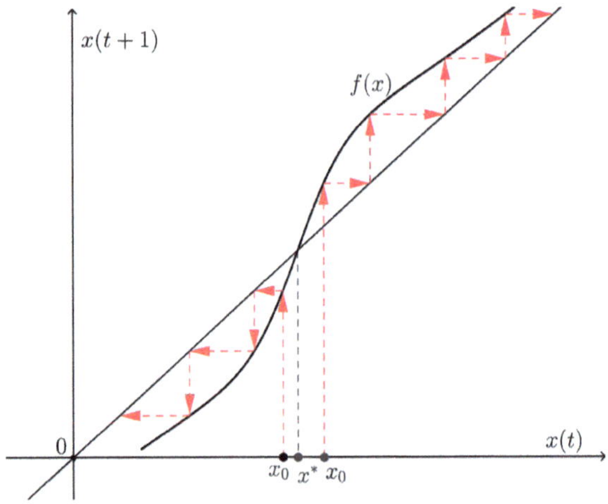

We conclude the present section with few and simple economic applications of difference equations.

1. A Cobweb Model.

This is a dynamical model which assumes that the demand of a good in a market depends on its current price, while the supply reacts to the price with a lag of one period (producers believe that the current price will hold also in the next period and so they start the new production according to the current price). Demand and supply are both linear functions of the price. In symbols:

$$D(t) = b - ap(t), \quad \text{with } a, b > 0;$$

$$S(t) = cp(t-1) - d, \text{ with } c, d > 0.$$

A last assumption is the market equilibrium assumption: the market determines the price in such a way that demand and supply are equal in each period, that is $D(t) = S(t)$, $t = 1, 2, \ldots$

$D(t) = S(t)$ implies $b - ap(t) = cp(t-1) - d$, that is $ap(t) + cp(t-1) = b + d$, $t = 1, 2, \ldots$ Equivalently, with a change of notations, we have

$$ap(t+1) + cp(t) = b + d, \quad t = 0, 1, 2, \ldots$$

so that

$$p(t+1) = -\frac{c}{a}p(t) + \frac{b+d}{a}, t = 0, 1, 2, \ldots \tag{7.21}$$

Let $p(t) = p^*$ be the equilibrium price; by substituting into (7.21) we have $p(t+1) = p^*$, $p(t) = p^*$ and hence

$$p^* = -\frac{c}{a}p^* + \frac{b+d}{a}$$

from which

$$p^* = \frac{b+d}{a+c}.$$

The general solution of the difference Eq. (7.21) is

$$p(t) = k\left(-\frac{c}{a}\right)^t + p^*, \ t = 0, 1, 2, \ldots \tag{7.22}$$

Setting $t = 0$ in (7.22), we obtain $k = p(0) - p^*$, so that the general solution is

$$p(t) = (p(0) - p^*)\left(-\frac{c}{a}\right)^t + p^*, \quad t = 0, 1, 2, \ldots$$

Let us note that if in a certain period t^* the price is fixed at the equilibrium value p^*, then the prices are constant over time, the demand is constant over time at the value $D(t) = D^* = b - ap(t) = b - ap^*$, and the supply is constant over time at the value $S(t) = S^* = cp^* - d$.

The equilibrium price is stable if and only if $\frac{c}{a} < 1$, that is $c < a$. This means that demand is more "reactive" than supply. The time path shows damped oscillations. If $c > a$ we have instability and explosive oscillations of the prices; if $c = a$, we have instability and finite oscillations of the prices. If we represent the solution for $c < a$, by means of a phase diagram, we obtain a sort of "cobweb" from which the name of the present model. See Fig. 7.8.

2. Interest Determination.

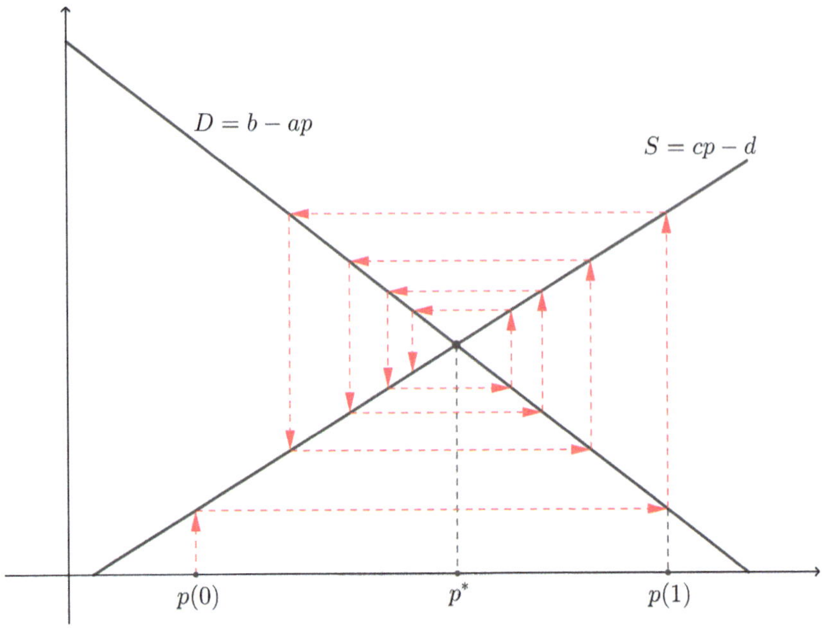

Fig. 7.8 A Cobweb model

We want to determine how much money one would have after t years upon investing a certain amount C at present ($t_0 = 0$) at an annual interest rate i, with a compound interest law, i.e. with accumulation and reinvestment of the interest payments.

Let $M(t)$ be the amount of money at period t. Then, the amount of money at period $(t + 1)$ will be

$$M(t + 1) = (1 + i)M(t), \quad t = 0, 1, 2, \ldots \tag{7.23}$$

The general solution of the first order linear homogeneous difference Eq. (7.23) is given by

$$M(t) = c(1 + i)^t, \quad t = 0, 1, 2 \ldots, \quad c \in \mathbb{R}.$$

Since at $t = 0$ we have $M(0) = C = c(1 + i)^0 = c \cdot 1 = c$, so that the total amount at the end of t years is given by

$$M(t) = C(1 + i)^t, \quad t = 0, 1, 2, \ldots$$

This computation is the inverse of the problem of determining the *present value* $A(t)$ of a sum of money C to be received after t years from now, when the interest rate is i per period and always under a common interest law. We must have

$$M(t) = A(t)(1+i)^t = C,$$

so that

$$A(t) = \frac{C}{(1+i)^t} = C(1+i)^{-t},\ t = 0, 1, 2, \ldots$$

Since $0 < i < 1$, we have $\frac{1}{(1+i)^t} < 1$, so that $A(t) < C$, that is the present value of a sum to be received in the future is lower than this sum. In Financial Mathematics the quantity $(1+i)^{-t}$ is referred as the *discount rate,* or better, the *discount factor.*

3. Present Value of a Stream Payments.

An individual makes annual payments at the end of each year, of amount R for t years, starting from the next year, with interest rate i per period. We want to determine the total amount after t payments and the present value of this total amount. Denote by $M(t)$ the total amount at period t. Then the total amount at the next period will be

$$M(t+1) = M(t) + iM(t) + R,\ \ t = 0, 1, 2, \ldots \tag{7.24}$$

Let $M(t) = k$ be the equilibrium value; by substituting in (7.24) we have

$$k = k + ik + R,$$

so that

$$k = -\frac{R}{i}.$$

The general solution of the first order linear non homogeneous difference Eq. (7.24) is given by

$$M(t) = c(1+i)^t - \frac{R}{i},\ \ t = 0, 1, 2, \ldots$$

Since at $t = 0$ we have $M(0) = 0 = c(1+i)^0 - \frac{R}{i} = c - \frac{R}{i}$, it results

$$c = \frac{R}{i},$$

so that the total amount at the end of t years is given by

$$M(t) = \frac{R}{i}\left((1+i)^t - 1\right),\ \ t = 0, 1, 2, \ldots$$

Taking into account that

$$A(t) = \frac{c}{(1+i)^t}, \quad t = 0, 1, 2, \dots$$

the present value of the stream of payments is given by

$$A(t) = \frac{\frac{R}{i}((1+i)^t - 1)}{(1+i)^t},$$

that is

$$A(t) = \frac{R}{i}(1 - \frac{1}{(1+i)^t}), \quad t = 0, 1, 2, \dots$$

4. Accelerator/Multiplier Model of P. A. Samuelson.

This classical economic model, due to P. A. Samuelson, is an attempt to explore the dynamic process of income determination when the acceleration principle is in operation along with the Keynesian multiplier. See Samuelson [15].

Consider an economic system observed over a number of consecutive years. Denote by $Y(t)$ the *national income* and by $C(t)$ the *consumption;* the model is based on the following three assumptions.

(a) $C(t)$ is a linear function of the income of the previous period:

$$C(t) = aY(t-1) + k \tag{7.25}$$

where a and k are propensity coefficients, $0 < a < 1$, and $k > 0$.

(b) The *investment* $I(t)$ is a linear function of the consumption variations and is given by the sum of two components, the first one is a constant and the second one is proportional to the consumption variations:

$$I(t) = A + b(C(t) - C(t-1)), \tag{7.26}$$

with $b > 0$.

(c) $Y(t)$ verifies the standard *balance condition:*

$$Y(t) = C(t) + I(t).$$

Inserting $I(t)$ from (7.26) and $C(t)$ from (7.25) we get

$$Y(t) - a(1+b)Y(t-1) + abY(t-2) = A + k. \tag{7.27}$$

Equation (7.27) is a difference equation: after shifting the time period forward by two periods, we get

$$Y(t+2) - a(1+b)Y(t+1) + abY(t) = A + k$$

which is a second order difference equation, with constant coefficients and non homogeneous.

This equation has a particular solution $Y(t) = \bar{Y}$ which solves

$$\bar{Y} - a(1+b)\bar{Y} + ab\bar{Y} = A + k.$$

We find

$$\bar{Y} = \frac{A+k}{1-a}.$$

The characteristic equation of the homogeneous equation is

$$\lambda^2 - a(1+b)\lambda + ab = 0$$

which has roots

$$\lambda_{1,2} = \frac{a(1+b) \pm \sqrt{a^2(1+b)^2 - 4ab}}{2}.$$

These roots are real and distinct if (and only if)

$$a > \frac{4b}{(1+b)^2}.$$

If we adopt this last assumption, we have that the general solution of the associated homogeneous equation is

$$Z(t) = c_1\lambda_1^t + c_2\lambda_2^t.$$

Thanks to Descartes Rule, the two roots obtained are both positive. The general solution of the non homogeneous equation is therefore

$$Y(t) = c_1\lambda_1^t + c_2\lambda_2^t + \frac{A+k}{1-a}.$$

Let us now examine the stability of the constant solution

$$\bar{Y} = \frac{A+k}{1-a}.$$

This solution is asymptotically stable if and only if the zero solution of the associated homogeneous equation is asymptotically stable. By rewriting the characteristic equation

$$\lambda^2 - a(1+b)\lambda + ab = 0,$$

we have asymptotic stability if and only if

$$ab < 1 \ \text{ and } a(1+b) < 1 + ab.$$

Since $0 < a < 1$, the coefficients have just to satisfy the condition

$$ab < 1. \tag{7.28}$$

The behaviour of the solution depends on the sign of the "discriminant" $a^2(1+b)^2 - 4ab$ or $a - \frac{4b}{(1+b)^2}$. If both conditions (7.28) and $a \geqq \frac{4b}{(1+b)^2}$ hold, any solution of the original non homogeneous equation evolves towards the equilibrium solution $\bar{Y}$ monotonically; if (7.28) holds, but $a < \frac{4b}{(1+b)^2}$, the evolution towards the equilibrium is oscillatory.

5. A Simple Dynamical Leontief model.

Also Leontief input-output models can be fitted into a discrete dynamical system. For example, [19] considered the following system of difference equations

$$\mathbf{x}(t+1) = A\mathbf{x}(t) + \mathbf{c},$$

where $A = \left[a_{ij}\right]$ is of order n, $A \geqq [0]$, and $x_j(t)$ is the output vector in period t. This system may be justified by the assumption that the demand for the i-th good by the j-th industry in period $t+1$, that is $x_{ij}(t+1)$, is proportional to this industry's sales (output) in period t, that is $x_j(t)$. Here $a_{ij} \equiv x_{ij}(t+1)/x_j(t)$.

Consider the constant solution $\mathbf{x}(t+1) = \mathbf{x}(t) = \mathbf{x}^*$ for all t (*equilibrium solution* or *steady state solution*). Then we have $\mathbf{x}^* = A\mathbf{x}^* + \mathbf{c}$, or $(I - A)\mathbf{x}^* = \mathbf{c}$. We can put two questions.

(i) Does $\mathbf{x}(t) \longrightarrow \mathbf{x}^*$, as $t \longrightarrow +\infty$?
(ii) When $(I - A)$ is nonsingular and $(I - A)^{-1} \geq [0]$?

The first question is the problem of stability, the second question is the problem of existence of nonnegative solutions. From Theorem 1.8.11 we know that $(I - A)$ is nonsingular and $(I - A)^{-1} \geq [0]$ if and only if $\lambda^*(A) < 1$, where $\lambda^*(A)$ is the Frobenius eigenvalue of A.

In order to treat the stability question, let us carry out the following iterations

$$\mathbf{x}(1) = A\mathbf{x}(0) + \mathbf{c}$$
$$\mathbf{x}(2) = A\mathbf{x}(1) + \mathbf{c} = (A)^2\mathbf{x}(0) + (I + A)\mathbf{c}$$

$$\mathbf{x}(3) = A\mathbf{x}(2) + \mathbf{c} = (A)^3\mathbf{x}(0) + (I + A + (A)^2)\mathbf{c}$$

$$\dots$$

$$\mathbf{x}(t) = A\mathbf{x}(t-1) + \mathbf{c} = (A)^t\mathbf{x}(0) + (I + A + \dots (A)^{t-1})\mathbf{c}.$$

Then we can say that $(A)^t \longrightarrow [0]$ as $t \longrightarrow +\infty$ if and only if $\lambda^*(A) < 1$ and that, under the same condition, $\sum_{t=0}^{+\infty}(A)^t\mathbf{c}$ is convergent and equal to $(I-A)^{-1}\mathbf{c} = \mathbf{x}^*$. Therefore $\mathbf{x}(t) \longrightarrow (I-A)^{-1}\mathbf{c}$ as $t \longrightarrow +\infty$ if and only if $\lambda^*(A) < 1$. In other words, the two problems, of stability and existence of nonnegative solutions, are in fact closely related.

In addition to the works cited throughout the chapter, the reader interested in further study can consult the books [1, 4, 7, 9].

References

1. R.P. Agarwal, *Difference Equations and Inequalities: Theory, Methods and Applications* (M. Dekker, New York, 1992)
2. J.S. Chipman, The multi-sector multiplier. Econometrica **18**, 355–374 (1950)
3. G. Debreu, I.N. Herstein, Nonnegative square matrices. Econometrica, **21**, 597–607 (1953)
4. S. Elaydi, *An Introduction to Difference Equations* (Springer, New York, 2005)
5. G. Gandolfo, *Economic Dynamics* (Springer, Berlin, 1997)
6. R.A. Horn, C.R. Johnson, *Matrix Analysis* (Cambridge University Press, Cambridge, 1985)
7. V. Lakshmikantham, D. Trigiante, *Theory of Difference Equations: Numerical Methods and Applications* (Academic Press, New York, 1988)
8. M. Livio, *The Golden Ratio* (Broadway Books, New York, 2003)
9. L.M. Milne-Thomson, *The Calculus of Finite Differences* (Chelsea Publishing Co., New York, 1981)
10. Y. Murata, *Mathematics for Stability and Optimization of Economic Systems* (Academic Press, New York, 1977)
11. H. Nikaido, *Convex Structures and Economic Theory* (Academic Press, New York, 1968)
12. H. Nikaido, *Introduction to Sets and Mappings in Modern Economics* (North Holland, Amsterdam, 1970)
13. E. Salinelli, F. Tomarelli, *Discrete Dynamical Systems* (Springer, Berlin, 2014)
14. S. Salsa, A. Squellati, *Dynamical Systems and Optimal Control. A Friendly Introduction* (Bocconi University Press, Milan, 2018)
15. P.A. Samuelson, Interactions between the multiplier analysis and the principle of accumulation. Rev. Econ. Stat. **21**, 75–78 (1939)
16. P.A. Samuelson, Conditions that the roots of a polynomial be less than unity in absolute value. Ann. Math. Stat. **12**, 360–364 (1941)
17. P.A. Samuelson, *Foundations of Economic Analysis* (Harvard University Press, Cambridge, 1947)
18. C.P. Simon, L. Blume, *Mathematics for Economists* (W. W. Norton & Co., New York, 1994)
19. R. Solow, On the structure of linear models. Econometrica, **20**, 29–46 (1952)
20. A. Takayama, *Mathematical Economics* (Cambridge University Press, Cambridge, 1985)

Chapter 8
Introduction to Dynamic Optimization

8.1 Introduction to the Calculus of Variations

In the present chapter we shall give some basic ideas on *problems of dynamic optimization:* in these kinds of problems we have not to find in a feasible set $K \subset \mathbb{R}^n$ a vector x^0 which maximizes or minimizes a given objective function $f : \mathbb{R}^n \longrightarrow \mathbb{R}$, a typical problem of *static optimization,* problems described in Chapters 4 and 5. In problems of dynamic optimization we have to choose, in a given class $\mathcal{U}$ of functions $x = x(t)$, $x : \mathbb{R} \longrightarrow \mathbb{R}^n$, hence $t \in \mathbb{R}$, an element $\bar{x} = \bar{x}(t)$ which maximizes (or minimizes) a given (real) *functional* $J(x)$, that is a mathematical relation which varies with the choice of $x(t)$ in $\mathcal{U}$, i.e. $J : \mathcal{U} \longrightarrow \mathbb{R}$. $\mathcal{U}$ is called "set of admissible functions". In other words, the objective to be maximized (or minimized) in these kinds of problems cannot be reduced to a function of a finite number of real variables: in general, we have a real function, whose argument is itself a function of one or several variables and these kinds of objectives are named *real functionals.* An example of a functional treated in these problems is the following one

$$J(x) = \int_{t_0}^{t_1} G(x, x', t)dt,$$

with $G(x, x', t)$ scalar function of the three arguments

$$x = x(t),\ x' = x'(t) = dx/dt,\ \ t,$$

t_0 and t_1 given constants (however it is not excluded $t_1 = +\infty$), and $x \in \mathcal{U}$, x real function (scalar or vector-valued) defined on the interval $T = [t_0, t_1]$.

The classical approach to dynamic optimization is the *Calculus of Variations,* which goes back to the 17th and 18th centuries and which was mainly developed by Euler and Lagrange. The basic *Euler equation* or *Euler condition* was obtained

G. Giorgi et al., *Lectures on Mathematics for Economic and Financial Analysis*,
https://doi.org/10.1007/978-3-031-83339-7_8

in 1744. The modern evolution of the Calculus of Variations is known as *Optimal Control Theory,* and is mainly due to the Russian mathematician L. S. Pontryagin and his collaborators (see Pontryagin et al. [30]). Another important contribution to dynamic optimization problems is *Dynamic Programming,* due to the American mathematician R. Bellman (see Bellman [7]).

All these topics are rather complex, from a mathematical point of view, hence we shall give only few basic ideas on the same, as a satisfactory treatment of the Calculus of Variations, Optimal Control Theory and Dynamic Programming is beyond the aims of the present book. However, the said topics are very important in recent economic literature, and students interested in dynamic economic analysis should therefore make a serious effort to learn the fundamental structures of these problems. In the present section we give the basic introductive ideas of the Calculus of Variations; in the second section the Optimal Control Theory (with the Pontryagin Maximum Principle) will be introduced and in the third final section, few hints on Dynamic Programming will be given. The willing reader is referred, for further information, to the following works: Bertsekas [8], Bellman [7], Athans and Falb [5], Caputo [10], Cesari [11], Chiang [12], Hadley and Kemp [19], Kamien and Schwartz [22], Hestenes [21], Lee and Markus [24], Leonard and Long [25], Gelfand and Fomin [17], Dreyfus [16], Pontryagin et al. [30], Seierstad and Sydsaeter [35], Tu [38, 39], and Salsa and Squellati [33]. Furthermore, in the References other books and papers are proposed, which may be useful, mainly in view of economic and management applications.

In order to introduce the essential features of the *Calculus of Variations,* we have, first of all, to specify the set $\mathcal{U}$ of functions $x : \mathbb{R} \longrightarrow \mathbb{R}^n$, $n \geqq 1$, defined on an interval $T = [t_0, t_1]$. These functions are called *admissible functions* and an application

$$J : \mathcal{U} \longrightarrow \mathbb{R}$$

is a *functional,* which is the objective to be maximized or minimized. Without loss of generality, we shall refer to a *maximization problem:* then the problem is to determine $\hat{x} \in \mathcal{U}$ such that

$$J(x) \leqq J(\hat{x}), \text{ for every } x \in \mathcal{U},$$

that is, to solve the problem

$$\max_{x \in \mathcal{U}} J(x).$$

We point out that a minimization problem can always be converted into a maximization problem by the relation

$$\min J(x) = \max(-J(x)).$$

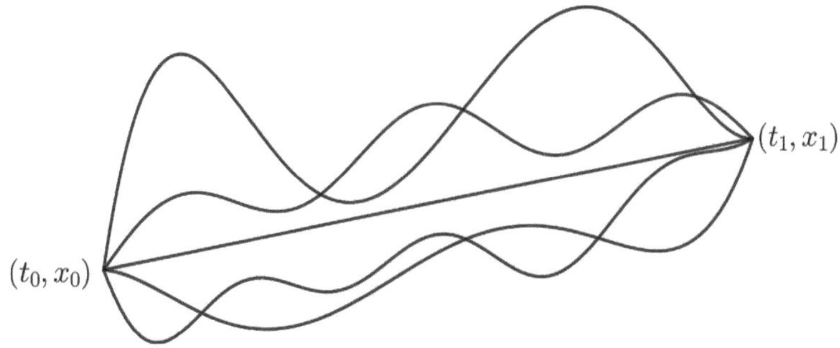

Fig. 8.1 Functions of class $\mathcal{C}^1$ joining points (t_0, x_0) and (t_1, x_1)

We call $\hat{x}$ (if there exists) a *maximizer* or an *optimal path.* Note that in the above formulation of the problem, there is no a priori restriction, neither on the nature and on the analytical expression of the functional J, nor on the class of admissible functions, however, the usual functionals of the Calculus of Variations are expressed by integrals.

For example, if we consider two points on the plane, with coordinates (t_0, x_0) and (t_1, x_1), with $t_0 < t_1$, and $\mathcal{U}$ is the set of all functions $x : \mathbb{R} \longrightarrow \mathbb{R}$ defined on $[t_0, t_1]$, of class $\mathcal{C}^1$ on the same interval and such that $x(t_0) = x_0$, $x(t_1) = x_1$, a problem of the Calculus of Variations could be to minimize the length of all possible paths which join the two points. See Fig. 8.1

The length of each path is given by

$$L(x) = \int_{t_0}^{t_1} \sqrt{1 + (x'(t))^2} dt.$$

This length is a real number which depends on the choice of $x(t)$: therefore it is a functional defined on $\mathcal{U}$. These considerations are the preliminaries of a famous problem: the queen Dido problem (see the next example).

Example 8.1.1 Mythology credits queen Dido (who appears in the famous poem "Aeneid" by Virgil) the following problem, known as the *Dido problem:* find the closed plane curve of a given length that encloses the greatest area. This is a typical problem of the Calculus of Variations, more precisely an *isoperimetric problem:* determine the curve $x(t)$, defined for $t \in [-a, a]$, of a given length $L > 2a$ and with $x(-a) = x(a) = 0$, such that the area of the part of the plane enclosed between the interval $[-a, a]$ and the curve, has the maximal area. See Fig. 8.2.

From a formal point of view we are requested to maximize the functional

$$J(x) = \int_{-a}^{a} x(t) dt$$

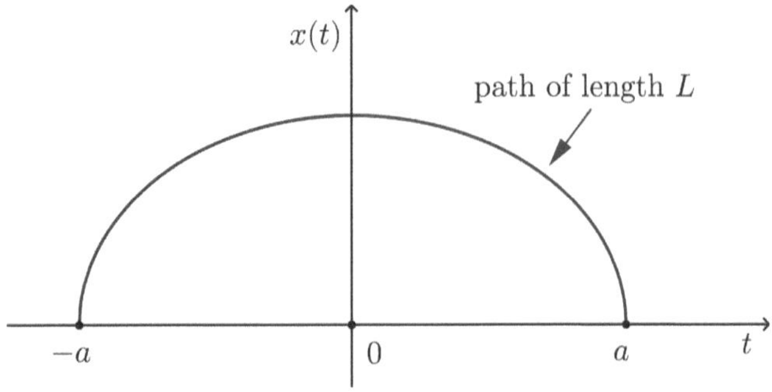

Fig. 8.2 Example 8.1.1, Dido problem

Fig. 8.3 Example 8.1.2, the brachistochrone problem

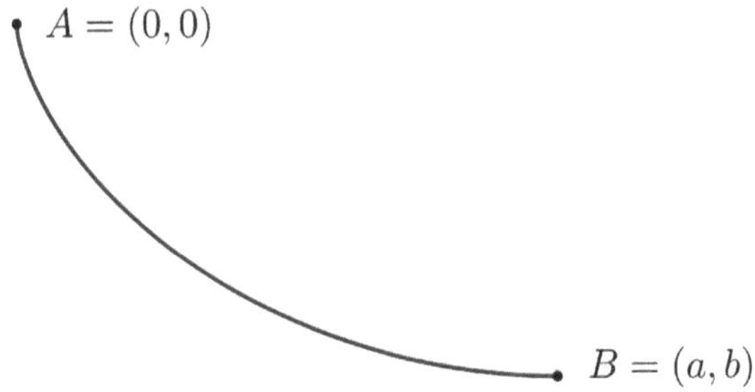

under the conditions $x(-a) = x(a) = 0$ and the constraint

$$\int_{-a}^{a} (1 + (x'(t))^2)^{1/2} dt = L.$$

Example 8.1.2 Another famous example of problems of the Calculus of Variations, which in fact gave rise to this part of Mathematics as an autonomous discipline, is the "brachistochrone problem" (from the two ancient Greek words *brachistos* and *chronos,* which mean "shortest time"), proposed in the second half of the seventeenth century by the Swiss mathematician Johann Bernoulli, and solved by the same (and also by other famous mathematicians, such as Newton and L'Hôpital). Given two points A and B in a vertical plane (A in a higher position with respect to B), the time required for a particle to slide along a curve from A to B under the sole influence of gravity, will depend on the shape of this curve. The problem is to find the curve along which the particle goes from A to B as quickly possible (curiously, this curve is *not* the straight line joining A to B). In 1606 Johann Bernoulli proved that the solution is part of the curve, called a *cycloid* (see Fig. 8.3).

Using elementary Physics one can show that the brachistochrone problem reduces to minimize the following real functional (A is the origin and B has coordinates (a, b), with $b < 0$)

$$\min J(x) = \int_0^a \sqrt{\frac{1 + (x'(t))^2}{2gx(t)}} dt$$

under the conditions $x(0) = 0$, $x(a) = b$, and where g is the acceleration of gravity.

Example 8.1.3 Another famous example of the Calculus of Variations is given by an economic problem, proposed by F. P. Ramsey (a student of Lord J. M. Keynes) in 1928. See Ramsey [32]; see also the textbooks by Blanchard and Fischer [9] and Barro and Sala-i-Martin [6]. Let us suppose that an individual gets his income from an exogenous wage $v = v(t)$ and from the interest $iK(t)$ coming from the capital $K(t)$, where i is the (constant) interest rate. The total income from interest and wage is allotted to consumption $C = C(t)$ and investment $K(t)$, according to the equation

$$v(t) + iK(t) = K'(t) + C(t).$$

Now, let $U = U(C)$ be an increasing and concave utility function (thus depending only on consumption):

$$U'(C) > 0, \quad U''(C) < 0.$$

Suppose we assign the initial and final value of the capital. We want to choose the consumption $C(t)$, that we need to maintain in the time period $[0, T]$, in order to maximize utility. Considering that the utility is discounted at a rate r, we need to maximize the following functional

$$J(K) = \int_0^T e^{-rt} U(C(t))dt = \int_0^T e^{-rt} U\left[v(t) + iK(t) - K'(t)\right] dt$$

subject to the conditions

$$K(0) = K_0 > 0, \quad K(T) = K_T,$$

where K_0 and K_T are fixed.

Now we consider a basic formulation of problems of the Calculus of Variations, a problem we call "the standard problem" or "the Lagrange problem" of the Calculus of Variations. Let $\mathcal{U}$ be the set of all functions $x : \mathbb{R} \longrightarrow \mathbb{R}^n$ of class C^1 on $T = [t_0, t_1]$. We have to find the function $x \in \mathcal{U}$ which maximizes (or minimizes) a functional of the type

$$J(x) = \int_{t_0}^{t_1} G(x, x', t)dt \tag{8.1}$$

over the class of *admissible functions*

$$\mathcal{U} = \left\{ x \in \mathcal{C}^1([t_0, t_1]) : x(t_0) = x^0,\ x(t_1) = x^1 \right\}.$$

For what concerns the function $G : \mathbb{R}^n \times \mathbb{R}^n \times \mathbb{R} \longrightarrow \mathbb{R}$, we assume the existence of its first and second order derivatives, with respect to all its variables.

The conditions $x(t_0) = x^0$, $x(t_1) = x^1$ are called *boundary conditions*, t_0 is the *initial time* (usually $t_0 = 0$) and t_1 is the *terminal time* ($t_1 = +\infty$ is not excluded), the components of x are the *state variables* (obviously if $n = 1$ we have only one state variable). This is a *fixed end points problem* (or *fixed boundary problem*). We wish to present a first order *necessary* optimality condition for this type of problems: the *Euler condition* or *Euler equation.* If we have a maximization problem, with $n = 1$, a geometric interpretation of the above problem is: among all well-behaved functions $x(t)$ that satisfy $x(t_0) = x^0$ and $x(t_1) = x^1$, find one making the integral

$$\int_{t_0}^{t_1} G(x, x', t)dt$$

as large as possible.

Before presenting the fundamental Euler equation, we need a previous lemma, given by Paul du Bois-Reymond in 1879.

Lemma 8.1.4 (Lemma of Du Bois-Reymond) *Let $x = x(t) : \mathbb{R} \longrightarrow \mathbb{R}^n$ be of class $\mathcal{C}^2$ on $T = [a, b]$. Let H be the set of all functions $h : \mathbb{R} \longrightarrow \mathbb{R}^n$ continuous on $[a, b]$ and such that $h(a) = h(b) = [0]$. If*

$$\int_a^b x(t)^\top h(t) = 0, \quad \forall h \in H,$$

then $x(t) = [0]$, $\forall t \in [a, b]$.

Proof We give the proof for $n = 1$. Absurdly suppose that there exists $\tau \in [a, b]$, with $x(\tau) \neq 0$, for instance $x(\tau) > 0$. As x is continuous, then it will hold $x(\tau) > 0$ in some interval $[\tau_0, \tau_1] \subset [a, b]$ and which contains τ. Now we choose $h(t) \in H$, with $h(t) > 0$, $\forall t \in [\tau_0, \tau_1]$ and $h(t) = 0$, $\forall t \notin [\tau_0, \tau_1]$. Then it holds

$$\int_a^b x(t)h(t)dt = \int_{\tau_0}^{\tau_1} x(t)h(t)dt > 0,$$

against the assumptions. In a similar way we obtain the thesis in the case $x(\tau) < 0$. □

Theorem 8.1.5 (Euler Equation or Euler-Lagrange Equation) *Let $\hat{x}$, $\hat{x} \in \mathcal{U}$, be a maximizer or a minimizer for the standard problem* (8.1) *of the Calculus of Variations:*

$$J(x) = \int_{t_0}^{t_1} G(x, x', t)dt,$$

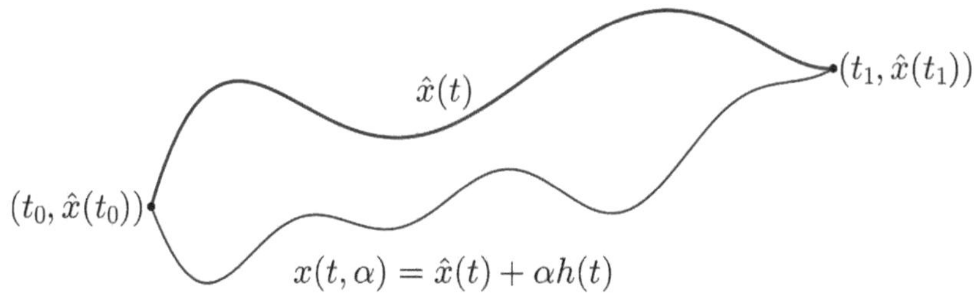

Fig. 8.4 Curve $\hat{x}(t)$ and perturbed curve $x(t, \alpha)$

with G possessing derivatives with respect to x, x' and the derivative with respect to t of the derivatives of G with respect to x'. Let $\hat{x}$ be of class $\mathcal{C}^2$ on $[t_0, t_1]$; then it holds

$$\frac{\partial G}{\partial x}(\hat{x}(t), \hat{x}'(t), t) - \frac{d}{dt}\left(\frac{\partial G}{\partial x'}(\hat{x}(t), \hat{x}'(t), t)\right) = [0], \ \forall t \in [t_0, t_1]. \tag{8.2}$$

Proof We give the proof for $n = 1$. Consider a functional $h \in \mathcal{U}$, with $h(t_0) = h(t_1) = 0$, and the function

$$x(t, \alpha) = \hat{x}(t) + \alpha h(t), \ \alpha \in \mathbb{R},$$

which modifies $\hat{x}(t)$ with the addendum $\alpha h(t)$, without changing the initial and the final points. See Fig. 8.4.

If the function $h(t)$ is kept fixed, then the values of J depend only on α; we denote by $\varphi(\alpha)$ this function, $\varphi : \mathbb{R} \longrightarrow \mathbb{R}$, which is differentiable, under our assumptions. As J assumes the maximum (or the minimum) for $x = \hat{x}$, φ will be maximized (or minimized) for $\alpha = 0$, which is an interior point of the domain of φ, hence $\varphi'(0) = 0$. Now we compute $\varphi'(\alpha)$:

$$\varphi'(\alpha) = \frac{d}{d\alpha}\left(\int_{t_0}^{t_1} G(\hat{x} + \alpha h, \hat{x}' + \alpha h', t)dt\right)$$

$$= \int_{t_0}^{t_1} \left[\frac{\partial G}{\partial x}(\hat{x} + \alpha h, \hat{x}' + \alpha h', t)\right] h dt + \int_{t_0}^{t_1} \left[\frac{\partial G}{\partial x'}(\hat{x} + \alpha h, \hat{x}' + \alpha h', t)\right] h' dt.$$

Therefore

$$\varphi'(0) = \int_{t_0}^{t_1} \frac{\partial G}{\partial x}(\hat{x}, \hat{x}', t) h dt + \int_{t_0}^{t_1} \frac{\partial G}{\partial x'}(\hat{x}, \hat{x}', t) h' dt.$$

Now we integrate "by parts" the second addendum:

$$\int_{t_0}^{t_1} \frac{\partial G}{\partial x'}(\hat{x}, \hat{x}', t)h' dt = \frac{\partial G}{\partial x'}(\hat{x}, \hat{x}', t)h(t)\Big|_{t=t_0}^{t=t_1} - \int_{t_0}^{t_1} \frac{d}{dt}\left(\frac{\partial G}{\partial x'}(\hat{x}.\hat{x}', t)\right)h(t)dt,$$

that is

$$\int_{t_0}^{t_1} \frac{\partial G}{\partial x'} h' dt = -\int_{t_0}^{t_1} \frac{d}{dt}\left(\frac{\partial G}{\partial x'}(\hat{x}.\hat{x}', t)\right)h(t)dt,$$

being $h(t_0) = h(t_1) = 0$.

Finally, as $\varphi'(0) = 0$, we have

$$\int_{t_0}^{t_1} \left(\frac{\partial G}{\partial x}(\hat{x}.\hat{x}', t) - \frac{d}{dt}\left(\frac{\partial G}{\partial x'}(\hat{x}.\hat{x}', t)\right)\right)h(t)dt = 0, \quad \forall t \in [t_0, t_1],$$

which, by Lemma 8.1.4, gives the Euler equation (8.2). □

We remark that in the proof of Theorem 8.1.5 the assumption that $\hat{x}(t) \in \mathcal{C}^2$ has not been directly used. Indeed, always for $n = 1$, if we compute the quantity

$$\frac{d}{dt}\left(\frac{\partial G}{\partial x'}(\hat{x}.\hat{x}', t)\right)$$

we find

$$\frac{d}{dt}\left(\frac{\partial G}{\partial x'}(\hat{x}.\hat{x}', t)\right) = \frac{\partial^2 G}{\partial x \partial x'}\hat{x}' + \frac{\partial^2 G}{\partial x' \partial x'}\hat{x}'' + \frac{\partial^2 G}{\partial t \partial x'}.$$

Hence $\frac{d}{dt}\left(\frac{\partial G}{\partial x'}(\hat{x}.\hat{x}', t)\right)$ exists if and only if $\hat{x}''$ exists. Inserting the last expression into the Euler equation and rearranging, the Euler equation becomes

$$\frac{\partial^2 G}{\partial x' \partial x'}(\hat{x}, \hat{x}', t) \cdot \hat{x}'' + \frac{\partial^2 G}{\partial x \partial x'}(\hat{x}, \hat{x}', t) \cdot \hat{x}' + \frac{\partial^2 G}{\partial t \partial x'}(\hat{x}, \hat{x}', t) - \frac{\partial G}{\partial x}(\hat{x}, \hat{x}', t) = 0,$$

which is a differential equation of the second order, if $\frac{\partial^2 G}{\partial x' \partial x'} \neq 0$.

The assumption that $\hat{x}'' \in \mathcal{C}^2$ can be avoided, by a more elaborate argument, by requiring only (see, e.g., Gelfand and Fomin [17]) that the admissible functions are of class $\mathcal{C}^1$. Indeed, it is possible to obtain the following result, called also *Euler equation in the form of Du Bois-Reymond* or in *integral form.*

Theorem 8.1.6 *If $\hat{x}(t)$ is a maximizer or a minimizer for J, in problem* (8.1) *and $\hat{x} \in \mathcal{C}^1$ and $n = 1$, there exists a constant $k \in \mathbb{R}$ such that*

$$\frac{\partial G}{\partial x'}(\hat{x}(t), \hat{x}', t) = \int_{t_0}^{t} \frac{\partial G}{\partial x}(\hat{x}(s), \hat{x}'(s), s)ds + k, \ \forall t \in [t_0, t_1]. \tag{8.3}$$

We deduce that (8.3) implies (8.2), but the vice-versa does not hold. However, it is possible to prove that if $\frac{\partial^2 G}{\partial x' \partial x'}$ exists and does not vanish in its domain, then the two necessary conditions (8.2) and (8.3) are equivalent. See. e.g., Salsa and Squellati [33]. In this case problem (8.1) is called *regular.* Also the assumption that x is of class C^1 can be relaxed, by assuming that x is continuously differentiable, except possible at a finite number of points, where a discontinuity jump is admitted. However, in this case we have to impose the so-called *Erdmann-Weierstrass conditions.* See further.

Now we present a *sufficient optimality condition* for problem (8.1). This condition has been given by Takayama [37] and is a special case of a more general theorem, due to Mangasarian [26] for optimal control problems.

Theorem 8.1.7 *Let the assumptions of Theorem 8.1.5 hold and let $\hat{x}(t)$ be a solution of the Euler equation* (8.2). *If $G(x, x', t)$ is concave (resp. convex) with respect to x and x', then $\hat{x}(t)$ maximizes (resp. minimizes)*

$$J(x) = \int_{t_0}^{t_1} G(x, x', t)dt,$$

in the respect of the boundary conditions $\hat{x}(t_0) = x^0$ and $\hat{x}(t_1) = x^1$.

Proof Again we give the proof for $n = 1$ and suppose that $G(x, x', t)$ is concave in (x, x'). Suppose further that $\hat{x} = \hat{x}(t)$ satisfies the Euler equation as well as the boundary conditions $\hat{x}(t_0) = x_0$ and $\hat{x}(t_1) = x_1$. Let $x = x(t)$ be an arbitrary admissible function in the problem. Because $G(x, x', t)$ is concave in (x, x'), we have (recall the characterization of differentiable concave and convex functions in Chapter 3)

$$G(x, x', t) - G(\hat{x}, \hat{x}', t) \leqq \frac{\partial G(\hat{x}, \hat{x}', t)}{\partial x}(x - \hat{x}) + \frac{\partial G(\hat{x}, \hat{x}', t)}{\partial x'}(x' - \hat{x}').$$

Using the Euler equation, reversing the above inequality yields (with simplified notation)

$$\begin{aligned}
\hat{G} - G &\geqq \frac{\partial \hat{G}}{\partial x}(\hat{x} - x) + \frac{\partial \hat{G}}{\partial x'}(\hat{x}' - x') \\
&= \left[\frac{d}{dt}\left(\frac{\partial \hat{G}}{\partial x'}\right)\right](\hat{x} - x) + \frac{\partial \hat{G}}{\partial x'}(\hat{x}' - x') \\
&= \frac{d}{dt}\left[\frac{\partial \hat{G}}{\partial x'}(\hat{x} - x)\right].
\end{aligned}$$

Because the last expression is valid for all t in $[t_0, t_1]$, integrating yields

$$\int_{t_0}^{t_1} (\hat{G} - G)dt \geqq \int_{t_0}^{t_1} \frac{d}{dt}\left[\frac{\partial \hat{G}}{\partial x'}(\hat{x} - x)\right] dt = \left.\frac{\partial \hat{G}}{\partial x'}(\hat{x} - x)\right|_{t_0}^{t_1} \tag{8.4}$$

However, the functions $\hat{x}(t)$ and $x(t)$ satisfy the boundary conditions $\hat{x}(t_0) = x(t_0) = x_0$ and $\hat{x}(t_1) = x(t_1) = x_1$. So, the last expression in (8.4) is equal to zero. It follows that

$$\int_{t_0}^{t_1} \left[G(\hat{x}, \hat{x}', t) - G(x, x', t)\right] dt \geqq 0$$

for every admissible function $x = x(t)$. This confirms that $\hat{x}(t)$ solves the maximization problem. The corresponding result for a minimization problem is easily obtained. □

It can be proved that if $G(x, x', t)$ is *strictly concave* (resp. *strictly convex*), the thesis of Theorem 8.1.7 holds in the version that $\hat{x}(t)$ is the *unique* maximizer (resp. the unique minimizer) for problem (8.1). We give two simple examples of the previous results.

Example 8.1.8 (This Example Is Taken from Sydsaeter et al. [36]) Solve the problem

$$\max \int_0^2 (4 - 3x^2 - 16x' - 4(x')^2)e^{-t} dt,$$

under the conditions $x(0) = -\frac{8}{3}$, $x(2) = \frac{1}{3}$.

Here $G(x, x', t) = (4 - 3x^2 - 16x' - 4(x')^2)e^{-t}$, so $\frac{\partial G}{\partial x} = -6xe^{-t}$ and $\frac{\partial G}{\partial x'} = (-16 - 8x')e^{-t}$. The Euler equation requires finding $\frac{d}{dt}\left[(-16 - 8x')e^{-t}\right]$. The product rule for differentiation gives

$$\frac{d}{dt}\left[(-16 - 8x')e^{-t}\right] = 16e^{-t} - 8x''e^{-t} + 8x'e^{-t};$$

so the Euler equation reduces to

$$-6xe^{-t} - 16e^{-t} + 8x''e^{-t} - 8x'e^{-t} = 0.$$

Cancelling the nonzero common factor $8e^{-t}$ yields

$$x'' - x' - \frac{3}{4}x = 2. \tag{8.5}$$

This is a linear differential equation of the second order with constant coefficients (see Chapter 6). The characteristic equation is

$$\lambda^2 - \lambda - \frac{3}{4} = 0,$$

with roots $\lambda_1 = -\frac{1}{2}$, $\lambda_2 = \frac{3}{2}$. The nonhomogeneous Eq. (8.5) has a particular solution $-\frac{8}{3}$ (given by $\frac{2}{\frac{-3}{4}}$), so the general solution is

$$x = x(t) = c_1 e^{-\frac{1}{2}t} + c_2 e^{\frac{3}{2}t} - \frac{8}{3},$$

where c_1 and c_2 are arbitrary real constants. The boundary conditions $x(0) = -\frac{8}{3}$ and $x(2) = \frac{1}{3}$ imply that $0 = c_1 + c_2$ and $c_1 e^{-1} + c_2 e^3 = 3$. It follows that $c_1 = -3/(e^3 - e^{-1})$ and $c_2 = -c_1$, so

$$x = x(t) = -\frac{3}{e^3 - e^{-1}} e^{-\frac{1}{2}t} + \frac{3}{e^3 - e^{-1}} e^{\frac{3}{2}t} - \frac{8}{3}.$$

The function $G(x, x', t) = (4 - 3x^2 - 16x' - 4(x')^2)$ is concave in (x, x'), as a sum of concave functions. We deduce that the function we have found is the solution of the problem.

Example 8.1.9 Solve the problem

$$\min \int_0^1 ((x')^2 + 2tx)dt,$$

under the conditions $x(0) = 1$, $x(1) = 2$. here we have $\frac{\partial G}{\partial x} = 2t$ and $\frac{\partial G}{\partial x'} = 2x'$. The Euler equation becomes $2t - 2x'' = 0$, that is $x'' = t$. This differential equation has the general solution given by $x = \frac{1}{6}t^2 + c_1 t + c_2$, with c_1 and c_2 arbitrary constants. As the boundary conditions are $x(0) = 1$ and $x(1) = 2$, we get $1 = c_2$ and $2 = \frac{1}{6} + c_1 + c_2$, hence $c_1 = \frac{5}{6}$. The unique function which satisfies the Euler equation is therefore

$$\hat{x} = \frac{1}{6}t^2 + \frac{5}{6}t + 1.$$

As $G(x, x', t)$ is a convex function with respect to x, x', we deduce that $\hat{x}$ is indeed the solution of the proposed problem.

Now we examine some simple formal variants of the functions of the problem of the Calculus of Variations (8.1) considered. These variants often give rise to considerable simplifications of the Euler equation. Again, we consider the case $n = 1$.

1. The integrand function G depends on x' only: $G = G(x')$. The Euler equation becomes $\frac{\partial^2 G}{\partial x' \partial x'} x'' = 0$. Hence $x'' = 0$ or $\frac{\partial^2 G}{\partial x' \partial x'} = 0$. In the first case we have $x(t) = c_1 t + c_2$, $c_1, c_2 \in \mathbb{R}$. In the second case, if the equation $\frac{\partial^2 G}{\partial x' \partial x'} = 0$ has one or more solutions $x' = z_s \in \mathbb{R}$, we have $x(t) = z_s t + c$, which is of the previous type of solutions.
2. G does not depend on $x : G(x', t)$. The Euler equation reduces to

$$\frac{d}{dt}\left[\frac{\partial G}{\partial x'}\right] = 0,$$

whence

$$\frac{\partial G}{\partial x'} = c \text{ (constant)}.$$

This is a first order differential equation.
3. G does not depend on $t : G(x, x')$. Also in this case it is possible to reduce the Euler equation to a first order equation. Indeed, we have

$$\begin{aligned} \frac{d}{dt}\left[G(x, x') - x' \frac{\partial G(x, x')}{\partial x'}\right] &= x' \frac{\partial G}{\partial x} + x'' \frac{\partial G}{\partial x'} - x'' \frac{\partial G}{\partial x'} - x' \frac{d}{dt}\frac{\partial G}{\partial x'} \\ &= x' \left[\frac{\partial G}{\partial x} - \frac{d}{dt}\left(\frac{\partial G}{\partial x'}\right)\right]. \end{aligned} \tag{8.6}$$

It follows that if the Euler equation is satisfied for all $t \in [t_0, t_1]$, then the expression in (8.6) is zero. That is, the derivative of $G - x' \frac{\partial G}{\partial x'}$ must be zero for all t. In this case the Euler equation implies that

$$G - x' \frac{\partial G}{\partial x'} = c \text{ (constant)}.$$

4. G does not depend on $x' : G(x, t)$. The Euler equation becomes $\frac{\partial G}{\partial x} = 0$, which is not a differential equation. As we have no arbitrary constants, the function $x(t)$ in general does not satisfy the boundary conditions $x(t_0) = x_0$ and $x(t_1) = x_1$.
5. G is linear with respect to $x' : G(x, x', t) = a(x, t) + b(x, t)x'$. The Euler equation reduces to

$$\frac{\partial a}{\partial x} - \frac{\partial b}{\partial t} = 0,$$

which is not a differential equation, hence the same considerations of the previous case hold.

The necessary optimality condition for problem (8.1) given by the Euler equation is perhaps the most "classical" condition for this type of problems of the Calculus of Variations. However, this condition is not the unique available; we point out, without entering into particular descriptions, other three necessary optimality conditions which, in a sense, improve the Euler condition.

(1) In the previous considerations we have required that $\hat{x}(t) \in C^2$, that is the path $\hat{x}(t)$ turns out to be smooth with no corners. The possibility that a trajectory may have corners is considered by the *Erdmann-Weierstrass conditions.* See, e.g., Hadley and Kemp [19], Hestenes [21].
(2) The Euler condition is essentially a first-order optimality condition; moreover, it does not differentiate between a maximum and a minimum. The *second-order conditions,* due independently to Legendre and Clebsch, have not this drawback, similarly to the second-order conditions for static optimization problems. For example, if $\hat{x}(t) \in \mathcal{U}$ is a maximizer for problem (8.1), then it holds ("Legendre condition" or "Clebsch condition"), with $n = 1$,

$$\frac{\partial^2 G}{\partial x' \partial x'}(\hat{x}(t), \hat{x}'(t), t) \leqq 0, \quad \forall t \in [t_0, t_1] .$$

If $n > 1$, the Hessian matrix $HG_{x'x'}(\hat{x}(t), \hat{x}'(t), t)$ must be negative semidefinite, $\forall t \in [t_0, t_1]$.
(3) A third necessary optimality condition for the classical problem (8.1) is based on the so-called *Weierstrass excess function,* defined as follows:

$$E(x, x', y', t) = G(x, y', t) - G(x, x', t) - \left(\frac{\partial G}{\partial x'}(x, x', t)\right)^{\top} (y' - x').$$

A necessary condition such that $\hat{x}$ maximizes J is:

$$E(\hat{x}.\hat{x}', y', t) \leqq 0, \quad \forall y \in \mathcal{U}, \ \forall t \in [t_0, t_1] .$$

See, e.g., Hadley and Kemp [19] and Hestenes [21].

In several applications we may encounter variants and generalizations of the standard problem (8.1) of the Calculus of Variations. For example, we may have the following problem: maximize or minimize the functional

$$J(x) + \varphi[x(t_1), t_1] = \int_{t_0}^{t_1} G(x, x', t)dt + \varphi[x(t_1), t_1],$$

where φ is a given function. In this case we speak of *problem of Bolza.* The *problem of Mayer* is the problem where we have to maximize or minimize $\varphi[x(t_1), t_1]$, that is, here G is everywhere equal to zero. These problems may appear to be different cases with respect to the standard problem of the Calculus of Variations, however, by suitable definition of the variables, the three formulations can be shown to be equivalent.

Since now we have considered *fixed end values problems,* i.e. problems where the boundary values are assigned. We shall give some hints on the case where $t_1 = +\infty$ ("infinite planning horizon") and the case where the values of $x(t)$ at the end points of the interval $[t_0, t_1]$ are free.

- *The case* $t_1 = +\infty$.

If we have to maximize (or minimize) an integral where $t_1 = +\infty$, it is usual that the problem imposes the condition

$$\lim_{t \longrightarrow +\infty} \hat{x}(t) = x^1, \ x^1 \text{ fixed}, \tag{8.7}$$

condition which substitutes the fixed value condition $\hat{x}(t_1) = x^1$. We may encounter also the more general condition

$$\lim_{t \longrightarrow +\infty} \left[\hat{x}(t) - y(t)\right] = [0], \tag{8.8}$$

with $y : \mathbb{R} \longrightarrow \mathbb{R}^n$ a given function. Conditions (8.7) or (8.8) are therefore modifications of the classical boundary conditions of the Euler equation. When no condition is imposed on $\hat{x}(t)$, for $t \longrightarrow +\infty$, we have, besides the usual necessary optimality conditions, also the following necessary condition:

$$\lim_{t \longrightarrow +\infty} \left(\frac{\partial G}{\partial x'}(\hat{x}(t), \hat{x}'(t), t)\right) = [0].$$

For the proof see. e.g., Hadley and Kemp [19].

- *Free end values problems.*

We consider the maximization of the functional

$$J(x) = \int_{t_0}^{t_1} G(x, x', t)dt$$

without boundary conditions, on the set of admissible functions $\mathcal{U} = \mathcal{C}^1([t_0, t_1])$. Since we deal with a wider class of admissible functions, an extremal has to satisfy some additional conditions, besides the usual Euler equations.

Theorem 8.1.10 *Let* $\hat{x} \in \mathcal{C}^1([t_0, t_1])$ *be a maximizer or a minimizer for* J *over* $\mathcal{U} = \mathcal{C}^1([t_0, t_1])$. *Then:*

(*a*) $\hat{x}$ *satisfies the Euler equation.*
(*b*) *The following transversality conditions hold at the end points.*

(*i*)

$$\frac{\partial G}{\partial x'}(\hat{x}(t_1), \hat{x}'(t_1), t_1) = [0] \quad (= 0 \textit{ if } n = 1).$$

(*ii*)

$$\frac{\partial G}{\partial x'}(\hat{x}(t_0), \hat{x}'(t_0), t_0) = [0] \quad (= 0 \textit{ if } n = 1).$$

If we have the boundary condition $x(t_0) = x^0$ (i.e. $x(t_1)$ is free), in Theorem 8.1.10 only relation (*i*) must hold. If we have the terminal condition $x(t_1) \geqq x^1$, x^1 fixed, besides the other condition $x(t_0) = x^0$, then in Theorem 8.1.10 relation (*ii*) is skipped, but relation (*i*) becomes, with $n = 1$, and in case of maximization,

$$\frac{\partial G}{\partial x'}(\hat{x}(t_1), \hat{x}'(t_1), t_1) \leqq 0,$$

with

$$\frac{\partial G}{\partial x'}(\hat{x}(t_1), \hat{x}'(t_1), t_1) = 0,$$

if $\hat{x}(t_1) > x^1$.

If $G(x, x', t)$ is concave with respect to (x, x'), then an admissible trajectory $\hat{x}(t)$ that satisfies both the Euler equation and the appropriate transversality condition, will solve the basic problem (8.1) of the Calculus of Variations.

If we have a minimization problem, the last transversality conditions become

$$\frac{\partial G}{\partial x'}(\hat{x}(t_1), \hat{x}'(t_1), t_1) \geqq 0,$$

with

$$\frac{\partial G}{\partial x'}(\hat{x}(t_1), \hat{x}'(t_1), t_1) = 0,$$

if $\hat{x}(t_1) > x^1$.

If t_1 is *free* and x^1 is *assigned,* we have the transversality condition ($n = 1$)

$$G(\hat{x}, \hat{x}', t_1) - \hat{x}'(t_1)\frac{\partial G}{\partial x'}(\hat{x}, \hat{x}', t_1) = 0.$$

Example 8.1.11 Find the solution of the problem

$$\max \int_0^1 (1 - x^2 - (x')^2 dt,$$

with $x(0) = 1$ and $x(1)$ free.

The Euler equation is easily seen to be $x'' - x = 0$, with general solution $x(t) = c_1 e^t + c_2 e^{-t}$. The condition $x(0) = 1$ gives $1 = c_1 + c_2$, so an optimal solution must be of the form $\hat{x}(t) = c_1 e^t + (1 - c_1)e^{-t}$, and thus $\hat{x}'(t) = c_1 e^t - (1 - c_1)e^{-t}$. Furthermore, $\partial G/\partial x' = -2x'$. Being $x(1)$ free, the related transversality condition requires $\hat{x}'(1) = 0$, so $c_1 e^1 - (1 - c_1)e^{-1} = 0$ and hence $c_1 = 1/(e^2 + 1)$. The only possible solution is therefore

$$\hat{x}(t) = \frac{1}{e^2 + 1}(e^t + e^2 e^{-t}).$$

Because $G(x, \hat{x}, t)$ is concave in (x, x'), the above trajectory is indeed the solution of the problem.

Example 8.1.12 We take again into consideration the model described in Example 8.1.3 (Ramsey model). Consider once more the maximization of the functional

$$J(K) = \int_0^T e^{-rt} U\left[v(t) + iK(t) - K'(t)\right] dt$$

subject to the conditions

$$K(0) = K_0 > 0; \quad K(T) = K_T.$$

We assume that K is assigned, but we require $K_T \geqq 0$ and let T be free. Therefore we have the following transversality conditions at T :

$$G(K, K', T) - K' \frac{\partial G}{\partial K'}(K, K', T) = 0,$$

namely

$$K'(T) = -\frac{U(C(T))}{U'(C(T))}$$

whence

$$K(T) \geqq 0, \quad U'(C(T)) \geqq 0, \quad K(T)U'[C(T)] = 0.$$

Analyzing the last equation, we obtain $K(T) = 0$, since $U' > 0$.

In several applications we encounter other generalizations of the basic problem of the Calculus of Variations (8.1), besides the various boundary conditions previously described. We give only some hints on two cases: problems of the Calculus of Variations with *equality constraints* and the so-called *isoperimetric problems.* In these types of problems an useful approach is the introduction of the so-called *Hamiltonian function,* or briefly, *Hamiltonian* (from the Irish mathematician W. R.

Hamilton, 1805–1865), which plays a role similar to the one of the Lagrangian function in static optimization problems.

• Problems of the Calculus of Variations with equality constraints.
Consider, for instance, the problem

$$\max_{x \in \mathcal{U}} \int_{t_0}^{t_1} G(x(t), x'(t), t)dt \tag{8.9}$$

under the constraints

$$x(t_0) = x^0; \quad x(t_1) = x^1,$$

$$g(x(t), x'(t), t) = [0],$$

where G and $g : \mathbb{R}^n \times \mathbb{R}^n \times \mathbb{R} \longrightarrow \mathbb{R}^m$ are of class $\mathcal{C}^2$ and $m < n$.
We define the *Hamiltonian function* or *Hamiltonian,* i.e. the function

$$H = H(x(t), x'(t), t, \lambda_0, \lambda(t)) = \lambda_0 G(x(t), x'(t), t) + \lambda(t)^\top \left[g(x(t), x'(t), t\right],$$

where $\lambda_0 \in \mathbb{R}$ and $\lambda(t) : \mathbb{R} \longrightarrow \mathbb{R}^m$. If the Jacobian matrix $\nabla_{x'} g$ is of full rank for every $t \in [t_0, t_1]$, then a necessary condition for $\hat{x}$ to be a solution of problem (8.9), is that the function H verifies the Euler equation. More precisely, there exist multipliers $\lambda_0, \lambda_1, \ldots, \lambda_m$, *not all zero,* such that

$$\nabla_x H(\hat{x}, \hat{x}', t) - \frac{d}{dt} H_{x'}(\hat{x}, \hat{x}', t) = [0], \quad \forall t \in [t_0, t_1].$$

In particular, it is possible to have $\lambda_0 = 0$ or $\lambda_0 = 1$. The problems where it is possible to choose $\lambda_0 = 1$ are said "normal problems". Note the parallelism with the regularity conditions for static optimization problems under equality constraints. Obviously, it is possible to consider also problems of the Calculus of Variations with inequality constraints; in this case we have conditions on the Hamiltonian similar to the Karush-Kuhn-Tucker conditions of mathematical programming, with all multipliers nonnegative and with the related complementary slackness conditions. We shall not be concerned with this case.

• Problems of the Calculus of Variations with isoperimetric constraints.
These are problems with integral constraints and take their name from the classical queen Dido problem, previously mentioned (Example 8.1.1. The problem may be formalized as follows:

$$\max_{x \in \mathcal{U}} \int_{t_0}^{t_1} G(x(t), x'(t), t)dt$$

$$(\text{or } \min_{x \in \mathcal{U}} \int_{t_0}^{t_1} G(x(t), x'(t), t)dt)$$

under the usual constraints

$$x(t_0) = x^0, \quad x(t_1) = x^1$$

and the integral constraints

$$\int_{t_0}^{t_1} F_i(x(t), x'(t), t)dt = c_i, \quad i = 1, \ldots, m,$$

where $F_i : \mathbb{R}^n \times \mathbb{R}^n \times \mathbb{R} \longrightarrow \mathbb{R}$, $i = 1, \ldots, m$, and G are of class C^2, c_i, $i = 1, \ldots, m$, are given constants and $m < n$. Here the Hamiltonian is

$$H(x(t), x'(t), t, \lambda_0, \lambda(t)) = \lambda_0 G(x(t), x'(t), t) + \lambda(t)^\top F(x(t), x'(t), t),$$

where $\lambda(t) : \mathbb{R} \longrightarrow \mathbb{R}^m$ and F is the vector function whose components are F_i, $i = 1, \ldots, m$.

A necessary condition such that $\hat{x}$ is a solution of the problem, is that there must exist multipliers $\lambda_0, \lambda_1, \ldots, \lambda_m$, *not all zero,* such that H satisfies the Euler equation.

Let us now suppose that there is only one integral constraint:

$$\int_{t_0}^{t_1} F(x(t), x'(t), t)dt = c.$$

Under certain additional conditions, we do not specify here, there is an interesting result, similar to duality results of mathematical programming, and called "reciprocity principle". Let us denote by J and I, respectively, the objective functional and the functional which appears in the constraint:

$$J(x) = \int_{t_0}^{t_1} G(x(t), x'(t), t)dt;$$

$$I(x) = \int_{t_0}^{t_1} F(x(t), x'(t), t)dt.$$

The reciprocity principle asserts that, under suitable conditions, if $\hat{x}$ maximizes J, under the given constraint I, then $\hat{x}$ minimizes I, under the given constraint J. For example, the closed curve of a given length, which maximizes the area bounded by the same curve (it is a circumference!), it is also that closed curve that, with the same given bounded area, results to be of minimal length.

8.2 Introduction to Optimal Control Problems

Optimal Control Theory is a modern extension of the classical Calculus of Variations. Whereas the Calculus of Variations goes back to XVII and XVIII centuries, the main results in Optimal Control Theory were obtained in the 1950s by a group of Soviet mathematicians, namely L. S. Pontryagin and his collaborators. See Pontryagin et al. [30]. The main result of Optimal Control Theory is the so-called "maximum principle": roughly speaking, this principle gives optimality conditions for a wide range of dynamic optimization problems. It seems that the famous American mathematician R. Bellman declared that "Optimal Control Theory is the corpse of the Calculus of Variations wrapped in the Soviet Union flag". If true, this sentence appears somewhat nasty and ungenerous, as it is true that Optimal Control Theory includes, as a particular case, the Calculus of Variations, but it can be applied to a significantly wider range of problems. Since 1960s, countless papers in economic analysis and management sciences have used Optimal Control Theory. Applications in other scientific fields are likewise countless.

In the previous section we have given the basic notions on the classical Calculus of Variations, that is the problem of finding, among the possible evolutions of a state variable or a state vector $x = x(t) \in \mathcal{U}$ of a system, that one which optimizes a given functional $J(x)$. In Optimal Control Theory we have dynamic optimization problems where it is possible to intervene through decisions, said *controls,* which affect directly the evolution of the state variables. This is a more general and fruitful approach than the one offered by the Calculus of Variations.

For the moment, consider, for simplicity, the case $n = 1$. Consider a system whose state at time t is characterized by a *state variable* $x(t)$. The process that causes $x(t)$ to change is controlled by a *control function* $u(t)$. We assume that the rate of change of $x(t)$ depends on $t, x(t)$ and $u(t)$. The state at the initial point x_0 is given by $x(t_0) = x_0$. Hence, the evolution of $x(t)$ is described by a differential equation of the type

$$x'(t) = g(x(t), u(t), t); \quad x(t_0) = x_0. \tag{8.10}$$

Then, we want, for example, to maximize the functional

$$J(u) = \int_{t_0}^{t_1} G(x(t), u(t), t)dt,$$

$u(t) \in (-\infty, +\infty)$, under conditions (8.10). Usually $x(t_1)$ is free, otherwise $x(t_1) = x_1$.

A pair $(x(t), u(t))$ that satisfies (8.10) is called an *admissible pair:* among all admissible pairs we search for an *optimal pair,* i.e. a pair of functions that maximizes the functional $J(u)$. This is a simple problem of Optimal Control with continuous time. It is also possible to consider problems with discrete time, but we shall not be concerned with this last case. More generally, with $n \geqq 1$, the elements of a problem of Optimal Control, with continuous time, are the following ones.

(*i*) The time t is a continuous variable defined on an interval $[t_0, t_1]$, with $t_1 \leqq +\infty$; t_0 is the *initial state* (fixed), whereas t_1 is the *final state* or *terminal state*, fixed or not.

(*ii*) The state of a system at every $t \in [t_0, t_1]$ is described by a function $x : \mathbb{R} \longrightarrow \mathbb{R}^n$, said *state variable* or *state vector* or *state trajectory.* Usually, this function is assumed to be $\mathcal{C}^1$ or, more generally, that its derivatives are *piecewise continuous* (a function is piecewise continuous if it has at most a finite number of discontinuities in each finite interval, with finite jumps at each point of discontinuity).

(*iii*) At every time $t \in [t_0, t_1]$ we have to choose k quantities $u_1(t), u_2(t), \dots, u_k(t)$, that is a vector

$$u(t) = [u_1(t), u_2(t), \dots, u_k(t)]^\top .$$

On the vector function $u : \mathbb{R} \longrightarrow \mathbb{R}^k$, said *admissible control trajectory* (the components of $u(t)$ are also said *admissible controls* or also *control variables*), we make the assumption of continuity or, more generally, of piecewise continuity. The control trajectory must be chosen in a given set $\Omega \subset \mathbb{R}^k$, open or closed, convex or not, but usually time invariant.

(*iv*) The law which governs the motion of the system is described by a system of differential equations

$$x'(t) = \varphi(x(t), u(t), t),$$

where $\varphi : \mathbb{R}^n \times \mathbb{R}^k \times \mathbb{R} \longrightarrow \mathbb{R}^n$ is usually assumed to be continuously differentiable. If φ does not depend explicitly on t, the equations of motion are called *autonomous.*

(*v*) The previous system is completed with a boundary condition $x(t_0) = x^0$. For what concerns t_1 and $x(t_1) = x^1$, these conditions may appear or not. In any case (unless $t_1 = +\infty$) it is always possible to suppose that we are given a set $Q \subset \mathbb{R}^{n+1}$, to which the pair (t_1, x^1) must belong. The set Q is also called *terminal surface.*

(*vi*) The last "ingredient" of an Optimal Control Problem is the *objective functional* to be maximized. Usually, it is of the type

$$J(u) = \int_{t_0}^{t_1} G(x(t), u(t), t)dt \tag{8.11}$$

where $G : \mathbb{R}^n \times \mathbb{R}^k \times \mathbb{R} \longrightarrow \mathbb{R}$ is assumed to be $\mathcal{C}^1$.

Instead of (8.11), sometimes the objective functional has the form

$$J(u) = F(t_1, x(t_1)) \tag{8.12}$$

or the form

$$J(u) = \int_{t_0}^{t_1} G(x(t), u(t), t)dt + F(t_1, x(t_1)) \tag{8.13}$$

with $F : \mathbb{R} \times \mathbb{R}^n \longrightarrow \mathbb{R}$, F continuously differentiable.

When the objective function is of the type (8.11) it is also called to be in the *canonical form* or that the problem is a *Lagrange problem;* when it is of the form (8.12), we have a *Mayer problem* and when it is of the type (8.13), we have a *Bolza problem.* Also here, as it is verified in the Calculus of Variations, it is possible to show that the three problems are substantially equivalent.

We speak of *terminal time problems,* when t_1 is given (but $x(t_1)$ is free); of *terminal state problems,* when also x^1 is given; of *generic terminal surface* when $x(t_1) = \phi(t_1)$, with $\phi : \mathbb{R} \longrightarrow \mathbb{R}^n$.

From what previously said, one of the versions of an Optimal Control Problem in the Lagrange form is

$$\max_{u(t)} \int_{t_0}^{t_1} G(x(t), u(t), t)dt \tag{8.14}$$

subject to:

$x'(t) = \varphi(x(t), u(t), t)$;
t_0, $x(t_0) = x^0$ given;
$(x(t), t) \in Q$ at $t = t_1$;
$u(t) \in \Omega$, $\forall t \in [t_0, t_1]$;
$u(t)$ continuous or even piecewise continuous;
$x(t) \in C^1([t_0, t_1])$;
$G, \varphi \in C^1(\mathbb{R}^{2n+1})$.

Remark 8.2.1 A problem of the Calculus of Variations can be considered a particular case of an Optimal Control Problem, where $u(t) = x'(t)$:

$$\max_{x(t)} \int_{t_0}^{t_1} G(x(t), x'(t), t)dt$$

subject to: $x(t_0) = x^0$, $x(t_1) = x^1$, $G \in C^2(\mathbb{R}^{2n+1})$; $x \in C^2([t_0, t_1])$. However, in a problem of the Calculus of Variations we maximize the functional J with respect to $x(t)$. It can be shown that the Euler equation can be obtained by maximizing, with respect to $x'(t)$, the Hamiltonian function of a corresponding Optimal Control Problem.

Remark 8.2.2 The objective functional depends only on the control trajectory u (indeed we have written $J(u)$). In other words, we choose $u \in \Omega$, and the state trajectory is uniquely determined by means of the system of differentiable equations $x'(t) = \varphi(x(t), u(t), t)$, with the boundary condition $x(t_0) = x^0$; hence it is uniquely determined also J.

Example 8.2.3 This example is taken from Chiang [12]. In a very interesting paper, the economist Robert Dorfman (see Dorfman [15]), considers the following economic model. A firm seeks to maximize its profits over the time interval $[0, T]$. There is a single state variable, capital stock K, and a single control variable u, representing some business decision that firm has to make at each moment of time (such as its advertising budget or inventory policy). The firm starts out at time $t = 0$ with capital K_0, but the terminal capital stock is left open. At any moment of time, the profit of the firm depends on the amount of capital it currently holds, as well as on the policy u it currently selects. Thus, the profit function is $p(K, u, t)$. But the policy selection u also bears upon the rate at which capital K changes over time, that is K' is affected by u. It follows that the optimal control problem is

$$\max J(u) = \int_0^T p(K, u, t)dt$$

subject to: $K' = f(K, u, t)$, $K(0) = K_0$, $K(t)$ free (K_0, T given).

Now we introduce the fundamental *Maximum Principle* (of Pontryagin) for an Optimal Control Problem. Also here it is convenient to define the *Hamiltonian function.* Let us introduce the following n variables depending on time:

$$\lambda(t)^\top = [\lambda_1(t), \lambda_2(t), \ldots, \lambda_n(t)].$$

In the present context this vector of multipliers is called *vector of costate variables* or *vector of auxiliary variables.* Then we introduce the function

$$H(x(t), u(t), \lambda(t), t) = G(x(t), u(t), t) + \lambda(t)^\top \varphi(x(t), u(t), t),$$

where $\lambda : \mathbb{R} \longrightarrow \mathbb{R}^n$. The function H is called the *Hamiltonian function* or simply the *Hamiltonian.* Some authors (e.g. Seierstad and Sydsaeter [35], Hadley and Kemp [19]) consider the Hamiltonian function of the form

$$H = \lambda_0 G(x(t), u(t), t) + \lambda^\top \varphi(x(t), u(t), t),$$

with $\lambda_0 \in \mathbb{R}$ and $\lambda = \lambda(t) : \mathbb{R} \longrightarrow \mathbb{R}^n$. When $\lambda_0 \neq 0$, obviously we can divide by λ_0 and get the Hamiltonian in which $\lambda_0 = 1$. In this case the problem is said to be *normal,* a condition which will be assumed in all what follows. In some "pathological" cases $\lambda_0 = 0$ is not excluded, however if in the related problem we have the free boundary condition, i.e. $x(t_1)$ free, the said pathological case cannot occur. See Hadley and Kemp [19].

For simplicity we consider first the case where $n = 1$ and $k = 1$. The necessary optimality conditions for a simplified version of problem (8.14) are given by the following set of conditions (Pontryagin Maximum Principle).

Theorem 8.2.4 (Pontryagin Maximum Principle) *Suppose that $(\hat{x}, \hat{u})$ is an optimal pair for problem* (8.14), *with $n = 1$, $k = 1$, initial state $x(t_1) = x_1$. Then, there exists a continuous function $\lambda(t)$ such that $\hat{u}$ solves the following problem*

$$\max_{u \in \Omega} H(\hat{x}, u, \lambda, t), \quad \forall t \in [t_0, t_1].$$

Moreover,

$$\hat{x}'(t) = \varphi(\hat{x}(t), \hat{u}(t), t) = \frac{\partial \hat{H}}{\partial \lambda};$$

$$\lambda' = -\frac{\partial \hat{H}}{\partial x};$$

$$x(t_0) = x_0; \; x(t_1) = x_1.$$

Note that the condition $\hat{x}'(t) = \varphi(\hat{x}(t), \hat{u}(t), t) = \frac{\partial \hat{H}}{\partial \lambda}$ is obvious and often omitted. Note moreover, that if $\hat{u}$ is in the interior of the control region Ω, the maximization of H with respect to u implies

$$\frac{\partial \hat{H}}{\partial u} = 0.$$

($\frac{\partial \hat{H}}{\partial u_i} = 0$, $i = 1, \ldots, k$, if $k > 1$).

In general Ω is not required to be open: for example, with $k > 1$, if $u_i(t)$ is restricted by $0 \leqq u_i(t) \leqq 1$ for all i, $\hat{u}_i(t)$ may be such that for all i

$$\hat{u}_i(t) = 0, \;\text{ for } t_0 \leqq t < \bar{t}$$

$$\hat{u}_i(t) = 1, \;\text{ for } \bar{t} \leqq t \leqq 1.$$

Such a solution is called a "bang-bang solution" and it appears in several practical applications.

In the present context the two last conditions of Theorem 8.2.4 are also called "transversality conditions", and they hold when both the terminal time t_1 and the terminal state x_1 are given. Otherwise, there are different transversality conditions.

1. When t_1 is given, but $x(t_1)$ is free, then $\lambda(t_1) = 0$.
2. When the terminal state x_1 is given, then $[H]_{t=t_1} = 0$.
3. When the terminal surface is given, $x(t_1) = \phi(t_1)$, then $\left[H - \lambda\phi'\right]_{t=t_1} = 0$.
4. $\{t_1$ given and $x(t_1) \geqq x_{\min}\}$. Then $\lambda(t_1) \geqq 0$, $x(t_1) \geqq x_{\min}$ and $(x(t_1) - x_{\min})\lambda(t_1) = 0$ (complementary slackness condition).

5. $\{x(t_1) = z,\ \text{given};\ t_1 \leqq t_{\max}\}$. Then $[H]_{t=t_1} \geqq 0$, $t_1 \leqq t_{\max}$ and $(t_1 - t_{\max})\,[H]_{t=t_1} = 0$ (complementary slackness condition).

Example 8.2.5 Consider the problem

$$\max_u \int_0^2 (2x - 3u)dt$$

subject to

$$x' = x + u,$$
$$x(0) = 4;\ u(t) \in [0, 2]\,.$$

The Hamiltonian function is

$$H(x, u, \lambda, t) = 2x - 3u + \lambda(x + u)$$

and, according to the Maximum Principle, we have to maximize H with respect to u :

$$\max_u\ 2x - 3u + \lambda(x + u).$$

H is an affine function with slope

$$\frac{\partial H}{\partial u} = \lambda - 3.$$

Hence, when $\lambda > 3$, H is maximized with respect to u when $u = 2$; when $\lambda \leqq 3$, H is maximized with respect to u when $u = 0$. So we have

$$\hat{u}(t) = \begin{cases} 2, & \hat{\lambda}(t) > 3 \\ 0, & \hat{\lambda}(t) \leqq 3. \end{cases}$$

Then, according to the Maximum Principle,

$$\lambda' = -\frac{\partial H}{\partial x} = -2 - \lambda.$$

In order to find $\hat{\lambda}(t)$ we have to solve the Cauchy problem

$$\begin{cases} \lambda' = -2 - \lambda \\ \hat{\lambda}(2) = 0. \end{cases}$$

We find $\hat{\lambda}(t) = ke^{-t} - 2$, $\hat{\lambda}(2) = ke^{-2} - 2 = 0$ and hence $k = 2e^2$ and then

$$\hat{\lambda}(t) = 2e^{2-t} - 2.$$

As $\hat{\lambda}(t)$ is a decreasing function, $\hat{\lambda}(0) = 2e^2 - 2 > 3$ and $\hat{\lambda}(2) = 0$, if we solve the equation in t

$$\hat{\lambda}(t) = 2e^{2-t} - 2 = 3$$

we obtain

$$\tau = 2 - \log \frac{5}{2}$$

and then

$$\hat{u}(t) = \begin{cases} 2, & 0 \leqq t < \tau \\ 0, & \tau \leqq t \leqq 2. \end{cases}$$

Always according to the Maximum Principle, we have

$$x'(t) = \frac{\partial H}{\partial \lambda} = \varphi(x(t), u(t), t) = x + u = \begin{cases} x + 2, & 0 \leqq t < \tau \\ x, & \tau \leqq t \leqq 2, \end{cases}$$

so that we must solve the Cauchy problems

$$\begin{cases} x' = x + 2 \\ x(0) = 4, \end{cases}$$

when $0 \leqq t < \tau$, and

$$\begin{cases} x' = x \\ x(\tau) = x_\tau, \end{cases}$$

when $\tau \leqq t \leqq 2$. In the first case $(0 \leqq t < \tau)$ we have

$$\begin{cases} x' = x + 2 \\ x(0) = 4, \end{cases}$$

$\hat{x}(t) = ke^{-t} - 2$, $\hat{x}(0) = k - 2 = 4$, i.e. $k = 6$, $\hat{x}(t) = 2(3e^t - 1)$.

In the second case $(\tau \leqq t \leqq 2)$ we have

$$\begin{cases} x' = x \\ x(\tau) = x_\tau, \end{cases}$$

$\hat{x}(t) = ke^t$ and, being $\hat{x}(t)$ continuous at $t = \tau$,

$$\hat{x}(t) = 2(3e^\tau - 1) = ke^\tau$$

and then

$$k = 2(3 - e^{-\tau}).$$

Then we have

$$\hat{x}(t) = \begin{cases} 2(3e^t - 1), & 0 \leqq t < \tau \\ 2(3 - e^{-\tau})e^t, & \tau \leqq t \leqq 2. \end{cases}$$

Example 8.2.6

$$\max_c \int_0^1 \log(4c(t)s(t))dt,$$

with $c(t)$ consumption per unit of output and $s(t)$ capital stock, under the constraints

$$\begin{aligned} s' &= 4s(t)(1 - c(t)), \\ s(0) &= 4, \\ s(1) &= e^2. \end{aligned}$$

The Hamiltonian function is

$$\begin{aligned} H(s, c, \lambda, t) &= \log(4cs) + \lambda(4s(1 - c)) \\ &= \log 4 + \log c + \log s + \lambda(4s(1 - c)) \end{aligned}$$

and according to the Maximum Principle we have to maximize H with respect to c, $c > 0$, i.e. Ω is an open interval:

$$\frac{\partial H}{\partial c} = \frac{1}{c} - 4\lambda s = 0,$$

that is

$$c = \frac{1}{4\lambda s}.$$

By the Maximum Principle we have also

$$\begin{cases} \lambda' = -\frac{\partial H}{\partial s} = -\frac{1}{s} - 4(1 - c)\lambda \\ s' = \frac{\partial H}{\partial \lambda} = 4s(1 - c), \end{cases}$$

that is, being $c = 1/(4\lambda s)$,

$$\begin{cases} \lambda' = -4\lambda \\ s' = 4s - \frac{1}{\lambda}. \end{cases}$$

So we have

$$\begin{cases} \lambda(t) = \lambda(0)e^{-4t} \\ s' = 4s - \frac{e^{4t}}{\lambda(0)} \end{cases}$$

and, multiplying both sides of the last equation by e^{-4t} :

$$s'e^{-4t} - 4se^{-4t} = \frac{1}{\lambda(0)}.$$

The left hand side of this equation is

$$\frac{\partial \left[se^{-4t}\right]}{\partial t} = s'e^{-4t} - 4se^{-4t}$$

and then

$$s'e^{-4t} - 4se^{-4t} = \frac{\partial \left[se^{-4t}\right]}{\partial t} = -\frac{1}{\lambda(0)}.$$

By integrating both sides we have

$$se^{-4t} = -\frac{t}{\lambda(0)} + k,$$

that is

$$s(t) = -\frac{te^{4t}}{\lambda(0)} + ke^{4t}.$$

As $s(0) = 1$, we have $k = 1$ and as $s(1) = e^2$, we have $\lambda(0) \cong 1.156$. So we obtain

$$\begin{cases} \hat{s}(t) = e^{4t} - 0.865te^{4t} \\ \hat{c}(t) = \frac{1}{4.624-4t}. \end{cases}$$

Example 8.2.7

$$\max_c \int_0^T e^{-\delta t} \log(c(t))dt,$$

where $c(t)$ is the consumption per unit of output, $s(t)$ is the capital stock and δ is the rate of discount. The constraints are:

$s'(t) = rs(t) - c(t)$, where r is the rate of interest; $s(0) = S_0$, $s(T) = S_T$.

The Hamiltonian function is

$$H(s, c, \lambda, t) = e^{-\delta t} \log c(t) + \lambda(t)(rs(t) - c(t))$$

and according to the Maximum Principle we have to maximize H with respect to $c > 0$ (Ω is an open interval):

$$\frac{\partial H}{\partial c} = e^{-\delta t}\frac{1}{c} - \lambda = 0.$$

By the Maximum Principle we have also

$$\begin{cases} \lambda' = -\frac{\partial H}{\partial s} = -r\lambda \Longrightarrow \hat{\lambda}(t) = \lambda(0)e^{-rt} \\ s' = \frac{\partial H}{\partial \lambda} = rs - c \Longrightarrow \hat{c}(t) = \frac{1}{\lambda(0)}e^{(r-\delta)t}. \end{cases}$$

We observe that:

$$r > \delta \Longrightarrow c \text{ is increasing}$$

$$r < \delta \Longrightarrow c \text{ is decreasing.}$$

From the above last equation we have

$$s' - rs = -\frac{1}{\lambda(0)}e^{(r-\delta)t}.$$

By multiplying both sides by e^{-rt} we get

$$s'e^{-rt} - rse^{-rt} = -\frac{1}{\lambda(0)}e^{-\delta t}.$$

Since the first term is the derivative of se^{-rt}, by integrating both sides we have

$$se^{-rt} = \frac{e^{-\delta t}}{\delta\lambda(0)} + K,$$

that is,

$$s(t) = \frac{e^{(r-\delta)t}}{\delta\lambda(0)} + Ke^{rt}.$$

The boundary conditions $s(0) = S_0$ and $s(T) = S_T$ imply the equations

$$S_0 = K + \frac{1}{\delta\lambda(0)}$$

$$S_T = Ke^{rT} + \frac{e^{(r-\delta)T}}{\delta\lambda(0)}.$$

Solving with respect to K and $\lambda^{-1}(0)$, we get

$$\frac{1}{\lambda(0)} = \frac{S_0 - S_T e^{-rT}}{\delta^{-1}(1 - e^{-\delta T})}$$

$$K = S_0 - \frac{1}{\delta\lambda(0)} = S_0 - \frac{S_0 - S_T e^{-rT}}{1 - e^{-\delta T}}.$$

We give only a sketch of the proof of the Maximum Principle, as a full proof requires some advanced results in the theory of differential equations which have not been given in the present book. Again we restrict our attention to problems with *one* state and *one* control variable; moreover, we assume that Ω is an open interval. Therefore we consider the following problem.

$$\max_{u\in\Omega} \int_{t_0}^{t_1} G(x(t), u(t), t)dt$$

subject to:

$$x' = \varphi(x, u, t),$$

$$x(t_0) = x_0, \quad x(t_1) \text{ free.}$$

• **Sketch of the proof of the Maximum Principle.**

Let us fix the control u and the corresponding state x and define the following function

$$V = \int_{t_0}^{t_1} G(x, u, t)dt + \int_{t_0}^{t_1} \lambda(t)\left[\varphi(x, u, t) - x'(t)\right]dt$$

$$= \int_{t_0}^{t_1} H(x, u, t, \lambda)dt - \int_{t_0}^{t_1} \lambda(t)x'(t)dt.$$

We integrate by parts the second integral and we obtain

$$\int_{t_0}^{t_1} \lambda(t)x'(t)dt = [\lambda(t)x(t)]_{t_0}^{t_1} - \int_{t_0}^{t_1} \lambda'(t)x(t)dt$$

and then

$$V = \int_{t_0}^{t_1} H(x, u, t, \lambda)dt - \lambda(t_1)x(t_1) + \lambda(t_0)x(t_0) + \int_{t_0}^{t_1} \lambda'(t)x(t)dt.$$

Let us consider the optimal control (solution of the problem) u^* and the corresponding optimal state x^*. Then we set

$$u = u^* + \varepsilon p$$
$$x = x^* + \varepsilon q$$

with q chosen such that $q(t_0) = 0$, so that $x(t_0) = x^*(t_0) = x_0$ and $x(t_1) = x^*(t_1) + \varepsilon \Delta x(t_1)$.

Since

$$V(\varepsilon) = \int_{t_0}^{t_1} \left[H(x^* + \varepsilon q, u^* + \varepsilon p, t, \lambda) + \lambda'(x^* + \varepsilon q)\right] dt$$
$$-\lambda(t_1)\left[x^*(t_1) + \varepsilon \Delta x(t_1)\right] + \lambda(t_0)x_0,$$

then

$$V'(\varepsilon) = \int_{t_0}^{t_1} \left[\frac{\partial H}{\partial x} q + \frac{\partial H}{\partial u} p + \lambda' q\right] dt - \lambda(t_1)\Delta x(t_1)$$

and, as u^* and x^* are optimal,

$$V'(0) = \int_{t_0}^{t_1} \left[(\frac{\partial H}{\partial x}(x^*, u^*, \lambda, t) + \lambda')q + \frac{\partial H}{\partial u}(x^*, u^*, \lambda, t)p\right]$$
$$\times dt - \lambda(t_1)\Delta x(t_1) = 0.$$

Being p, q and $\Delta x(t_1)$ arbitrarily chosen, this last equality implies that, at $u = u^*$ and $x = x^*$ ($\varepsilon = 0$),

$$\lambda' = -\frac{\partial H}{\partial x}$$
$$\frac{\partial H}{\partial u} = 0$$
$$\lambda(t_1) = 0. \ \square$$

The conditions in the Maximum Principle are necessary, but generally not sufficient for optimality. We consider two different kinds of sufficient optimality conditions:

1. **The Mangasarian condition** (Mangasarian [26]).
2. **The Arrow condition** (Arrow [3] and Arrow and Kurz [4]).

Theorem 8.2.8 (Mangasarian) *Suppose that $(x^*(t), u^*(t))$ is an admissible pair with a corresponding costate variable $\lambda(t)$ such that the conditions of Theorem 8.2.4 are satisfied, with $n = 1$, $k = 1$. If $x(t_0) = x_0$, $x(t_1) = x_1$, no specific transversality conditions are required; if $x(t_0) = x_0$ and $x(t_1)$ is free, then the transversality conditions is $\lambda(t_1) = 0$. Suppose further that the control region Ω is convex (therefore Ω is an interval of $\mathbb{R}$) and that $H(x, u, \lambda(t), t)$ is concave in (x, u) for every $t \in [t_0, t_1]$. Then $(x^*(t), u^*(t))$ is an optimal pair. If $H(x, u, \lambda(t), t)$ is strictly concave with respect to (x, u), then the pair $(x^*(t), u^*(t))$ is the unique solution of the problem.*

Note that if G and φ are both concave with respect to (x, u) and $\lambda(t) \geqq 0$ for all $t \in [t_0, t_1]$, then the concavity conditions on H hold true. If G is concave, with respect to (x, u), and φ is *linear affine,* with respect to (x, u), then H is concave, independently of the sign of $\lambda(t)$.

Most of the control problems can be solved by means of the Mangasarian condition, however, in several economic models the Hamiltonian is not concave. Arrow [3] has suggested a weakening of the Mangasarian condition. The following result appears without proof in Arrow [3]; it appears with proof in Arrow and Kurz [4]. For a quite complete and more exact treatment see Seierstad and Sydsater [34].

Define the function

$$H^0(x, \lambda, t) = \max_{u \in \Omega} H(x, u, \lambda, t)$$

and assume that the maximum value is attained. The function $H^0(x, \lambda, t)$ is called the *maximized Hamiltonian.* Note that the concept of H^0 is different from the optimal Hamiltonian H^*, since H^* denotes the Hamiltonian evaluated along all the optimal paths, that is evaluated at $x^*(t)$, $u^*(t)$ and $\lambda^*(t)$. In contrast, H^0 is evaluated along $u^*(t)$ only, so that $H^0 = H^0(x, \lambda, t)$ is still a function with three arguments.

Theorem 8.2.9 (Arrow) *Assume that the requirements on $(x^*(t), u^*(t))$ of Theorem 8.2.8 holds true and suppose further that*

$$H^0(x, \lambda, t) \text{ is concave in } x \text{ for every } t \in [t_0, t_1].$$

Then $(x^*(t), u^*(t))$ solves the Optimal Control Problem.

Example 8.2.10 See Chiang ([12], Example 1, page 171): "shortest distance between a point and a line". Consider the problem

$$\max_u \int_0^T -((1 + u^2)^{\frac{1}{2}} dt$$

subject to:

$$x'(t) = u$$
$$x(0) = x_0, \quad x(T) \text{ free.}$$

The necessary conditions are given by the Maximum Principle:

$$\max_u H(x.u, \lambda, t) = \max_u \left\{ -(1+u^2)^{\frac{1}{2}} + \lambda u \right\}, \quad \forall t \in [0, T];$$

$$x'(t) = \varphi(x(t), u(t), t) = \frac{\partial H}{\partial \lambda} = u;$$

$$\lambda' = -\frac{\partial H}{\partial x} = 0;$$

$$x(0) = x_0; \quad \lambda(T) = 0.$$

As, $\forall t \in [0, T]$, we have

$$\lambda' = 0; \ \lambda(t) = \lambda \text{ and } \lambda(T) = 0,$$

this implies $\lambda(t) = 0, \forall t \in [0, T]$.

Then we solve

$$\max_u \left\{ -(1+u^2)^{\frac{1}{2}} \right\},$$

that is $u(t) = 0, \ \forall t \in [0, T]$.

Then

$$x'(t) = u(t) = 0, \quad \forall t \in [0, T]$$

implies

$$x(t) = c, \quad \forall t \in [0, T],$$

and as $x(0) = x_0$, we have

$$x(t) = c = x_0, \quad \forall t \in [0, T].$$

So that we have, $\forall t \in [0, T]$:

$$u^*(t) = 0; \quad \lambda^*(t) = 0; \quad x^*(t) = x_0.$$

Let us check the Mangasarian condition on the functions

$$G(x, u, t) = -(1 + u^2)^{\frac{1}{2}}$$
$$\varphi(x, u, t) = u.$$

As G does not depend on x, we have to check the convexity of G with respect to u, i.e. we must have $G''_{\mathbf{uu}} \leqq 0$.

$$G'_u = -\frac{1}{2}(1 + u^2)^{-\frac{1}{2}} 2u = -u(1 + u^2)^{-\frac{1}{2}}$$

$$G''_{\mathbf{uu}} = -(1 + u^2)^{-\frac{1}{2}} + \frac{1}{2}u(1 + u^2)^{-\frac{3}{2}} 2u = (1 + u^2)^{-\frac{1}{2}} \left[\frac{u^2}{u^2 + 1} - 1\right].$$

As $u^*(t) = 0, \ \forall t \in [0, T]$, then

$$G''_{\mathbf{uu}} = -1 < 0, \ \ \forall t \in [0, T],$$

and the Mangasarian condition is therefore satisfied.

Example 8.2.11

$$\max_u \int_0^T 4u\sqrt{x}e^{-\delta t} dt$$

subject to:

$$x'(t) = -u^4$$
$$x(0) = x_0 > 0$$
$$x(T) = x_T > 0.$$

The Hamiltonian function is

$$H(x, u, \lambda, t) = 4u\sqrt{x}e^{-\delta t} - \lambda u^4$$

and

$$G(x, u, t) = 4u\sqrt{x}e^{-\delta t}.$$

The function G is not concave with respect to (x, u), as

$$G'_x = 2ux^{-\frac{1}{2}}e^{-\delta t};$$
$$G'_u = 4\sqrt{x}e^{-\delta t};$$

$$G''_{xx} = -ux^{-\frac{3}{2}}e^{-\delta t};$$
$$G''_{uu} = 0;$$
$$G''_{ux} = G''_{xu} = 2x^{-\frac{1}{2}}e^{-\delta t}.$$

The Hessian matrix of G with respect to (x, u) is

$$\nabla^2_{xu} G = \begin{bmatrix} 0 & 2x^{-\frac{1}{2}}e^{-\delta t} \\ 2x^{-\frac{1}{2}}e^{-\delta t} & -ux^{-\frac{3}{2}}e^{-\delta t} \end{bmatrix}$$

and $\det(\nabla^2_{xu} G) < 0$.

The Mangasarian condition is not satisfied. Let us check the Arrow condition:

$$H^0(x, \lambda, t) = u^0(x, \lambda, t) = x^{\frac{1}{6}}\lambda^{-\frac{1}{3}}e^{-\frac{\delta t}{3}} = \sqrt[3]{\sqrt{x}\lambda^{-1}e^{-\delta t}}.$$

As

$$\frac{\partial H}{\partial u} = 4\sqrt{x}e^{-\delta t} - 4\lambda u^3 = 0,$$

we have

$$u^3 = \frac{\sqrt{x}e^{-\delta t}}{\lambda}$$

$$u^0 = \sqrt[3]{\frac{\sqrt{x}e^{-\delta t}}{\lambda}} = \sqrt[3]{\sqrt{x}\lambda^{-1}e^{-\delta t}}.$$

Hence

$$H^0(x, \lambda, t) = 4\sqrt[3]{\sqrt{x}\lambda^{-1}e^{-\delta t}}\,\sqrt{x}e^{-\delta t} - \lambda\left(\sqrt[3]{\sqrt{x}\lambda^{-1}e^{-\delta t}}\right)^4$$
$$= 4x^{\frac{1}{6}}x^{\frac{1}{2}}\lambda^{-\frac{1}{3}}e^{-\frac{\delta t}{3}}e^{-\delta t} - \lambda\lambda^{-\frac{4}{3}}x^{\frac{4}{6}}e^{-\frac{4\delta t}{3}}$$
$$= 4x^{\frac{2}{3}}\lambda^{-\frac{1}{3}}e^{-\frac{4\delta t}{3}} - x^{\frac{2}{3}}\lambda^{-\frac{1}{3}}e^{-\frac{4\delta t}{3}} = 3x^{\frac{2}{3}}\lambda^{-\frac{1}{3}}e^{-\frac{4\delta t}{3}}.$$

This function is therefore strictly concave with respect to x when $\lambda > 0$.

One of the few papers which extend to generalized convex functions the sufficient optimality conditions of Mangasarian for an Optimal Control Problem is the paper of Mond and Smart [27].

• Economic interpretation of the costate variables

As before, we consider the simple case $n = 1$, $k = 1$.

We give some considerations on the meaning of the costate variable ("multiplier") λ, which shows a similarity with the interpretation of Lagrange multipliers (or Karush-Kuhn-Tucker multipliers) of mathematical programming problems. Consider the "classical" Optimal Control Problem

$$\max_u \int_{t_0}^{t_1} G(x, u.t)dt$$

subject to:

$$x'(t) = \varphi(x(t), u(t), t),$$
$$x(t_0) = x_0; \quad x(t_1) = x_1.$$

Suppose that it has a unique optimal solution $(x^*(t), u^*(t))$, with a unique costate variable $\lambda^*(t)$. The corresponding value of the objective function will depend on x_0, x_1, t_0 and t_1, so it is natural to consider the *(optimal) value function*

$$V(x_0, x_1, t_0, t_1) = \int_{t_0}^{t_1} G(x^*(t), u^*(t), t)dt.$$

Now, suppose that x_0 is changed slightly. Then, both $x^*(t)$ and $u^*(t)$ will change over the whole interval $[t_0, t_1]$. If V is differentiable (and this is not a trivial question; see, e.g., Seierstad and Sydsaeter [35]), then

$$\frac{\partial V}{\partial x_0}(x_0, x_1, t_0, t_1) = \lambda(t_0).$$

This can be justified as follows (not a full proof!). We have

$$V^* = \int_{t_0}^{t_1} G(x^*, u^*, t)dt = \int_{t_0}^{t_1} \left[G(x^*, u^*, t) + \lambda(t)\left(\varphi(x^*, u^*t) - (x^*)'(t)\right)\right] dt.$$

Integrating by parts $\lambda(t)(x^*)'(t)$, we have

$$\int_{t_0}^{t_1} \lambda(t)(x^*)'(t)dt = \left[\lambda(t)x^*(t)\right]_{t_0}^{t_1} - \int_{t_0}^{t_1} \lambda'(t)x^*(t)dt$$

and then

$$V^* = \int_{t_0}^{t_1} \left[H(x^*, u^*, \lambda, t) + \lambda' x^*\right] dt - \lambda(t_1)x^*(t_1) + \lambda(t_0)x^*(t_0).$$

If V^* is differentiable, we have

$$\frac{\partial V^*}{\partial x_0} = \int_{t_0}^{t_1} \left[\frac{\partial H}{\partial x}\frac{\partial x^*}{\partial x_0} + \frac{\partial H}{\partial u}\frac{\partial u^*}{\partial x_0} + \lambda' \frac{\partial x^*}{\partial x_0} \right] dt + \lambda^*(t_0).$$

As $\lambda = \lambda^*$ and $H^* = H(x^*, u^*, \lambda, t)$, by the Maximum Principle we have $\frac{\partial H^*}{\partial u} = 0$, $(\lambda^*)' = -\frac{\partial H^*}{\partial x}$, and substituting in the last equality, we have

$$\frac{\partial V^*}{\partial x_0} = \int_{t_0}^{t_1} \left[-(\lambda^*)' \frac{\partial x^*}{\partial x_0} + (\lambda^*)' \frac{\partial x^*}{\partial x_0} \right] dt + \lambda^*(t_0) = \lambda^*(t_0).$$

Therefore the number $\lambda^*(t_0)$ measures the marginal change in the optimal value function, as x_0 varies. If, as it is quite usual in several problems of economic nature, x has the meaning of quantities and the objective function has the meaning of values (costs, profits, etc.), then $\lambda(t_0)$ assumes the meaning of a price, price often called, as in the case of mathematical programming problems, "shadow price".

- **Current value formulations**

Many control problems have the following structure (for simplicity $n = 1$, $k = 1$):

$$\max_{u \in \Omega \subset \mathbb{R}} \int_{t_0}^{t_1} G(x, u, t) e^{-rt} dt \tag{8.15}$$

subject to:

$$x' = \varphi(x, u, t);$$
$$x(t_0) = x_0;$$
$$x(t_1) = x_1 \text{ or } x(t_1) \text{ free.}$$

In other words, we have in the objective function also a "discount factor" e^{-rt}. For this reason it is convenient to redefine the Hamiltonian function. The ordinary Hamiltonian for the above problem is

$$H = G(x, u, t) e^{-rt} + \lambda \varphi(x, u, t).$$

Multiplying the same by e^{rt}, we obtain the so-called *current value Hamiltonian*

$$H^c = He^{rt} = G(x, u, t) + e^{rt} \lambda \varphi(x, u, t).$$

By setting $p = e^{rt}\lambda$ ("current value shadow price"), we can write

$$H^c(x, u, t, p) = G(x, u, t) + p\varphi(x, u, t).$$

Note that if $p = e^{rt}\lambda$, then $p' = re^{rt}\lambda + e^{rt}\lambda' = rp + e^{rt}\lambda'$ and hence $\lambda' = e^{-rt}(p' - rp)$. Also, $H^c = He^{rt}$ implies $\partial H^c/\partial x = e^{rt}(\partial H/\partial x)$. So $\lambda' = -\partial H/\partial x$ takes the form $p' - rp = -\partial H^c/\partial x$. The following result is taken from Sydsaeter et al. [36].

Theorem 8.2.12 *Suppose that the admissible pair* $(x^*(t), u^*(t))$ *solves problem* (8.15) *and let* H^c *be its related current value Hamiltonian. Then there exists a continuous function* $p(t)$ *such that for all* $t \in [t_0, t_1]$:

$$u = u^*(t) \text{ maximizes } H^c(x^*(t), u, p(t), t) \text{ for all } u \in \Omega;$$

$$p'(t) - rp(t) = -\frac{\partial H^c(x^*(t), u^*(t), p(t), t)}{\partial x}.$$

Moreover, if $x(t_1) = x_1$, *no condition is required on* $p(t_1)$; *if* $x(t_1)$ *is free, then* $p(t_1) = 0$.

- **Infinite horizon problems:** $t_1 = +\infty$

In several optimal control problems appearing in economic analysis the planning horizon is assumed to be of infinite length, even if one of the most quoted sentences of J. M. Keynes is: “In the long run we are all dead”. A problem of optimal control with $t_1 = +\infty$ could be of the form ($n = 1$, $k = 1$):

$$\max_{u \in \Omega \subset \mathbb{R}} \int_{t_0}^{+\infty} G(x(t), u(t), t)dt$$

subject to:

$$x'(t) = \varphi(x(t), u(t), t),$$
$$x(t_0) = x_0; \quad \lim_{t \longrightarrow +\infty} x(t) = x_1 \ (x_1 \text{ fixed number}).$$

Because our objective functional is an improper integral, we *assume* that the said integral is convergent for all admissible pairs (there exist several sufficient conditions for the convergence of improper integrals: see any textbook on Mathematical Analysis). For this problem the Maximum Principle of Pontryagin still holds. Moreover, if we replace the condition

$$\lim_{t \longrightarrow +\infty} x(t) = x_1$$

with

$$\lim_{t \longrightarrow +\infty} x(t) \geqq x_1 \text{ or } \lim_{t \longrightarrow +\infty} x(t) \text{ free,}$$

then all necessary conditions in the Maximum Principle hold, *except the transversality condition* $\lambda(t_1) = 0$. With no transversality condition we have too many solution candidates. There is a famous example, due to Halkin [20], which shows that if we substitute the condition $\lambda(t_1) = 0$ with the "natural" transversality condition $\lim_{t \longrightarrow +\infty} \lambda(t) = 0$, we impose a condition which is *not a necessary one;* the same is true for the condition $\lim_{t \longrightarrow +\infty} \lambda(t)x(t) = 0$. Halkin [20] shows that, for infinite horizon problems, there are none, in general, transversality conditions. For the development of necessary and sufficient optimality conditions in the case of infinite horizon problems, the reader may consult Caputo [10], Seierstad and Sydsaeter [35] and Sydsaeter et al. [36].

For what concerns multidimensional optimal control problems, i.e. the previous problems with $n > 1$, $k > 1$, only formal variants are needed. Again the reader is referred to the books quoted at the beginning of the present chapter. Finally, we give some insights on Optimal Control Problems with *inequality constraints,* of the form $g(x, u, t) \geqq [0]$. These constraints are often referred, in the present context, as *mixed constraints,* in opposition to the constraints of the type $g(x, t) \geqq [0]$, called *pure state constraints.*

We consider, following Sydsaeter et al. [36], the following mixed constraints problem

$$\max_u \int_{t_0}^{t_1} G(x, u, t)dt \tag{8.16}$$

subject to

$$\begin{aligned} x' &= \varphi(x, u, t) \\ x(t_0) &= x^0 \\ g(x, u, t) &\geqq [0] \text{ for all } t \end{aligned}$$

and the terminal conditions

$$\begin{aligned} &(a)\ \ x_i(t_1) = x_i^1, \ \ i = 1, \dots, \ell; \\ &(b)\ \ x_i(t_1) \geqq x_i^1, \ \ i = \ell + 1, \dots, m; \\ &(c)\ \ x_i(t_1) \text{ free}, \ \ i = m + 1, \dots, n. \end{aligned}$$

Here $x \in \mathbb{R}^n$, $u \in \mathbb{R}^k$, $G : \mathbb{R}^n \times \mathbb{R}^k \times \mathbb{R} \longrightarrow \mathbb{R}$, $\varphi : \mathbb{R}^n \times \mathbb{R}^k \times \mathbb{R} \longrightarrow \mathbb{R}^n$, while g is an s-dimensional vector function, so that the inequality $g(x, u, t) \geqq [0]$ represents the s inequalities

$$g_r(x(t), u(t), t) \geqq 0, \ \ r = 1, \dots, s.$$

It is assumed, in addition to the usual requirements, that g is a C^1 function with respect to x, u, t. We take from Sydsaeter et al. [36] the following sufficient optimality conditions for the above problem. The necessary optimality conditions given by Takayama ([37], Theorem 8.C.1) seem to be not fully correct: see Seierstad and Sydsaeter [34]. A correct treatment of necessary optimality conditions for this type of problems is given by Hestenes [21] and Seierstad and Sydsaeter [35].

Define the *Lagrangian function* or*generalized Hamiltonian function*

$$\mathcal{L}(x, u, \lambda, q, t) = H(x, u, \lambda, t) + q^\top g(x, u.t),$$

with the Hamiltonian given, as usually, by

$$H(x, u, \lambda, t) = G(x, u, t) + \lambda^\top \varphi(x, u, t).$$

Theorem 8.2.13 (Sufficient Optimality Conditions for (8.16)) *Suppose that* $(x^*(t), u^*(t))$ *is an admissible pair in the mixed constraints problem* (8.16). *Suppose further that there exist functions* $\lambda(t)^\top = (\lambda_1(t), \ldots, \lambda_n(t))$ *and* $q(t)^\top = (q_1(t), \ldots, q_s(t))$, *where* $\lambda(t)$ *is continuous, whereas* $\lambda'(t)$ *and* $q(t)$ *are piecewise continuous, such that the following conditions are satisfied:*

$$\frac{\partial \mathcal{L}^*}{\partial u_j} = 0, \quad j = 1, \ldots, k;$$

$$q_r(t) \geqq 0 \text{ and } q_r = 0 \text{ if } g_r(x^*(t), u^*(t), t) > 0, \ r = 1, \ldots, s;$$

$$\lambda_i'(t) = -\frac{\partial \mathcal{L}^*}{\partial x_i}, \text{ at all continuity points of } u^*(t), \ i = 1, \ldots, n;$$

$$\text{no condition on } \lambda_i(t_1), \ i = 1, \ldots, \ell;$$

$$\lambda_i(t_1) \geqq 0 \text{ and } \lambda_i(t_1) = 0 \text{ if } x_i^*(t_1) > x_i^1, \ i = \ell + 1, \ldots, m;$$
$$\lambda_i(t_1) = 0, \quad i = m + 1, \ldots, n.$$

$$H(x, u, \lambda, t) \text{ is concave in } (x, u);$$

$$g_r(x, u, t) \text{ is quasiconcave in } (x, u), \quad r = 1, \ldots, s.$$

Then $(x^*(t), u^*(t))$ *solves problem* (8.16).

The above theorem may be considered an extension of the results of Mangasarian [26] and is proved in Seierstad and Sydsaeter [35], where also the related necessary optimality conditions are discussed.

8.3 Some Hints on Dynamic Programming

As previously mentioned, besides the Optimal Control Theory of Pontryagin et al., the other modern approach to dynamic optimization is the *Dynamic Programming* of the American mathematician R. Bellman. This approach is suitable both for discrete optimization problems, and for continuous optimization problems. We shall give only few basic notions, as these topics involve advanced mathematical notions, namely the theory of partial differential equations. The method of Bellman has some advantages with respect to the method of Optimal Control: one is that it yields sufficient optimality conditions, without requiring concavity (or convexity) assumptions, nor the belonging of the state vector to finite-dimensional spaces. Moreover, in the case of stochastic optimal control problems, that is decisions problems under random assumptions, the Dynamic Programming of Bellman is in practice the unique general method for investigating these kinds of problems. On the other hand, the most relevant difficulty of the method of Bellman, a difficulty not to be underestimated, is the enormous quantity of computations often required.

The method of Dynamic Programming is based on a principle, formulated by Bellman himself, which is named *Bellman's Principle of Optimality.* This principle states that "an optimal policy has the property that, whatever the initial state and decision are, the remaining decisions must constitute an optimal policy with regard to the state resulting from the first decision" (Bellman [7]). This principle is valid for a vast range of systems, whose evolution in the future is determined from the present state, that is, it is independent from the past "history".

We give only few notions. We introduce the subject by means of a simple discrete model, taken from Salsa and Squellati [33]. Another discrete production model which describes the salient features of Dynamic Programming is given by Chiang [12]. Take into consideration Fig. 8.5.

Imagine starting from node 0 and reaching one of the $3.k$ nodes, where $k = 1, \ldots, 8$. Each segment corresponds to a score: we want to select the path with the maximum total score. Since there are only 8 possible paths, the total score of each path can be easily computed to find the optimal one. However, if every node had 6 exits, instead of 2, and the number of levels were more than 20, instead of 4, the computation would require a powerful computer. Thus, we will proceed using the Dynamic Programming method to solve this problem, as prototype. Suppose we are in one of the following nodes: 2.1, 2.2, 2.3, 2.4. The Bellman principle require choosing the maximum score path, for each one we gain

35 points going from 2.1 to 3.2;
40 points going from 2.2 to 3.3;
38 points going from 2.3 to 3.5;
31 points going from 2.4 to 3.8.

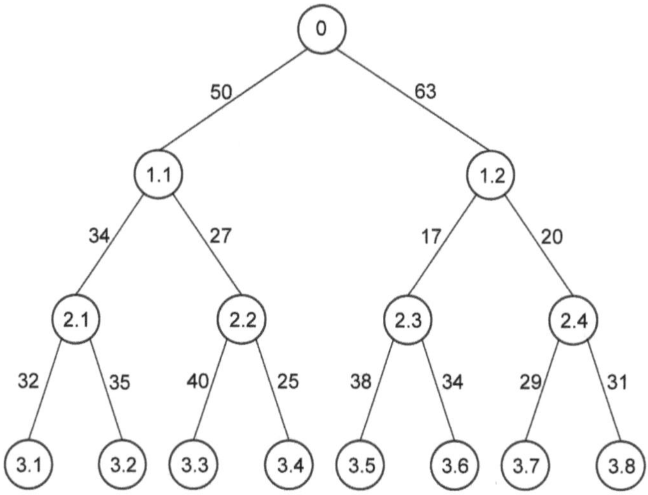

Fig. 8.5 Example to solve by using the Dynamic Programming method

Now, we proceed backward, placing ourselves at one of the nodes 1.1, 1.2. The Bellman principle suggests selecting the remaining path in order to have the maximum total score. On account of our previous choice, we get

$$\begin{cases} \text{going from } 1.1 \text{ to } 2.1, \text{ and then to } 3.2, \text{ the score is } 34 + 35 = 69. \\ \text{going from } 1.1 \text{ to } 2.1, \text{ and then to } 3.3, \text{ the score is } 27 + 40 = 67. \end{cases}$$

$$\begin{cases} \text{going from } 1.2 \text{ to } 2.3 \text{ and then to } 3.5, \text{ the score is } 17 + 38 = 55. \\ \text{going from } 1.2 \text{ to } 2.4, \text{ and then to } 3.8, \text{ the score is } 20 + 31 = 51. \end{cases}$$

Thus, the best paths starting from 1.1 and 1.2 are
- from 1.1 to 2.1, and then to 3.2, with score 69;
- from 1.2 to 2.3, and then to 3.5, with score 55.

Finally:
- if one goes from 0 to 1.1, then to 2.1 and 3.2, the score is $50 + 69 = 119$;
- if one goes from 0 to 1.2, then to 2.3 and 3.5, the score is $63 + 55 = 118$.

In conclusion, the maximum score is 119, obtained by the path

$$0 \longrightarrow 1.1 \longrightarrow 2.1 \longrightarrow 3.2.$$

We remark that as a first step, the Bellman principle *does not* require choosing the best score of the final path, 40 in our example, but rather it requires determining the paths giving the best scores for each node at the penultimate level. On account of these results, the second step is to choose of the paths giving the best scores for each node at the subsequent lower level, and so on.

The Bellman optimality principle, introduced above with a simple discrete problem of choice, can be used in building efficient algorithms to solve more complex problems, and, as we have anticipated, it is suitable to be applied also to continuous dynamic problems. With reference to this last case, we consider the following simplified version of a dynamic maximization problem, i.e. with $n = 1$, $k = 1$:

$$\max_{u\in\Omega} J(u) = \int_{t_0}^{t_1} G(x, u, t)dt + \psi(x(t_1))$$

under the conditions

$$x' = \varphi(x, u, t)$$
$$x(t_0) = x_0, \; t_1 \text{ given, } x(t_1) \text{ free.}$$

As usual, we assume that the scalar functions G, φ and ψ ($n = 1$, $k = 1$) are continuously differentiable on their domains, with respect to their arguments; Ω is the set of admissible controls, in the interval $[t_0, t_1]$. We denote by $V(t_0, x_0)$ the *optimal value function* of J with respect to the initial condition $x(t_0) = x_0$, that is

$$V(t_0, x_0) = \max_{u\in\Omega}\left[\int_{t_0}^{t_1} G(x, u, t)dt + \psi(x(t_1))\right] = \int_{t_0}^{t_1} G(x^*, u^*, t)dt + \psi(x^*(t_1)).$$

It is then possible to prove the following fundamental result.

Theorem 8.3.1 *Let the above assumptions hold and, furthermore, assume that V is differentiable with respect to x and t. Then it satisfies the following partial differential equation*

$$-V_t(t, x) = \max_{u\in\Omega}\{G(x, u, t) + V_x(t, x)\varphi(x, u, t)\}, \tag{8.17}$$

together with the final condition

$$V(t_1, x(t_1)) = \psi(x(t_1)).$$

Here V_t denotes the gradient of V with respect to t and V_x the gradient of V with respect to x.

Equation (8.17) is known as the *Bellman-Hamilton-Jacobi equation* and is one of the main results of Dynamic Programming theory. Note that the expression to be maximized in the right hand side of (8.17) is similar to the Hamiltonian function, with V_x in place of λ. Actually, since the multiplier λ represents the marginal value of the state variable x, we conclude that

$$\lambda(t) = V_x(t, x(t)).$$

Thus, the maximum problem in the right hand side of (8.17) coincides with the *maximum principle* in the variational method. We conclude here our few hints on Dynamic Programming and refer the reader to the works of Bellman [7], Bertsekas [8], Dreyfus [16], and Hadley [18].

Besides the works quoted along the chapter, we refer to the reader interested in the topics of this chapter to the references [1, 2, 13, 14, 23, 28, 29, 31].

References

1. G.C. Bliss, *Lectures on the Calculus of Variations* (University of Chicago Press, Chicago, 1946)
2. M. Aoki, *Optimal Control and System Theory in Dynamic Economic Analysis* (North Holland, Amsterdam, 1976)
3. K.J. Arrow, Applications of control theory to economic growth, in *Mathematics of the Decision Sciences*, ed. by G.B. Dantzig, A.F. Veinott, Part 2 (American Mathematical Society, Providence, 1968), pp. 85–119
4. K.J. Arrow, M. Kurz, *Public Investment, the Rate of Return, and Optimal Fiscal Policy* (John Hopkins University Press, Baltimore, 1970)
5. M. Athans, P.L. Falb, *Optimal Control: An Introduction to the Theory and Its Applications* (McGraw-Hill, New York, 1966)
6. R.J. Barro, X. Sala-i-Martin, *Economic Growth* (McGraw-Hill, New York, 1995)
7. R. Bellman, *Dynamic Programming* (Princeton University Press, Princeton, 1957)
8. D.P. Bertsekas, *Dynamic Programming and Optimal Control* (Athena Scientific, Belmont, 2000)
9. O. Blanchard, S. Fischer, *Lectures on Macroeconomics* (The MIT Press, Cambridge, 1989)
10. M.R. Caputo, Foundations of dynamic economic analysis. *Optimal Control Theory and Applications* (Cambridge University Press, Cambridge, 2005)
11. L. Cesari, *Optimization Theory and Applications* (Springer Verlag, Berlin/New York, 1983)
12. A.C. Chiang, *Elements of Dynamic Optimization* (McGraw-Hill, New York, 1992)
13. A.C. Chiang, K. Wainwright, *Fundamental Methods of Mathematical Economics* (McGraw-Hill, New York, 2004)
14. A.K. Dixit, *Optimization in Economic Theory* (Oxford University Press, Oxford, 1990)
15. R. Dorfman, An economic interpretation of optimal control theory. Am. Econ. Rev. **59**, 817–831 (1969)
16. S.E. Dreyfus, *Dynamic Programming and the Calculus of Variations* (Academic Press, New York, 1965)
17. I.M. Gelfand, S.V. Fomin, *Calculus of Variations* (Prentice-Hall, Englewood Cliffs, 1963)
18. G. Hadley, *Nonlinear and Dynamic Programming* (Addison-Wesley Publishing Company, Reading, 1964)
19. G. Hadley, M.C. Kemp, *Variational Methods in Economics* (North Holland, Amsterdam, 1971)
20. H. Halkin, Necessary conditions for optimal control problems with infinite horizons. Econometrica, **42**, 267–272 (1974)
21. M.R. Hestenes, *Calculus of Variations and Optimal Control Theory* (John Wiley & Sons, New York, 1966)
22. M.I. Kamien, N.I. Schwartz, *Dynamic Optimization: The Calculus of Variations and Optimal Control in Economics and Management*, 2nd edn. (North Holland, Amsterdam, 1991)
23. H.W. Kuhn, G.P. Szegö, *Mathematical Systems. Theory and Economics* (Springer, Berlin, 1969)
24. E.B. Lee, L. Markus, *Foundations of Optimal Control Theory* (John Wiley & Sons, New York, 1967)

25. D. Leonard, N.V. Long, *Optimal Control Theory and Static Optimization in Economics* (Cambridge University Press, Cambridge, 1992)
26. O.L. Mangasarian, Sufficient conditions for the optimal control of nonlinear systems. SIAM J. Control, **4**, 139–152 (1966)
27. B. Mond, I. Smart, Duality and sufficiency in control problems with invexity. J. Math. Anal. Appl. **136**, 325–333 (1988)
28. H. Nikaido, *Introduction to Sets and Mappings in Modern Economics* (North Holland, Amsterdam, 1970)
29. J.D. Pitchford, S. Turnovsky (eds.), *Applications of Control Theory to Economic Analysis* (North Holland, Amsterdam, 1977)
30. L.S. Pontryagin, V.G. Boltyanskii, R.V. Gamkrelidze, E.F. Mischenko, *The Mathematical Theory of Optimal Processes* (Interscience, New York, 1962)
31. R. Rammanathan, *Introduction to the Theory of Economic Growth* (Springer Verlag, Berlin, 1982)
32. F.P. Ramsey, A mathematical theory of saving. Econ. J. **38**, 543–559 (1928)
33. S. Salsa, A. Squellati, *Dynamical Systems and Optimal Control. A Friendly Introduction* (Bocconi University Press, Milan, 2018)
34. A. Seierstad, K. Sydsaeter, Sufficient conditions in optimal control theory. Int. Econ. Rev. **18**, 367–391 (1977)
35. A. Seierstad, K. Sydsaeter, *Optimal Control Theory with Economic Applications* (North Holland, Amsterdam, 1987)
36. K. Sydsaeter, P. Hammond, A. Seierstad, A. Strom, *Further Mathematics for Economic Analysis*, 2nd edn. (Prentice Hall, London/New York, 2008)
37. A. Takayama, *Mathematical Economics* (Cambridge University Press, Cambridge, 1985)
38. P. Tu, *Introductory Optimization Dynamics: Optimal Control with Economic and Management Applications* (Springer Verlag, Berlin, 1991)
39. P. Tu, *Dynamical Systems* (Springer Verlag, Berlin, 1994)

Index

G. Giorgi et al., *Lectures on Mathematics for Economic and Financial Analysis*,
https://doi.org/10.1007/978-3-031-83339-7

U

V

W

MIX
Papier aus verantwortungsvollen Quellen
Paper from responsible sources
FSC® C105338

If you have any concerns about our products,
you can contact us on
ProductSafety@springernature.com

In case Publisher is established outside the EU,
the EU authorized representative is:
Springer Nature Customer Service Center GmbH
Europaplatz 3, 69115 Heidelberg, Germany

Printed by Libri Plureos GmbH
in Hamburg, Germany